S0-EJV-224

ADVANCES IN VIRAL ONCOLOGY

Volume 4

Advances in Viral Oncology

Volume 4

Mechanisms of Neoplastic Transformation at the Cellular Level

Editor

George Klein, M.D., D.Sc.
Department of Tumor Biology
Karolinska Institutet
Stockholm, Sweden

Raven Press ■ New York

Raven Press, 1140 Avenue of the Americas, New York, New York 10036

Library of Congress Cataloging in Publication Data
Main entry under title:

Mechanisms of neoplastic transformation at the cellular
level.

(Advances in viral oncology ; v. 4)
Includes bibliographical references and index.
1. Viral carcinogenesis. 2. Cancer cells.
3. Oncogenes. I. Klein, George, 1925– . II. Series.
[DNLM: 1. Cell transformation, Neoplastic. W1 AD888 v.4 /
QZ 202 M4856]
RC268.57.M43 1984 616.99'4071 83-26930
ISBN 0-89004-977-7

Made in the United States of America

The material contained in this volume was submitted as previously unpublished material, except in the instances in which credit has been given to the source from which some of the illustrative material was derived.

Great care has been taken to maintain the accuracy of the information contained in the volume. However, Raven Press cannot be held responsible for errors or for any consequences arising from the use of the information contained herein.

Preface

The impact of the "new biology" on cancer research already has revolutionized the study of oncogenic or potentially oncogenic DNA sequences. However, the regulation of oncogene expression in normal and malignant cells is much less well understood. Their protein products are only partly known and their transforming effects are largely obscure, with the exception of some glaring bright spots. They may lead the way or turn out to be nothing more than will-o'-the-wisp.

The long and fruitless chase after *the* mechanism of malignant transformation is now widely regarded as illusory, a few die-hards notwithstanding. Neoplastic behavior is a phenotypic property, not a mechanism. Who would ever think of finding a single mechanism for the escape of bacteria from penicillin or of prisoners from a well-guarded camp? Clearly, all available mechanisms will be utilized although some will occur more frequently than others in each cell–host combination.

This volume presents some examples of such mechanisms that are now just becoming discernible. Oncogene products that associate with the inner surface of the plasma membrane often possess tyrosine kinase activity. But what is their normal substrate and how does phosphorylation affect cellular behavior?

Other oncogene products are nuclear DNA binding proteins. Nuclear and membrane-associated oncogene products may complement each other, as shown by the recent double transfection experiments of Land et al. (3).

One of the most crucial missing pieces concerns the role of the oncogenes in normal growth and development. The evolutionary analysis of Shilo is a step in this direction. Another crucial unknown is the first identification of an oncogene *(sis)* product with a known growth factor (PDGF) (1,4). The search continues for other examples of "autocrine" stimulation by oncogene products that may act as growth factors for the same cell.

In the DNA virus field, Doerfler takes a look at the role of DNA methylation for viral expression, transformation, and persistence. Retroviruses lacking their own transforming genes that may or may not act by inserting 3′-terminal promoter or enhancer sequences in the neighborhood of a cellular oncogene are represented by murine leukemia virus and are discussed by Ihle et al.; mammary tumor virus is discussed by Fanning and Cardiff, and the human T-cell leukemia virus is the focus of the chapter by Popovic et al. The action of a virally transduced oncogene is analyzed by Rohrschneider and Gentry as well as by Weber. The transformed phenotype is discussed by Perbal, Pollack et al., and Puck. Nonviral transformation is the subject of the chapter by Potter et al. that discusses the story of mouse plasmacytomas. This chapter is an outstanding example of oncogene activation by chromosomal translocation.

Oncogene products that transform hemopoietic cells appear to meddle with differentiation. Appropriate differentiation-inducer signals may overrule their action in some leukemic cells but not in others. This is a complex interplay between growth regulatory signals, cellular receptors, and growth and differentiation responses. The chapter by Sachs on murine myeloid leukemia describes one of the most extensively analyzed systems. "Constitutive switch-on" is a frequently occurring motto. The recent discovery that specific chromosomal translocations can contribute to the transformation process by transposing oncogenes into highly active chromatin regions illustrates one of several mechanisms whereby oncogenes can be activated permanently in cells of the proper differentiation type (2).

Last, the chapters by Sachs and Puck also include studies on the reversibility of the transformed phenotype. This is a reality in many systems, although its mechanism may be as diversified as the mechanism of transformation itself.

All three major areas of oncogene activation—via retroviral transduction, retroviral insertion, and chromosomal translocation—conclude with the same basic questions. Is transformation due to a purely quantitative change, a constitutive switch-on of a normal gene product expressed in the wrong cell, at the wrong place or at the wrong time, or does it require a qualitative change, like a point mutation, or more extensive modifications at the DNA level? Examples exist for both models, alone and combined in various ways.

As this volume emphasizes, the term "constitutive expression" may be an enduring key word. Normal household genes that control important cellular functions must lead a strictly regulated life. Dysfunction of the switch-off mechanism may lead to permanent disaster.

This book will be of interest to those workers involved in cancer research and/ or viral oncology who wish to be informed about recent developments in this area.

George Klein

REFERENCES

1. Doolittle, R. F. et al. (1983): *Science*, 221:275–277.
2. Klein, G. (1983): *Cell*, 32:311–315.
3. Land, H. (1983): *Nature*, 304:596–602.
4. Waterfield, M. D. et al. (1983): *Nature*, 304:35–39.

Contents

Contributors

Robert D. Cardiff
Department of Pathology
University of California, Davis
Davis, California 95616

Suzie Chen
Department of Biological Sciences
Columbia University
New York, New York 10027

Walter Doerfler
Institute of Genetics
University of Cologne
5000 Köln 41 Germany

Thomas G. Fanning
Department of Pathology
University of California, Davis
Davis, California 95616

R. C. Gallo
Laboratory of Tumor Cell Biology
National Cancer Institute
National Institutes of Health
Bethesda, Maryland 20205

L. E. Gentry
Fred Hutchinson Cancer Research Center
Tumor Virology Program
Seattle, Washington 98104

James N. Ihle
Laboratory of Viral Immunobiology
National Cancer Institute
Frederick Cancer Research Facility, LBI
Basic Research Program
Frederick, Maryland 21701

Richard Mural
Laboratory of Viral Immunobiology
National Cancer Instutute
Frederick Cancer Research Facility, LBI
Basic Research Program
Frederick, Maryland 21701

J. Frederic Mushinski
Laboratory of Genetics
National Cancer Institute
National Institutes of Health
Bethesda, Maryland 20205

Bernard Perbal
Institut Curie
Section de Biologie
Centre Universitaire
91405 Orsay Cedex, France

Robert Pollack
Department of Biological Sciences
Columbia University
New York, New York 10027

M. Popovic
Laboratory of Tumor Cell Biology
National Cancer Institute
National Institutes of Health
Bethesda, Maryland 20205

Michael Potter
Laboratory of Genetics
National Cancer Institute
National Institutes of Health
Bethesda, Maryland 20205

Scott Powers
Cold Spring Harbor Laboratory
Cold Spring Harbor, New York 11224

Theodore T. Puck
Department of Biochemistry, Biophysics and Genetics and Department of Medicine
University of Colorado Health Sciences Center
Denver, Colorado 80262

Alan Rein
Laboratory of Molecular Virology and Carcinogenesis
National Cancer Institute
Frederick Cancer Research Facility, LBI
Basic Research Program
Frederick, Maryland 21701

L. R. Rohrschneider
Fred Hutchinson Cancer Research Center
Tumor Virology Program
Seattle, Washington 98104

Leo Sachs
The Weizmann Institute of Science
Rehovot 76100, Israel

P. S. Sarin
Laboratory of Tumor Cell Biology
National Cancer Institute
National Institutes of Health
Bethesda, Maryland 20205

Ben-Zion Shilo
Department of Virology
Weizmann Institute of Science
Rehovot, Israel

Michael Verderame
Department of Biological Sciences
Columbia University
New York, New York 10027

Michael Weber
Department of Microbiology
University of Virginia School of Medicine
Charlottesville, Virginia 22908

Francis Wiener
Department of Tumor Biology
Karolinska Institute
S-104 01
Stockholm, Sweden

F. Wong-Staal
Laboratory of Tumor Cell Biology
National Cancer Institute
National Institutes of Health
Bethesda, Maryland 20205

Mechanisms of Neoplastic Transformation at the Cellular Level

Advances in Viral Oncology, Volume 4,
edited by George Klein. Raven Press,
New York © 1984.

Transformation Mechanisms at the Cellular Level

Robert Pollack, Suzie Chen, *Scott Powers,
and Michael Verderame

Department of Biological Sciences, Columbia University, New York, New York 10027; and
**Cold Spring Harbor Laboratory, Cold Spring Harbor, New York 11224*

A major question in biology today is: What are the differences in behavior between normal cells and cells that either come from or can become tumors? A partial answer is forthcoming from somatic tissue cells grown in dishes and flasks. Even this success may be surprising, since cell cultures contain an element of artifact: insofar as they are growing, often at lower densities than found *in vivo*, they may be intentionally different from the cells in the tissue they mimic (50,109).

In the adult body, many cells are not growing; rather, they are sitting in a matrix of other cells and their mutually secreted products in assemblies called tissues. A typical tissue has epithelia and mesodermal and fibroblast cells separated by an acellular basement membrane (Fig. 1). The epithelial side is the topologic outside of the tissue. All epithelial cells are in contact with either neighboring cells or a collagen-based matrix. Blood vessels are on the mesodermal side, along with a nondividing cell called the fibroblast. Fibroblasts normally are quiescent.

GROWTH CONTROL MECHANISMS

A typical tissue in an adult is not growing in size; thus net tissue cell growth must be kept to zero. Normal cell growth in a tissue is regulated differently by cells on the two sides of the basement membrane. The first physiologic mechanism is simplest: some cells do not normally divide. The second is more complex: some cells divide regularly, but one of their two daughters is destined always to die by differentiation.

Remarkably, these two mechanisms are topologically distinct in the body. The basement membrane separates them, with the "inside" cells showing the simpler control and the "outside" ones the more complex. In the case of the epithelium, molecular biology must explain how one cell divides into two different cells, only one of which will divide again (7,39,45). In the case of the stromal cells, we must explain what signals these cells require in order to divide (4,32,49).

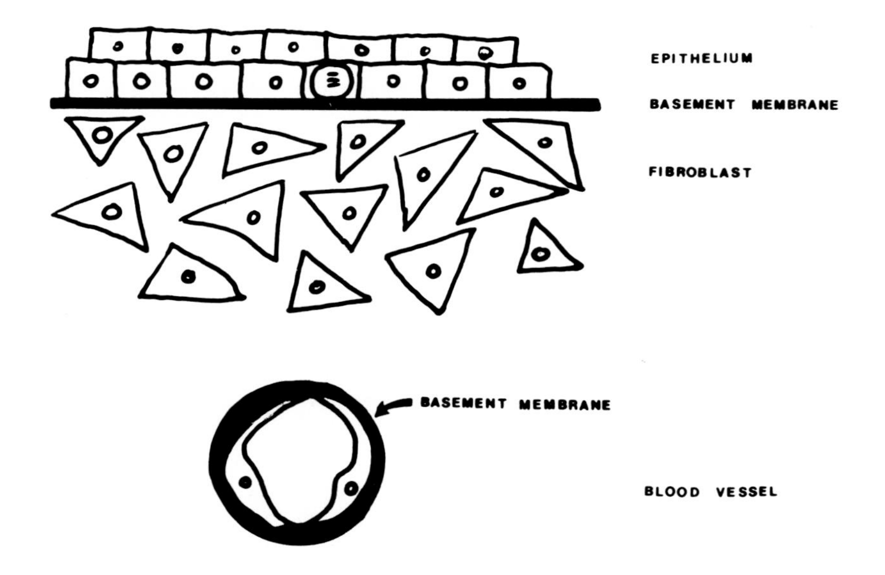

FIG. 1. This schematic diagram shows the relative positions of epithelial cells, fibroblasts, blood vessel endothelial cells, and the interposing basement membranes. Epithelial and fibroblastic cells are separated by a basement membrane.

WOUNDS AND THE GROWTH CONTROL OF FIBROBLASTS

Epithelia are the origin of the majority of tumors. Apparently, the more complex mechanism is the one that most often goes awry; more is known about the simpler case of stromal cell growth regulation. We have a good idea of what the signals are in the body for normal fibroblast growth, because we can make these cells grow by wounding any tissue. The consequence of wounding is to cause fibroblasts to grow, to secrete a collagen scar, and then to stop growing as the epithelial layer fills in over the collagen. Thus a wound provides all the signals necessary to reversibly activate the growth controls of a normal fibroblast (85).

What then are the signals provided by a wound? A wound is an opportunity for blood to exit a capillary or a larger blood vessel. The blood then clots. Unclotted blood is a suspension of lymphocytes, macrophages, erythrocytes, and platelets in plasma. Plasma is composed of fibrinogen, hormones, and a vast assortment of soluble molecules. Upon wounding, blood contacts the collagen of the cut vessel and the collagen-based matrix in which the vessel was embedded. This leads to a string of events. First, the platelets lyse to release a mesh of actin and myosin; they also release a local high concentration of soluble, platelet-derived hormones. At the same time, the fibrinogen of the plasma is converted to an insoluble fibrin mesh by collagen activation of the plasma enzyme thrombin. These events culminate as the fibrin, actin, and myosin become a contractile clot, which squeezes out a fluid called serum. The presence of serum and a fibrin clot are the necessary and sufficient signals for fibroblasts to begin dividing and secreting large amounts of collagen in a feedback loop that reestablishes the structural integrity of a tissue.

Recently, we have learned that serum contains a set of polypeptide hormones (49), including the platelet-derived growth factor PDGF (106). PDGF stimulates fibroblasts to secrete their own mitogenic hormone, the insulin-like growth factor somatomedin C (19). Somatomedin C, also known as insulin-like growth factor-I (IGF-I), is the final hormone in the hypothalamic loop stimulating normal body growth through growth hormone (Fig. 2). At the same time as IGF-I appears, the clot with collagen provides an anchoring structure on which a fibroblast can spread and organize its cytoskeleton into large stress fibers containing intracellular actomyosin (42,71,126). Receptors for such hormones as IGF, PDGF, epidermal growth factor (EGF), and insulin are integral membrane proteins (63) (Fig. 3), and the insulin and EGF receptors have an associated tyrosine kinase activity (3,13,54,60). Tyrosine kinase activity is associated with tumor virus transforming gene products (9). Evidence has accumulated that hormone binding is followed by clustering of these hormone receptors (59), and that such lateral, directed receptor movement requires a well-organized cytoskeleton (6,31).

Cell culture techniques thus permit the growth of tissue fibroblasts by providing necessary growth signals. These come in the form of serum and a dish on which the fibroblasts can anchor and spread. Given these signals, tissue fibroblasts in a dish will proliferate for a few dozen divisions, as if filling in a giant wound. Eventually, they will cease dividing, dying in a crisis (50) that mimics aging. If crisis were inevitable, there would be little hope of doing reproducible molecular biology with the cultures. Often, however, a process called establishment (2,44,123) occurs, to provide populations of cells in culture with an abnormal but desirable extra degree of autonomy.

ESTABLISHMENT

Experimental protocols, especially those requiring recovery of variant cell types, depend on the capacity of cells to form separate colonies on a dish (47). Colony formation demands that tissue fibroblasts show a novel degree of autonomy, initiating colonies at a distance from one another that exceeds by many orders of magnitude the distance wound healing cells get from their nearest neighbors in any tissue (44,109,130). Most tissue cells cannot form colonies, and these colonies are even less often capable of throwing off colonies in turn. Variants that survive crisis and readily form colonies thereafter are called established (123). Although they

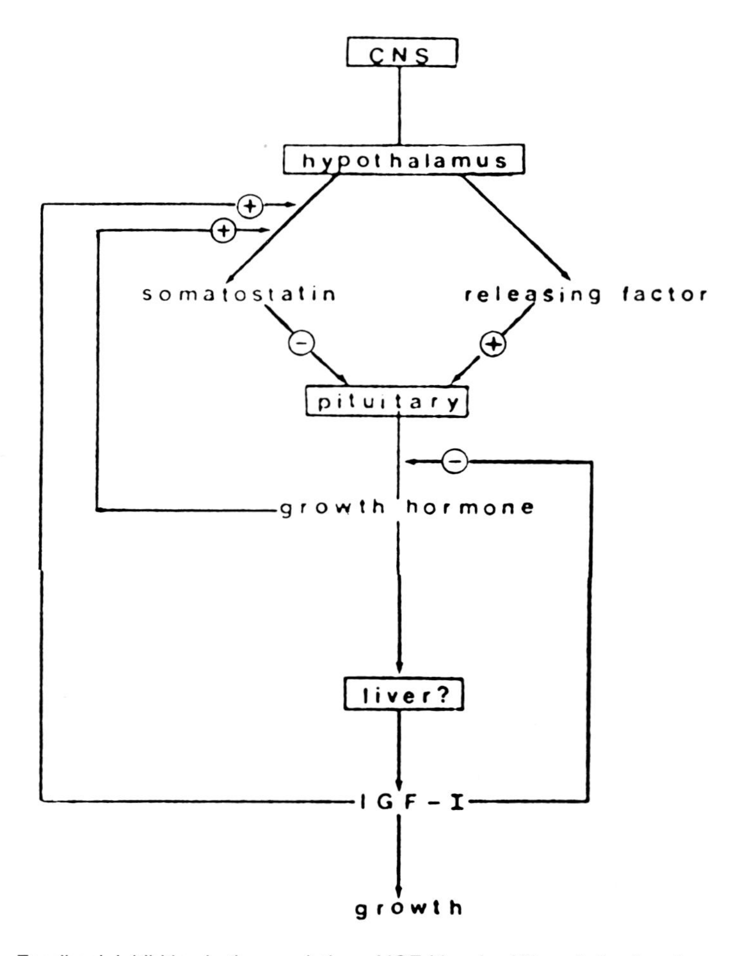

FIG. 2. Feedback inhibition in the regulation of IGF-I levels. Although the liver is currently the suspected site of IGF-I production, this issue has never been conclusively settled. The existence of a releasing factor has recently been discovered (11). +, positive stimulation; −, negative stimulation. (Adapted from ref. 8.)

are often not abnormal enough to make tumors, they have clearly become genetically different from normal tissue fibroblasts.

Obviously, ignoring the difference between established cells and tissue cells introduces artifact; nevertheless, the choice to do so provides a great boon. Colonies are the descendents of single cells, while tissues are a complex architecture of many cell types. Homogeneity of material is a prerequisite to the molecular analysis of mechanism.

TUMOR CELLS AND TRANSFORMATION

A sarcoma is a tumor of fibroblast origin. It will grow whether or not a wound or clot is present; it is no longer in need of these signals. The loss of a need for

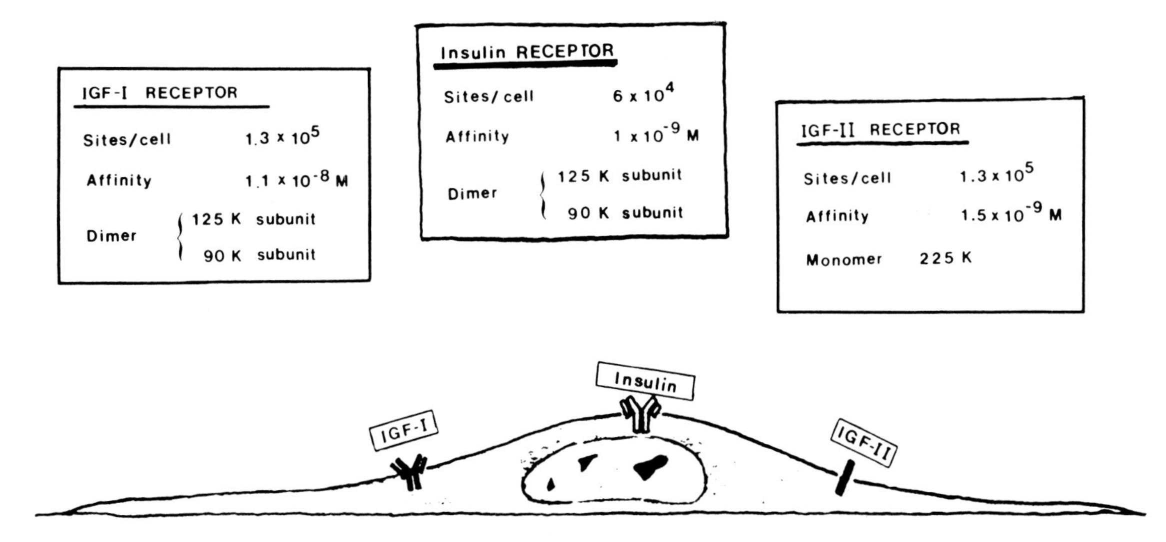

FIG. 3. Insulin-hormone family of receptors on 3T3 cells. Each hormone has a distinct receptor. Note that the receptors for insulin and IGF-I are multiple polypeptide chains held together by disulfide bonds, while the IGF-II receptor is a single polypeptide chain (M. Czech, *personal communication*).

one or more of the signals involving hormones or anchorage defines the loss of growth control in the cells of this form of tumor (33,57). Established normal cell lines can be converted by a number of agents into a state in which they can form tumors upon injection into an appropriate normal tissue (35,61,75,119,122,124). Thus cell culture provides two different kinds of transformed cells: from the tumor and from the transformation (Fig. 4).

Once a culture is established from a single cell, it can be grown into a large amount of genetically and biochemically homogeneous material for further study. This material, whether from a tumor, a transformant, or a normal cell culture, will be suitable for the tracking of biochemical differences in a manner not possible using biopsy material.

What then are the biochemical and phenotypic differences between normal and tumor cells, and how do they occur? Because it is inappropriate to approach the question of mechanism by the accumulation of differences (Table 1), we concentrate only on these differences that are characteristic of both cultured tumor cells and transformed cells (Fig. 5). In this light, three phenotypic changes are common to both sarcomas and transformed fibroblasts: (a) the ability to grow without anchorage (28,32–34,75,104,115), (b) the inability to organize the actomyosin cytoskeleton properly (1,51,52,65,80,89,116,126), and (c) the abnormal capacity to secrete proteases (72,87,100,102,103).

Selections designed to separate out the parts of this complex of phenotypic changes have generated a set of cell lines that permit linkage of these phenotypes to tumorigenicity (4,75,104,122,124). The wound teaches us that normal fibroblasts must receive at least two different signals before they can divide. These are, again, the hormones found in serum and the substrate on which to spread. Tumor cells in culture often grow equally well in the presence or absence of these signals (4,25,28,40,49,53). Since both kinds of cells grow well in culture, we can construct assay conditions that directly select for the cells that no longer need these signals (Fig. 6). In such an assay, cells are intentionally deprived of one or more of the signals necessary for growth. They remain in suspended animation. A transforming agent (e.g., virus or chemical) may convert some of the cells so that they can grow despite the absence of the signal. When the conversion is genetically stable, a clone

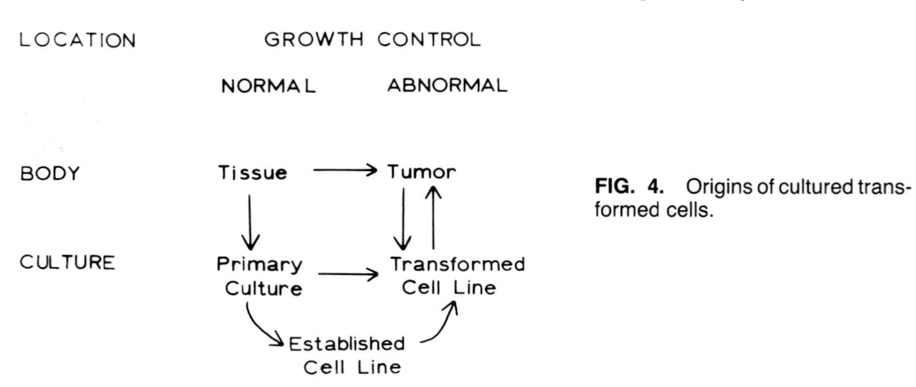

FIG. 4. Origins of cultured transformed cells.

TABLE 1. *Properties of cells transformed by SV40 or polyoma virus*[a,b]

Growth
High or indefinite saturation density[c]
Different, usually reduced serum requirement[c]
Growth in agar or Methocel suspension, anchorage independence[c]
Tumor formation upon injection into susceptible animals
Not susceptible to contact inhibition of movement
Growth in a less oriented manner[c]
Growth on monolayers of normal cells[c]
Surface
Increased agglutinability by plant lectins[c]
Changes in composition of glycoproteins and glycolipids
Tight junctions missing
Fetal antigens revealed
Virus-specific transplantation antigen
Different staining properties
Increased rate of transport of nutrients
Increased secretion of proteases or activators[c]
Intracellular
Disruption of the cytoskeleton
Changed amounts of cyclic nucleotides
Evidence of virus
Virus-specific antigenic proteins detectable
Viral DNA sequences detected
Viral mRNA present
Virus can be rescued in some cases

[a]From ref. 125.
[b]Transformed cells show many, if not all, of these properties, which are not shared by untransformed parental cells.
[c]Several of these properties have formed the basis of selection procedures for isolating transformants.

of transformed cells will grow out against a background of nongrowing normal cells.

Surprisingly, the transformants isolated in these different assays are not the same (104,105,127,128). Each selection generates a cell type insensitive to the growth signal that is absent in the selective assay; for example, serum transformants arise from the serum-starvation assay and anchorage transformants from the anchorage-deprivation assay. However, other transformation phenotypes accompanying the one that served as a basis for selection usually occur in an ordered way (Fig. 7). Serum-transformed clones lack the capacity to grow in the anchorage-deprivation assay, while tumorigenic anchorage transformants always also grow in low serum (34). Also, cells of tumors arising from a normal hamster line after chemical mutagenesis are both serum and anchorage independent, even though the transformants themselves are initially only anchorage independent (83). Apparently, while both the serum and the anchorage requirements must be lost before tumorigenicity is possible, the anchorage assay has a much higher probability of generating tumorigenic transformants (90,126). In biochemical terms, the concomitants of the anchorage requirement are likely to be most directly related to tumorigenicity (90).

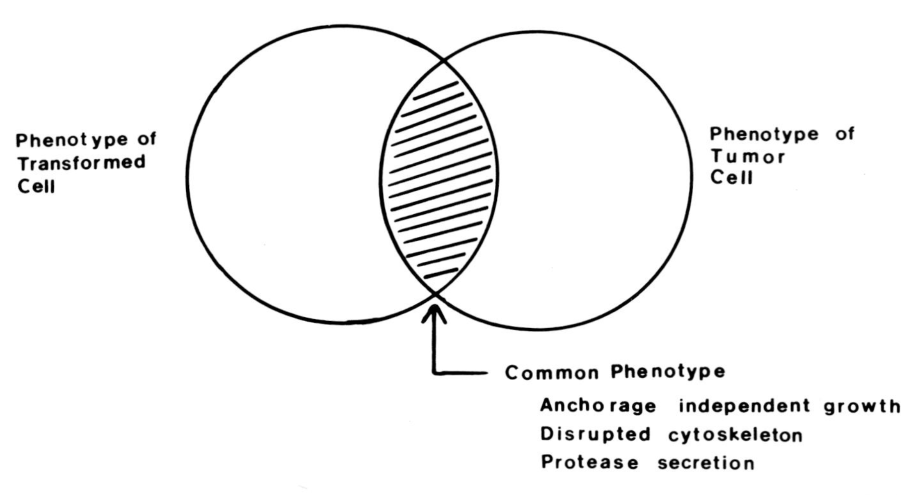

FIG. 5. Overlap of the phenotypic differences in transformation and in tumor cell growth. Each circle in this Venn diagram represents the complete set of phenotypes for the transformed cell **(left)** or the tumor cell **(right)**. A common strategy of modern research is to focus on those properties common to both sets (here represented by the hatched area).

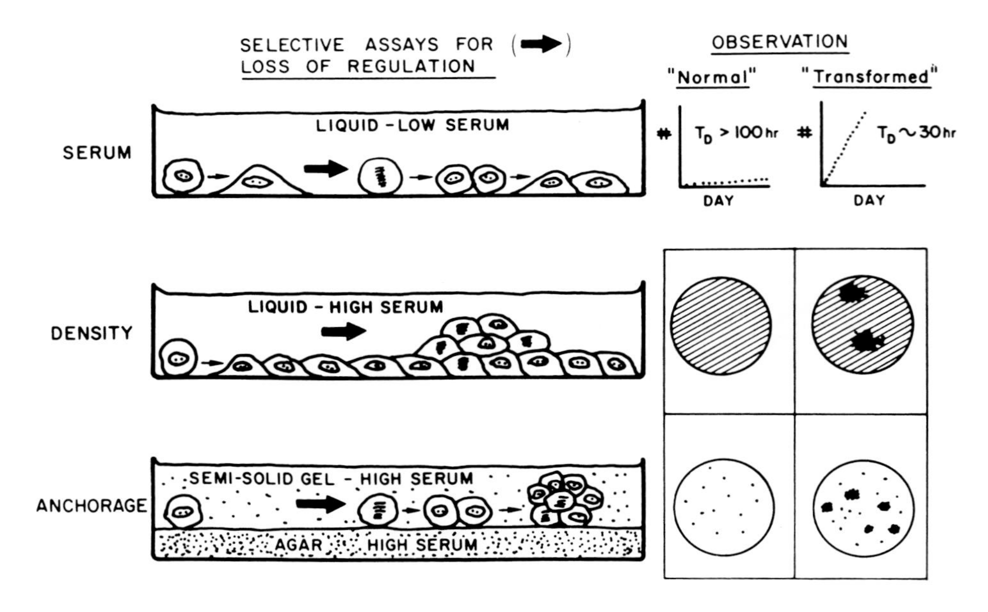

FIG. 6. Three selective transformation assays. In each assay, transformed cells and their descendents grow, whereas untransformed cells, due to one or another restrictive environmental signals, remain alive but do not increase in number. *Heavy arrow*, "transformation." Note that only in the anchorage assay can the transformant continuously traverse the cell cycle without ever spreading out. (Adapted from ref. 90.) Other transformation assays include hormone deprivation (94) and calcium reduction (79).

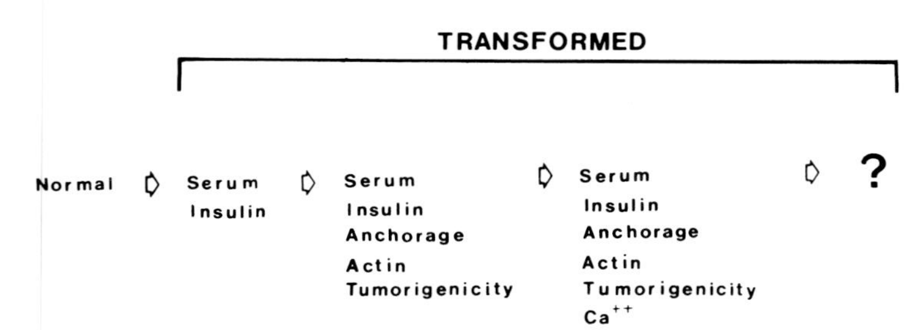

FIG. 7. Transformations occur in an ordered series. Cells that are serum transformed are also transformed with respect to the hormone insulin or IGF-I. Cells that are transformed with respect to anchorage are also transformed with respect to serum and the other phenotypes indicated; also, they usually are tumorigenic. Transformants selected for their ability to grow in low calcium (79) are transformed with respect to all the other phenotypes indicated.

INSULIN TRANSFORMATION

We recently carried the serum assay one step further toward molecular resolution. By growing cells in the absence of serum, differences in hormone requirement between normal and transformed cells became apparent (94,95). Figure 8 shows the growth requirement curves for insulin and IGF-I of a normal, established cell line. Insulin inefficiently mimics IGF-I (62); in fact, the physiologic IGF-I requirement is one of the major hormone requirements for established normal fibroblasts (118). Insulin itself is unlikely to have any direct growth-promoting activity through its own receptors on these cells (62).

Figure 9 shows the curves for a serum-transformed derivative line. Note that the requirements for both hormones are shifted to the left by at least a factor of 10. Apparently, serum transformation involves the reduction of the IGF-I requirement.

ANCHORAGE TRANSFORMATION

Because anchorage transformation most strongly correlates with the capacity to grow as a tumor in a susceptible animal (34,90), we concentrated on asking which biochemical events are most closely correlated with anchorage independence in fibroblasts. First, anchorage-transformed cells elaborate proteases, in particular the protease that activates the clot-dissolving plasma proenzyme plasminogen to plasmin (88,101) (Table 2). Second, we found that anchorage-transformed cells specifically lack the capacity to organize their cytoskeleton into large, actin-containing stress fibers, and that plasmin can efficiently disrupt the cytoskeleton of the normal cell (87,89,126). The cytoskeletal change is most easily seen by comparing the organization of normal and transformed cell cytoskeletons with a dye, such as phalloidin, that specifically stains the actin of the cell (104) (Fig. 10A). The cables run to the edge of the spread cell and are large and many in number. In most anchorage-transformed cells, but not serum-transformed cells, this organization is replaced by a markedly diffuse one (Fig. 10B). The red nucleus is stained at the same time with antibodies that pick out the presence of a transforming virus protein, SV40

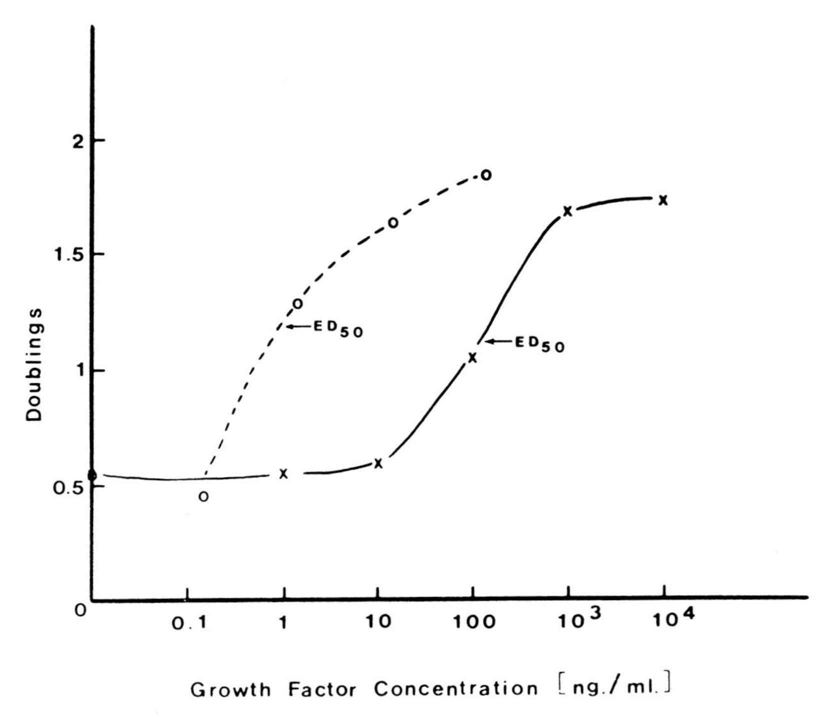

FIG. 8. Dose response curves of 3T3 with insulin and IGF-I. Cells were grown in serum-free medium, including either insulin (x) or IGF-I (o). Without either of the hormones, normal 3T3 cells do not even complete one doubling in 4 days. Note the shifted response curve for IGF-I as compared to insulin (94).

T antigen. Figure 10C is a serum-transformed, anchorage-requiring line. Note the correlation of cytoskeletal retention with the maintenance of the anchorage requirement in this cell, despite the presence of the SV40 T antigen (126).

EPITHELIAL CELLS AND THE TRANSFORMED PHENOTYPE

The majority of *in vitro* analyses concerning cellular growth control have been conducted with fibroblasts because of the relative ease with which these cells grow out in culture from any tissue biopsy. Recent progress in plating epithelial cells in culture allows short-term survival of these cells (30,38,82,96,97,99,114,131). Given the advantages of *in vitro* growth outlined above, studies with these populations should lead to significant observations concerning the mechanisms of both growth and differentiation—two intimately related phenomena in tissues that give rise to the majority of human malignant neoplasms.

The long-standing observation that tumorigenic fibroblasts have a disrupted actin cytoskeleton (89,126) provides a starting point for the examination of colonies of epithelial cells derived from colon biopsies of patients with frank adenocarcinomas,

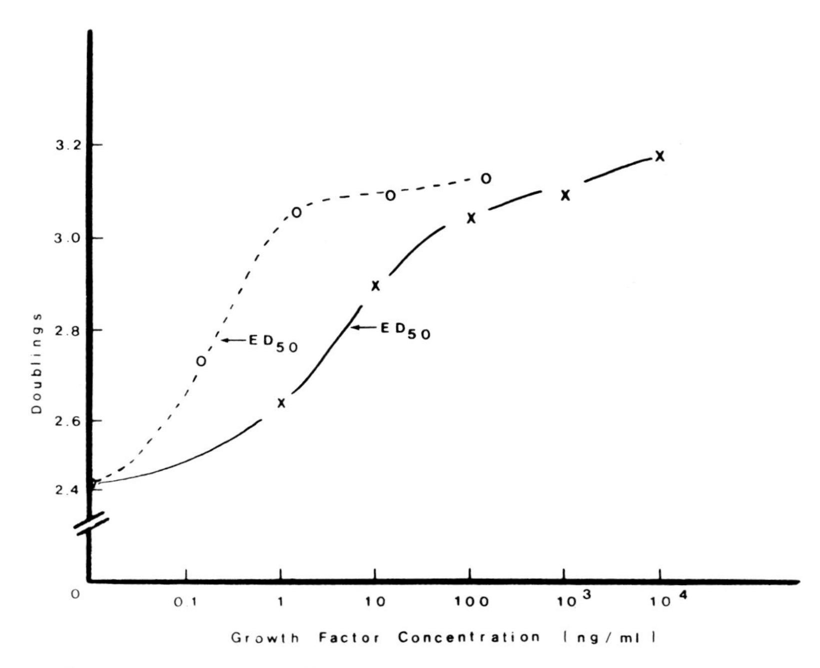

FIG. 9. Dose response curves of SV101 with insulin and IGF-I. These transformed cells were grown in serum-free medium, including insulin (x) or IGF-I (o). Note that even without added insulin or IGF-I, SV101 cells grow considerably (~2.5 doublings in a 4-day assay). Also note the shifted response curve for IGF-I as compared to insulin (94).

TABLE 2. *Correlation among* in vitro *growth properties and tumorigenicity of Syrian hamster cell lines*[a]

Growth parameter	Coefficient of rank correlation
Cloning efficiency, semisolid agar	0.986[b]
Fibrinolytic activity	0.754[c]
Generation time, 1% serum	0.754[c]
Cloning efficiency, liquid medium	0.638[c]
Organization of intracellular actin	0.570[c]
Saturation density, 10% serum	0.246
Generation time, 10% serum	0.062

[a]From ref. 5.
[b]Significant at the 95% confidence level.
[c]Significant at the 90% confidence level.

premalignant conditions of various states, or apparently normal tissue. Unlike fibroblasts, which have an elaborate stress fiber network in normal cells that is lost upon transformation to the tumorigenic state, primary epithelial colonies cultured for a short time *in vitro* exhibit several distinct classes of organization differing from fibroblasts (37).

In normal colon, the epithelium is organized into finger-like indentations called crypts of Lieberkuhn. Only the cells at the base of the crypt are dividing (78). Cells *in vitro* derived from normal epithelial tissue are well spread. Electron micrographs have shown that the normal cells overlap one another like roofing shingles (36). In some colonies, the cells are well separated from their neighbors (i.e., not touching one another). In all cases, normal biopsy colonic epithelial cells contain some actin cables, but not as abundant an amount as found in fibroblasts. Instead, there is some weak staining at the perimeter of the cells (37) (Fig. 11A, B).

Premalignant adenomas *in situ* are characterized by a breakdown in the normal organization of the crypts usually seen *in vivo* (78). Dividing cells can be found all the way up the crypt wall, not confined to the bottom as in normal tissue (74). This elongates the crypt, until it forms a polyp which extends into the lumen of the gut. Colonies derived from early stage (tubular) adenomous tissue show a remarkable alteration in their *in vitro* phenotype as well (37). The cells are always organized into sheets of well-spread cells, in which each cell is in intimate contact with its neighbor along its entire perimeter (Fig. 11C, D). There is usually an intense staining lining the plasma membrane of each cell. Actin stress fibers can be seen scattered throughout some of the cells. A small number of colonies from premalignant adenomas exhibit an elaborate system of stress fibers, which seem to extend intercellularly, essentially blanketing the colony.

Late stage (villous) adenomas show a similar organization to early stage adenomas (37). Cells from late stage adenomas can be distinguished from early stage adenomas by their response to the potent tumor promoter 12-O-tetradecanoyl phorbol-13-acetate (TPA). In addition to increasing the secretion of the protease plasminogen activator (36,72), TPA causes the late stage adenomas to round up, separate from one another, and completely lose all actin cables (37). In contrast, early stage adenomas are completely unaffected by this treatment.

Adenocarcinomas are characterized by a loss of tissue organization and inappropriate cellular proliferation (78). Biopsies containing these cells are placed in culture, and the cells are stained for actin; the cells are readily distinguished from either normal or premalignant cells by their rounded appearance. They can occur either patches of cells touching each other or as single cells, but in neither case

FIG. 10. Cells stained for F-actin and for SV40 large T-antigen. These cells have been double stained with the fluorescent compound Fl-phalloidin and with antibodies to SV40 large T-antigen (126). **Top:** A normal cell with its characteristic numerous F-actin bundles, shown here in green. **Middle:** A typical transformed cell. Note the presence of SV40 T-antigen (indicated by the red nucleus) and the absence of large actin bundles. **Bottom:** A typical anchorage revertant. Despite the continued expression of the SV40 T-antigen, the F-actin bundles are obvious, demonstrating the correlation of anchorage-dependent growth and an organized cytoskeleton.

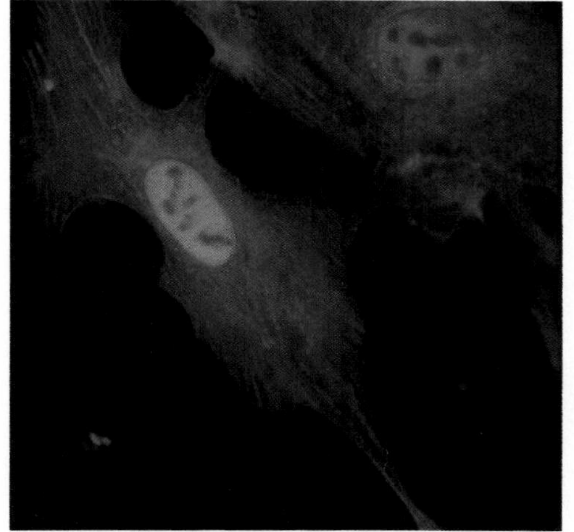

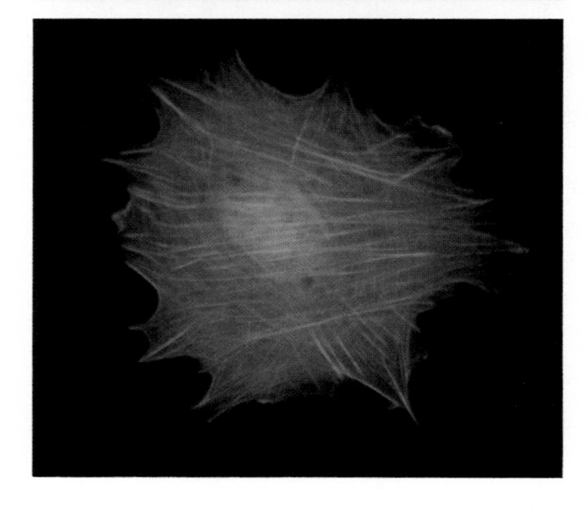

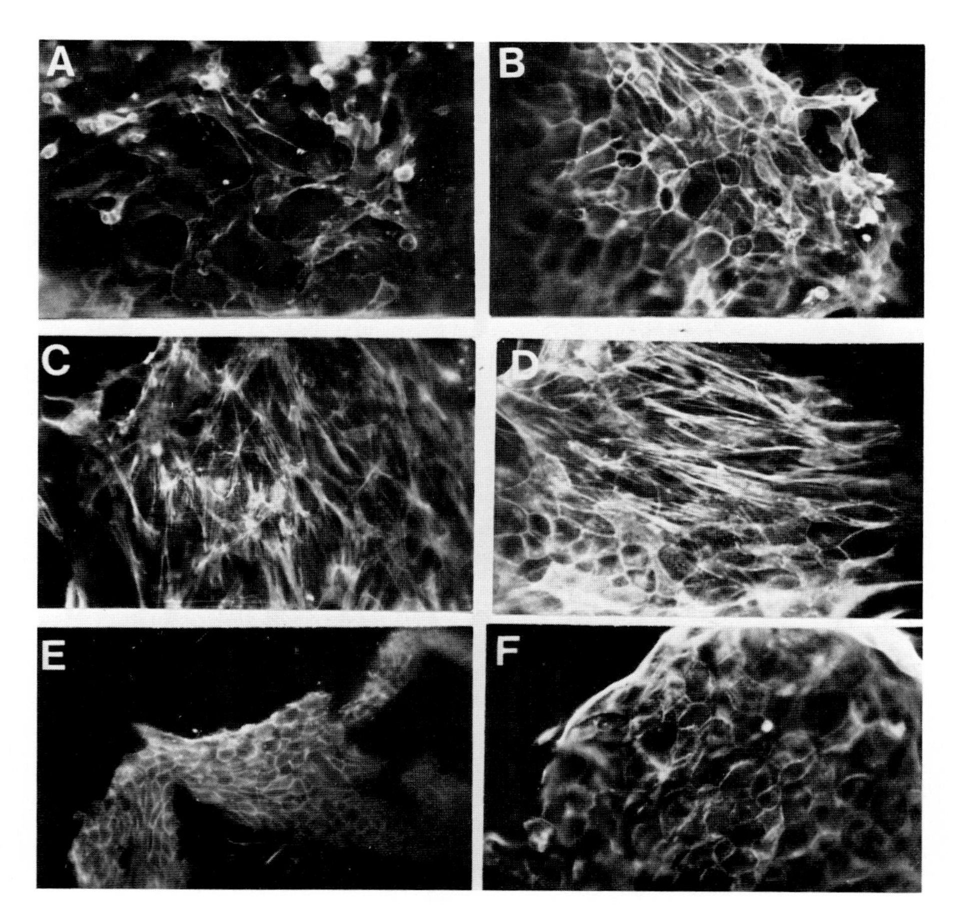

FIG. 11. Epithelial cells derived from colon tissue and stained for F-actin. Primary epithelial cells from an explant were cultured for 1 to 4 days and stained for F-actin with the compound Fl-phalloidin (37). **A** and **B:** Typical examples of cells from normal tissue. **C** and **D:** Typical examples of cells from benign adenomas. **E** and **F:** Examples of cells derived from adenocarcinomas. There is only a moderate amount of actin bundles in the normal cells **(A,B)**. The cells from the adenomas, on the other hand, are blanketed by actin cables **(C,D)**. Note the complete absence of detectable cables in the cells from the adenocarcinomas **(E,F)**.

do they exhibit any stress fibers (Fig. 11E, F) (37). These invasive populations usually display actin at their periphery, but they are easily distinguished from adenoma cells, which are much more angular at their edges. Interestingly, late stage adenomas that have been treated with TPA in culture are similar in appearance to adenocarcinoma cells, perhaps because either plasminogen activator or plasmin is disrupting their cytoskeletons (36,103). Because TPA can disrupt the cytoskeleton even in serum-free media, it is possible that plasminogen activator is directly responsible for the effect (36).

We have tentatively concluded that cultured epithelial cells from the colon lose their organized cytoskeleton upon becoming tumorigenic. Compared to normal skin fibroblasts, however, normal colonic epithelial cells apparently have only a moderately organized skeleton (Fig. 11A, B) (37). The cytoskeleton is dramatically more prominent in premalignant colonic cells and is completely lost upon the transition to malignant tissue. The molecular mechanism underlying cytoskeletal changes in both fibroblast and epithelial cells remains a mystery; nevertheless, models exist.

MODELS OF MECHANISM

Three distinct models of mechanism have been proposed to link growth signals, cytoskeletal changes, and protease activations to each other in a biochemical pathway (85,86,91,92,101).

Automimicry of an External Signal

The postulate is that the transformed cell is secreting the equivalent of a growth hormone or serum, or clustering the receptors for that hormone as if the hormone had been present and bound to those receptors. Secretion of IGF is indeed a normal response by fibroblasts to PDGF or growth hormone. Clearly, either stimulation of PDGF receptors or inherited loss of a requirement for that hormone would lock in repeated cell division. In the first case, this would happen by a cell stimulating itself into repeated divisions; in the second case, it would occur without any further hormone requirement.

Clustering of hormone receptors through protease disruption of the cytoskeleton would be a simple way for the cell to accomplish this bypass. Such a model links cytoskeletal changes with reduced hormone requirements and is consistent with the presence in culture fluid from transformed cells of growth-stimulating factors (85).

Regulatory Mutation Derepressing a Cellular Gene That is Normally Activated Only by a Growth Signal

The cell is presumed to bypass the need for environmental signals completely by making a cellular gene product that is normally never made except as a consequence of full hormone and anchorage signaling in a normal growing tissue. This model predicts that due to gene activation, a transformed cell will express intracellular or surface antigens common to rapidly growing normal cells. If such gene activation were sufficient to generate tumorigenic potential in a cell, certain hierarchically ancient genes should be found expressed both in embyros and in a disparate collection of tumors. Monoclonal antibodies are indeed turning up such antigens.

A major example is a cellular phosphoprotein of 54,000 (54K) daltons. This protein was detected almost simultaneously by workers in two separate fields. Searchers for tumor antigens reported that mice with methylcholanthrene-induced tumors had antibodies to this cellular protein (23). Virologists working with the DNA tumor virus SV40 reported that the 54K cellular protein could be immunoprecipitated with sera obtained from animals with tumors induced by SV40-transformed cells (68,73). The same 54K cellular protein subsequently was found in cells transformed by Abelson murine leukemia virus and adenovirus, in spontaneous teratocarcinomas, and in a variety of other tumor-derived cell lines of murine, human, and other species (58,108). Small amounts of this protein have been detected in normal mouse tissues and cell cultures (46,73) and larger amounts in early mouse embryos (14).

Recently, we have used inbred mouse strains to examine endogenous cellular 54K protein in the absence of any transforming agent (18). We have found that the amount of endogenous cellular 54K protein in early precrisis mouse cells varies in a strain-specific manner. Strains that can produce a high level of endogenous 54K protein in late passage do not further increase 54K expression after SV40 transformation (Table 3).

Structural Mutation Generating an Abnormal Gene Product Involved in Hormone and Anchorage Signaling

A cell would bypass the environment by making an abnormal gene product whose activity is no longer subject to normal cell cycle regulation. This would be the intracellular equivalent of surface PDGF receptor clustering. Examples of such host proteins with altered activity are not yet available. However, transcriptionally active variant alleles of human oncogenes are found in cultured cells of many tumors, and genes for these variant alleles are capable of transfecting anchorage independence (12,120; M. Wigler, *personal communication*).

TUMOR VIRUSES AND MECHANISMS OF TRANSFORMATION

These three mechanisms need not be mutually exclusive, and any is likely to be at best only partly correct. Because they are small, viruses offer the possibility of a resolution of the problem of mechanism. The smallest ones encode only one protein, which by itself can overcome all the growth requirements of a fibroblast and convert it into a sarcomagenic transformed line (125). The transforming gene product of the RNA retroviruses is typically a protein with a hydrophobic C-terminus that places it in the plasma membrane of the cell (67). This protein usually has kinase activity, transferring phosphate to tyrosine residues of a small number of cellular proteins (20,22,55). A close similarity to hormone receptors and their immediate substrates at the inner membrane probably is not coincidental (6,26,31,59,60,107).

The transforming proteins of the DNA tumor viruses are of two sorts: small proteins, which are claimed to have the capacity to mimic growth hormones (125), and large phosphoproteins, which bind to viral and cellular DNA (48) and the 54K cellular phosphoprotein (70). Recently, we have found that SV40 transformation

TABLE 3. *Quantitation of [35S]-methionine-labeled host 54K protein in primary mouse cells*[a]

Cells	Passage	Amounts of 54K protein immunoprecipitated with monoclonal 122 antibody[b]
C57L	p3	0.27
C57L	p12	0.30
SVC57L	[c]	5.75
NFS	p3	0.30
NFS	p12	6.75
SVNFS	[c]	6.38
A[d]	p2	0.36
A	p12	2.70
BALB/C	p2	0.81
BALB/C	p12	2.92
SVBALB/C[d]	[c]	5.73
C3H	p2	0.26
C3H	p12	0.20
SVC3H	[c]	5.58
(C57LxNFS)F1	p2	0.38
(C57LxNFS)F1	p12	4.25

[a]Adapted from ref. 18.
[b]Total trichloracetic acid precipitable radioactivity ([35S]-methionine) was determined in each cell extract, and immunoprecipitation was done using equal amounts of protein. Amounts of 54K protein were determined from fluorograms by scanning on a Gilford 250 Spectrophotometer equipped with a Gilford 2520 Gel Scanner (arbitrary units).
[c]After SV40 transformation, phenotypes are passage independent.
[d]Female mice only.

frequency varies coordinately with endogenous 54K expression (Table 4). This suggests that the genes regulating 54K expression may predispose mice to have a greater or lesser chance of developing at least some forms of cancer.

With tumor viruses and cellular transformation assays, it should be possible to pose the molecular question: How does this one gene product work, and by interaction with which cellular molecules, to overthrow the requirement of the cell for growth signals? A major limitation of the tumor virus strategy so far lies in the use of established normal cells. We have already pointed out that normal tissue cells can be passaged in culture as autonomous microorganisms. Upon passage, the cell population changes, since passage itself selects for the most autonomous of the cells in the biopsied tissue.

Presumably, a set of structural gene mutations must accompany stable establishment. For example, expression of the 54K phosphoprotein is increased in established lines, since this protein is found in many "normal" established lines as well as in many tumors (73); tumor viruses in most laboratories are expected to transform abnormal but anchorage-requiring cells into anchorage-transformed lines. There-

TABLE 4. *Covariation of endogenous 54K protein and susceptibility to SV40 transformation in various strains of mice*

Amount of 54K protein	SV40 transformation frequency	
	High	Low
High	BALB/C NFS (C57L x NFS) A Female (late passage)	None
Low	None	C57L C3H A Female (early passage)

fore, viral gene products are usually interacting with a set of cellular gene products that are different from those found in a normal tissue cell. This is confirmed by a recent study of the polyoma gene products necessary to transform cells. Polyoma middle T was sufficient to anchorage-transform the established rat line FR3T3. However, both polyoma middle T and large T were necessary to anchorage-transform precrisis fibroblasts (98).

The strategy of transforming already abnormal cells may generate information only about a necessary but not sufficient part of oncogenic transformation. Direct approaches will also be necessary to map and understand the mechanism by which a normal cell elaborates genetically stable established variants. Indeed, it is possible that the changes from normal to established cell are part of the changes that must occur before a tissue cell becomes oncogenic. If so, then this second set of changes will be of equal importance to anchorage independence in our understanding of the origins of cancer (10,18,29,56,93).

TRANSFORMATION BY GENES FROM SPONTANEOUS TUMORS

Recently, a new strategy has permitted an old question to be asked about transformation: Which genes in a spontaneously transformed cell or putatively nonviral tumor are capable of transforming a normal cell?

In the past 10 years, a new technology has grown for the direct manipulation of specific DNA sequences. The enzymatic and biochemical tools are now available to isolate, amplify, and sequence a gene and to insert it into a cell, where it will integrate and express in that cell and its descendents (21,43,64,129). Furthermore, the sequence transforming a cell can be found at a later time in the cellular genome and recovered for another cycle of gene sequencing, site-specific mutagenesis, and reimplantation in the genome of another recipient cell. With this set of tools, the search has begun in tumors for the cellular genes that can transform normal cell lines. These are called oncogenes.

The protocol is simple. DNA from a transformed cell is transfected into normal recipient cells. The normal cell cultures are then treated as if they were infected with any other transforming agent. Transformants for anchorage or density are recovered, and the causative sequences of DNA from the donor are tracked and isolated (21,66,69,84,111–113). These studies have led to two remarkable results: First, the genes that have been followed out in detail are homologous to the transforming genes found in Kirsten and Harvey sarcoma viruses, even though the donor DNA may come from a cell line derived from a "spontaneous" human tumor (16,17,81). Second, these tumor oncogenes differ only slightly from ones found in a conserved way in every mammalian species. Expression from the "normal" alleles of these genes is tightly down-regulated in normal cells (12,120). For example, the normal *ras* gene and the tumor *ras* gene can both anchorage transform NIH3T3. However, these transformants showed significant difference in p21 *ras* production. Much more normal *ras* protein than tumor *ras* protein was made in the transformants (121). Apparently, a small amount of tumor p21 *ras* is as effective as a normal amount of normal p21 *ras*.

HYPOTHESES TO EXPLAIN ONCOGENES

Model 1: Tumor oncogenes are oncogenic because they are mutant forms of necessary growth regulatory genes, which play a role in rapid cell growth, perhaps during embryogenesis.

Model 2: Tumor oncogenes are normal genes; that is, the oncogene of the donor tumor may be a member of a polyallelic family of growth regulatory genes. In this family of genes, all alleles would yield normal growth control in tissue, while some would generate the transformed state against the changed genetic background of an established cell. The donor tumor itself would then have arisen as a result of the establishment of its parental cell, perhaps by mutation of the sort that converts normal tissue cells to established cell lines. Upon transfection into a recipient cultured cell that is also established, such a variant allele of a growth regulatory gene would again be able to generate the full phenotype of transformation.

If model 1 is the case, then these mutant cellular oncogenes should make products whose functions are sufficient to generate a tumor from a tissue cell. If model 2 is true, then certain alleles of these host oncogenes may be markers for a higher than normal cellular propensity to develop a tumor. This would imply that the oncogene set may persist in any one of a set of multiple allelic forms in the human genome without inevitably generating a tumor. If a tumor arises, it would be the result of a combination of an inherited "bad" allele of an oncogene in all cells of the body and somatic mutation leading to establishment in a single cell. This second mutation is a good candidate for environmental mutagenesis.

According to model 2, environmental mutation would be responsible for the conversion of cells from normal to established phenotypes in the tissue. Such mutated established cells would not produce tumors unless they arose in a host carrying the sort of bad allele now detected as a "transforming gene" in gene transfer assay. In the presence of a bad allele, however, environmental mutagenesis would result in a fully transformed phenotype, and the cell would grow into a tumor.

One of the oncogenes cloned from a human bladder tumor cell line is reported to be closely related to the oncogenes of the rat-derived Harvey sarcoma virus, as well as a cellular homolog present in normal human DNA (15,24,41,81,110). This particular oncogene is distinguished from its normal cellular counterpart by the loss of a specific restriction endonuclease cleavage site and a single amino acid difference at that site. However, only the normal allele of c-ras_1^{Ha} has been reported to be present in at least one bladder tumor (77).

At least some tumor oncogenes are mutated in one or more amino acid, compared with their normal homologs. Enhanced transcription of c-onc sequences in colon carcinoma (27), however, and similar examples of regulatory changes in oncogenes need not be the result of oncogene mutation. Indeed, changing the local microenvironment of early mouse embryo cells by placing them under the testes capsule of an adult male mouse is sufficient to generate a highly malignant, undifferentiated population of teratocarcinoma cells (117). Even after 8 years of being maintained as a transplantable tumor, teratocarcinoma cells can be entrained to normality by a subsequent change in the microenvironment to the inner cell mass of a normal early mouse embryo (56,76).

Time will tell whether current human sequences called oncogenes are actually agents of carcinogenesis or markers of individual susceptibility to environmental carcinogenesis, or both, or neither.

REFERENCES

1. Albrecht-Buehler, G. (1977): Phagokinetic tracks of 3T3 cells: Parallels between the orientation of track segments and of cellular structures which contain actin or tubulin. *Cell*, 12:333–339.
2. Augusti-Tocco, G., and Sato, G. (1969): Establishment of functional clonal lines of neurons from mouse neuroblastoma. *Proc. Natl. Acad. Sci. USA*, 64:311–315.
3. Baldwin, G. S., Knesel, J., and Monckton, J. M. (1983): Phosphorylation of gastrin-17 by epidermal growth factor-stimulated tyrosine kinase. *Nature*, 301:435–437.
4. Balk, S. D., Whitfield, J. F., Youdale, T., and Braun, A. (1973): Roles of calcium, serum, plasma and folic acid in the control of proliferation of normal and rous sarcoma virus-infected chicken fibroblasts. *Proc. Natl. Acad. Sci. USA*, 70:675–679.
5. Barrett, C. J., Crawford, B. D., Mixter, L. O., et al. (1979): Correlation of in vitro growth properties and tumorigenicity of Syrian hamster cell lines. *Cancer Res.*, 39:1504–1510.
6. Ben-Ze'ev, A., Duerr, A., Solomon, F., and Penman, S. (1979): The outer boundary of the cytoskeleton: A lamina derived from plasma membrane proteins. *Cell*, 17:859–865.
7. Bennett, D. C. (1980): Morphogenesis of branching tubules in cultures of cloned mammary epithelial cells. *Nature*, 285:657–659.
8. Berelowitz, M., Szabo, M., Frohman, L. A., Firestone, S., Chu, L., and Hintz, R. L. (1981): Somatomedin-C mediates growth hormone negative feedback by effects on both the hypothalamus and the pituitary. *Science*, 212:1279–1281.
9. Bishop, J. M., and Varmus, H. (1982): Functions and origins of retroviral transforming genes. In: *RNA Tumor Viruses*, edited by R. Weiss, N. Teich, H. Varmus, and J. Coffin, ch. 9., pp. 999–1108. Cold Spring Harbor Laboratories, Cold Spring Harbor.
10. Boone, C. W. (1975): Malignant hemangioendotheliomas produced by subcutaneous inoculation of Balb/3T3 cells attached to glass beads. *Science*, 188:68–70.
11. Brazeau, P., Ling, N., Bohlen, P., Esch, F., Ying, S. Y., and Guillemin, R. (1982): Growth hormone releasing factor, somatocrinin, releases pituitary growth hormone *in vitro*. *Proc. Natl. Acad. Sci. USA*, 79:7907–7913.
12. Capon, D. J., Chen, E. Y., Levinson, A. D., Seeburg, P. H., and Goedel, D. V. (1983): Complete nucleotide sequences of the T24 human bladder carcinoma oncogene and its normal homologue. *Nature*, 302:33–37.
13. Carpenter, G., King, L., and Cohen, S. (1978): Epidermal growth factor stimulates phosphorylation in membrane preparations in vitro. *Nature*, 276:409–410.

14. Chandrasekaran, K., McFarland, V. W., Simmons, D., Dziadek, M., Gurney, E., and Mora, P. T. (1981): Quantitation and characterization of a species specific and embryo stage-dependent 55 kilodalton phosoprotein also present in cells transformed by simian virus 40. *Proc. Natl. Acad. Sci. USA*, 78:6953–6957.
15. Chang, E. H., Gonda, M. A., Ellis, R. W., Scolnick, E. M., and Lowy, A. R. (1982): Human genome contains four genes homologous to transforming genes of Harvey and Kirsten murine sarcoma viruses. *Proceedings of the National Academy of Sciences (79)*, 4848–4852.
16. Chang, E., Furth, M. E., Scolnick, E. M., and Lowy, D. R. (1982): Tumorigenic transformation of mammalian cells induced by a normal human gene homologous to the oncogene of Harvey murine sarcoma virus. *Nature*, 297:479–483.
17. Chang, E. H., Gonda, M. A., Ellis, R. W., Scolnick, E. M., and Lowy, D. R. (1982): Human genome contains four genes homologous to transforming genes of Harvey and Kirsten murine sarcoma viruses. *Proc. Natl. Acad. Sci. USA*, 79:4848–4852.
18. Chen, S., Blanck, G., and Pollack, R. (1983): Early passage of pre-crisis mouse cells shows strain specific covariation in expression of 54k phosphoprotein and in susceptibility to transformation by SV40. *Proc. Natl. Acad. Sci. USA*, 80:5670–5674.
19. Clemmons, D. R., Underwood, L. E., and Van Wyk, J. J. (1981): Hormonal control of immunoreactive somatomedin production by human fibroblasts. *J. Clin. Invest.*, 67:10–17.
20. Collett, M., and Erikson, R. (1978): Protein kinase activity associated with the avian sarcoma virus *src* gene product. *Proc. Natl. Acad. Sci. USA*, 75:2021–2024.
21. Cooper, G. M., Okenquist, S., and Silverman, L. (1980): Transforming activity of DNA of chemically transformed and normal cells. *Nature*, 284:418–421.
22. Courtreidge, S. A., Levinson, A. D., and Bishop, J. M. (1980): The protein encoded by the transforming gene of avian sarcoma virus ($pp60^{src}$) and a homologous protein in normal cells ($pp60^{proto\text{-}src}$) are associated with the plasma membrane. *Proc. Natl. Acad. Sci. USA*, 77:3783–3787.
23. DeLeo, A. B., Appella, E., DuBois, G. C., Law, L. W., and Old, L. J. (1979): Detection of a transformation-related antigen in chemically induced sarcomas and other transformed cells of the mouse. *Proc. Natl. Acad. Sci. USA*, 76:2420–2424.
24. Der, C. J., Krontivis, J. G., and Cooper, G. M. (1982): Transforming genes of human bladder and lung carcinoma cell lines are homologous to the ras genes of Harvey and Kirsten sarcoma viruses. *Proc. Natl. Acad. Sci. USA*, 79:3637–3640.
25. Eagle, H. (1955): Nutritional needs of mammalian cells in tissue culture. *Science*, 122:501–504.
26. Eckhart, W., Dulbecco, R., and Burger, M. (1971): Temperature-dependent surface changes in cells infected or transformed by a thermosensitive mutant of polyoma virus. *Proc. Natl. Acad. Sci. USA*, 68:283–286.
27. Eva, A., Robbins, K. C., Andersen, P. R., Srinivasan, A., Tronick, S. R., Reddy, E. P., Ellmore, N. W., Galen, A. T., Lautenberger, J. A., Papas, T. S., Westin, E. H., Wong-Staal, F., Gallo, R. C., and Aaronson, S. A. (1982): Cellular genes analogous to retroviral *onc* genes are transcribed in human tumour cells. *Nature*, 295:116–119.
28. Farmer, S. R., Ben-Ze'ev, A., Benecke, B., and Penman, S. (1978): Altered translatability of messenger RNA from suspended anchorage-dependent fibroblasts: Reversal upon cell attachment to a surface. *Cell*, 15:627–637.
29. Fidler, I. J., and Kripke, M. L. (1977): Metastasis results from preexisting variant cells within a malignant tumor. *Science*, 197:893–895.
30. Fischer, S. M., Viaje, A., Harris, K. L., Miller, D. R., Bohrman, J. S., and Slaga, T. J. (1980): Improved conditions for murine epidermal cell culture. *In Vitro*, 16:180–188.
31. Flanagan, J., and Koch, G. L. E. (1978): Cross-linked surface Ig attaches to actin. *Nature*, 273:278–281.
32. Folkman, J., and Moscona, A. (1978): Role of cell shape in growth control. *Nature*, 273:345–349.
33. Freedman, V., and Shin, S. (1974): Cellular tumorigenicity in *nude* mice: Correlation with cell growth in semi-solid medium. *Cell*, 3:355–359.
34. Freedman, V., Shin, S., Risser, R., and Pollack, R. (1975): Tumorigenicity of virus transformed cells in nude mice is correlated specifically with anchorage independent growth *in vitro*. *Proc. Natl. Acad. Sci. USA*, 72:4335–4339.

35. Freeman, A. E., Black, P. H., Wolford, R., and Huebner, R. (1967): Adenovirus type 12-rat embryo transformation system. *J. Virol.*, 1:362–367.
36. Freidman,·E. A. (1981): Differential response of premalignant epithelial cell classes to phorbol ester tumor promoters and to deoxycholic acid. *Cancer Res.*, 41:4588–4599.
37. Friedman, E., Verderame, M., and Pollack, R. (1983): Loss of actin cytoskeletal organization accompanies the benign to malignant tumor transition in colonic epithelium examined *in vitro*. *Can. Res. (submitted).*
38. Friedman, E. A., Higgins, P. J., Lipkin, M., Shinya, H., and Gelb, A. M. (1981): Tissue culture of human epithelial cells from benign colonic tumors. *In Vitro*, 17:632–644.
39. Friend, C., Scher, W., Holland, J. G., and Sato, T. (1971): Hemoglobin synthesis in murine virus-induced leukemic cells *in vitro*: Stimulation of erythroid differentiation by dimethyl sulfoxide. *Proc. Natl. Acad. Sci. USA*, 68:378–382.
40. Gey, G. O., Coffman, W. D., and Kubicek, M. T. (1952): Tissue culture studies of the proliferative capacity of cervical carcinoma in normal epithelium. *Cancer Res.*, 12:264–265.
41. Goldfarb, M., Shimizu, C., Perucho, M., and Wigler, M. (1982): Isolation and preliminary characterization of a human transforming gene from T24 bladder carcinoma cells. *Nature*, 296:404–409.
42. Goldman, R. D., and Follett, E. A. C. (1969): The structure of the major cell processes of isolated BHK21 fibroblasts. *Exp. Cell Res.*, 57:263–276.
43. Graessman, M., Graessman, A., and Mueller, C. (1977): The biological activity of different early SV40 DNA fragments. *INSERM*, 69:233–240.
44. Green, H., and Meuth, M. (1974): An established pre-adipose cell line and its differentiation in culture. *Cell*, 3:127–133.
45. Green, H., Rheinwald, J. G., and Sun, T. (1977): *Properties of an Epithelial Cell Type in Culture: The Epidermal Keratinocyte and its Dependence on Products of the Fibroblast.* Liss, New York.
46. Gurney, E., Harrison, R. O., and Fenno, J. (1980): Monoclonal antibodies against simian virus 40 T antigens: Evidence for distinct subclasses of large T antigen and for similarities among nonviral T antigens. *J. Virol.*, 34:752–763.
47. Hamilton, W. G., and Ham, R. (1977): Clonal growth of Chinese hamster cell lines in protein-free media. *In Vitro*, 13:537–547.
48. Hand, R. (1981): Functions of T antigens of SV40 and polyoma virus. *Biochim. Biophys. Acta*, 651:1–24.
49. Hayashi, I., and Sato, G. (1976): Replacement of serum by hormones permits growth of cells in a defined media. *Nature*, 259:132–134.
50. Hayflick, L., and Moorehead, P. S. (1961): The serial cultivation of human diploid cell strains. *Exp. Cell Res.*, 25:585–621.
51. Heggeness, M., Wang, K., and Singer, S. J. (1977): Intracellular distributions of mechanochemical proteins in cultured fibroblasts. *Proc. Natl. Acad. Sci. USA*, 74:3883–3887.
52. Heuser, J. E., and Kirscher, M. W. (1980): Filament organization revealed in platinum replicas of freeze-dried cytoskeletons. *J. Cell Biol.*, 86:212–234.
53. Kohler, G., and Milstein, C. (1975): Continuous cultures of fused cells secreting antibody of predefined specificity. *Nature*, 256:495–497.
54. Hunter, T., and Cooper, J. A. (1981): Epidermal growth factor induces rapid tyrosine phosphorylation of proteins in A431 human tumor cells. *Cell*, 24:741–752.
55. Hunter, T., and Sefton, B. M. (1980): The transforming gene product of rous sarcoma virus phosphorylates tyrosine. *Proc. Natl. Acad. Sci. USA*, 77:1311–1315.
56. Illmensee, I., and Mintz, B. (1976): Totipotency and normal differentiation of single teratocarcinoma cells cloned by injection into blastocyts. *Proc. Natl. Acad. Sci. USA*, 73:549–553.
57. Israel, M., Chan, H. W., Martin, M. A., and Rowe, W. P. (1979): Molecular cloning of polyoma virus DNA in *Escherichia coli*: Oncogenicity testing in hamsters. *Science*, 205:1140–1142.
58. Luka, J., Jornvall, H., and Klein, G. (1979): Purification and biochemical characterization of the Epstein-Barr virus determined nuclear antigen and an associated protein with a 53,000 dalton. *J. Virol.*, 35:592–602.
59. Kahn, C. R., Baird, K., Jarrett, D., and Flier, J. (1978): Direct demonstration that receptor crosslinking of aggregation is important in insulin action. *Proc. Natl. Acad. Sci. USA*, 75:4209–4213.
60. Kasuga, M., Zick, Y., Blithe, D. L., Crettaz, M., and Kahn, C. R. (1982): Insulin stimulates tyrosine phosphorylation of the insulin receptor in a cell-free system. *Nature*, 298:667–669.
61. Kennedy, A. R., Fox, M., Murphy, G., and Little, J. (1980): Relationship between X-ray exposure and malignant transformation in C3h 10T1/2 cells. *Proc. Natl. Acad. Sci. USA*, 77:7262–7266.

62. King, G., and Kahn, C. (1981): Non-parallel evolution of metabolic and growth-promoting functions of insulin. *Nature*, 292:644–646.
63. King, L. E., Carpenter, G., and Cohen, S. (1980): Characterization by electrophoresis of epidermal growth factor stimulated phosphorylation using A431 membranes. *Biochemistry*, 19:1524–1528.
64. Klobutcher, L. A., and Ruddle, F. H. (1979): Phenotype stabilization and integration of transferred material in chromosome-mediated gene transfer. *Nature*, 280:657–660.
65. Kopelovich, L., Blattner, W. A., Fraumeni, J. F., Lipkin, M., and Pollack, R. (1980): Organization of cytoskeletal actin in cultured skin fibroblast from patients at high risk of colon cancer. *Int. J. Cancer*, 26:301–308.
66. Krontiris, T. G., and Cooper, G. M. (1981); Transforming activity of human tumor DNAs. *Proc. Natl. Acad. Sci. USA*, 78:1181–1184.
67. Krueger, J. G., Wang, E., and Goldberg, A. R. (1980): Evidence that the *src* gene product of rous sarcoma virus is membrane associated. *Virology*, 101:25–40.
68. Lane, D. P., and Crawford, L. V. (1979): T antigen is bound to a host protein in SV40 transformed cells. *Nature*, 278:261–263.
69. Lane, M., Sainten, A., and Cooper, G. M. (1981): Activation of related transforming genes in mouse and human mammary carcinomas. *Proc. Natl. Acad. Sci. USA*, 78:5185–5189.
70. Lane, D., and Crawford, L. (1979): T antigen is bound to a host protein in SV40 transformed cells. *Nature*, 278:261–263.
71. Lazarides, E., and Weber, K. (1974): Actin antibody: The specific visualization of actin filaments in non-muscle cells. *Proc. Natl. Acad. Sci. USA*, 71:2268–2272.
72. Lee, L., and Weinstein, I. B. (1978): Epidermal growth factor, like phorbol esters, induces plasminogen activator in HeLa cells. *Nature*, 274:696–697.
73. Linzer, D. I. H., and Levine, A. J. (1979): Characterization of a 54K dalton cellular SV40 tumor-antigen present in SV40-transformed cells and uninfected embryonal carcinoma cells. *Cell*, 17:43–52.
74. Lipkin, M., and Deschner, E. (1976): Early proliferative changes in intestinal cells. *Cancer Res.*, 36:2665–2668.
75. Macpherson, I., and Montagnier, L. (1964): Agar suspension culture for the selective assay of cells transformed by polyoma virus. *Virology*, 23:291–294.
76. Mintz, B., and Illmensee, K. (1975): Normal genetically mosaic mice produced from malignant teratocarcinoma cells. *Proc. Natl. Acad. Sci. USA*, 72:3585–3589.
77. Muschel, R. J., Kouhry, G., Lebowitz, P., Koller, R., and Dahr, R. (1983): Retraction of data on the human oncogene c-ras_1^H. *Science*, 220.
78. Muto, T., Bussey, H. J. R., and Morson, B. C. (1975): The evolution of cancer of the colon and rectum. *Cancer*, 36:2251–2270.
79. Nicholson, N. B., and Pollack, R. (1983): SV40 induces a reduced calcium requirement in mouse cells. *Mol. Cell Biol., (submitted).*
80. Nicholson, N., Verderame, M., Lipkin, M., and Pollack, R. (1981): F-actin patterns quantitated with Fl-phalloidin in skin fibroblasts of individuals genetically predisposed to colon cancer. In: *International Cell Biology 1981–82*, edited by H. Schweiger, pp. 331–335. Springer-Verlag, Berlin.
81. Parada, L. F., Tabin, C. J., Shih, C., and Weinberg, R. A. (1982): Human EJ bladder carcinoma oncogene is homologue of harvey sarcoma virus *ras* gene. *Nature*, 297:474–478.
82. Peehl, D. M., and Ham, R. G. (1980): Growth and differentiation of human keratinocytes without a feeder layer or conditioned medium. *In Vitro*, 16:516–525.
83. Perez-Rodriguez, R., Chambard, J. C., Van Obberghen-Schilling, E., Franchi, A., and Pouyssegur, J. (1981): Emergence of hamster fibroblast tumors in nude mice—evidence for in vivo selection leading to loss of growth factor requirement. *J. Cell. Physiol.*, 109:387–396.
84. Perucho, M., Goldfarb, M., Shimizu, K., Lama, C., Fogh, J., and Wigler, M. (1981): Human tumor-derived cell lines contain common and different transforming genes. *Cell*, 27:467–476.
85. Pollack, R. (1981): *Hormones, Anchorage and Oncogenic Cell Growth*. Grune & Stratton, New York.
86. Pollack, R., and Hough, P. V. C. (1974): The cell surface and malignant transformation. *Annu. Rev. Med.*, 25:431–446.
87. Pollack, R., and Rifkin, D. (1975): Actin-containing cables within anchorage-dependent rat embryo cells are dissociated by plasmin and trypsin. *Cell*, 6:495–506.
88. Pollack, R., and Rifkin, D. (1976): Maintenance of mammalian cell shape: Redistribution of

intracellular actin by SV40, proteases, cytochalasin B and dimethylsulfoxide. In: *Cell Motility*, edited by J. Rosenbaum, T. Pollard, and R. Goldman, pp. 389–401. Cold Spring Harbor, New York.

89. Pollack, R., Osborn, M., and Weber, K. (1975): Patterns of organization of actin and myosin in normal and transformed cells. *Proc. Natl. Acad. Sci. USA*, 72:994–998.
90. Pollack, R., Risser, R., Conlon, S., Freedman, V., Rifkin, D., and Pollack, R. (1975): Production of plasminogen activator and colonial growth in semisolid medium are *in vitro* correlates of tumorigenicity. In: *Proteases and Biological Control*, edited by E. Reich, D. Rifkin, and E. Shaw, pp. 885–899. Cold Spring Harbor, New York.
91. Pollack, R., Prives, C., and Manley, J. (1983): Do variant SV40 sequences have a role in the maintenance of the oncogenic transformed phenotype? In: *Genes and Proteins in Oncogenesis*, edited by H. Vogel and I. B. Weinstein, pp. 295–302. Academic Press, New York.
92. Pollack, R. (1977): Viral and cellular contributions to expression of transformed state. In: *Neoplastic Transformation*, edited by H. Kaprowski, pp. 169–179. Dahlem Konferenzen, Berlin.
93. Pollack, R. (1983): Letter. *Science*, 218:4577.
94. Powers, R. S., Chen, S., and Pollack, R. (1983): SV40 transformation in hormonally defined medium: Direct isolation of insulin-independent transformed lines. *Science (submitted)*.
95. Powers, S., Alcorta, D., Nicholson, N., and Pollack, R. (1982): Effects of calcium and hormones on the cytoskeleton and cell proliferation. In: *Cold Spring Harbor Conferences on Cell Proliferation*, pp. 243–258. Cold Spring Harbor, New York.
96. Price, F. M., Camalier, R. F., Grant, R., Taylor, W. G., Smith, H. G., and Sanford, K. K. (1980): A new culture medium for human skin epithelial cells. *In Vitro*, 16:147–158.
97. Quaroni, A., Wands, J., Trelstad, R. L., and Isselbacher, K. I. (1979): Epitheloid cell cultures from the small intestine. Characterization by morphological and immunological criteria. *J. Cell Biol.*, 80:248–265.
98. Rassoulzadegan, M., Cowie, A., Carr, A., Glaichenhaus, N., Kamen, R., and Cuzin, F. (1982): The roles of individual polyoma virus early proteins in oncogenic transformation. *Nature*, 300:713–718.
99. Rheinwald, J. G., and Green, H. (1975): Serial cultivation of human epidermal keratinocytes: The formation of keratinizing colonies from single cells. *Cell*, 6:331–334.
100. Rifkin, D., and Pollack, R. (1977): The production of plasminogen activator by established cell lines of mouse origin. *J. Cell Biol.*, 73:47–55.
101. Rifkin, D., and Pollack, R. (1976): Proteases produced by normal and malignant cells in culture. In: *Proteases and Physiological Regulation*, edited by D. W. Ribbons and K. Brew, pp. 263–285. Academic Press, New York.
102. Rifkin, D. B., Beal, L. P., and Reich, E. (1975): Macromolecular determinants of plasminogen activator synthesis. In: *Proteases and Biological Control, Cold Spring Harbor Conferences on Cell Proliferation*, edited by E. Reich, D. B. Rifkin, and E. Shaw, pp. 841–847. Cold Spring Harbor, New York.
103. Rifkin, D., Crowe, R., and Pollack, R. (1979): Tumor promoters induce changes in cytoskeletal organization in chick embryo fibroblasts. *Cell*, 18:1504–1510.
104. Risser, R., and Pollack, R. (1974): A nonselective analysis of SV40 transformation of mouse 3T3 cells. *Virology*, 59:477–489.
105. Risser, R., and Pollack, R. (1974): Biological analysis of clones of SV40-infected mouse 3T3 cells. In: *Control of Proliferation in Animal Cells*, edited by B. Clarkson and R. Baserga, pp. 125–138. Cold Spring Harbor, New York.
106. Ross, R., and Vogel, A. (1978): The platelet-derived growth factor. *Cell*, 14:203–210.
107. Roth, R., and Cassell, D. (1983): Insulin receptor: Evidence that it is a protein kinase. *Science*, 219:299–301.
108. Rotter, V., Witte, O., Coffman, R., and Baltimore, D. (1980): Abelson murine leukemia virus-induced tumors elicit antibodies against a host cell protein P50. *J. Virol.*, 36:547–555.
109. Rous, P., and Jones, F. S. (1916): A method for obtaining suspensions of living cells from the fixed tissues, and for the plating out of individual cells. *J. Exp. Med.*, 23:549–555.
110. Shih, C., and Weinberg, R. A. (1982): Isolation of a transforming sequence from a human bladder carcinoma cell line. *Cell*, 29:161–169.
111. Shih, C., Shilo, B., Goldfarb, M. P., Dannenberg, A., and Weinberg, R. A. (1979): Passage of phenotypes of chemically transformed cells via transfection of DNA and chromatin. *Proc. Natl. Acad. Sci. USA*, 76:5714–5718.

112. Shih, C., Padhy, L. C., Murray, M., and Weinberg, R. A. (1981): Transforming genes of carcinomas and neuroblastomas introduced into mouse fibroblasts. *Nature*, 290:261–264.
113. Shilo, B., and Weinberg, R. A. (1981): Unique transforming gene in carcinogen-transformed mouse cells. *Nature*, 289:607.
114. Smith, H. S., Lan, S., Ceriani, R., Hackett, A. J., and Stampfer, M. R. (1981): Clonal proliferation of cultured non-malignant and malignant human breast epithelium. *Cancer Res.*, 41:4637–4643.
115. Steinberg, B. M., Rifkin, D., Shin, S., Boone, C., and Pollack, R. (1979): Tumorigenicity of revertants from an SV40 transformed line. *J. Supramol. Struct.*, 11:539–546.
116. Steinberg, B., Smith, K., and Pollack, R. (1980): Viral transformation diminishes cellular contractivity in native collagen gels. *J. Cell Biol.*, 87:304–308.
117. Stevens, L. C. (1970): The development of transplantable teratocarcinomas from intratesticular grafts of pre- and postimplantation mouse embryos. *Dev. Biol.*, 21:364–382.
118. Stiles, C. D., Capone, G. T., Scher, C. D., Antoniades, H. N., Van Wyk, J. J., and Pledger, W. J. (1979): Dual control of cell growth by somatomedins and platelet-derived growth factor. *Proc. Natl. Acad. Sci. USA*, 76:1279–1283.
119. Sutherland, B. M., Cimino, J. S., Delihas, N., Shih, A. G., and Oliver, R. P. (1980): Ultraviolet light-induced transformation of human cells to anchorage-independent growth. *Cancer Res.*, 40:1934–1939.
120. Tabin, C. J., Bradley, S. M., Bargmann, C. I., Weinberg, R. A., Papageorge, A. G., Scolnick, E. M., Dahr, R., Lowy, D. R., and Chang, E. H. (1982): Mechanism of activation of a human oncogene. *Nature*, 300:143–149.
121. Taparowski, E., Suard, Y., Fasano, O., Shimizu, K., Goldfarb, M., and Wigler, M. (1982): Activation of the T24 bladder carcinoma gene is linked to a single amino acid change. *Nature*, 300:762–764.
122. Temin, H. M., and Rubin, H. (1958): Characteristics of an assay for Rous sarcoma virus and Rous sarcoma cells in tissue culture. *Virology*, 6:669–688.
123. Todaro, G., and Green, H. (1963): Quantitative studies of the growth of mouse embryo cells in culture and their development into established lines. *J. Cell Biol.*, 17:299–313.
124. Todaro, G., and Green, H. (1964): An assay for cellular transformation by SV40. *Virology*, 23:117–119.
125. Topp, W., Lane, D., and Pollack, R. (1980): Transformation by polyoma and SV40. In: *DNA Tumor Viruses*, edited by J. Tooze, pp. 205–296. Cold Spring Harbor, New York.
126. Verderame, M., Alcorta, D., Egnor, M., Smith, K., and Pollack, R. (1980): Cytoskeletal F-actin patterns quantitated with fluorescein-isothiocynate phalloidin in normal and transformed cells. *Proc. Natl. Acad. Sci. USA*, 77:6624–6628.
127. Vogel, A., and Pollack, R. (1974): Isolation and characterization of revertant cell lines. VI. Susceptibility of revertants to retransformation by simian virus 40 and murine sarcoma virus. *J. Virol.* 14:1404–1412.
128. Vogel, A., Oey, J., and Pollack, R. (1974): Two classes of revertants isolated from SV40 transformed 3T3 mouse cells. In: *Control of Proliferation in Aminal Cells*, edited by B. Clarkson and R. Baserga, pp. 125–138. Cold Spring Harbor, New York.
129. Wigler, M., Pellicer, A., Silverstein, S., and Axel, R. (1978): Biochemical transfer of single-copy eukaryotic genes using total cellular DNA as donor. *Cell*, 14:725–731.
130. Yaffe, D. (1968): Retention of differentiation potentialities during prolonged cultivation of myogenic cells. *Proc. Natl. Acad. Sci. USA*, 61:477–483.
131. Yang, J., Richards, J., Gluzman, R., Imagawa, W., and Nandi, S. (1980): Sustained growth in primary culture of normal mammary epithelial cells embedded in collagen gels. *Proc. Natl. Acad. Sci. USA*, 77:2088–2092.

Advances in Viral Oncology, Volume 4,
edited by George Klein. Raven Press,
New York © 1984.

Evolution of Cellular Oncogenes

Ben-Zion Shilo

Department of Virology, Weizmann Institute of Science, Rehovot, Israel

The vertebrate genome contains a set of cellular genes which can be converted into transforming genes (oncogenes). Activation can occur as a result of infection with retroviruses, leading to recombination of the oncogene with the viral genome (7,8), or following integration of the infecting virus in the proximity of the cellular gene (28). Alternatively, nonviral agents were shown to activate the genes by inducing point mutations (50,67) or by causing transpositions (49) and chromosomal translocations (17,39,55,68). The target genes for these modifications are referred to as cellular oncogenes (c-*onc*) or proto oncogenes. These are somewhat misleading terms, since they indicate that conversion to potent oncogenes is their ultimate function. In fact, it is more likely that the c-*onc* fulfill basic roles in the physiology and metabolism of the normal cell or tissue, and that the oncogenic activation represents an exceptional situation. One of the major justifications for this argument is the observation that c-*onc* sequences are highly conserved in evolution. They can be identified not only in the organism in which the tumor was originally isolated but also in distantly related species and even in other phyla where they are not known to be involved in transformation. The detection of expression of c-*onc* in nontransformed cells, which in some cases is tissue specific or restricted to defined developmental stages, is also a compelling indication of their crucial role in the organism.

Only a limited number of papers consider oncogene evolution directly. From the rapidly accumulating information on oncogenes, interesting conclusions, ideas, or hints bearing on the evolution of the genes have begun to emerge. This chapter highlights these points and focuses on the distribution and structure of c-*onc* in different species and phyla. The implications on the function of the gene products, on the essential domains of the c-*onc* proteins, and on the interaction of oncogenes with other cellular proteins are discussed.

The existing data on oncogene structure in various species allow us to begin to trace the evolutionary process of these genes. Two parallel courses can be identified. On one hand, a static impression emerges due to the high degree of conservation of each oncogene; on the other hand, the identification of families of oncogenes of common ancestry reveals a dynamic situation in which multiple duplication events of oncogenes have occurred during the evolution of the species.

The two major families of oncogenes are discussed, with emphasis on structural similarities and differences among the members of each family.

STRUCTURAL CONSERVATION OF CELLULAR ONCOGENES IN VERTEBRATES

A variety of methods have been utilized to study the structural conservation of c-*onc*, including liquid DNA:DNA hybridization, Southern blotting, cloning, DNA sequencing, and immunoprecipitation. The structural conservation of c-*onc* across a wide spectrum of vertebrate species has been demonstrated for most oncogenes that were thoroughly tested. It provides a compelling argument for the cellular origin of viral oncogenes and points to the central role these genes are likely to play in vertebrates. The ability to identify regions within each gene that are more conserved is a powerful method for defining the functional domains of the protein. This chapter presents the results that bear on the structural conservation of each of the oncogenes that has been studied in detail.

src

The first oncogene to be identified and studied was the *src* gene of the Rous avian sarcoma virus. The ability to make hybridization probes that are specific for this oncogene (65) has allowed testing for the presence of homologous sequences in uninfected cells. The probe hybridized to cellular DNA of uninfected cells following kinetics of single copy genes (66). Moreover, under stringent hybridization conditions, homology could also be detected in the DNA of distant avian species, such as duck and emu. Similar experiments that were performed under nonstringent conditions facilitating pairing of partially matched DNA sequences have detected sequences homologous to *src* in human, calf, mouse, and salmon DNA (64). The identification of sequences homologous to the *src* oncogene in distantly related organisms has far-reaching implications for the evolution of this gene. According to the fossil record, birds, mammals, and teleosts diverged more than 400 million years ago (52,76); thus the *src* gene evolved prior to this period. The wide distribution of *src* sequences in contrast to the rather narrow host range of retroviruses provides one of the most convincing arguments for the cellular origin of retroviral oncogenes (7).

The gene product of the *src* oncogene was the first to be identified and studied in detail. It has a molecular weight of 60,000 daltons (9) (hence its name $pp60^{src}$) and is phosphorylated *in vivo*. The identification of a protein kinase activity of this protein and its unique tyrosine-phosphorylating properties (15,31) provide a valuable clue to the function of the protein. In view of the structural conservation of the *src* DNA sequence, it is not surprising that antibodies generated against the viral gene product were able to immunoprecipitate the homologous protein in distantly related species. Rabbit antiserum prepared against the viral $pp60^{src}$ was able to immunoprecipitate a c-*src* protein of similar size from chicken, rat, frog, and human cells (44,51). Not all the rabbit tumor antisera displayed complete cross reactivity with pp60 from uninfected vertebrate cells, indicating that only some of the antigenic determinants of the protein are conserved. The products of partial protease digestion of the immunoprecipitated protein from chicken, rat, and human cells were indistinguishable, again demonstrating the high degree of structural conservation. The pp60 from vertebrate cells also displayed functional homology to the viral $pp60^{src}$. *In vitro* phosphorylation of the immunoglobulin heavy chain was observed with the cellular pp60 from different species. In addition, the pp60 was shown to be phosphorylated in all cases, similar to the viral $pp60^{src}$.

mos

A more detailed view on the structural conservation of oncogenes was obtained when the homologous genes in different species were sequenced. The ability to deduce the amino acid sequence allows a more accurate determination of homology, in positions where third base changes may have caused mismatch of the nucleotide sequence. Furthermore, the domains within the protein which are more highly conserved can be identified from sequence comparison. These domains are likely to be crucial for the function of the c-*onc*.

Mouse and human c-*mos* sequences were sequenced and aligned (71,77). The two sequences are 77% homologous at the nucleotide level and 75% homologous at the amino acid level. Specific regions of functional importance are suggested by even greater homology (e.g., 85% within the middle of the polypeptide chain). Mouse and human genes terminate at the same position, but their initiation positions are different. The putative first ATG in the human c-*mos* was aligned with the 48th internal codon of the mouse sequence. The suggestion that the first codons of the mouse c-*mos* are not essential for its function gains further support from the observation that in viral *mos* (v-*mos*) the first 20 codons of the cellular gene are not present (71). A rearranged c-*mos* has recently been identified as the transforming gene of a mouse plasmocytoma cell line (49). In this case, a foreign sequence has been inserted into the coding region of c-*mos* in codon position 87, again suggesting that the NH_2 terminus of the protein is not functionally essential.

fes-fps

The oncogenes *fes* and *fps* were independently isolated from cats and chickens, respectively, where hybridization and immunologic cross-reactivity experiments suggested that they represent the same gene (4,6,56). Sequence comparison of the two viral genes identified conservation of 70% of the deduced amino acid sequence (27,57). This gene has been altered at a rate that is significantly slower than that of c-*mos*, in which a similar degree of conservation was observed for mouse and human genes (which separated more than 200 million years after the separation of cat and chicken).

abl

Hybridization experiments with a probe prepared from the Abelson murine leukemia virus oncogene have detected homologous bands in the DNA of rat, hamster, rabbit, chicken, and human (26). The human c-*abl* gene has recently been cloned. Similar to the mouse gene, it contains multiple large introns and spans more than 30 kilobases (kb) (34).

has

Probes prepared from the *has* sequence have detected homologous sequences in the DNA of rat, chicken, mouse, and human (22). Cloning of the cellular homologs of *has* from rat and human DNA has demonstrated the presence of two copies of the cellular gene: one contains three introns, and the other has none (10,18). The positions of the three introns within the spliced mouse and human genes are similar. Moreover, heteroduplex analysis has shown that the homology between the mouse and human genes is not restricted to the coding region but spans part of the introns as well (10).

The gene product of the *has* oncogene was studied in detail. It has a molecular weight of 21,000 daltons (hence termed p21) (60). The protein is present both in a phosphorylated and a nonphosphorylated form *in vivo*. A single amino acid substitution in the 12th residue of this protein was shown to be responsible for its conversion to a potent oncogene in a human bladder carcinoma (50,67). By using rat antiserum prepared against the p21 protein of Harvey murine sarcoma virus, a protein of similar size was immunoprecipitated from uninfected vertebrate cells, including rat, mouse, cat, dog, monkey, human, and turkey cells (37). Partial protease digestion has demonstrated the structural similarity among the viral p21 protein and its cellular homologs in a variety of vertebrate species. The phosphorylated form of the p21 protein was not detected, however, in uninfected vertebrate cells.

myc

The cellular gene homologous to the oncogene of avian myelocytomatosis virus has been isolated from chicken and human DNA. Characterization of the clones has shown that the organization of the c-*myc* gene in human and chicken DNA is similar. In both cases, two coding regions separated by an intron of similar size have been identified (19,42,74). Comparison of the nucleotide sequence of the human c-*myc* gene to the v-*myc* (which originated from the chicken c-*myc* gene) revealed partial homology of the predicted amino acid sequence in the first exon and a 71% homology for the second (1,14). Both proteins terminate in the same position. It is interesting to note that in addition to the c-*myc* gene, the v-*myc* probe has identified a set of sequences in human DNA which display only partial homology to the probe (19). They may represent distantly related genes or pseudogenes.

MULTIGENE FAMILIES OF ONCOGENES IN VERTEBRATES

The preceding chapter emphasizes the structural conservation of c-*onc*. Selective pressures for conservation of the genes, however, are not the only factors that guide their evolution. In parallel, modifications in the structure or regulation of the c-*onc* must occur for adjustment of the genes to the changing physiology of the evolving species. The two conflicting requirements are settled by gene duplication events which allow a variety of selective forces to act on the different copies of the duplicated gene. This chapter discusses the presence of two multigene families of oncogenes that are a consequence of duplication events: the *src* and the *ras* gene families.

The *src* Gene Family

More than a dozen types of distinct oncogenes were isolated by the generation of acutely transforming retroviruses (8,12). The v-*onc* sequences were defined as distinct genes by lack of cross-homology in nucleic acid hybridization experiments and by their hybridization to different cellular restriction fragments in Southern blots of genomic DNA. With the accumulation of DNA sequence data for a variety of oncogenes, the deduced amino acid sequences could be used to search for structural relationships among different oncogenes. It became apparent that at the amino acid level, several oncogenes exhibited structural homology to *src*. The *yes* oncogene showed the strongest homology: 82% of the 436 amino acids in the COOH-terminus of $pp60^{src}$ were identical to *yes* (33). The *fes* and *fps* oncogenes exhibited homology of 45% of the amino acids with $pp60^{src}$ (27,57). A smaller degree of homology was observed between *mos* and *src* (72). The genes that are related to this family are not confined to oncogenes. Bovine cyclic AMP-dependent protein kinase was also shown to contain structural homology to *src* (5). Of particular interest is the conservation of lysine (residue 71 in the bovine protein) in all members of the *src* family. This residue is the proposed ATP binding site. Identification of additional cellular proteins with known functions as members of the *src* family may prove to be instrumental in understanding the role and regulation of the *src* oncogenes.

The different members of the *src* family were found not only to be structurally similar but also to possess common functional features. The gene products of *src*, *yes*, *fes*, and *fps* were all shown to be protein kinases, phosphorylating tyrosine on target proteins (3,15,23,31,32,43,73). The possibility that other oncogenes (*abl* and *ros*), which are also protein kinases (78), may be members of the *src* family has arisen. By the utilization of *Drosophila* sequences homologous to v-*src* and to v-*abl*, it has been shown that *abl* indeed is a member of the *src* family (29). These results are discussed in detail in the following chapter. DNA sequencing of v-*abl* has recently confirmed its relationship to the *src* gene (D. Baltimore and S. A. Aaronson, *personal communication*). Another feature common to the tyrosine kinase oncogene family is their localization to the cytoplasmic surface of the plasma membrane.

A cellular protein with the typical properties of the *src* family has been identified. The EGF receptor was shown to be a membrane protein with tyrosine kinase activity (13,70). It remains to be shown whether it is indeed structurally related to the *src* family.

The phosphorylated tyrosine residue of $pp60^{src}$ has been identified as residue 416 (63). This residue was shown to be conserved in *yes*, *fes-fps*, *mos*, and *abl*. The availability of evolutionarily related proteins with similar phosphorylation properties makes it possible to study the structural features contributing to the specificity of the tyrosine phosphorylation site. Comparison of the amino acid sequence surrounding the tyrosine in the *src*, *yes*, *fes*, and *fps* proteins was performed. In all cases, a basic amino acid (arginine or lysine) is located seven residues to the NH_2 terminus from the phosphorylated tyrosine, and a glutamic acid residue is located four residues to the NH_2 terminus or in close proximity to the tyrosine (47). The similar specificities of the tyrosine phosphorylation sites have been demonstrated by the ability of a decapeptide fragment of $pp60^{src}$ containing the residues around tyrosine 416 to inhibit the *in vitro* autophosphorylation by the proteins *src*, *yes*, and *fps* (79).

The active site for phosphotransfer resides in the COOH-terminal half of $pp60^{src}$. A fragment containing this half of the protein was generated following partial cleavage with trypsin and was shown to be even more active as a protein kinase than the native $pp60^{src}$ (38). The highest degree of amino acid homology among *src*, *yes*, *fes-fps*, and *mos* was also localized to the same part of the protein; this is a further indication of the crucial role this domain plays in the different members of the *src* family. The smaller degree of sequence conservation at the NH_2-terminus of the proteins of the *src* family may be a result of the smaller selective pressures to maintain that sequence. Alternatively, the possibility of evolution by "exon shuffling," in which the COOH-terminus of the molecule was attached to different genes, remains open (25).

Evolutionary relationships among distinct oncogenes with similar enzymatic activities may simplify the elucidation of the function of c-*onc*. If members of the same family have similar catalytic properties, the number of different mechanisms leading to transformation may be smaller than previously thought. However, if members of the *src* family have similar roles, why are there multiple genes in the group? The finding that each of the members is independently conserved in evolution (as outlined in the previous chapter) emphasizes that despite the similarities, each of these genes fulfills a distinct and essential role in the organism. The duplication of genes within the *src* family has maintained the utilization of the same enzymatic activity (e.g., the kinase function) while broadening the spectrum of substrate specificities, modes of regulation, tissue distribution, or cellular localization.

From the available sequences of the different members of the family, it is difficult to draw conclusions regarding the time of duplication which led to the generation of each gene. The differences in sequence between two genes are a consequence not only of the time elapsed from their separation but also of the rate at which each of them evolves. For example, a comparison of the nucleotide sequences of *fes-fps* and of *mos* in different species indicates that stronger selective forces are acting to conserve the structure of the *fes-fps* gene.

The *ras* Gene Family

The oncogenes *has* and *kis* have been independently isolated by the generation of different acutely transforming retroviruses following infection of rats with Moloney murine leukemia virus. Subsequent experiments have shown that the two oncogenes are related to each other: The cloned genes cross-hybridize, the p21 proteins encoded by the two genes are serologically related (75), and about two-thirds of the tryptic peptides of both proteins are similar (22). A nucleotide sequence of v-*has* and v-*kis* has allowed a detailed comparison of the sequences in common (21,69). The NH_2-terminal halves of both proteins show extensive homology (110 of 120 amino acids). In the COOH-terminal region, however, only three of 22 amino acids are conserved. The overall homology of amino acids is 81% (33 changes).

The high degree of conservation suggests that the two proteins would be similar in their transformation mechanisms and possibly in the function of their cellular counterparts. Yet the duplication event that has generated the genes has not occurred recently. Each of the probes detects a different set of cellular sequences in chicken, mouse, and human DNA (22). Thus duplication of the genes preceded the separation of avian and mammalian species more than 300 million years ago. The question arises as to the critical sequence differences that are responsible for the distinct functions of the two genes. These differences could lie within the coding region, for example, at the COOH-terminus, where little homology is observed, or at the NH_2-terminus in the 12th position, in which sequence changes were shown to have a direct effect on the transforming role of the protein (50,67). Alternatively, sequence differences in the noncoding region could be responsible for differential regulation of the two genes. Detailed analysis of the cellular genes in distant species may be useful in the identification of sequence differences crucial for the distinct function of *has* and *kis*, since these sequences are likely to be conserved.

A third member of the *ras* family has recently been identified. The oncogene of a human neuroblastoma cell line was molecularly cloned (62). This gene hybridizes weakly with both v-*has* and v-*kis* probes, suggesting that it is not identical but structurally related to both.

The *has* and *kis* genes hybridize to more than one band in cellular rat and human DNA (22). It should be pointed out that the presence of multiple bands hybridizing with a probe in the genomic blot does not necessarily reflect the presence of more than one gene. It is possible that a single cellular gene would be fragmented by the restriction enzymes used, especially if that gene is large and contains many introns. The presence of multiple copies of a given gene can be conclusively demonstrated only by structural analysis of the cloned cellular fragments. Cloning the cellular sequences homologous to *has* from rat and human DNA has demonstrated the presence of two cellular genes in both cases: c-*has* 1 is composed of four exons and three introns, while c-*has* 2 is colinear with the viral *has* (10,18). The spliced c-*has* 1 genes in rats and humans are homologous by the heteroduplex analysis not only in the coding region but in certain regions of the introns as well, thus demonstrating the direct evolutionary relationship between them. Since the non-spliced c-*has* 2 gene was identified in rat and human cells, the duplication event most likely occurred prior to the divergence of the two species. Ligation of the rat c-*has* 1 and c-*has* 2 to a long terminal repeat (LTR) and transfection of NIH/3T3 cells resulted in the induction of transformed foci in both cases (18). Therefore, both cellular genes have an intact coding region and the capacity to be functional.

With the availability of the nucleotide sequences of the *ras* genes, it was possible to search for sequences of other cellular proteins which would display structural homology to the *ras* family. For example, the β-chain of bovine mitochondrial proton translocating ATPase exhibited short sequence homology to the NH_2-terminus of human c-*has* (24). The conserved region is thought to be the ATP binding site of the ATPase and could be responsible for the GTP binding properties of the p21 protein of Harvey murine sarcoma virus (54).

CONSERVATION OF CELLULAR ONCOGENES IN *DROSOPHILA*

The accumulating experimental data summarized in the previous chapters demonstrate that c-*onc* are widely distributed and highly conserved throughout vertebrates. It was of interest to determine whether oncogenes can be detected in organisms of other phyla. This approach should generate a broader view on the evolution of oncogenes and provide a system to analyze the structure of these genes in organisms in which the number and size of introns are generally smaller. The detection of c-*onc* in organisms in which genetic manipulations are available would provide a powerful tool to study their function.

Genomic DNAs of a variety of organisms, including *Drosophila*, yeast, and slime molds, were hybridized with a set of oncogene DNA probes (61). Nonstringent conditions of hybridization were used to allow pairing of partially matched sequences. No hybridization with the DNAs of yeast or slime molds was observed. However, several probes detected homologous sequences in *Drosophila*. Among them were *abl*, *has*, *src*, and *fes*. Two features of the hybridization patterns to *Drosophila* should be pointed out. First, only *abl* hybridized with a single band, while the other probes detected two or three homologous fragments. Second, the intensity of hybridization of the probes to the *Drosophila* bands was equal to or even stronger than the hybridization to the genomic mouse DNA (although the probes were derived from vertebrates). This is a result of the genomic complexity of *Drosophila*, which is 25 times lower than that of mouse; thus more genome equivalents are loaded in each lane of *Drosophila* DNA. The identification of sequences homologous to vertebrate oncogenes in *Drosophila* demonstrates that they evolved more than 800 million years ago, prior to the divergence of the Annelid-Arthropoid and the Echinoderm-Chordate superphyla. The central roles played by c-*onc* appear not to be restricted to vertebrates and may be shared by all metazoan organisms.

The approach of using vertebrate probes to detect homologous sequences in distant organisms by nonstringent hybridization may prove to be misleading in cases in which the DNA sequence of the probe possesses special features, such as a high GC content or stretches of repeated sequences. It remains to be shown whether the *Drosophila* sequences that hybridize to the v-*myc* probe [which is highly enriched for GC residues (1)] indeed have a structural homology to the *myc* amino acid sequence.

Detailed analysis was performed with clones of the *Drosophila* sequences homologus to the v-*abl* and v-*src* probes (29). A single clone was isolated with the v-*abl* probe and mapped by *in situ* hybridization of polytene chromosomes to chromosomal position 73B on the left arm of chromosome 3. Two types of clones were isolated by the v-*src* probe, one mapped to position 64B (on chromosome 3L), and the other mapped again to 73B. Further analysis showed that the same cloned *Drosophila* fragment hybridizes with both v-*abl* and v-*src* probes. The region of homology to the two probes is confined to a 220 bp fragment, while the homology to v-*abl* is more extensive, extending over 700 bp. Hybridization of the two vertebrate probes to the same *Drosophila* DNA fragment demonstrated the relationship of *abl* to the *src* gene family. This conclusion was confirmed by sequencing the v-*abl* gene (D. Baltimore and S. A. Aaronson, *personal communication*) and its *Drosophila* homolog (30). The vertebrate *abl* and *src* genes do not cross-hybridize, while the Drosophila *abl* gene reacted readily with v-*src*. It may reflect the smaller degree of alteration this gene has undergone in *Drosophila*.

Sequence analysis of the *Drosophila abl* and *src* clones was performed to determine whether the *Drosophila* genes have conserved regions which are known to be crucial for the kinase function of the vertebrate genes. In the region of nucleotide homology, each of the two *Drosophila* genes has only one open reading frame. The predicted amino acid sequences in that frame show significant homology to the vertebrate viral counterparts. The homology is concentrated in the carboxy terminus of the $pp60^{src}$ protein, a region previously shown to be conserved for other members of the *src* family. In view of the high degree of conservation in this region, some batches of antisera prepared against the viral oncogenes are expected to cross-react with the *Drosophila* proteins.

A consensus sequence of the ancestor of the *src* and *abl* genes can be deduced by the alignment of the amino acid sequences of v-*src*, v-*abl*, and the *Drosophila* Dsrc and Dash (the homolog of v-*abl* which can also hybridize with *src*). In positions where three or four of the four sequences display the same amino acid, this is taken as the consensus sequence. In more than 140 positions, such a sequence could be deduced; it includes most notably the tyrosine residue, which was shown to be phosphorylated in $pp60^{src}$, and the lysine residue, which is probably the ATP binding site in the cyclic AMP-dependent protein kinase. The consensus sequence defines a set of highly conserved residues which are likely to be crucial for the general functions of the proteins of the *src* family and may relate to their role as kinases.

In more than 110 positions in which no consensus sequence was deduced, the sequences of v-*abl* and v-*src* differ, while one or both of them display an identical residue to its closest *Drosophila* homolog (Dash or Dsrc, respectively). These amino acids appear to be essential for the specialized roles that the *abl* and *src* proteins have assumed. Furthermore, the presence of those conserved positions demonstrates that the *abl* and *src* cellular genes have diverged prior to the divergence of the Annelid-Arthropoid and Echinoderm-Chordate superphyla. Their divergence and subsequent conservation in metazoa indicate that the distinct and specialized functions of *abl* and *src* are shared by a wide spectrum of multicellular organisms.

The other multigene families identified in *Drosophila* with the v-*has* and v-*fes* probes are also not likely to be a result of a recent gene duplication event. The three members of the *has* family in *Drosophila* have been cloned (B.-Z. Shilo, *unpublished*). They exhibit only weak cross-hybridization with each other, thus demonstrating the accumulation of multiple structural differences among them. The ability to detect multiple genes in *Drosophila* with probes that recognize only a single gene in vertebrates (such as v-*src* or v-*fes*) is probably a result of the higher sensitivity of the Southern blots of *Drosophila* DNA due to the lower genomic complexity and the nonstringent hybridization conditions. It reflects the dynamic evolution and duplication of c-*onc* which are general to metazoa.

The identification of *Drosophila* genes which are structurally similar to vertebrate oncogenes provides a new approach to look for the function of these genes. Utilization of the sophisticated genetic manipulations available in *Drosophila* should indicate whether the gene products are essential during development, and in which tissues they are required.

FUNCTIONAL CONSERVATION

The accumulating evidence on the structural conservation of c-*onc* strongly suggests that they fulfill analogous functions in the different vertebrate and invertebrate species. One of the most convincing pieces of data is the demonstration that c-*src* proteins in a variety of vertebrate species all have *in vitro* kinase activity (44,51). Other experimental systems test indirectly the functional conservation of c-*onc* products. The oncogenes were originally isolated by virtue of their ability to induce tumors in animals or to generate foci of transformed cells in culture. The transforming potential of oncogenes is likely to be tightly linked to the function of their nontransforming cellular counterparts, in view of the high degree of sequence homology between the transforming form of the gene and the normal c-*onc*, observed for a variety of oncogenes (8). The transformation assay, therefore, is a useful criterion for the conservation of normal c-*onc* function.

Data from a variety of systems demonstrate that the same c-*onc* sequence can be converted to a transforming oncogene in different species. This is a reflection of the similar function it is likely to fulfill in its normal form in the various species. Several examples have been documented: (a) The oncogene *fes-fps* was independently isolated from cat and chicken, respectively, following its recombination with the infecting retrovirus and induction of tumors in the infected host animal (4,6,27,56,57). (b) The oncogene *has-bas* was independently isolated from rat and mouse in the same manner (2). Moreover, this oncogene was also shown to be responsible for transformation in human tumors of nonviral origin, such as bladder carcinoma (20,46,53). (c) The *kis* oncogene originally isolated from rats as a recombinant retrovirus was also shown to be responsible for a variety of human carcinomas (20,48). (d) Finally, the *myc* oncogene isolated from chicken in a recombinant retrovirus was recently shown to be involved in the transformation of mouse plasmacytomas and human Burkitt lymphomas (17,39,55,68).

The development of an assay for transformation of mouse NIH/3T3 cells following DNA transfection allows testing of the ability of oncogenes from various species to function as transforming genes in the mouse cells. A successful transfer of the transformed phenotype by interspecies DNA transfection implies not only that the oncogene itself has been conserved, but also that the cellular mouse components that interact with it are sufficiently conserved to allow manifestation of the transformed state. Several examples for successful interspecies DNA transfection have been reported: (a) The v-*src* oncogene from chicken elicited foci formation in NIH/3T3 mouse cells following transfection (16). (b) Activation of the human c-*has* 1 oncogene, by its ligation to the viral LTR promoter, proved to be potent in the transformation of mouse cells (11). (c) The utilization of the transfection of NIH/3T3 cells to detect oncogenes in tumors of nonviral origin (58) led to the identification of new oncogenes. DNAs extracted from a variety of human tumor cell lines and solid tumors (35,36,41,48,59) were shown to be positive in the transfection assay. They include bladder, colon, pancreas, and lung carcinomas (35,41,48,59), embryonal rhabdomyosarcoma, promyelocytic leukemia (41), B- and T-lymphocyte neoplasms (36), and neuroblastoma (62).

The availability of the transfection assay in NIH/3T3 cells has made it possible to obtain structural information on the oncogenes before they are molecularly cloned. Treatment of the tumor DNA with restriction enzymes prior to DNA transfection identifies enzymes that cleave within the oncogene and thus abolish its transforming activity. The spectrum of enzymes that inactivate or maintain the oncogenic potential of the transfected DNA is a direct consequence of the sequence of the oncogene. DNAs extracted from mouse and human B- and T-cell lymphomas were tested for their restriction enzyme sensitivity (36). For tumors of a similar type, the pattern of enzyme sensitivity of mouse and human DNA was similar. Such results reflect a structural conservation of these oncogenes. It is not possible to determine the degree of sequence conservation, however, since the restriction enzymes tested recognize only limited portions of the gene.

The control of the expression of c-*onc* genes is not well understood. Intricate controls must exist to allow the genes to fulfill their specific roles at the tissue, cellular, or developmental level. Several experiments have shown that enhancement of the expression of some c-*onc* genes, such as c-*mos* and c-*has*, results in cellular transformation (18,45). Signals that regulate the normal level of expression of these genes must exist to maintain it below the critical threshold level of transformation. These regulatory elements in the DNA are also likely to be conserved throughout evolution. A way to identify such regulatory elements may be to compare the noncoding sequence of a given cellular oncogene in distantly related species.

Transcription of c-*onc* genes has been widely studied. A more detailed set of data exists for the genes c-*src*, c-*myc*, c-*erb*, c-*myb*, and c-*has*. Each is transcribed in a variety of tissues and in every species that has been satisfactorily examined (8,40). It is striking that several c-*onc* genes give rise to distinctive RNAs whose sizes have been shown to be identical in different species and various cell types (8). The size of the c-*src* RNA is 3.9 kb in chicken, mouse, and rat, while that of the c-*myc* RNA in the same species is 2.5 kb. Since the RNA size is much larger than its coding capacity, the constancy of its size in the different species testifies to selective pressures that conserve not only the coding sequences but the position of regulatory signals for transcription initiation and termination.

CONCLUDING REMARKS

Study of the evolution of c-*onc* is intimately linked to some of the major issues in the field of oncogenes: The cellular origin of viral oncogenes was deduced from the wide distribution of c-*onc* sequences. The conservation of oncogenes within vertebrates and in other phyla point to their universal role in multicellular organisms. What these functions are remains one of the central puzzles. The ability to identify families of oncogenes which are an evolutionary consequence of gene duplication events confirms distinctive features among proteins that share structural and functional similarities.

In the future, a detailed comparison of oncogene sequences in distantly related species should prove to be useful in the identification of amino acid sequences that are essential for the function of the gene products. The identification of c-*onc* in *Drosophila* provides genetic tools to study those functions. New oncogenes of nonviral origin have recently been identified and are currently being isolated. A study of their distribution and conservation may provide an important clue to their function. Finally, despite failures to detect c-*onc* in unicellular organisms until now, the utilization of new and more sensitive techniques may yet allow their identification. Success in this direction will provide a spectrum of useful organisms for studying the function of c-*onc*.

ACKNOWLEDGMENTS

I thank K. Berns, H. Hoffman-Falk, and F. S. Neumann for critical reading of the manuscript. This work was supported by grants from the Leukemia Research Foundation and the Charles H. Revson Foundation. The views expressed, however, are solely the responsibility of the author.

REFERENCES

1. Alitalo, K., Bishop, J. M., Smith, D. H., Chen, E. Y., Colby, W. W., and Levinson, A. D. (1983): Nucleotide sequence of the v-*myc* oncogene of avian retrovirus MC29. *Proc. Natl. Acad. Sci. USA*, 80:100–104.
2. Andersen, P. R., Devare, S. G., Tronick, S. R., Ellis, R. W., Aaronson, S. A., and Scolnick, E. M. (1981): Generation of BALB-MuSV and Ha-MuSV by type C virus transduction of homologous transforming genes from different species. *Cell*, 26:129–134.

3. Barbacid, M., Beemon, K., and Devare, S. G. (1980): Origin and functional properties of the major gene product of the Snyder-Theilen strain of feline sarcoma virus. *Proc. Natl. Acad. Sci. USA*, 77:5158–5162.
4. Barbacid, M., Breitman, M. L., Lauver, A. V., Long, L. K., and Vogt, P. K. (1981): The transformation specific proteins of avian (Fujinami and PRC-II) and feline (Snyder-Theilen and Gardner-Arnstein) sarcoma viruses are immunologically related. *Virology*, 110:411–419.
5. Barker, W. C., and Dayhoff, M. O. (1982): Viral *src* gene products are related to the catalytic chain of mammalian c-AMP-dependent protein kinase. *Proc. Natl. Acad. Sci. USA*, 79:2836–2839.
6. Beemon, K. (1981): Transforming proteins of some feline and avian sarcoma viruses are related structurally and functionally. *Cell*, 24:145–153.
7. Bishop, J. M. (1981): Enemies within: The genesis of retrovirus oncogenes. *Cell*, 23:5–6.
8. Bishop, J. M., and Varmus, H. E. (1982): Functions and origins of retroviral transforming genes. In: *RNA Tumor Viruses*, edited by R. Weiss, N. Teich, H. Varmus, and J. Coffin, pp. 999–1108. Cold Spring Harbor, New York.
9. Brugge, J. S., and Erikson, R. L. (1977): Identification of a transforming-specific antigen induced by an avian sarcoma virus. *Nature*, 269:346–348.
10. Chang, E. H., Gonda, M. A., Ellis, R. W., Scolnick, E. M., and Lowy, D. R. (1982): Human genome contains four genes homologous to transforming genes of Harvey and Kirsten murine sarcoma virus. *Proc. Natl. Acad. Sci. USA*, 79:4848–4852.
11. Chang, E. H., Furth, M. E., Scolnick, E. M., and Lowy, D. R. (1982): Tumorigenic transformation of mammalian cells induced by a normal human gene homologous to the oncogene of Harvey murine sarcoma virus. *Nature*, 297:479–483.
12. Coffin, J. M., Varmus, H. E., Bishop, J. M., Essex, M., Hardy, W. D., Martin, G. S., Rosenberg, N. E., Scolnick, E. M., Weinberg, R. A., and Vogt, P. K. (1981): A proposal for naming host cell-derived inserts in retrovirus genomes. *J. Virol.*, 40:953–957.
13. Cohen, S., Carpenter, G., and King, L. (1980): Epidermal growth factor-receptor-protein kinase interactions. *J. Biol. Chem.*, 255:4834–4842.
14. Colby, W. W., Chen, E. Y., Smith, D. H., and Levinson, A. D. (1983): Identification and nucleotide sequence of a human locus homologous to the v-*myc* oncogene of avian myelocytomatosis virus MC29. *Nature*, 301:722–725.
15. Collett, M. S., Purchio, A. F., and Erikson, R. L. (1980): Avian sarcoma virus-transforming protein, $pp60^{src}$, shows protein kinase activity specific for tyrosine. *Nature*, 285:167–169.
16. Copeland, N. G., Zelenetz, A. D., and Cooper, G. M. (1979): Transformation of NIH/3T3 mouse cells by DNA of Rouse sarcoma virus. *Cell*, 17:993–1002.
17. Crews, S., Barth, R., Hood, L., Prehn, J., and Calame, K. (1982): Mouse c-*myc* oncogene is located on chromosome 15 and translocated to chromosome 12 in plasmocytomas. *Science*, 218:1319–1321.
18. DeFeo, D., Gonda, M. A., Young, H. A., Chang, E. H., Lowy, D. R., Scolnick, E. M., and Ellis, R. W. (1981): Analysis of two divergent rat genomic clones homologous to the transforming gene of Harvey murine sarcoma virus. *Proc. Natl. Acad. Sci. USA*, 78:3328–3332.
19. Della Favera, R., Gelmann, E. P., Martinotti, S., Franchini, G., Papas, T. S., Gallo, R., and Wong-Staal, F. (1982): Cloning and characterization of different human sequences related to the *onc* gene (v-*myc*) of avian myelocytomatosis virus (MC29). *Proc. Natl. Acad. Sci. USA*, 79:6497–6501.
20. Der, C. J., Krontiris, T. G., and Cooper, G. M. (1982): Transforming genes of human bladder and lung carcinoma cell lines are homologous to the *ras* genes of Harvey and Kirsten sarcoma viruses. *Proc. Natl. Acad. Sci. USA*, 79:3637–3640.
21. Dhar, R., Ellis, R. W., Shih, T. Y., Oroszlan, S., Shapiro, B., Maizel, J., Lowy, D., and Scolnick, E. M. (1982): Nucleotide sequence of the p21 transforming protein of Harvey murine sarcoma virus. *Science*, 217:934–937.
22. Ellis, R. W., DeFeo, D., Shih, T. Y., Gonda, M. A., Young, H. A., Tsuchida, N., Lowy, D. R., and Scolnick, E. M. (1981): The p21 *src* genes of Harvey and Kirsten sarcoma viruses originate from divergent members of a family of normal vertebrate genes. *Nature*, 292:506–511.
23. Feldman, R. A., Hanafusa, T., and Hanafusa, H. (1980): Characterization of protein kinase activity associated with the transforming gene product of Fujinami sarcoma virus. *Cell*, 22:757–765.

24. Gay, N. J., and Walker, J. E. (1983): Homology between human bladder carcinoma oncogene product and mitochondrial ATP-synthase. *Nature*, 301:262–264.
25. Gilbert, W. (1978): Why genes in pieces? *Nature*, 271:501.
26. Goff, S. P., Gilboa, E., Witte, O. N., and Baltimore, D. (1980): Structure of the Abelson murine leukemia virus genome and the homologous cellular gene: Studies with cloned viral DNA. *Cell*, 22:777–785.
27. Hampe, A., Laprevotte, I., Galibert, F., Fedele, L. A., and Sherr, C. J. (1982): Nucleotide sequences of feline retroviral oncogenes (v-*fes*) provide evidence for a family of tyrosine-specific protein kinase genes. *Cell*, 30:775–785.
28. Hayward, W. S., Neel, B. G., and Astrin, S. M. (1981): Activation of a cellular *onc* gene by promoter insertion in ALV-induced lymphoid leukosis. *Nature*, 290:475–480.
29. Hoffman-Falk, H., Einat, P., Shilo, B.-Z., and Hoffmann, F. M. (1983): *Drosophila melanogaster* DNA clones homologous to vertebrate oncogenes: Evidence for a common ancestor to the *src* and *abl* cellular genes. *Cell*, 32:289–298.
30. Hoffmann, F. M., Fresco, L. D., Hoffman-Falk, H., and Shilo, B.-Z. (1983): Nucleotide sequences of the *Drosophila src* and abl homologs: Conservation and variability in the src-family oncogenes. *Cell (in press)*.
31. Hunter, T., and Sefton, B. (1980): Transforming gene product of Rous sarcoma virus phosphorylates tyrosine. *Proc. Natl. Acad. Sci. USA*, 77:1311–1315.
32. Kawai, S., Yoshida, M., Segawa, K., Sugiyama, H., Ishizaki, R., and Toyoshima, K. (1980): Characterization of Y73, an avian sarcoma virus: A unique transforming gene and its product, a phosphoprotein with protein kinase activity. *Proc. Natl. Acad. Sci. USA*, 77:6199–6203.
33. Kitamura, N., Kitamura, A., Toyoshima, K., Hirayama, Y., and Yoshida, M. (1982): Avian sarcoma virus Y73 genome sequence and structure similarity of its transforming gene product to that of Rous sarcoma virus. *Nature*, 297:205–208.
34. de Klein, A., van Kessel, A. G., Grosveld, G., Bastram, C. R., Hagemeijer, A., Bootsma, D., Spurr, N. K., Heisterkamp, N., Groffen, J., and Stephenson, J. R. (1982): A cellular oncogene is translocated to the Philadelphia chromosome in chronic myelocytic leukemia. *Nature*, 300:765–767.
35. Krontiris, T. G., and Cooper, G. M. (1981): Transforming activity in human tumor DNAs. *Proc. Natl. Acad. Sci. USA*, 78:1181–1184.
36. Lane, M. A., Sainten, A., and Cooper, G. M. (1982): Stage-specific transforming genes of human and mouse B- and T-lymphocyte neoplasms. *Cell*, 28:873–880.
37. Langbeheim, H., Shih, T. Y., and Scolnick, E. M. (1980): Identification of a normal vertebrate cell protein related to the p21 *src* of Harvey murine sarcoma virus. *Virology*, 106:292–300.
38. Levinson, A. D., Courtneige, S. A., and Bishop, J. M. (1981): Structural and functional domains of the Rous sarcoma virus transforming protein ($pp60^{src}$). *Proc. Natl. Acad. Sci. USA*, 78:1624–1628.
39. Marcu, K. B., Harris, L., Stanton, L. W., Erikson, J., Watt, R., and Croce, C. M. (1983): Transcriptionally active c-*myc* oncogene is contained within NIARD, a DNA sequence associated with chromosome translocations in B-cell neoplasia. *Proc. Natl. Acad. Sci. USA*, 80:519–523.
40. Müller, R., Slamon, D. J., Trembly, J. M., Cline, M. J., and Verma, I. M. (1982): Differential expression of cellular oncogenes during pre- and postnatal development of the mouse. *Nature*, 299:640–644.
41. Murray, M. J., Shilo, B.-Z., Shih, C., Cowing, D., Hsu, H. W., and Weinberg, R. A. (1981): Three different human tumor cell lines contain different oncogenes. *Cell*, 25:355–361.
42. Neel, B. G., Gasic, G. P., Rogler, C. E., Skalka, A. M., Ju, G., Hishinuma, F., Papas, T., Astrin, S., and Hayward, W. S. (1982): Molecular analysis of the c-*myc* locus in normal tissue and in avian leukosis virus-induced lymphomas. *J. Virol.*, 44:158–166.
43. Neil, J. C., Ghysdael, J., and Vogt, P. K. (1981): Tyrosine-specific protein kinase activity associated with p105 of avian sarcoma virus PRCII. *Virology*, 109:223–228.
44. Oppermann, H., Levinson, A. D., Varmus, H. E., Levintow, L., and Bishop, J. M. (1979): Uninfected vertebrate cells contain a protein that is closely related to the product of the avian sarcoma virus transforming gene *(src)*. *Proc. Natl. Acad. Sci. USA*, 76:1804–1808.
45. Oskarsson, M., McClements, W. L., Blair, D. G., Maizel, J. V., and Vande-Woude, G. F. (1980): Properties of a normal mouse cell DNA sequence (sarc) homologous to the src sequence of Moloney sarcoma virus. *Science*, 207:1222–1224.

46. Parada, L. F., Tabin, C. J., Shih, C., and Weinberg, R. A. (1982): Human EJ bladder carcinoma oncogene is homologue of Harvey sarcoma virus *ras* gene. *Nature*, 297:474–478.
47. Patchinsky, T., Hunter, T., Esch, F. S., Cooper, J. A., and Sefton, B. M. (1982): Analysis of the sequence of amino acids surrounding sites of tyrosine phosphorylation. *Proc. Natl. Acad. Sci. USA*, 79:973–977.
48. Pulciani, S., Santos, E., Lauver, A. V., Long, L. K., Aaronson, S., and Barbacid, M. (1982): Oncogenes in solid human tumours. *Nature*, 300:539–542.
49. Rechavi, G., Givol, D., and Canaani, E. (1982): Activation of a cellular oncogene by DNA rearrangement: Possible involvement of an IS-like element. *Nature*, 300:607–611.
50. Reddy, E. P., Reynolds, R. K., Santos, E., and Barbacid, M. (1982): A point mutation is responsible for the acquisition of transforming properties by the T24 human bladder carcinoma oncogene. *Nature*, 300:149–152.
51. Rohrschneider, L. R., Eisenman, R. N., and Leitch, C. R. (1979): Identification of a Rous sarcoma virus transformation-related protein in normal avian and mammalian cells. *Proc. Natl. Acad. Sci. USA*, 76:4479–4483.
52. Romer, A. S. (1966): *Vertebrate Paleontology*. University of Chicago Press, Chicago.
53. Santos, E., Tronick, S. R., Aaronson, S. A., Pulciani, S., and Barbacid, M. (1982): T24 human bladder carcinoma oncogene is an activated form of the normal human homologue of BALB- and Harvey-MSV transforming genes. *Nature*, 298:343–347.
54. Scolnick, E. M., Papageorge, A. G., and Shih, T. Y. (1979): Guanine nucleotide-binding activity as an assay for *src* protein of rat-derived murine sarcoma viruses. *Proc. Natl. Acad. Sci. USA*, 76:5355–5359.
55. Shen-Ong, G. L. C., Keath, E. J., Piccoli, S. P., and Cole, M. D. (1982): Novel *myc* oncogene RNA from abortive immunoglobulin-gene recombination in mouse plasmacytomas. *Cell*, 31:443–452.
56. Shibuya, M., Hanafusa, T., Hanafusa, H., and Stephenson, J. R. (1980): Homology exists among the transforming sequences of avian and feline sarcoma viruses. *Proc. Natl. Acad. Sci. USA*, 77:6536–6540.
57. Shibuya, M., and Hanafusa, H. (1982): Nucleotide sequence of Fujnami sarcoma virus: Evolutionary relationship of its transforming gene with transforming genes of other sarcoma viruses. *Cell*, 30:787–795.
58. Shih, C., Shilo, B.-Z., Goldfarb, M. P., Dannenberg, A., and Weinberg, R. A. (1979): Passage of phenotypes of chemically transformed cells via transfection of DNA and chromatin. *Proc. Natl. Acad. Sci. USA*, 76:5714–5718.
59. Shih, C., Padhy, L. C., Murray, M., and Weinberg, R. A. (1981): Transforming genes of carcinomas and neuroblastomas introduced into mouse fibroblasts. *Nature*, 290:261–264.
60. Shih, T. Y., Weeks, M. O., Young, H. A., and Scolnick, E. M. (1979): Identification of a sarcoma virus coded phosphoprotein in nonproducer cells transformed by Kirsten or Harvey murine sarcoma virus. *Virology*, 96:64–79.
61. Shilo, B.-Z., and Weinberg, R. A. (1981): DNA sequences homologous to vertebrate oncogenes are conserved in *Drosophila melanogaster*. *Proc. Natl. Acad. Sci. USA*, 78:6789–6792.
62. Shimizu, K., Goldfarb, M., Perucho, M., and Wigler, M. (1983): Isolation and preliminary characterization of the transforming gene of a human neuroblastoma cell line. *Proc. Natl. Acad. Sci. USA*, 86:383–387.
63. Smart, J. E., Oppermann, H., Czernilofsky, A. P., Purchio, A. F., Erikson, R. L., and Bishop, J. M. (1981): Characterization of sites for tyrosine phosphorylation in the transforming protein of Rous sarcoma virus ($pp60^{v\text{-}src}$) and its normal cellular homologue ($pp60^{c\text{-}src}$). *Proc. Natl. Acad. Sci. USA*, 78:6013–6017.
64. Spector, D. H., Varmus, H. E., and Bishop, J. M. (1978): Nucleotide sequences related to the transforming gene of avian sarcoma virus are present in DNA of uninfected vertebrates. *Proc. Natl. Acad. Sci. USA*, 75:4102–4106.
65. Stehelin, D., Guntaka, R. V., Varmus, H. E., and Bishop, J. M. (1976): Purification of DNA complementary to nucleotide sequences required for neoplastic transformation of fibroblasts by avian sarcoma viruses. *J. Mol. Biol.*, 101:349–365.
66. Stehelin, D., Varmus, H. E., Bishop, J. M., and Vogt, P. K. (1976): DNA related to the transforming gene(s) of avian sarcoma viruses is present in normal avian DNA. *Nature*, 260:170–173.
67. Tabin, C. J., Bradley, S. M., Bargmann, C. I., Weinberg, R. A., Papageorge, A. G., Scolnick,

E. M., Dhar, R., Lowy, D. R., and Chang, E. H. (1982): Mechanism of activation of a human oncogene. *Nature*, 300:143–149.
68. Taub, R., Kirsch, T., Morton, C., Lenoir, G., Swan, D., Tronick, S., Aaronson, S., and Leder, P. (1982): Translocations of the c-*myc* gene into immunoglobulin heavy chain locus in human Burkitt lymphoma and murine plasmacytoma cells. *Proc. Natl. Acad. Sci. USA*, 79:7837–7841.
69. Tsuchida, N., Ryder, T., and Ohtsubo, E. (1982): Nucleotide sequence of the oncogene encoding the p21 transforming protein of Kirsten murine sarcoma virus. *Science*, 217:937–938.
70. Ushiro, H., and Cohen, S. (1980): Identification of phosphotyrosine as a product of epidermal growth factor-activated protein kinase in A-431 cell membranes. *J. Biol. Chem.*, 255:8363–8365.
71. Van Beveren, C., Van Straaten, F., Galleshaw, J. A., and Verma, I. M. (1981): Nucleotide sequence of the genome of a murine sarcoma virus. *Cell*, 27:97–108.
72. Van Beveren, C., Galleshaw, J. A., Jonas, V., Berns, A. J. M., Doolittle, R. F., Donoghue, D. J., and Verma, I. M. (1981): Nucleotide sequence and formation of the transforming gene of a mouse sarcoma virus. *Nature*, 289:258–262.
73. Van de Ven, W. J. M., Renolds, F. H., and Stephenson, J. R. (1980): The nonstructural components of polyproteins encoded by replication-defective mammalian transforming retroviruses are phosphorylated and have associated protein kinase activity. *Virology*, 101:185–197.
74. Vennstrom, B., Sheiness, D., Zabielski, J., and Bishop, J. M. (1982): Isolation and characterization of c-*myc*, a cellular homolog of the oncogene (v-*myc*) of avian myelocytomatosis virus strain 29. *J. Virol.*, 42:773–779.
75. Young, H. A., Shih, T. Y., Scolnick, E. M., Rasheed, S., and Gardner, M. B. (1979): Different rat-derived transforming retroviruses code for an immunologically related intracellular phosphoprotein. *Proc. Natl. Acad. Sci. USA*, 76:3523–3527.
76. Young, J. Z. (1962): *The Life of Vertebrates*. University of Chicago Press, Chicago.
77. Watson, R., Oskarsson, M., and Vande-Woude, G. F. (1982): Human DNA sequence homologous to the transforming gene (mos) of Moloney murine sarcoma virus. *Proc. Natl. Acad. Sci. USA*, 79:4078–4082.
78. Witte, O. N., Dasgupta, A., and Baltimore, D. (1980): Abelson murine leukemia virus protein is phosphorylated *in vitro* to form phosphotyrosine. *Nature*, 283:826–831.
79. Wong, T. W., and Goldberg, A. R. (1981): Synthetic peptide fragment of *src* gene product inhibits the *src* protein kinase and crossreacts immunologically with avian *onc* kinases and cellular phosphoproteins. *Proc. Natl. Acad. Sci. USA.*, 78:7412–7416.

Note added in proof: The first ATG codon of the mouse c-*mos* was recently shown to be located at the same position as the initiation codon of the human c-*mos*. The NH_2-terminus of the sequence is thus conserved between the two species.

Advances in Viral Oncology, Volume 4,
edited by George Klein. Raven Press,
New York © 1984.

Biology of Human T-Cell Leukemia/ Lymphoma Virus Transformation of Human T-Cells *In Vivo* and *In Vitro*

M. Popovic, F. Wong-Staal, P. S. Sarin, and R. C. Gallo

Laboratory of Tumor Cell Biology, National Cancer Institute, National Institutes of Health, Bethesda, Maryland 20205

It is well known that a variety of neoplasias (leukemias and lymphomas, sarcomas, and occasionally, carcinomas) in animals are induced by retroviruses. In many instances, the retroviruses have been identified as the prime factor in the pathogenesis of naturally occurring leukemias and lymphomas; therefore, many are called leukemia viruses. Despite decades of accumulated evidence showing retroviruses to be involved in leukemogenesis in a variety of animal and subhuman primate species, evidence for an etiologic role of retroviruses in human neoplasias was not available until recently. There was a prevailing bias that these viruses might not be present in humans as infecting agents, because earlier studies in animals, particularly of avian and murine leukemias, showed an association of the disease with abundant virus replication, while such viruses were not seen in primary human tumors. In addition, most mammalian retroviruses shared common antigenic determinants of at least some of their proteins, so that immunologic reagents of one virus could be used to detect the presence of others. Therefore, failure to detect viral markers in human leukemias using reagents of animal retroviruses was considered as evidence against the existence of human retroviruses.

Two animal model systems, feline and bovine leukosis, provided several contrasting and important findings. For the first time, Jarret et al. in Glasgow (35) and, subsequently, Essex and colleagues in the United States (17,31) firmly established feline leukemia virus (FeLV) as the cause of spontaneous leukemia of domestic cats, thus clearly identifying a retrovirus cause of a naturally occurring leukemia in nonlaboratory animals. In addition, FeLV was shown to be readily detectable in most spontaneous leukemias and lymphomas of cats, yet in 30 to 40% of leukemic cats, the leukemic cells have no detectable virus or viral components. These "virus-negative" tumors are clinically and histologically indistinguishable from the virus-positive tumors; they occur in cats exposed to FeLV, suggesting that FeLV may still cause these leukemias (10,17,41,42). Moreover, Rojko et al. (76)

found FeLV in some bone marrow cells of these cats. There is no evidence for integration of exogenous FeLV sequences in the virus-negative tumors, suggesting that if FeLV causes the virus-negative leukemias, a novel mechanism in tumor induction by a retrovirus may be involved.

Some other observations relevant to the recent isolation and detection of human retroviruses came from studies in bovine leukemia. Bovine leukemia virus (BLV) is the causative agent of two related lymphoproliferative diseases in cattle: (a) the enzootic forms of lymphoid leukemia or lymphosarcoma, and (b) persistent lymphocytosis (19). The latter is a benign, polyclonal, persistent lymphoid proliferation that sometimes develops into a monoclonal lymphosarcoma. Although the role of BLV in causation of the enzootic form of bovine leukemias/lymphomas is now well established, fresh bovine tissues frequently do not have detectable virus, viral proteins, or viral mRNA (40,57). Full expression of the virus was found in tumor cells which have been successfully grown *in vitro*. BLV lacks the homology detectable in other mammalian or avian retroviruses (9,16,26). Therefore, prior to the isolation of BLV, evidence for a virus involved in bovine leukemias and lymphomas could not be obtained using probes derived from the previously isolated animal retroviruses. As shown later, these observations are similar to our results with human T-cell leukemia/lymphoma virus (HTLV), clearly indicating the importance of long-term *in vitro* culture of appropriate human target cells and the necessity of development of reagents prepared against a human retrovirus for subsequent detection of related human retroviruses.

DETECTION AND ISOLATION OF HTLV

The discovery of T-cell growth factor (TCGF) in 1976 enabled the development of the technology to grow normal, mature T-cells *in vitro* for considerable periods of time (54,77). Using purified TCGF, it was also possible to grow T-cells that exhibited several characteristics of primary tumor cells from patients with mature T-cell malignancies (62,66). A number of human T-cell lines were established from patients with these malignancies. Many of these cultured T-cell lines release a type-C retrovirus, which we have named HTLV. Morphologically, HTLV is a typical type-C retrovirus as seen by electron microscopy (EM) (61,63). Like other retroviruses, HTLV contains a reverse transcriptase and a high molecular weight RNA genome of approximately 9 kb (70S) and viral core *gag* proteins, composed of similar units as in the other retroviruses (38,69,71). HTLV has been shown to be an exogenous human virus, since HTLV-related sequences are not present in the DNA of normal, uninfected human cells but are readily detected in DNA or RNA from HTLV-positive tumor cells (68–70). In addition, HTLV sequences were not found in the normal Epstein-Barr virus-infected B-cells from a HTLF-positive T-cell lymphoma patient; only his neoplastic T-cells contained HTLV-specific sequences (22,68).

Since the first isolation of HTLV reported in our laboratory several years ago (61,63), we have obtained many additional isolates from cell lines established from patients with mature T-cell malignancies and some normal family members from different parts of the world (22,23,39,66,81). Recently, HTLV has also been isolated from patients with the acquired immune deficiency syndrome (AIDS) (24). The virus isolates obtained or studied in our laboratory are summarized in Table 1. All cell lines have karyotype and HLA patterns that match those of the primary (fresh) donor cells, and the HLA profiles are distinct in all cell lines. The HTLV-positive T-cell lines were established from seven patients from the United States, one from Israel, two from the West Indies, and five from Japan. Four isolates were obtained from a family from the northwest part of Honshu Island, Japan. In this family, the patient (S. K.) has acute T-cell leukemia (ATL), and both his parents and brother are virus positive. The father (H. K.) and brother (B. K.) are clinically healthy, while the mother (T. K.) exhibited lymphocytosis; some of the lymphocytes had abnormal, convoluted nuclei, a feature characteristic of the leukemic T-cells of patients with ATL. The cell lines derived from patients with mature T-cell malignancies, including a cell line established from a patient (M. O.) with hairy cell leukemia (T-cell variant), and from family members of the patient S. K. with ATL were analyzed for the presence of HTLV by highly sensitive competition radioimmunoprecipitation assay (RIPA) for the major core protein p24 (38), indirect immunofluorescence assay (IFA) using highly specific monoclonal antibody for p19 (75), and reverse transcriptase activity (RTA) in culture fluids. The presence of type-C virus particles was examined by EM. As shown in Table 1, HTLV was fully expressed in all established T-cell lines. Representative examples of the EMs of HTLV particles in six established T-cells are shown in Fig. 1.

TABLE 1. *HTLV isolates in T-cell lines derived from patients with T-cell malignancies and healthy relatives*

Patient/ cell line	Diagnosis[a]	Patient origin	Protein (% positive) p24	p19	RT	EM	Ref.
CR	CTCL, MF	Alabama	+	60	+	+	61
MJ	CTCL, Sezary	Massachusetts	+	85	+	+	66
WA	T-cell lymphoma	Georgia	+	78	+	+	66
PL	T-DML	Florida	+	23	+	+	66
OB	T-cell lymphoma	Georgia	+	50	+	+	66
MO	Hairy cell leukemia	Washington	+	80	+	ND	39
EP	AIDS	New York	+	9	+	+	24
MB	CTCL, Sezary	West Indies	+	5	+	+	63
MI	T-LCL	West Indies	+	63	+	+	66
UK	PTCL	Israel	+	71	+	+	66
CC	AIDS	France	+	12	+	ND	24
SD	ATL	Japan	+	90	+	+	32
SK	ATL	Japan	+	39	+	+	66,81
TK	PL (mother)	Japan	+	54	+	+	66,81
HK	Healthy (father)	Japan	+	38	+	+	66,81
BK	Healthy (brother)	Japan	+	8	+	+	81

[a]CTCL, cutaneous T-cell lymphoma; MF, mycosis fungoides; T-DML, diffuse mixed lymphoma of T-cells; PTCL, peripheral T-cell lymphoma (diffuse histiocytic lymphoma); PL, persistent lymphocytosis.

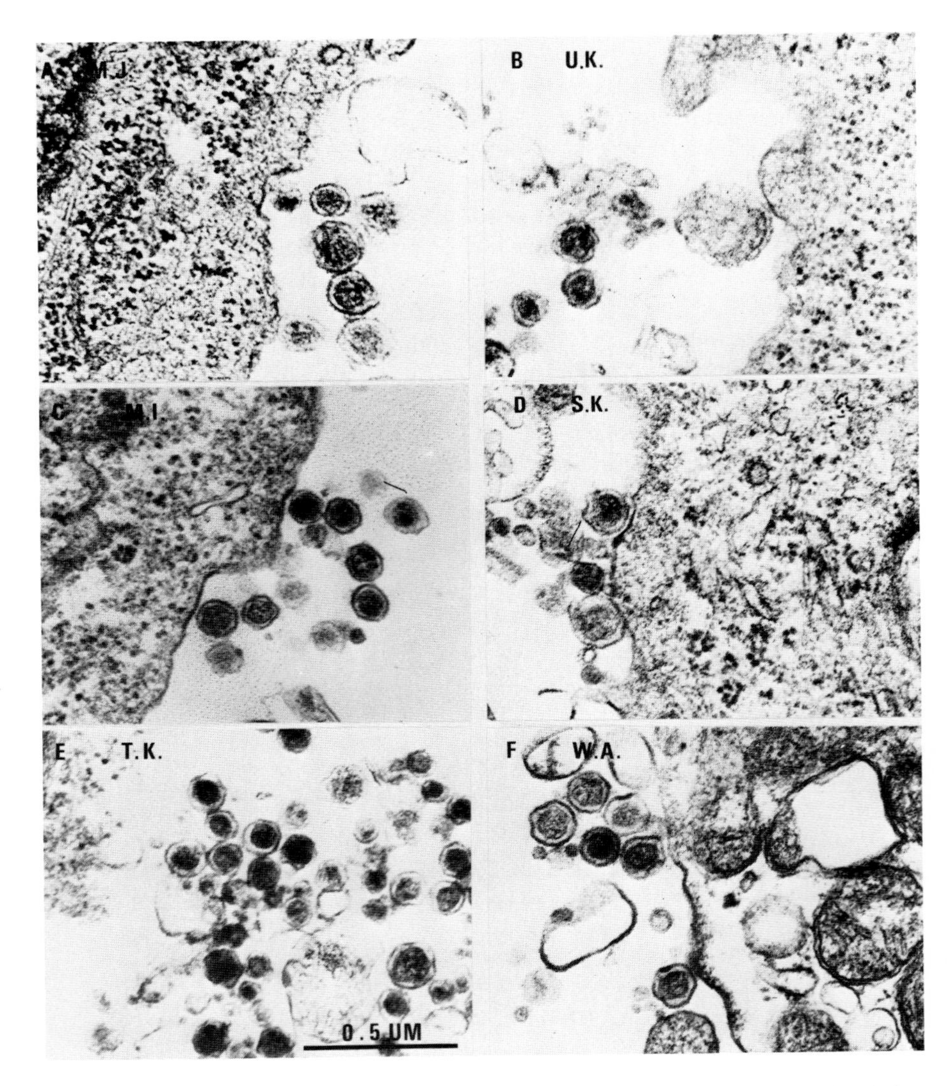

FIG. 1. Electron micrographs of T-cell lines established from patients with T-cell leukemia/lymphoma from **(A)** Boston, Massachusetts; **(B)** Jerusalem, Israel; **(C)** Granada, West Indies; **(D)** Honshu Island, Japan; **(E)** Honshu Island, Japan (mother of patient in **D**); **(F)** Augusta, Georgia.

Human retrovirus isolates have been independently reported from other laboratories (2,32,51,85,92). In Japan, the human retrovirus isolates have been named adult T-cell leukemia virus (ATLV). However, the virus isolates described by these investigators are clearly members of the HTLV family, and the few that have been analyzed are closely related or identical to the same strain (HTLV-1) originally isolated from the United States (65).

SUBGROUPS OF HTLV

As mentioned above, serologic and nucleic acid hybridization studies demonstrated that original isolates from the United States (C. R., M. B., and M. J.) and Japan (MT-1 and MT-2) belong to the same group of HTLV (65,70). We have compared the different HTLV isolates from the United States, Japan, the Caribbean, and elsewhere by serologic reactions to the viral core proteins p24 and p19 and by the homology of the virion genomic RNA. The data are summarized in Table 2. It was found that all virus isolates except $HTLV_{MO}$, detected (39) and isolated (M. Popovic et al., *submitted*) in our laboratory from a cell line derived from a patient with hairy cell leukemia of a T-cell type, and the new isolate recently described by Barre-Sinoussi et al. (2) in France are virtually indistinguishable by these assays (38,39,68,70). In addition, we have recently compared the HTLV proviruses of these isolates as present in leukemic cells and observed that all isolates, except $HTLV_{MO}$, are closely related and have highly conserved genomes, as determined by cleavage sites of several restriction endonucleases. On the other hand, $HTLV_{MO}$ competes poorly in the p24 assays (39) and nucleic acid sequence homology with $HTLV_{CR}$ was detected only under nonstringent hybridization conditions (E. P. Gelmann et al., *submitted*).

In addition, it was recently shown by using two different biologic assays (syncytia induction and vesicular stomatitis virus pseudotypes) that the $HTLV_{MO}$ isolate has markedly different envelope antigens compared to other HTLV isolates (55; R. A. Weiss, *personal communication*). The new isolate obtained from a French patient with lymphadenopathy (2) also has minimal homology with the prototype HTLV. However, similar to other HTLV isolates, including $HTLV_{MO}$, the lymphadenopathy-

TABLE 2. *Comparative properties of HTLV subgroups I, II, and III*[a]

HTLV	Cell line	RNA[b]	Core protein p24	Core protein p19
Subgroup I				
CR	HUT-102	..	..	..
MB	CTCL-2	+ +	+ +	+ +
ATLV	MT-1	+ +	+ +	+ +
ATLV	MT-2	+ +	+ +	+ +
UK	UK	+ +	+ +	+ +
MI	MI	+ +	+ +	+ +
MJ	MJ	+ +	+ +	+ +
SD	SD	+ +	+ +	+ +
Subgroup II				
MO	MO	±	+	+ +
Subgroup III				
RUB (France)	RUB	ND	–	–

[a]From refs. 65,66,68, and 70.

[b]In relation to CR, the prototype HTLV, the virus was highly related or indistinguishable (+ +), related but readily distinguishable (+), slightly related (±), and not related (–). ND, not done.

associated virus is characterized by T-cell tropism, biochemical properties (ionic and template requirement) of its reverse transcriptase, and common cell surface antigens with HTLV-infected T-cells. The virus did not cross react with type-specific antisera against viral structural proteins (p19 and p24) of HTLV. Moreover, preliminary nucleic acid hybridization analysis indicates that this virus is clearly different from other HTLV isolates, including M.O. isolate (F. Wong-Staal, et al., *unpublished results*). Therefore, these two virus isolates, M.O. and the lymphadenopathy virus, may form distinct subgroups in the HTLV family. We propose to group these as HTLV-I_{CR}, HTLV-I_{MB}, and HTLV-I_{MJ}, versus HTLV-II_{MO} and HTLV-III from France (subscripts designate patients' initials). Further studies may reveal other members of subgroups II and III and more subgroups of HTLV. Even within the HTLV-I subgroup, there are differences between individual viral isolates based on nucleic acid analysis of cloned proviruses (G. M. Shaw et al., *submitted*), which could represent subtype differences.

UNIQUENESS OF HTLV AND ITS RELATIONSHIP TO OTHER RETROVIRUSES

HTLV does not cross react in immunologic assays with known retroviruses (38,69,71), and there is only little nucleic acid sequence homology with BLV (E. P. Gelmann et al., *unpublished results)*. Although it is clear that HTLV and BLV are different viruses, there is significant similarity in *gag* proteins of both (56). Extensive amino acid sequence analysis of major core protein p24 revealed correspondence among amino acids between the first 25 amino terminal residues of HTLV p24 and BLV p24. Similar correspondence was found up to the first 150 residues so far sequenced; this has been considered statistically significant correlation. A homology between HTLV p15 and BLV p12 was found, including a conservation of the nucleic acid binding domain (T. D. Copeland et al., *submitted*).

These data suggest that these homologous proteins of HTLV and BLV may have originated from common ancestral molecules in the distant past. The extent of amino acid sequence homology also underscores the substantial dissimilarities between the two viruses and, in fact, the uniqueness of HTLV among mammalian retroviruses. As noted above, no immunologic cross reaction was found between HTLV and any other retroviruses, including BLV, in conventional radioimmunoassays in either homologous or several broadly cross reactive heterologous systems (38).

Recently, complete nucleotide sequence analysis of cloned HTLV (ATLV) provirus composed of 9,032 bases was determined (82). These results suggest the following organization of HTLV genome. The provirus DNA contains two long terminal repeats (LTRs), one at each side, consisting of 755 bases, which could be arranged into a unique secondary structure making possible the transcriptional termination within the 3′ LTR but not in the 5′ LTR. The nucleotide sequence of the provirus contains three open reading frames, which are capable of coding for proteins at 48K, 99K, and 54K daltons from the 5′ end of the viral genome and presumably code for *gag*, *pol*, and *env* genes, respectively. On the 3′ side of these three open frames is a region composed of four short open reading frames, which may code for 10K, 11K, 12K, and 27K dalton polypeptides. Although DNA of this region has no homology with the normal human genome, some of these polypeptides may have transforming activity. These data on HTLV genome organization represent further evidence for the uniqueness of HTLV.

SEROEPIDEMIOLOGY OF HTLV

Natural HTLV infection frequently elicits serum antibodies to the viral antigens (34,36,37,67,74). Seroepidemiologic surveys of patients with T-cell leukemia/lymphoma, their relatives, and normal donors for antibodies to HTLV were conducted in order to determine the relationship with particular human malignancies. Three different assays have been employed, including: (a) a solid phase radioimmunoassay using whole disrupted viral particles of HTLV, (b) a RIPA using homogeneously purified p24, and (c) IFA against virus-infected cells. Positive results in one single assay system were always cross checked with another assay to provide confirmation of the results. A large number of patients with a variety of T-cell malignancies were screened for evidence of HTLV infection because of T-cell tropism of HTLV (21,68).

As shown in Table 3, HTLV occurs sporadically in the United States. However, there is a form of ATL which is more common in Japan and apparently in the Caribbean. Clinical epidemiology data of Japanese ATL had indicated a geographic clustering in the southwestern islands of Kyushu and Shikoku (34,85). Our sero-

TABLE 3. *Prevalence of natural antibodies to HTLV in sera of patients with malignancies of mature T-cells, their healthy relatives, and random normal donors*[a]

Serum donors	HTLV antibodies[b]	
	No. positive/no. tested	Percent positive
Japanese ATL patients	40/46	87
Healthy relatives of ATL patients	19/40	48
Random healthy donors		
Nonendemic area	9/600	2
Endemic area	50/419	12
Healthy relatives of U.S. patients with HTLV-associated malignancy	2/12	17
Unrelated healthy donors		
Washington, D.C.	1/185	<1
Georgia	3/158	2
Caribbean T-LCL patients	20/20	100
Healthy relatives of Caribbean patients	3/16	19
Random healthy donors, Caribbean	12/337	4

[a]From ref. 72.
[b]Antibodies were detected by RIP or HTLV p24 or by solid-phase radioimmunoassay.

epidemiologic studies have demonstrated that nearly 90% of all Japanese ATL patients have serum antibodies to HTLV, thus indicating HTLV association with this disease (37,74). Another sera where HTLV-associated malignancies are endemic is the Caribbean basin (3,4,11). Antibodies to HTLV were found in all 20 West Indian patients with T-cell lymphosarcoma cell leukemia (T-LCL), a disease similar to Japanese ATL with an aggressive course. Other cases of T-cell malignancies associated with HTLV have also been found in the Boston and Seattle areas, and particularly in the southeastern United States, Alaska, Central and South America, Africa, and Israel (21,72).

Sera of random donors in the United States and Europe generally do not contain antibodies to HTLV. However, HTLV endemic regions have been identified in several areas of the world. The prevalence of serum antibodies in the normal Japanese population varied considerably and correlated closely with the geographic distribution of ATL (34,37,72,74). Thus the incidence was highest, about 16%, in the Nagasaki and Kagoshima areas of Kyushu Island, about 9% in the Uwajima area of Shihoku Island, but was only about 2% in the Honshu Island and even lower in Hakkaido Island (37,72,74). Clustering of T-LCL in regions of the Caribbean basin also occurs (4). The normal Caribbean population screened for the presence of HTLV antibodies exhibited positively in 4%, indicating that HTLV infection may be generally widespread in this region as well (3–5).

Since HTLV is an exogenous virus, close family members of virus-positive patients should be the most likely group to have high incidence of serum antibodies to HTLV proteins. The sera of families that we have analyzed were from the United States, the Caribbean basin, and Japan (73). The single significant normal group positive for HTLV-specific antibodies in the United States comprises the family members of HTLV-positive patients (Table 3). Similarly, Caribbean families were found to be significantly antibody-positive. The highest incidence was found among relatives of Japanese ATL patients. Although the possibility of infection outside family contacts could occur in the endemic regions, family members of virus-positive patients exhibit a much higher incidence of infection in all cases from any given area than do healthy donors who are not related to HTLV-positive patients. This was assessed by serum antibodies to HTLV or the expression of virus antigens upon culturing their T-cells, further strengthening the evidence of HTLV transmission by horizontal infection.

MOLECULAR EPIDEMIOLOGY OF HTLV-RELATED DISEASES

Molecular clones of HTLV-I and HTLV-II genomes have been obtained in our laboratory (47); E. P. Gelmann et al., *submitted*). These clones as well as subclones representing different parts of HTLV genomes have been used as probes to survey fresh leukemic cells and tumor tissues from patients with different lymphoid and myeloid malignancies for DNA sequences related to HTLV (8). These studies show that in some patients with mature T-cell malignancies, including all patients with ATL studied, cells containing one or few copies of HTLV provirus are present.

Cells from other types of malignancies involving immature T-, B-, or myeloid cells are generally negative for the presence of HTLV sequences. Although the results of molecular hybridization and of serologic surveys correlate closely, in a few instances, patients negative for circulating antibodies against HTLV antigens were positive for HTLV provirus. Moreover, it was shown that tumor cells are clonal expansions of single infected cells.

CHARACTERISTICS OF HTLV-ASSOCIATED DISEASES

The isolation of a human type-C retrovirus, HTLV, from the cell line of a patient originally believed to have a cutaneous T-cell lymphoma (61), subsequent isolation of HTLV from other patients with T-cell leukemia/lymphoma in the United States, Japan, and the Caribbean (21,66), and the epidemiologic surveys for antibodies to HTLV have led to the recognition of a distinct form of malignant T-cell leukemia/lymphoma (5,7,21). Clinical and laboratory features of patients positive for HTLV from Japan, the Caribbean, the United States, and elsewhere share common characteristics (6–8,11). The clinical features of Japanese patients with ATL, described by Uchiyama et al. in 1977 (84), included high white blood counts, often skin lesions, hepatosplenomegaly, lymphadenopathy, and a rapidly fatal course. Most of the patients developed opportunistic infections, and many had hypercalcemia during their clinical course. The syndrome was clinically distinct from T-cell lymphoplastic lymphoma, as none of the ATL patients had mediastinal masses. Many of the Japanese patients had birthplaces that clustered in the southern Japanese islands of Kyushu and Shikoku (32,72,85).

Initial morphologic studies of ATL cells circulating in peripheral blood revealed that the malignant cells were pleomorphic, of variable size, had indented or lobulated nuclei, coarsely clumped nuclear chromatin, and scant cytoplasm. The malignant cell of ATL had T-cell surface markers (85). Phenotypically, the cell was best characterized as a helper cell on the basis of the OKT4 and Leu 3 surface markers, but functional assays showed that these cells suppress immunoglobulin synthesis *in vitro*. This observation has recently been confirmed and extended (86,88). In addition, fresh primary cells from ATL patients reacted with a monoclonal antibody, termed anti-TAC, detecting TCGF receptors (86) and a monomorphic anti-HLA-Dr antibody (3.1) (D. Mann, et al., *in preparation*).

Subsequently, in patients from the Caribbean, T-LCL was recognized as a pathologic entity that was indistinguishable from ATL (11). Japanese and Caribbean patients with T-cell leukemia/lymphoma have been shown to have antibodies to HTLV in the majority of instances (3–5,21,34,37,74). Similarly, the cases from the southeastern part of the United States with mature T-cell malignancies that were associated with HTLV closely resemble clinical and laboratory features of those described in Japan and the Caribbean (4,6–8,11,21,74). However, there were additional virus-positive patients who had atypical presentations and diagnoses, usually with less aggressive clinical courses and atypical demographic and laboratory features (3,21,74). The presence of HTLV serum antibodies in cases of ATL (with hypercalcemia and circulating malignant cells) appears to define a distinct clinicopathologic entity, which may occur in geographic clusters.

HTLV also may be associated with AIDS, but it is too early to know if it plays an important role in the disease. It is a T-lymphotropic retrovirus (21,44,62), and the target cells for the putative AIDS agent may be the T-cell or a T-cell subset (28). HTLV is endemic in the Caribbean (3–5,11), and AIDS has been well documented in Haitians (60). Also, there is precedence for retroviruses causing profound cellular and humoral immunosuppression in animals (1,30,59,83). Patients with AIDS include homosexuals (12,20,28), intravenous drug users (12), hemophiliacs (14), heterosexual contacts of members of other high risk groups (49), and, as mentioned, Haitians (13,60). The disease is frequently manifested with opportunistic infection, predominantly pneumocystis carcinii pneumonia (12,28), and/or with Kaposi's sarcoma (12–14,20) in previously healthy persons. Studies of cell-mediated immunity in patients with AIDS have demonstrated generalized impairment of T-lymphocyte functions, including lymphopenia, cutaneous anergy, and reduced helper T-lymphocyte ($OKT4^+$) subpopulations. This results in reversed ratios of helper-to-suppressor T-lymphocyte ($OKT4^+/OKT8^+$), poor lymphocyte responsiveness to mitogens, and, in some cases, decreased natural killer cell activity (28).

The epidemiology of this syndrome, that is, the increasing incidence and clustering of cases, particularly in New York City (NY) and Los Angeles and San Francisco (California), suggests the involvement of a transmissible agent (12–14). Patients with AIDS are often chronically infected with cytomegalovirus and/or hepatitis B virus (12). There is evidence that serum antibodies of many patients with AIDS react with membrane proteins from HTLV-infected cells (18), and HTLV sequences have been found in DNA from two of 33 AIDS cases (25). HTLV has been isolated and partially characterized from two patients, one from the United States with AIDS (24) and a second from France, a patient with lymphadenopathy and other clinical signs and symptoms preceding AIDS (2). HTLV antigens were detected in cultured T-lymphocytes of six other cases of AIDS from France and the United States (M. Popovic, F. Wong-Staal, P. S. Sarin, and R. C. Gallo, *unpublished results*). HTLV or a related retrovirus, therefore, might be important in this disease, but it could also be another opportunistic infection.

IN VITRO TRANSMISSION OF HTLV

Seroepidemiologic and nucleic acid hybridization studies have demonstrated that HTLV is an exogenous virus. A possible *in vitro* infectivity of HTLV was reported by Miyoshi et al. (51), when the virus was isolated from cord blood used as a feeder layer during attempts to establish a cell line (MT-2) from an ATL patient in Japan. However, the cord blood cells used in these cocultivation experiments were not examined for the presence of HTLV. Thus it was possible that these cells were not virus-free. Transmission of MT-2 virus into blood leukocytes has recently been reported by Yamamoto et al. (90). In order to prove infectivity of HTLV, we have carried out extensive transmission studies with each of the new HTLV isolates on target (recipient) cells that were tested and shown to be negative for HTLV (66). Donor cells were X-irradiated or treated with mitomycin-C and cocultivated with HTLV-negative recipient cord blood T-cells of the opposite sex (male × female, and vice versa). After 4 to 5 weeks of cocultivation, the cultures were analyzed for expression of HTLV antigens, extracellular virus, karyotype, and HLA profiles.

To firmly establish the transmission of HTLV, two different approaches were used (Table 4). In the first series of experiments, transmission of the virus (from the cell line derived from M.J.) into human umbilical cord blood T-cells from four different newborns was carried out. On each successive transmission, the recipient cord blood T-cells became the donor for infection of the new cord blood T-cells. Cells of the M.J. line were treated with mitomycin-C and cocultured with cord blood cells (C_1) from a newborn female. The same procedures were used to transmit HTLV-I_{MJ} from C1 to C2 and from C2 to C3 cord blood cells. Analysis of p19, p24, RT, and EM examinations showed that HTLV was fully expressed in all three cord blood T-cell lines. These three cell lines exhibited distinct HLA profiles, thereby confirming the karyologic data. When C3/MJ cord blood T-cells with a female karyotype were cocultured with C5 cord blood cells from a male, the recipient cells showed male karyotype, and HLA profiles matched the recipient cord blood cells (C5). HTLV was fully expressed in these T-cells. Cord blood T-cells cocultured with HTLV-negative T-cells from peripheral blood of a normal donor or PHA-stimulated cell cultures of the cord blood recipient cells were consistently negative for p24 and p19. Sera from cord blood donors were negative for HTLV antibodies as well.

In a second series of experiments, thymus T-cells from a fetus with Klinefelter's syndrome characterized by the presence of XXY sex chromosomes were used as target cells. Two isolates of HTLV (M.J. and T.K.) were transmitted by cocultivation into these fetal thymus T-cells. HTLV was fully expressed in the infected cells, and chromosomal analysis confirmed the presence of the additional marker (sex) chromosome in the recipient thymic T-cells. Nine other isolates (Table 4) of HTLV, including HTLV-II obtained from a patient with a hairy cell leukemia (M.O.), were transmitted into fresh human cord blood T-cells, resulting in productive infection, with the exception of C6/WA (66; M. Popovic et al., *submitted*). The cord blood T-cells infected with the WA isolate were positive for both HTLV-specific proteins p19 and p24, but culture fluids and cells were consistently negative for particulate RTA and type-C particles, respectively. The restricted HTLV expression in infected T-cells has been observed not only in the cells infected *in vitro* but also in T-cells derived from some patients with mature T-cell malignancies (29,80).

Thus the results from transmission studies demonstrate that more than a dozen of the new HTLV isolates of both subgroups can infect and replicate in human T-cells. Recent transmission studies from our laboratory using only concentrated (cell-free) virus of HTLV-I (78; M. Markham et al., *submitted*) or HTLV-II (M. Popovic et al., *submitted*) for infection of T-cells have also been successful.

TABLE 4. *Transmission of HTLV into human cord blood and thymus derived T-cells by cocultivation with HTLV-positive T-cell lines*[a]

Cocultured cells[b] (recipient/donor)	Sex chromosome[b] of fresh and cultured cells		Expression of HTLV proteins			
	Recipient	Donor	p19 (% positive cells)	p24 (ng/mg)	RT (pmoles/ml extract)	EM
				Successive transmission		
C1/MJ	XX	XY	90	385	3.3	+
C2/MJ	XX	XX	85	930	ND	+
C3/MJ	XX	XX	95	690	5.1	+
C5/MJ	XY	XX	90	540	8.1	+
				Thymus (Klinefelter's syndrome)		
FTH[b]/TK	XXY	XX	90	600	ND	+
FTH[b]/MJ	XXY	XY	90	700	ND	+
				Individual transmission		
C4/UK	XX	XY	81	740	34.1	+
C21/MI	XY	XX	47	235	60.4	+
C6/WA	XX	XY	53	502	0	−
C91/PL	XY	XX	71	500	35.6	+
C126/0B	XY	XX	30	140	ND	ND
C8/SK	XX	XY	47	1,000	8.7	+
C7/TK	XY	XX	65	685	84.2	+
C90/HK	XX	XY	46	500	30.3	+
C3-44/MO[c]	XX	XY	90	ND[d]	16.5	+
C2-18/MO[c]	XX	XY	87	ND[d]	21.8	+

[a]From ref. 66.

[b]Thymus cells from fetus with Klinefelter's syndrome associated with a characteristic abnormality of sex chromosomes.

[c]Two isolates of HTLV-II subgroup used for infection of cord blood T-cells.

[d]Positive for p24 in heterologous assay using radiolabeled p24 of HTLV-I and patient (M.O.) serum.

Sex chromosomes and HLA antigens were used as markers for HTLV-infected T-cells. Control or PHA- or allogenic (human T-cells from peripheral blood)-stimulated recipient cord blood T-cells were processed simultaneously and tested for HTLV. Prior to the transmission experiments, cord blood samples were consistently negative for HTLV. ND, not done.

However, seroepidemiologic as well as *in vitro* transmission studies of HTLV indicate that this human retrovirus is far less infectious than the majority of avian or mammalian retroviruses.

T-CELL TROPISM OF HTLV

Detection of HTLV in neoplastic T-cells and not in B-cells of the same patient (C. R.) suggests that the virus is exogenous and T-cell tropic (21,22,44,51,61,66,69).

TABLE 5. *T-cell tropism of HTLV*

Source of T- and B-cells used for infection	HTLV-I[a]	HTLV-II[b]
Human T-cells from		
Fetal thymus	+	+
Fetal spleen	+	ND
Newborn cord blood	+	+
Adolescent nasopharyngeal tonsils[c]	+	+
Adult peripheral blood	+	+
Adult bone marrow	+	+
Adult spleen	+	ND
Adult liver	+	ND
Human B-cells from		
Newborn cord blood	−	ND
Adolescent nasopharyngeal tonsils[c]	−	ND
Adult peripheral blood	−	ND
Marmoset T-cells from		
Peripheral blood	+	+

[a]HTLV-I_{TK} and HTLV-I_{MJ} isolates were used for infection.
[b]In the case of HTLV-II, MO and MO-F isolates from the same patient (J.M.) were used; +, positive for p19, p24, and RT; −, negative for p19 and RT; ND, not done.
[c]In coinfection (HTLV + EBV) experiments, HTLV-positive T-cells were obtained only in the presence of TCGF and EBV-positive B-cells in the absence of TCGF.

This has been documented by seroepidemiologic and nucleic acid studies and by establishment of HTLV-positive T-cell lines from more than 15 cases of patients with mature T-cell malignancies and their close relatives. Fresh or *in vitro* cultured cells revealed markers of mature T- but not B-cells in every instance (44,66). Recently, Yamamoto et al. (89) reported the establishment of a unique cell line from an ATL patient with B-cell characteristics and positive for both HTLV and Epstein-Barr virus (EBV). We have extended our transmission studies of HTLV into cells derived from fetal thymus and spleen, cord blood of newborns, bone marrow, peripheral blood, liver and spleen of adults, and nasopharyngeal tonsils of an adolescent. The results are summarized in Table 6. In all cases, the HTLV-positive T-cell population was originally at least partially dependent on exogenous TCGF; most of them, as previously demonstrated (44,66), revealed markers of mature T-cells (OKT4$^+$ and Leu 3$^+$). None of the HTLV-positive cells possessed B-cell markers. Attempts to infect B-cells by HTLV were unsuccessful. In coinfection experiments (HTLV + EBV) of cells derived from nasopharyngeal lymphoid tissues in the presence of TCGF, only HTLV-positive T-cells with T-cell markers grew. In contrast, growth of EBV-positive cells with B-cell markers developed in cell cultures simultaneously infected with HTLV and EBV and cultured in the absence of TCGF. Thus the HTLV positivity of fetal thymus T-cells suggests that less mature T-cells can be infected with HTLV than previously believed but not B-cells susceptible to EBV infection. T-cells from primates, as demonstrated here

TABLE 6. *Comparison of properties of HTLV-positive human neoplastic T-cells with normal, uninfected, and HTLV-infected human cord blood T-cells*

Property	HTLV-positive neoplastic T-cell lines	Cord blood T-cells	
		HTLV-infected	Mitogen-stimulated
In vitro growth	>180 days (Immortal)	>180 days (Immortal)	<60 days (Temporary)
Requirement for exogenous TCGV (v/v)	0–5%	0–5%	10–12%
TCGF receptor (TAC)[a]	+++	+++	+
Transferrin receptor	+++	+++	+
E-rosette	+++	+++	+++
S-IgG[b], EBNA[c], TdT[d]	–	–	–
Cell phenotype			
Inducer/helper (OKT4, Leu3)[d]	Most or all	Most or all	Most or all
Suppressor/cytotoxic (OKT8, Leu2)[a]	Few or none	Few or none	Few or none
Cell morphology			
Presence of multinucleated giant cells	+	+	–
Presence of lobulated nuclei	+	+	–
Constitutive lymphokine production	+	+	–
HLA modification			
Expression of additional HLA antigens	+	+	–
Expression of HLA-Dr[a]	+	+	–
Expression of HT-3 sequences	+++	+++	+
HTLV detection			
p19, p24, and RT expression	+	+	–
Type-C virus particles (EM)	+	+	–

[a]Determined by cell sorter using monoclonal antibodies.
[b]S-IgG, cell surface immunoglobulins.
[c]EBNA, Epstein-Barr nuclear antigen.
[d]TdT, terminal deosynucleotidyl transferase.

in the case of marmoset T-cells, and rodent T-cells have also been shown to be susceptible to infection by HTLV (52,53). It is possible that a precursor cell of B- and T-cell lineage can be infected by HTLV which, after maturation, becomes susceptible to EBV infection. This might explain the development of a cell line from a patient with ATL harboring both HTLV and EBV, as described by Yamamoto et al. (89).

TRANSFORMATION OF T-CELLS BY HTLV

HTLV-infected T-cells exhibit several characteristic features which distinguish them from uninfected T-cells and show similarities to T-cell lines established from patients with T-cell malignancies (23,51,62,64). As shown in Table 6, a comparison of the properties of HTLV-infected T-cells with normal (uninfected) and HTLV-positive neoplastic T-cells derived from patients with mature T-cell malignancies indicates that *in vitro* infected T-cells with HTLV are transformed. Like primary neoplastic T-cells, the HTLV in *in vitro* transformed T-cells exhibit indefinite growth potential, show a decreased or complete independence of requirement for TCGF, have generally helper/inducer ($OKT4^{+}/Leu3^{+}$) phenotype (23,44,66,79), become a constitutive producer of lymphokines, and show morphologic and cell surface alterations (see below).

Morphologic Characteristics of Normal and HTLV-Infected Cord Blood T-Cells

Normal (mitogen-stimulated) T-cells from cord blood grow transiently in culture, predominantly as single cell suspensions with some small clumps (Fig. 2A). Morphologically, they are a homogenous population of lymphoblastoid cells (Fig. 2B). The HTLV-infected T-cells from cord blood initially grew as single cells but soon displayed growth in large clumps (Fig. 2C), which was a characteristic feature of those HTLV-infected T-cells that became less dependent on or independent of exogenous TCGF. The cultures of transformed T-cells were also considerably less uniform in appearance than uninfected normal T-cells in culture. Examination of smears from these infected cultures after Wright-Giemsa stain showed that the cell size and number of nuclei per cell were variable (Fig. 2D). Mono- and binucleated cells were frequently found around multinucleated giant cells. The multinucleated cells (0.7 to 10% of total cell population) were always present in the HTLV-infected T-cell lines. Both the growth pattern and morphology of HTLV-infected cord blood T-cells are similar to the neoplastic T-cell lines derived from patients with mature T-cell malignancies (23,62,64). An example of growth in large clumps and of cell polymorphism of a neoplastic T-cell line established *in vitro* from an HTLV-positive patient with cutaneous T-cell lymphoma is shown in Figs. 2E and F.

The cell morphology of *in vitro* transformed T-cells and T-cells from primary T-cell malignancies was examined by EM. Electron micrographs of HTLV-infected and mitogen-stimulated cord blood T-cells are shown in Fig. 3. The HTLV-transformed T-cells frequently exhibited convoluted nuclei with one or two nucleoli. Although the nuclear convolution was not as prominent as in the case of neoplastic T-cells, it was consistently observed in the HTLV-infected cord blood T-cells. The nuclei of the HTLV-infected cells differ from the spheroidal shaped nuclei of the uninfected cord blood T-cell population.

Immortalization of HTLV-Infected T-Cells and TCGF Dependence

HTLV infection of normal T-cells can induce long-term growth of the infected cells (23,51,64,78). This was demonstrated in our laboratory with the various isolates of both subgroups of HTLV and by Japanese investigators with the MT-2 isolate (23,64,90). Long-term growth of mitogen-stimulated (uninfected) and HTLV-infected T-cells from cord blood was followed simultaneously in many separate experiments. As shown in Fig. 4, the uninfected T-cells consistently entered a "crisis" period, in which they failed to increase in cell number. In the absence of HTLV infection, no permanent TCGF dependence or independent T-cell line from human cord blood was obtained, and all exhibited limited lifespan *in vitro*. In contrast, no such crisis period was observed in any of the HTLV-infected T-cells, and all (40 infected T-cell lines) grew indefinitely.

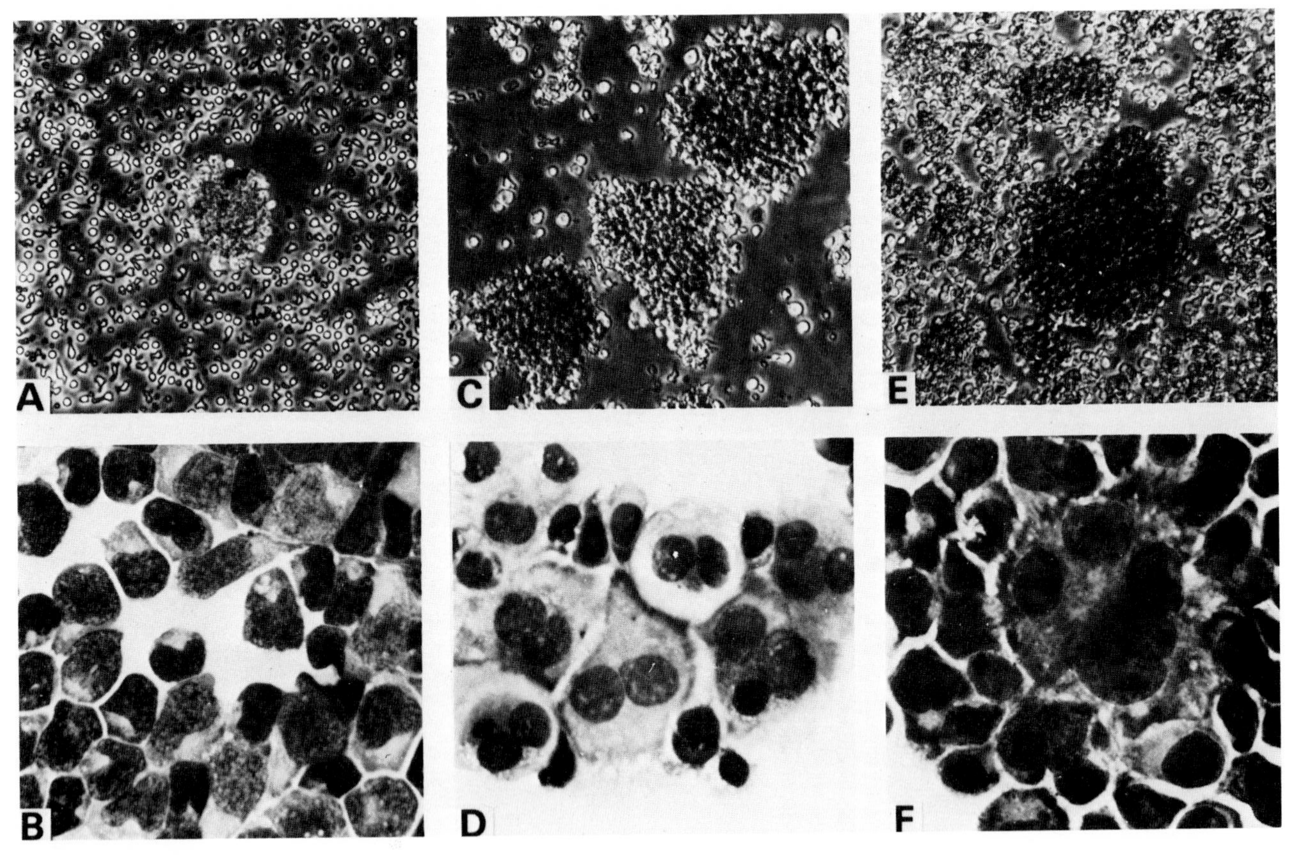

FIG. 2.

Activated normal T-cells depend on exogenous TCGF for growth *in vitro*. T-cells originating from clinical specimens obtained from patients with HTLV-positive mature T-cell malignancies differ from their normal counterparts in at least two respects: (a) they contain TCGF receptors and, therefore, do not require lectin activation to respond to TCGF (23,62,66); and (b) they often have a capacity for indefinite growth (23,66). Some become constitutive producers of TCGF, perhaps explaining why they become independent of exogenous TCGF (27). Therefore, the cellular requirements for exogenous TCGF of HTLV-infected cord blood T-cells were compared to HTLV-positive neoplastic T-cells derived from patients with mature T-cell malignancies. The requirements for exogenous TCGF of both cord blood cells newly infected with HTLV and of HTLV-positive primary tumor cells were two- to 10-fold less than for the normal (uninfected) T-cells from adult peripheral blood or from cord blood (Table 6). With cultures initiated in the presence of TCGF, four of 16 HTLV-infected cord blood T-cells are now completely independent of the requirement for exogenous TCGF for their growth. Under culture conditions in which TCGF-independent, HTLV-infected T-cells are preferentially selected, however, the frequency of establishment of HTLV transformed cell lines independent of exogenous TCGF can be substantially higher: more than 90% of newly infected T-cell lines (48).

Phenotypic Alterations of HTLV Transformed Cord Blood T-Cells: TCGF, Transferrin Receptors, and HLA

As pointed out previously, the fresh neoplastic T-cells from people with HTLV-positive malignancies express TCGF receptors, in contrast to the majority of normal T-cells. This was suggested by the fact that these cells grew in direct response to TCGF (27,62,66) and bound TCGF to cell surface (27). It was further demonstrated by a monoclonal antibody (anti-TAC) recognizing these receptors (86). In the case of cultured neoplastic T-cells of HTLV-positive cell lines, more than 70% of the cells reacted with anti-TAC monoclonal antibody (66). Moreover, it was shown that more than 90% of the cultured T-cells from adults possess HLA-Dr determinants (50), which mitogen-stimulated cord blood T-cells cultured *in vitro* failed to express (91). It was of interest to determine whether umbilical cord blood T-cells infected with HTLV develop or have both TCGF receptors and HLA-Dr determinants.

FIG. 2. Growth pattern (×100) and morphology (Wright-Giemsa ×460) of cultured normal (uninfected) and HTLV transformed T-cells from cord blood, and neoplastic T-cells from a patient (M.J.) with a HTLV-positive cutaneous T-cell lymphoma. **A:** Phase contrast microscopy of (uninfected) cord blood T-cells cultured in the presence of 10% TCGF showing growth of cells as single cells in suspension. **B:** Light microscopy of Giemsa stained normal (uninfected) cord blood T-cells grown with 10% TCGF showing uniform morphology. **C:** Phase contrast microscopy of cord blood T-cells transformed by the M.J. isolate of HTLV showing cells growing in large clumps. **D:** Light microscopy of Giemsa stained HTLV transformed cord blood T-cells showing some of bi- and multinucleated cells. **E:** Phase contrast microscopy of the neoplastic T-cells from patient M.J. **F:** Light microscopy of Giemsa stained M.J. neoplastic T-cells. Note the presence of multinucleated cells in this culture. These changes and the clumping of cells shown in **E** are similar to the HTLV *in vitro* transformed cord blood T-cells.

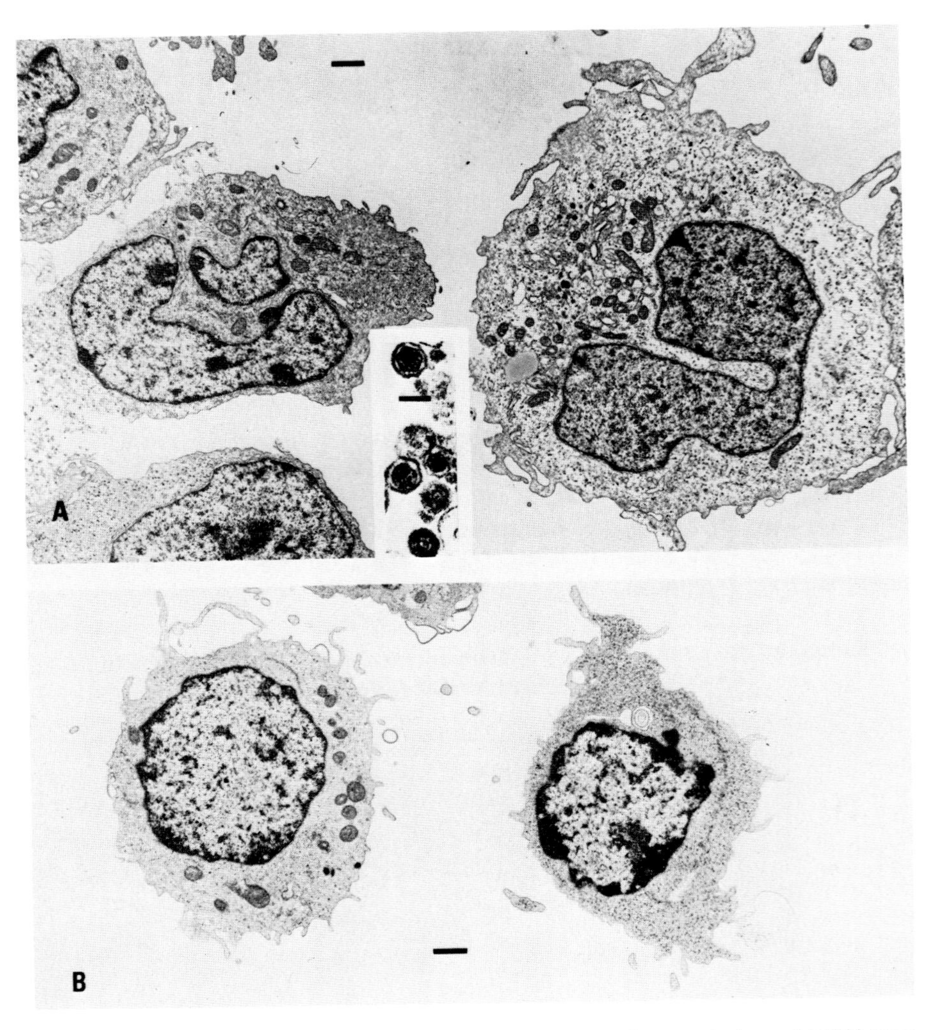

FIG. 3. Electron micrographs showing differences between the morphology of HTLV transformed cord blood T-cells **(A)** and normal cord blood T-cells grown with TCGF **(B)** (bar = 1 μm). Note the high proportion of cells with lobulated nuclei in the HTLV transformed cell culture. *Inset*, extracellular type-C virus (bar = 0.1 μm).

In parallel experiments, HTLV transformed cord blood T-cells and their counterparts (mitogen-stimulated T-cells) from the same newborns were analyzed for the presence of these antigens. The percentage of positive cells and density of TCGF receptors and HLA-Dr determinants were determined by FACS analysis using the anti-TAC monoclonal antibody and a monomorphic anti-HLA-Dr antibody (3.1), respectively. All HTLV-infected T-cells were highly positive for the expression of both TCGF receptors and HLA-Dr, and they were approximately 50 times

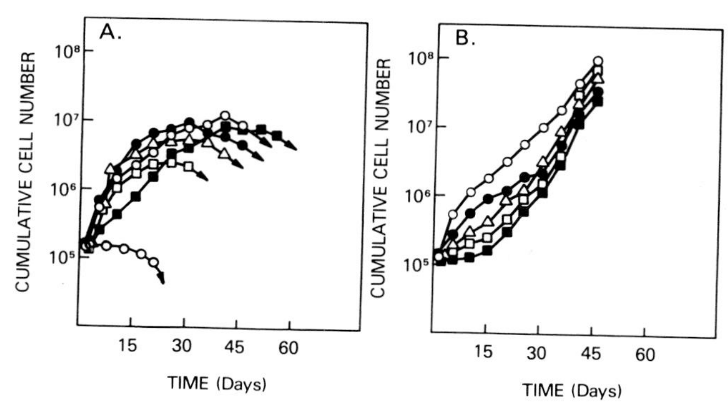

FIG. 4. Cumulative growth curves of **(A)** mitogen-stimulated T-cells derived from newborn umbilical cord human blood and **(B)** HTLV infected cord blood T-cells from newborns in long-term cultures.

greater than mitogen-stimulated normal human cord blood T-cells. This estimate of the number of TCGF receptors is comparable to neoplastic HTLV-positive T-cells obtained from patients with T-cell malignancies. Thus the expression of both TCGF receptors and HLA-Dr determinants may play an important role in the proliferation of cord blood T-cells infected by HTLV. In addition, as in the case of neoplastic T-cells characterized by rapid growth, the density of transferrin receptors was relatively high in the HTLV transformed T-cells.

HLA typing of the peripheral blood lymphocytes and HTLV-positive cultured T-cells from patients with mature T-cell malignancies suggested the expression of additional HLA-A and -B locus antigens on the HTLV-positive cultured T-cells that were not present on the EBV transformed B-cells or on the fresh peripheral blood lymphocytes from the same patient. Cultured T-cells positive for HTLV derived from patients or transformed *in vitro* showed parallel expression of both HTLV and altered HLA alloantigen (22,45). The meaning of this consistent finding is not yet clear. However, using molecular clones of HTLV and human major histocompatibility antigen DNA, Clarke and co-workers (15) demonstrated in Southern blots a homology between the envelope gene region of HTLV and the region of an HLA-B locus gene which codes for the extracellular portion of a class I histocompatibility antigen. Some interesting implications of the finding of the *env* protein of HTLV relationship to certain HLA antigens biologically may help clarify the mechanism of T-cell infection and its transformation. For instance, HTLV might use these HLA antigen(s) as a viral receptor, or inappropriate expression of these HLA antigen(s) on infected T-cells could impair or abrogate their normal function.

Constitutive Lymphokine Production by HTLV Transformed T-Cells

The central role of T-cells in humoral and cell-mediated immunity is well established. In these highly complex and precise interactions of humoral factors and cells, T-cells operate as both effector and modulator cells. Proliferation and several T-cell functions are mediated by the release of soluble, biologically active factors termed lymphokines, and some can also influence the behavior of other cells. More than 10 such active factors released by T-cells have been reported (79). Previous studies clearly demonstrated that HTLV transformed T-cells are not only capable of responding efficiently to TCGF (27,62,66), but some become constitutive TCGF producers (27).

Subsequently, a survey of a constitutive production of various lymphokines by HTLV transformed T-cell lines was performed (79). Conditioned media from 21 T-cell cultures, established either from the peripheral blood of HTLV-positive patients with T-cell malignancies or from human umbilical cord blood or bone marrow T-cells transformed *in vitro* by HTLV, were assayed for biologic activity. The data clearly demonstrated that all HTLV transformed T-cell lines became constitutive producers of several lymphokines, including macrophage migration-inhibitory factor, leukocyte-inhibitory factor, migration-enhancement factor, macrophage-activating factor, differentiation-inducing factor, colony-stimulating factor, eosinophil growth-maturation activity, interleukin 3, fibroblast-activating factor, B-cell growth factor(s), and gamma-interferon. In some HTLV transformed T-cell lines, low levels of TCGF production were also detected. In conclusion, the decreased requirement for TCGF, cell surface alterations, and constitutive production of various lymphokines represent essential features of HTLV transformed T-cells, which may result in their impaired function and unrestrained growth.

Possible Molecular Mechanism of T-Cell Transformation by HTLV

The analysis of HTLV-positive leukemic T-cells showed that the cells are of clonal origin with respect to provirus integration sites (29,87). Analysis of cloned complete HTLV genomes indicates that HTLV does not contain a cell-derived *onc* gene (82; F. Wong-Staal et al., *unpublished results*). The absence of a cell-derived *onc* gene and the monoclonality of the tumors suggest that HTLV is a chronic leukemia virus, despite its *in vitro* transforming activity. Several chronic leukemia viruses are known to induce leukemia by activating cellular *onc* genes (e.g., *myc* in B-cell lymphomas) (33). This occurs from provirus integration in the proximity of these genes. Activation of these genes probably is induced by providing either a viral promotor or viral enhancer sequence (43,58).

We have examined some of the primary HTLV-infected neoplastic T-cells and cord blood T-lymphocytes infected *in vitro* with HTLV for possible rearrangement of known *onc* genes and expression of these genes. Thus far, there is no evidence for such rearrangements or any abnormal expression of any *onc* gene, with the possible exception of low level *sis* expression in some cells. This phenomenon is under study. Since HTLV specifically transforms mature T-cells, it is likely to affect expression of genes that are important in T-cell proliferation and not related to known *onc* genes.

To study genes that are turned on by HTLV infection, we constructed a cDNA library from mRNA or a HTLV-positive tumor cell line and screened the library differentially with cDNA probes made from the same cell line and a closely analogous T-cell lymphoma line negative for HTLV (46). From these experiments, a gene called HT-3 was identified and isolated. It is expressed at high levels in all HTLV-positive neoplastic T-cells and in normal cord blood T-cells after infection with HTLV (Table 6). A survey of a variety of human hematopoietic cell lines showed that this gene is usually expressed at low or undetectable levels, but several sources of T-cells, including normal lymphocytes and the cell lines Jurkat and HSB-2, express high levels of this gene after lectin activation. These results suggest that this gene may code for a product intimately involved in blastic transformation of T-cells and their proliferation. At present, experiments are being carried out to determine whether activation of this gene is a direct consequence of HTLV provirus integration (e.g., insertion of viral promotor or enhancer sequences). If so, this gene would be equivalent to an *onc* gene in its role in HTLV-induced neoplastic transformation.

We have previously proposed that HTLV-transformed cells synthesize both TCGF and TCGF receptor, resulting in autostimulation and uncontrolled proliferation (23,27). We have now cloned the gene for TCGF (S. C. Clark et al., *submitted*) and can directly test this model by using the clone to study expression of TCGF mRNA in different HTLV-infected cells (S. K. Arya et al., *submitted*). Many of the transformed cord blood cell lines do not express TCGF, even though they are independent of exogenous TCGF for growth. Therefore, they have either completely bypassed the TCGF-TCGF receptor system and utilize an entirely different growth factor (e.g., one of the many factors that these cell lines produce) or, alternatively, the TCGF receptors in these cells may have been altered so that they behave as if they are bound to TCGF. These possibilities are currently being investigated.

SUMMARY AND FUTURE STUDIES

HTLV is a unique, exogenous retrovirus which is not related to any known retrovirus, except for a distant relationship with BLV. Epidemiologic studies indicate that HTLV is endemic in parts of Japan, the Caribbean Islands, Central and South America, and Africa. A number of important questions remain to be answered, including: (a) the mechanism of transmission from individual to individual, (b) the mechanism by which HTLV transforms mature T-cells, whether this may involve a mechanism of an appropriate HLA, expression of specific viral gene products, and/or an integration near a specific gene or genes, and (c) whether early HTLV isolates represent a restricted group of retroviruses associated with a specific disease or belong to a family of human retroviruses associated with other diseases. HTLV is not readily infectious and may require prolonged close contact for transmission. This might occur in relatives of patients or by sexual contact, blood transfusion, or insect vector transmission. Further studies will determine the involvement of HTLV and its worldwide distribution in other human malignancies, including the recently described AIDS and Kaposi's sarcoma.

ACKNOWLEDGMENT

We thank A. Mazzuca for her excellent editorial assistance.

REFERENCES

1. Anderson, L. J., Jarret, W. F. H., Jarret, O., and Laird, H. M. (1971): Feline leukemia-virus infection of kittens: Mortality associated with atrophy of the thymus and lymphoid deplition. *J. Natl. Cancer Inst.*, 47:807–817.
2. Barre-Sinoussi, F., Chermann, J. C., Rey, F., Nugeyre, M. T., Chamaret, S., Gruest, J., Dauguet, C., Axler-Blin, C., Vezinet-Brun, F., Rouzious, C., Rozenbaum, W., and Montagnier, L. (1983): Isolation of a T-lymphotropic retrovirus from a patient at risk for acquired immune deficiency syndrome (AIDS). *Science*, 220:868–871.
3. Blattner, W. A., Blayney, D. W., Robert-Guroff, M., Sarngadharan, M. G., Kalyanaraman, V. S., Sarin, P. S., Jaffe, E. S., and Gallo, R. C. (1983): Epidemiology of human T-cell leukemia/lymphoma virus. *J. Infect. Dis.*, 147:406–416.
4. Blattner, W. A., Gibbs, N. W., Saxinger, C., Robert-Guroff, M., Clark, J., Loftus, W., Hanchard, B., Cambell, M., and Gallo, R. C. (1983): Human T-cell leukemia/lymphoma virus-associated lymphoreticular neoplasia in Jamaica. *Lancet*, 8341:61–64.
5. Blattner, W. A., Kalyanaraman, V. S., Robert-Guroff, M., Lister, T. A., Galton, D. A. G., Sarin, P. S., Crawford, M. H., Catovsky, D., Greaves, M., and Gallo, R. C. (1982): The human type-C retrovirus, HTLV, in blacks from the Caribbean region and relationship to adult T-cell leukemia/lymphoma. *Int. J. Cancer*, 39:257–264.
6. Blayney, D. W., Blattner, W. A., Robert-Guroff, M., Jaffe, E. S., Fischer, R. I., Bunn, P. A., Patton, M. G., Rarick, H. R., and Gallo, R. C. (1983): The human T-cell leukemia/lymphoma virus (HTLV) in the Southeastern United States. *JAMA*, 250:1048–1052.
7. Blayney, D. W., Jaffe, E. S., Schechter, G. P., Cossman, J., Robert-Guroff, M., Kalyanaraman, V. S., Blattner, W. A., and Gallo, R. C. (1983): The human T-cell leukemia/lymphoma virus, lymphoma, lytic bone lesions, and hypercalcemia. *Ann. Intern. Med.*, 98:144–151.
8. Bunn, P. A., Jr., Schechter, G. P., Jaffe, E., Blayney, D., Young, R. C., Matthews, M. J., Blattner, W., Broder, S., Robert-Guroff, M., and Gallo, R. C. (1983): Retrovirus associated with adult T cell lymphoma in the United States: Staging evaluation and management. *N. Engl. J. Med.*, 309:257–264.
9. Burny, A., Bruck, C., Chantrenne, H., Cleuter, Y., Dekezel, D., Ghysdael, J., Kettmann, R., Leclercq, M., Leunen, J., Mammerickx, M., and Portelle, D. (1980): Bovine leukemia virus: Molecular biology and epidemiology. In: *Viral Oncology*, edited by G. Klein, pp. 231–280. Raven Press, New York.
10. Casey, J. W., Roach, A., Mullins, J. I., Burck, K. B., Nicolson, M. O., Gardner, M. B., and Davidson, S. (1981): The U3 portion of the feline leukemia virus identifies horizontally acquired provirus in leukemic cats. *Proc. Natl. Acad. Sci. USA*, 78:7778–7782.
11. Catovsky, D., Greaves, M. F., Rose, M., Galton, D. A. G., Goolden, A. W. G., McCluskey, D. R., White, J. M., Lampert, I., Bourikas, G., Ireland, R., Brownell, A. I., Bridges, J. M., Blattner, W. A., and Gallo, R. C. (1982): Adult T-cell lymphoma-leukaemia in blacks from the West Indies. *Lancet* i(8372):639–642.
12. Center for Disease Control (1982): Task force on Kaposi's sarcoma and opportunistic infections. *Morbidity Mortality Weekly Report*, 31:249.
13. Center for Disease Control (1982): Opportunistic infections and Kaposi's sarcoma among Haitians in the United States. *Morbidity Mortality Weekly Report*, 31:353–354.
14. Center for Disease Control (1982): Pneumocystis carcinii pneumonia among persons with hemophilia A. *Morbidity Mortality Weekly Report*, 31:365–367.
15. Clarke, M. F., Gelmann, E. P., and Reitz, M. S. (1983): Homology of the human T-cell leukemia virus envelope gene with a class-I major histocompatibility antigen gene. *Nature*, 305:60–62.
16. Deschamp, J., Kettmann, R., and Burny, A. (1981): Experiments with cloned complete tumor-derived bovine leukemia virus information prove that the virus is totally exogenous to its target animal species. *J. Virol.*, 40:605–609.
17. Essex, M. (1975): Horizontally and vertically transmitted oncornaviruses of cats. *Adv. Cancer Res.*, 21:175–264.
18. Essex, M., McLane, M. F., Lee, T. H., Falk, L., Howe, C. W. S., Mullins, J. I., Cabradilla, C.,

and Francis, D. P. (1983): Antibodies to cell membrane antigens associated with human T-cell leukemia virus in patients with AIDS. *Science*, 220:859–862.

19. Ferrer, J. F., Abt, D. A., Bhatt, D. M., and Marshak, R. R. (1974): Studies on the relationship between infection with bovine C-type virus, leukemia and persistent lymphocytes in cattle. *Cancer Res.*, 34:893–898.
20. Friedman-Klein, A. E., Laubeustein, L. J., Rubinstein, P., Buimovie-Klein, E., Marmor, M., Stahl, R., Spigland, I., So, I., and Zolla-Pazner, S. (1982): Disseminated Kaposi's sarcoma in homosexual men. *Ann. Intern. Med.*, 96:693–700.
21. Gallo, R. C., Kalyanaraman, V. S., Sarngadharan, M. G., Sliski, A., Vonderheid, E. C., Maeda, M., Nakao, Y., Yamada, K., Ito, Y., Gutensohn, N., Murphy, S., Bunn, P. A., Jr., Catovsky, D., Greaves, M. F., Blayney, D. W., Blattner, W., Jarrett, W. F. H., zur Hausen, H., Seligmann, M., Brouet, J. C., Haynes, B. F., Jegasothy, B. V., Jaffe, E., Cossman, J., Broder, S., Fisher, R. I., Golde, D. W., and Robert-Guroff, M. (1983): The human type-C retrovirus: Association with a subset of adult T-cell malignancies. *Cancer Res.*, 43:3892–3899.
22. Gallo, R. C., Mann, D., Broder, S., Ruscetti, F. W., Maeda, M., Kalyanaraman, V. S., Robert-Guroff, M., and Reitz, M. S. (1982): Human T-cell leukemia-lymphoma virus (HTLV) is in T- but not B-lymphocytes from a patient with cutaneous T-cell lymphoma. *Proc. Natl. Acad. Sci. USA*, 79:4680–4683.
23. Gallo, R. C., Popovic, M., Lange-Wantzin, G., Wong-Staal, F., and Sarin, P. S. (1983): Stem cells, leukemia viruses, and leukemia of man. In: *Haemopoietic Stem Cells*, edited by Sv.-Aa. Killmann, E. P. Cronkite, and C. N. Muller-Berat, pp. 155–170. Munksgaard, Copenhagen.
24. Gallo, R. C., Sarin, P. S., Gelmann, E. P., Robert-Guroff, M., Richardson, E., Kalyanaraman, V. S., Mann, D., Sidhu, G. D., Stahl, R. E., Zolla-Pazner, S., Leibowitch, J., and Popovic, M. (1983): Isolation of human T-cell leukemia virus in acquired immune deficiency syndrome (AIDS). *Science*, 220:865–867.
25. Gelmann, E. P., Popovic, M., Blayney, D., Masur, H., Sidhu, G., Stahl, E. R., and Gallo, R. C. (1983): Proviral DNA of a retrovirus, human T-cell leukemia virus in two patients with AIDS. *Science*, 200:862–865.
26. Gilden, R. V., Long, C. W., Hanson, M., Toni, R., Charman, H. P., Oroszlan, S., Miller, J. M., and Van Der Maaten, M. J. (1975): Characteristics of the major internal protein and RNA dependent DNA polymerase of bovine leukemia virus. *J. Gen. Virol.*, 29:305–314.
27. Gootenberg, J. E., Ruscetti, F. W., Mier, J. W., Gazdar, A., and Gallo, R. C. (1981): Human cutaneous T cell lymphoma and leukemia cell lines produce and respond to T cell growth factor. *J. Exp. Med.*, 154:1403–1418.
28. Gottlieb, M. S., Sehroff, R., Shauker, M. M., Weisman, I. D., Fan, P. T., Wolf, R. A., and Saxon, A. (1981): Pneumocystic carinii pneumonia and mucosal candidiasis in previously healthy homosexual men: Evidence of a new acquired cellular immunodeficiency. *N. Engl. J. Med.*, 305:1425–1430.
29. Hahn, B., Manzari, V., Colombini, S., Franchini, G., Gallo, R. C., and Wong-Staal, F. (1983): Common site of integration of HTLV in cells of three patients with mature T-cell leukaemia-lymphoma. *Nature*, 303:253–256.
30. Hardy, W. D., Jr., Hess, P. W., MacEwen, E. G., McClelland, A. J., Zuckerman, E. E., Essex, M., and Cotter, S. M. (1976): The biology of feline leukemia virus in the natural environment. *Cancer Res.*, 36:582–588.
31. Hardy, W. D., Jr., Old, L. J., Hess, P. W., Essex, M., and Cotter, S. M. (1973): Horizontal transmission of feline leukemia virus. *Nature*, 244:266–269.
32. Haynes, B. F., Miller, S. E., Moore, T. O., Dunn, P. H., Bolognesi, D. P., and Metzgar, R. S. (1983): Identification of human T cell leukemia virus in a Japanese patient with adult T cell leukemia and cutaneous lymphomatous vasculitis. *Proc. Natl. Acad. Sci. USA*, 80:2054–2058.
33. Hayward, W. S., Neel, B. G., and Astrin, S. M. (1981): Activation of a cellular *onc* gene by promoter insertion in ALV-induced lymphoid leukosis. *Nature*, 209:475–480.
34. Hinuma, Y., Nagata, K., Misoka, M., Nakai, M., Matsumoto, T., Kinoshita, K. I., Shirakawa, S., and Miyoshi, I. (1981): Adult T-cell leukemia: Antigen in an ATL cell line and detection of antibodies to the antigen in human sera. *Proc. Natl. Acad. Sci. USA*, 78:6476–6480.
35. Jarrett, W. F. H., Crawford, E. M., Martin, W. B., and Davie, F. (1964): A virus-like particle associated with leukemia (lymphosarcoma). *Nature*, 202:567–570.
36. Kalyanaraman, V. S., Sarngadharan, M. G., Bunn, P. A., Minna, J. D., and Gallo, R. C. (1981):

Antibodies in human sera reactive against an internal structural protein of human T-cell lymphoma virus. *Nature*, 294:271–273.

37. Kalyanaraman, V. S., Sarngadharan, M. G., Nakao, Y., Ito, Y., Aoki, T., and Gallo, R. C. (1982): Natural antibodies to the structural core protein (p24) of the human T-cell leukemia (lymphoma) retrovirus (HTLV) found in sera of leukemic patients in Japan. *Proc. Natl. Acad. Sci. USA*, 79:1653–1657.
38. Kalyanaraman, V. S., Sarngadharan, M. G., Poiesz, B. J., Ruscetti, F. W., and Gallo, R. C. (1981): Immunological properties of a type C retrovirus isolated from cultured human T-lymphoma cells and comparison to other mammalian retroviruses. *J. Virol.*, 38:906–913.
39. Kalyanaraman, V. S., Sarngadharan, M. G., Robert-Guroff, M., Blayney, D., Golde, D., and Gallo, R. C. (1982): A new subtype of human T-cell leukemia virus (HTLV-II) associated with a T-cell varient of hairy cell leukemia. *Science*, 218:571–573.
40. Kettmann, R., Deschamps, J., Cleuter, Y., Couez, D., Burny, A., and Marbaix, G. (1984): Leukemogenesis by bovine leukemia virus: Proviral DNA integration and lack of RNA expression of viral long terminal repeat and 3′ proximate cellular sequences. *Proc. Natl. Acad. Sci. USA (in press)*.
41. Koshy, R., Gallo, R. C., and Wong-Staal, F. (1980): Characterization of the endogenous feline leukemia virus related DNA sequences in cats and attempts to identify exogenous viral sequences in tissues of virus-negative leukemic animals. *Virology*, 103:434–445.
42. Koshy, R., Wong-Staal, F., Gallo, R. C., Hardy, W., and Essex, M. (1979): Distribution of FeLV sequences in DNA of normal and leukemic domestic cats. *Virology*, 99:135–144.
43. Levinson, B., Khoury, G., Vande Woude, G., and Gruss, P. (1982): Activation of SV40 genome by 72 base pair tandem repeats by Moloney sarcoma virus. *Nature*, 295:568–572.
44. Mann, D. L., Popovic, M., Murray, C., Neuland, C., Strong, D. M., Sarin, P. S., Gallo, R. C., and Blattner, W. A. (1983): Cell surface antigen expression in newborn cord blood lymphocytes infected with HTLV. *J. Immunol.*, 131:2021–2024.
45. Mann, D. L., Popovic, M., Sarin, P., Murray, C., Neuland, C., Strong, D. M., Haynes, B. F., Gallo, R. C., and Blattner, W. A. (1984): Cell lines producing human T-cell lymphoma virus (HTLV) have altered HLA expression. *Nature*, 305:58–60.
46. Manzari, V., Gallo, R. C., Franchini, G., Westin, E., Ceccherini-Nelli, L., Popovic, M., and Wong-Staal, F. (1983): Abundant transcription of a cellular gene in T-cells infected with human T-cell leukemia-lymphoma virus (HTLV). *Proc. Natl. Acad. Sci. USA*, 80:11–15.
47. Manzari, V., Wong-Staal, F., Franchini, G., Colombini, S., Gelmann, E. P., Oroszlan, S., Staal, S. P., and Gallo, R. C. (1983): Human T-cell leukemia-lymphoma virus, HTLV: Molecular cloning of an integrated defective provirus and flanking cellular sequences. *Proc. Natl. Acad. Sci. USA*, 80:1574–1578.
48. Markham, P., Salahuddin, Z., Kalyanaraman, V. S., Popovic, M., Sarin, P., and Gallo, R. C. (1983): Infection and transformation of fresh human umbilical cord blood cells by multiple sources of human T-cell leukemia/lymphoma virus (HTLV). *Int. J. Cancer*, 31:413–420.
49. Masur, H., Michelis, M. A., Wormser, G. P., Lewin, S., Gold, J., Tapper, M. L., Giron, J., Letner, C. W., Armstrong, D., Sefia, U., Sender, J. A., Siebken, R. S., Nicholas, P., Arlen, Z., Maayan, S., Ernst, J. A., Siegal, F. P., and Cunningham-Rundles, S. (1982): Opportunistic infection in previously healthy women. *Ann. Intern. Med.*, 97:533–539.
50. Metzgar, R. S., Bertoglio, I., Anderson, J. K., Bonnard, G. B., and Ruscetti, F. W. (1979): Detection of HLA-DRw (Ia-like) antigens on human T lymphocytes grown in tissue culture. *J. Immunol.*, 122:949–953.
51. Miyoshi, I., Kubonishi, I., Yoshimot, S., Akagi, T., Ohtsuki, Y., Shiraishi, Y., Nagato, K., and Hinuma, Y. (1982): Type C virus particles in a cord T-cell line derived by co-cultivating normal human cord leukocytes and human leukemic T-cells. *Nature*, 294:770–771.
52. Miyoshi, I., Taguchi, H., Fujishita, M., Yoshimoto, S., Kuboishi, I., Ohtsuki, Y., Shiraishi, Y., and Akagi, T. (1982): Transformation of monkey lymphocytes with adult T-cell leukemia virus. *Lancet*, i:1016.
53. Miyoshi, I., Yoshimoto, S., Taguchi, H., Kubonishi, I., Fujishita, M., Ohtsuki, Y., and Shiraishi, Y. (1983): Transformation of rabbit lymphocytes with T-cell leukemia virus. *Gann*, 74:1–4.
54. Morgan, D. A., Ruscetti, F. W., and Gallo, R. C. (1976): Selective *in vitro* growth of T-lymphocytes from normal human bone marrow. *Science*, 193:1007–1008.
55. Nagy, K., Clapham, P., Cheingsong-Popov, R., and Weiss, R. A. (1983): Human T-cell leukemia virus type I: Induction of syncytia and inhibition by patients' sera. *Int. J. Cancer*, 32:321–328.

56. Oroszlan, S., Sarngadharan, M. G., Copeland, T. D., Kalyanaraman, V. S., Gilden, R. V., and Gallo, R. C. (1982): Primary structure analysis of the major internal protein p24 of human type-C T cell leukemia virus. *Proc. Natl. Acad. Sci. USA*, 79:1291–1294.
57. Paul, P. A., Pomeroy, K. A., Johnson, D. W., Muscoplat, C. C., Handwerger, B. S., Soper, F. S., and Sorensen, D. K. (1977): Evidence for the replication of bovine leukemia virus in the B-lymphocytes. *Am. J. Vet. Res.*, 38:873–876.
58. Payne, G., Bishop, M., and Varmus, H. (1982): Multiple arrangements of viral DNA and an activated host oncogene in bursal lymphomas. *Nature*, 295:209–214.
59. Perryman, L. E., Hoover, D. S., and Yohn, J. (1972): Immunologic reactivity of the cat: Immunosuppression in experimental feline leukemia. *J. Natl. Cancer Inst.*, 49:1357–1365.
60. Pitchenik, A., Fischl, M. A., Dickinson, G. M., Becker, P. M., Fournier, A. M., O'Connell, M. T., Colton, R. M., and Spird, T. J. (1983): Opportunistic infections and Kaposi's sarcoma among Haitians: Evidence of a new acquired immunodeficiency state. *Ann. Intern. Med.*, 98:277–284.
61. Poiesz, B. J., Ruscetti, F. W., Gazdar, A. F., Bunn, P. A., Minna, J. D., and Gallo, R. C. (1980): Isolation of type-C retrovirus particles from cultured and fresh lymphocytes of a patient with cutaneous T-cell lymphoma. *Proc. Natl. Acad. Sci. USA*, 77:7415–7519.
62. Poiesz, B. J., Ruscetti, F. W., Mier, J. W., Woods, A. M., and Gallo, R. C. (1980): T-cell lines established from human T-lymphocyte neoplasias by direct response to T-cell growth factor. *Proc. Natl. Acad. Sci. USA*, 77:6815–6819.
63. Poiesz, B. J., Ruscetti, F. W., Reitz, M. S., Kalyanaraman, V. S., and Gallo, R. C. (1981): Isolation of a new type C retrovirus (HTLV) in primary uncultured cells of a patient with Sezary T-cell leukaemia. *Nature*, 294:268–271.
64. Popovic, M., Lange-Wentzin, G., Sarin, P. S., Mann, D., and Gallo, R. C. (1983): Transformation of human umbilical cord blood T-cell leukemia/lymphoma virus (HTLV). *Proc. Natl. Acad. Sci. USA*, 80:5402–5406.
65. Popovic, M., Reitz, M. S., Jr., Sarngadharan, M. G., Robert-Guroff, M., Kalyanaraman, V. S., Nakao, Y., Miyoshi, I., Minowada, J., Yoshida, M., Ito, Y., and Gallo, R. C. (1982): The virus of Japanese adult T-cell leukaemia is a member of the human T-cell leukaemia virus group. *Nature*, 300:63–66.
66. Popovic, M., Sarin, P., Robert-Guroff, M., Kalyanaraman, V. S., Mann, D., Minowda, J., and Gallo, R. C. (1983): Isolation and transmission of human retrovirus (human T-cell leukemia virus). *Science*, 219:856–859.
67. Posner, L. E., Robert-Guroff, M., Kalyanaraman, V. S., Poiesz, B. J., Ruscetti, F. W., Fossieck, B., Bunn, P. A., Jr., Minna, J. D., and Gallo, R. C. (1981): Natural antibodies to the human T cell lymphoma virus in patients with cutaneous T cell lymphomas. *J. Exp. Med.*, 154:333–346.
68. Reitz, M. S., Jr., Kalyanaraman, V. S., Robert-Guroff, M., Popovic, M., Sarngadharan, M. G., Sarin, P. S., and Gallo, R. C. (1983): Human T-cell leukemia/lymphoma virus: The retrovirus of adult T-cell leukemia/lymphoma. *J. Infect. Dis.*, 147:399–405.
69. Reitz, M. S., Poiesz, B. J., Ruscetti, F. W., and Gallo, R. C. (1981): Characterization and distribution of nucleic acid sequences of a novel type C retrovirus isolated from neoplastic human T lymphocytes. *Proc. Natl. Acad. Sci. USA*, 78:1887–1891.
70. Retiz, M. S., Jr., Popovic, M., Haynes, B. F., Clark, S. C., and Gallo, R. C. (1983): Relatedness by nucleic acid hybridization of new isolates of human T-cell leukemia-lymphoma virus (HTLV) and demonstration of provirus in uncultured leukemic blood cells. *Virology*, 126:688–692.
71. Rho, H. M., Poiesz, B. J., Ruscetti, F. W., and Gallo, R. C. (1981): Characterization of the reverse transcriptase from a new retrovirus (HTLV) produced by a human cutaneous T-cell lymphoma cell line. *Virology*, 112:355–358.
72. Robert-Guroff, M., and Gallo, R. C. (1983): Establishment of an etiologic relationship between the human T-cell leukemia/lymphoma virus (HTLV) and adult T-cell leukemia. *Blut*, 46:1–12.
73. Robert-Guroff, M., Kalyanaraman, V. S., Blattner, W. A., Popovic, M., Sarngadharan, M. G., Maeda, M., Blayney, D., Catovsky, D., Bunn, P. A., Shibata, A., Nakao, Y., Ito, Y., Aoki, T., and Gallo, R. C. (1983): Evidence for human T cell lymphoma-leukemia virus infection of family members of human T cell lymphoma-leukemia virus positive T cell leukemia-lymphoma patients. *J. Exp. Med.*, 157:248–258.
74. Robert-Guroff, M., Nakao, Y., Notake, K., Ito, Y., Sliski, A., and Gallo, R. C. (1982): Natural antibodies to human retrovirus HTLV in a cluster of Japanese patients with adult T-cell leukemia. *Science*, 215:975–978.

75. Robert-Guroff, M., Ruscetti, F. W., Posner, L. E., Poiesz, B. J., and Gallo, R. C. (1981): Detection of the human T-cell lymphoma virus p19 in cells of some patients with cutaneous T-cell lymphoma and leukemia using a monoclonal antibody. *J. Exp. Med.*, 154:1957–1965.
76. Rojko, J. L., Hoover, E. A., Finn, B. L., and Olsen, G. R. (1981): Determinants of susceptibility and resistance to feline leukemia virus infection. II. Susceptibility of feline lymphocytes to productive feline leukemia. *J. Natl. Cancer Inst.*, 67:899–909.
77. Ruscetti, F. W., and Gallo, R. C. (1981): Human T-lymphocyte growth factor: The second signal in the immune response. *Blood*, 57:379–393.
78. Ruscetti, F. W., Robert-Guroff, M., Ceccherini-Nelli, L., Minowada, J., Popovic, M., and Gallo, R. C. (1983): Persistent *in vitro* infection by human T-cell leukemia-lymphoma virus (HTLV) of normal human T-lymphocytes from blood relatives of patients with HTLV-associated mature T-cell neoplasms. *Int. J. Cancer*, 31:171–183.
79. Salahuddin, S. Z., Markham, P. D., Lindner, S., Gootenberg, J., Popovic, M., Hemmi, H., Sarin, P. S., and Gallo, R. C. (1983): Constitutive lymphokine production by HTLV-positive T-cell lines *Science. (Submitted.)*
80. Salahuddin, S. Z., Markham, P. D., Wong-Staal, F., Franchini, G., Kalyanaraman, V. S., and Gallo, R. C. (1983): Restricted expression of human T-cell leukemia-lymphoma virus (HTLV) in transformed human umbilical cord blood lymphocytes. *Virology*, 129:51–64.
81. Sarin, P., Aoki, T., Shibata, A., Ohnishi, Y., Aoyagi, Y., Miyakoshi, H., Emura, I., Kalyanaraman, V. S., Robert-Guroff, M., Popovic, M., Sarngadharan, M. G., Nowell, P. C., and Gallo, R. C. (1983): High incidence of human type-C retrovirus (HTLV) in family members of an HTLV-positive Japanese T-cell leukemia patient. *Proc. Natl. Acad. Sci. USA*, 80:2370–2374.
82. Seiki, M., Hattori, S., Hirayama, Y., and Yoshida, M. (1983): Human adult T-cell leukemia virus: Complete nucleotide sequence of the provirus genome integrated in leukemia cell DNA. *Proc. Natl. Acad. Sci. USA*, 88:3618–3622.
83. Trainin, Z., Wernicke, D., Ungar-Waron, H., and Essex, M. (1983): Suppression of the humoral antibody response in natural retrovirus infections. *Science*, 220:858–859.
84. Uchiyama, T., Yodoi, J., Sagawa, K., Takatsuki, K., and Uchino, H. (1977): Adult T-cell leukemia: Clinical and hematological features of 16 cases. *Blood*, 50:481–503.
85. Vyth-Dreese, F. A., and DeVries, J. E. (1982): Human T-cell leukemia virus in lymphocytes from T-cell leukemia patient originating from Surinam. *Lancet*, i:993.
86. Waldmann, T., Broder, S., Greene, W., Sarin, P. S., Saxinger, C., Blayney, D. W., Blattner, W. A., Goldman, C., Frost, K., Sharrow, S., Depper, J., Leonard, W., Uchiyama, T., and Gallo, R. C. (1983): A functional and phenotropic comparison of human T-cell leukemia/lymphoma virus (HTLV): Positive adult T-cell leukemia with HTLV negative Sezary leukemia and their distinction using anti-TAC, a monoclonal antibody identifying the human receptor for T-cell growth factor. *Clin. Res., (in press).*
87. Wong-Staal, F., Hahn, B., Manzari, V., Colombini, S., Franchini, G., Gelmann, E. P., and Gallo, R. C. (1983): A survey of human leukemias for sequences of a human retrovirus. *Nature*, 302:626–628.
88. Yamada, Y. (1983): Phenotypic and functional analysis of leukemic cells from 16 patients with adult T-cell leukemia/lymphoma. *Blood*, 61:192–199.
89. Yamamoto, N., Matsumoto, T., Koyanagi, T., Tamaka, Y., and Hinuma, Y. (1982): Unique cell lines harboring both Epstein-Barr virus and adult T-cell leukemia virus established from leukemia patients. *Nature*, 299:367–369.
90. Yamamoto, N., Okada, M., Koyanagi, Y., Kannagi, M., and Hinuma, Y. (1982): Transformation of human leukocytes by cocultivation with an adult T-cell leukemia virus. *Science*, 217:737–739.
91. Yokoi, T., Miyawaki, T., Yachie, A., Ohzeki, S., and Taniguchi, N. (1982): Discrepancy in expression ability of TAC antigen and Ia determinants defined by monoclonal antibodies on activated or cultured cord blood T lymphocytes. *J. Immunol.*, 129:1441–1445.
92. Yoshida, M., Miyoshi, I., and Hinuma, Y. (1982): Isolation and characterization of retrovirus from cell lines of human adult T-cell leukemia and its implication in the disease. *Proc. Natl. Acad. Sci. USA*, 79:2031–2035.

Advances in Viral Oncology, Volume 4,
edited by George Klein. Raven Press,
New York © 1984.

Alterations of Mouse Mammary Tumor Virus DNA During Mammary Tumorigenesis

Thomas G. Fanning and Robert D. Cardiff

Department of Pathology, University of California, Davis, California 95616

The mouse mammary tumor virus (MuMTV) is one of the few retroviruses associated with epithelial malignancy. Therefore, an understanding of the role of MuMTV in murine mammary tumorigenesis could provide insight into other epithelial malignancies. Although MuMTV infection of susceptible mice may result in mammary cancer (8,25,103), the mechanism(s) underlying this carcinogenic effect has remained elusive.

Biologic studies indicate that productive infection with MuMTV is not an essential feature of the malignant phenotype. For example, the virion is produced in normal mammary cells but not found in many tumor cells (8,25,26). Since MuMTV-induced tumors arise only after a prolonged latent period, it is unlikely that MuMTV carries an oncogene analogous to those of the acutely transforming retroviruses (14). Instead, the data point to a more subtle type of host-virus interaction in mammary tumorigenesis.

The development of the restriction endonuclease/Southern blot technique has permitted analysis of this interaction at the molecular level. During the last several years, these studies have identified several events associated with the neoplastic state. These phenomena, namely (a) MuMTV amplification, (b) proviral DNA hypomethylation, and (c) identification of favored insertion regions, are highlighted in this brief review.

The biology of MuMTV, the role of hormones in MuMTV transcription, and the molecular biology of the MuMTV life cycle have been reviewed in detail (6, 8,25,26,89,90,103,131,132,140). In addition, a virtual encyclopedia of retrovirology has recently been compiled (136). This chapter primarily covers reports published in the last decade and correlates new data on premalignant and malignant stages of mammary neoplasia.

BACKGROUND

MuMTV is a RNA tumor virus that has been placed in the family Retraviridae as the prototype virus for the genus *Oncornavirus B.* The generic name comes from the distinctive morphologic structure of the virion, which was named the B-particle in Bernhard's (10) classification of virus-like particles associated with malignancy.

The B-particle is causatively associated with mammary tumorigenesis (6,8,25). When a low tumor incidence "virus-free" strain, such as Balb/c, is given MuMTV, tumor incidence increases (6,8,25,26). A direct relationship exists between the amount of MuMTV injected, the amount of MuMTV shed, the proportion of cells infected, and the increase in tumor incidence (25,26). Furthermore, the introduction of MuMTV into the Balb/c mouse leads to an increase in the incidence of premalignant hyperplastic alveolar nodules (HAN).

Two major stages have been identified in mouse mammary tumorigenesis. The first stage involves the transformation of normal mammary epithelial cells into nodule cells. The nodule cells emerge from the normal population as discrete HAN, a process referred to as nodulogenesis (89,90). The second stage occurs when nodule cells undergo a malignant transformation to emerge from the nodule population as a mammary adenocarcinoma. This second process is termed tumorigenesis (24,89,90).

The HAN can be identified in the mammary gland, isolated, and transplanted into gland-free mammary fat pad where it forms a hyperplastic outgrowth (HPO) (38). Transplanted HAN develop tumors at a higher frequency than transplanted normal ducts (38,90). This provides an operational definition of premalignancy as a lesion that has a high risk of developing malignant tumors (24). As a result, HAN and HPO are regarded as high risk preneoplasias or high risk premalignant neoplasias.

The fact that B-type particles can be found in abundance in "normal" mammary epithelial cells of many mouse strains represents the major difference between B- and C-type retroviruses. In low leukemia incidence strains of mice, C-type viral expression tends to occur chiefly in embryogenesis and in neoplasms (125). In the B-type system, the virus is found throughout the life of virus-infected animals in normal mammary epithelium (8,26,103). In fact, it has been suggested that lobuloalveolar (epithelium) differentiation is necessary for virus replication, since poorly differentiated tumors frequently do not produce MuMTV (85).

The original description of the "Bittner agent" (MuMTV) emphasized transmission through the milk (15). In the middle 1960s, Bentvelzen (5) demonstrated that the virus also can be transmitted genetically. Subsequent studies using nucleic acid hybridization have demonstrated MuMTV DNA sequences in all laboratory strains of mice (100), and approximately equal numbers of proviruses have been found in both high and low cancer incidence strains (100). Feral mice lacking MuMTV sequences have been described (21,33,35) and provide a convincing demonstration that MuMTV DNA is not essential to any phase of mouse development.

The MuMTV RNA genome sediments at approximately 35S, corresponding to a size of about 9 kilobases (kb) (47). Each virion contains two copies of 35S RNA, which are presumed to be identical in sequence. Once inside the cell, the RNAs are reverse transcribed by a virion-encoded enzyme giving rise to a double-stranded, unintegrated circular DNA (132). Several different unintegrated forms of proviral DNA have been found in MuMTV infected cells: 10 kb circular molecules containing two long terminal repeats (LTRs) and 9 kb circular molecules with a single

LTR (120). In addition, linear molecules have also been detected (120). At present, it is unknown which form integrates into the host genome (131), but integrated MuMTV proviruses are bracketed by LTRs (131) (Fig. 1).

MuMTV genetic information has been grouped into three domains: the *gag*, *pol*, and *env* regions (41). The *gag* proteins (p27, p21, p14, and p10) are structural proteins found in the interior of the virion. The *pol* gene encodes the MuMTV RNA-dependent reverse transcriptase (p100). Two major proteins are encoded by the *env* region (gp52 and gp36) and form the virion glycoprotein envelope. Transcription of these genes is initiated in the 5′-LTR and results in two major products: a 35S unspliced RNA encoding all the viral proteins and a 24S spliced RNA which encodes only the *env* proteins (48,61,118,119) (Fig. 1).

MuMTV expression is under hormonal control, a particularly intriguing aspect, since the mammary gland is an endocrine-responsive end organ (9). Two hormones related to mammary development, estrogen and prolactin, have also been recognized as necessary for mammary tumor development (102). These same hormones, as well as insulin and glucocorticoids, also enhance MuMTV transcription (116,117,141). Direct evidence for hormone induction of MuMTV has come primarily from experiments using glucocorticoid hormones in cell cultures (116,117). The glucocorticoid effect in culture appears to work at a transcriptional level (117). Stimulation is correlated with the presence of glucocorticoid receptors (122), and the net result is a rapid accumulation of MuMTV RNA in the cytoplasm of the host cell (117).

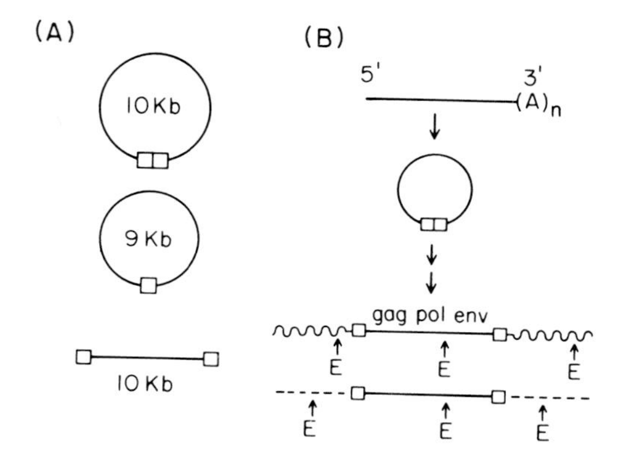

FIG. 1. Proviral forms of MuMTV. **A:** Intracellular, unintegrated MuMTV proviral DNAs: 9 and 10 kb circles and 10 kb linear forms (120). The boxes represent the proviral LTRs. **B:** Integration of MuMTV into the host genome. *Upper line*, 9 kb (35S) MuMTV RNA. Proviral DNA results from a copying process in which two RNAs play a role (131). Although the circular proviral form with two LTRs is shown integrating into the host DNA, it is presently unknown which form shown in **A** is actually inserted into the host genome. *Lower two lines*, proviruses integrated at two different locations in the host DNA (wavy and dotted lines). When cleaved with a restriction enzyme (E), four fragments are generated whose sizes depend on the distances of the enzyme sites from the proviruses. *gag*, *pol*, and *env*, MuMTV genes encoding the group-specific antigens, RNA-dependent DNA polymerase, and envelope proteins.

AMPLIFICATION

Introduction

An increased number of proviruses per cell was one of the first bona fide molecular changes observed in MuMTV-induced tumors at the DNA level. However, gene amplification is not restricted to tumor viruses (or to viruses in general), nor is it found only in malignant cells. Many DNA sequences, both large and small, functional and nonfunctional, are known to be reiterated within the eukaryotic genome (11,19). Reiteration frequencies vary from several copies to many thousands of copies, depending on the sequence (11,19). Saltatory replication (17) and unequal crossing over (123) are widely accepted mechanisms for forming tandem DNA repeats, while transposition (59) and integration of "processed" genes (75,121) are mechanisms that may explain dispersed repetitive DNAs. In addition, amplification events involving large amounts of DNA (e.g., partial or complete chromosomal trisomy) are found in many malignancies and have been reviewed in detail elsewhere (77).

Two mechanisms of gene amplification, transposition and integration of processed genes, are especially cogent in the present context. Transposition of DNA sequences was originally observed in microorganisms (22), although the process is also known to occur in eukaryotes (59). Many eukaryotic organisms are thought to harbor mobile genetic elements (transposable elements) in their genomes. In *Drosophila*, for example, 15% of the genome may be composed of families of mobile elements (139). The similarities between proviruses and mobile elements have been discussed in detail (124); we return to the topic of transposition later in this chapter.

Processed genes are those in which mRNAs are reverse transcribed into double-stranded DNA molecules which integrate into the host genome. These genes normally lack introns, have an AT-rich region at one boundary, and are flanked by short, direct repeats (75,121). Processed genes have been described in mice, rats, and humans and share several features in common with proviruses, in particular the reverse transcription step and the direct repeats flanking the integrated DNA (75,121).

The exact mechanism giving rise to processed genes is unknown at present but may be simpler than that for retroviruses, since no duplications of the mRNA (i.e., LTRs) are found flanking these genes (75,121). Nevertheless, the striking similarities between proviruses and processed genes suggest an evolutionary relationship (75,121). Reverse transcription and integration of cellular mRNAs appear to be capricious events of low probability. However, a sufficiently strong selection pressure, acting on mRNA encoding a RNA-dependent DNA polymerase, could have led to the evolution of retroviruses (124) which maintained the RNA–DNA–RNA procession as part of their life cycle.

MuMTV Amplification in Mammary Tumors

Every inbred mouse strain contains an endogenous, genetically transmitted complement of MuMTV proviral DNA sequences. MuMTV-induced mammary tumors are characterized by an increased number of MuMTV proviral sequences relative to the endogenous number. This phenomenon has been referred to as MuMTV amplification.

In this chapter, we use the term amplification to mean any increase in proviral DNA copy number detected by liquid hybridization of Southern blot analysis. As defined here, amplification does not distinguish between integrated and nonintegrated proviruses, nor does it distinguish between a homogeneous or heterogeneous cell mass (two terms which we define below).

Amplification in virus-induced tumors is generally considered the result of infection of the cells by virions originating outside the animal. This is certainly true for the first exogenous MuMTV provirus inserted into the cell of a low tumor incidence mouse strain, such as Balb/c. However, several other mechanisms of amplification are theoretically possible. Infection of the tissue could result from the spread of MuMTV virions released inside the animal. This situation might be particularly important in strains that carry an oncogenic endogenous virus, e.g., GR or C_3Hf.

In addition to viral infection, several possibilities exist for proviral amplification within the infected cell: reverse transcription of a newly synthesized MuMTV mRNA followed by reintegration of the DNA copy, and direct transposition of MuMTV DNA via mobile elements. Although no compelling evidence for these two processes has been presented, the presence of subgenomic proviral fragments in the mouse genome (32) suggests that a processed gene mode of amplification occasionally may occur.

Early results using liquid hybridization convincingly demonstrated MuMTV proviral amplification in the mammary tumors of most inbred strains of mice. Mammary tumors from Balb/cfC_3H (87,100), C_3H (62,92,94), and RIII (92,100) inbred mouse strains all exhibited significant increases in MuMTV proviral DNA copy number. Tumors from GR mice appeared to be an exception, since no proviral amplification was detected (92,100).

The Southern blotting technique generally verified the results obtained by liquid hybridization and at the same time allowed a more detailed examination of host-proviral interactions. Restriction enzyme digests of DNA extracted from the tissues of Balb/cfC_3H mice, when Southern blotted and probed with labeled MuMTV cDNA, yielded a wealth of information (31). Cohen et al. (31) were able to distinguish the acquired MuMTV(C_3H) from the endogenous Balb/c proviruses with the enzyme Pst I. This permitted the detection of acquired MuMTV(C_3H) proviruses in Balb/cfC_3H DNA from lactating mammary gland and mammary tumors (31). Using a restriction endonuclease (Eco RI), which cuts most proviruses only once, multiple additional MuMTV restriction fragments were detected in tumors but not in lactating mammary glands. None of the tumors used in this study had the same size acquired MuMTV restriction fragments (31). These results established two widely held tenets of MuMTV-induced carcinogenesis: (a) proviruses integrate at many sites in the mouse genome; and (b) tumors are more homogeneous cell populations than normal, virus-infected mammary tissue.

Restriction enzyme analyses of MuMTV-induced mouse mammary tumor DNA usually give rise to patterns in which numerous MuMTV proviral DNA fragments are found in addition to those produced by the endogenous proviruses (Fig. 2). The enzymes routinely employed (e.g., Eco RI, Hind III) cleave most MuMTV proviral DNA once near the middle of the genome and other cleavages occur in

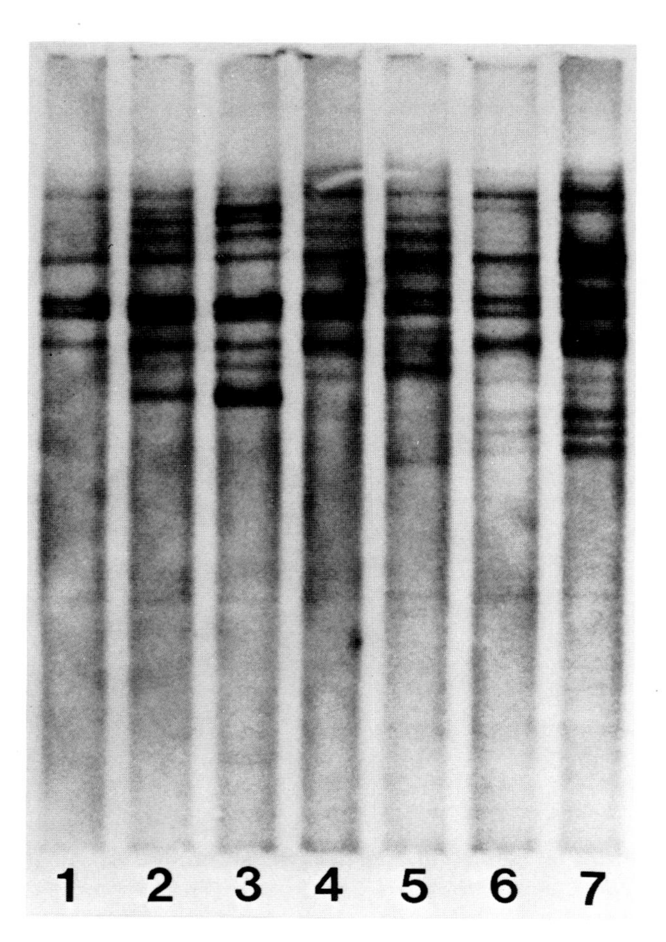

FIG. 2. Restriction enzyme/Southern blot analysis of MuMTV proviruses in Balb/cfC_3H normal, premalignant, and malignant tissue DNAs. The DNAs were obtained from Z-series animals (*4*), digested with Eco RI, Southern blotted, and probed with labeled MuMTV-specific cDNA. The lanes contain DNA from *(1)* spleen, *(2)* Z 4 outgrowth, *(3)* Z 4 tumor, *(4)* Z 5c outgrowth, *(5)* Z 5c tumor, *(6)* Z 5c_1 outgrowth, and *(7)* Z 5c_1 tumor.

mouse DNA flank both sides of the provirus. Since many additional bands are produced by each tumor sample, and most tumors produce few bands in common (Fig. 2), the Southern blots suggest that MuMTV proviruses are flanked by a wide variety of mouse DNAs; in other words, MuMTV can integrate into numerous sites in the mouse genome (31).

Both tumors and virus-infected, lactating mammary gland contain acquired MuMTV proviruses, yet restriction enzymes that readily generate additional proviral fragments with tumor DNA fail to do so with mammary gland DNA (31). Since MuMTV proviruses appear to integrate at many sites, the absence of acquired proviral fragments is easily explained by assuming the virus-infected gland consists of a large collection of cells, each of which contains acquired proviral DNA at different sites. If the DNA from a single virus-infected mammary cell could be examined by restriction analysis, the acquired proviruses would be readily apparent. This is exactly what one finds when tumor DNAs are examined (Fig. 2), suggesting that tumors represent the proliferation of a select number of infected cells (31).

Since the restriction enzyme/Southern blotting definition of tumor homogeneity is strictly operational (31), it is not surprising to find that the concept of homogeneity varies from one laboratory to another. In this chapter, we designate a tissue as homogeneous if acquired proviral-mouse flanking fragments can be detected after restriction enzyme/Southern blotting analysis (Fig. 2). This definition is a phenomenologic one and does not imply a mechanism for the origin or evolution of tumors (e.g., monoclonality with divergent evolution versus polyclonality). These topics are reviewed in the general context of neoplasia (57,105) and, specifically, for MuMTV-induced tumorigenesis (23a).

The initial observations with Balb/cfC$_3$H mice (31) were followed by a host of reports demonstrating MuMTV amplification and tumor homogeneity in viral-induced mammary tumors. Balb/cfC$_3$H (1,101), C$_3$H (46,62,71,106), C$_3$Hf (34,51,113), Balb/cNIV (113), and GR (34,52,71,81,98) mammary tumors have all been demonstrated to contain acquired MuMTV proviruses by the Southern blotting technique. The GR strain is interesting in this respect, since liquid hybridization had failed to detect proviral amplification in GR mammary tumors (92,100). The Southern blotting data revealed the reason for this: the degree of MuMTV amplification in GR tumors is small, amounting to only several proviral copies on average per cell.

In addition to the small number of acquired proviruses in GR mammary tumors, these tumors are unusual in another respect: They appear to be much less homogeneous than tumors arising in other inbred strains of mice. Early reports suggested that acquired proviral DNA fragments were either submolar or completely lacking in GR mammary tumors (71). Recently, studies designed to follow the fate of MuMTV proviruses during the progression of GR mammary tumors from hormone dependence to hormone independence have confirmed that GR mammary tumors consist of numerous cell populations (81,98). Results demonstrating that most Balb/cfC$_3$H tumors contain different subpopulations of virus-infected cells have also been presented (101; R. D. Cardiff, D. W. Morris, and L. J. Young, *unpublished observations*).

No apparent correlations exist between viral oncogenicity and proviral amplification or between mouse susceptibility and proviral amplification. Tumors induced by the low oncogenic MuMTV(C_3Hf) and MuMTV(NIV) appear to contain as many acquired proviruses as tumors induced by the highly oncogenic MuMTV(C_3H). The number of acquired MuMTV proviruses in tumors induced by these low and high oncogenic MuMTVs normally exceeds the number found in tumors induced by the high oncogenic MuMTV(GR). In addition, tumors from an early tumor incidence Balb/cfC_3H line and a late tumor incidence Balb/cfC_3H line were indistinguishable in terms of numbers of acquired proviruses (1).

MuMTV(GR) is both genetically inherited and horizontally transmitted in GR mice (5,6). Genetic experiments designed to breed out the high tumor incidence "gene" in GR mice (127) resulted in a strain termed GR-*Mtv*-2^-, which contained fewer MuMTV proviral DNA sequences than the parental GR strain (93). Subsequent experiments demonstrated that only a single MuMTV provirus (GR-MTV-2) was missing in the GR-*Mtv*-2^- strain (52,71,97). Restriction enzyme digests of GR mammary tumors, as well as preneoplastic tissues (54), showed that GR-MTV-2 was the provirus amplified during tumorigenesis. In addition, the RNA genomes of MuMTV(GR) virions have been shown to be transcribed from the GR-MTV-2 provirus (53,68). Thus mammary tumorigenesis in GR mice hinges on the presence or absence of a particular endogenous provirus, and GR-*Mtv*-2^- mice represents an example in which cancer has been "cured" by classic genetic manipulation (127).

The long-standing dogma that MuMTV is solely involved in mammary tumorigenesis may be in need of revision. The recent discovery of MuMTV proviral amplification in T-cell lymphomas of male GR mice suggests that the oncogenic spectrum of MuMTV may be broader than supposed (69,99). However, as Michalides et al. (99) conclude, until more information is accumulated, a cause and effect relationship between MuMTV proviral amplification and leukemogenesis is only suggested by the result and has not been definitely established.

Restriction mapping suggests that not all mouse mammary tumors are associated with MuMTV proviral amplification. Hormonally induced (86,87,96), spontaneous (46,94), and chemically induced (46,95) mammary tumors appear to contain only the endogenous complement of MuMTV proviruses. However, tacit assumptions, which may not be correct, are present in many of these reports. These include the following: (a) the tumors are homogeneous; and (b) additional proviruses have restriction maps which differ from those of the endogenous proviruses. Let us assume, for example, that an endogenous provirus amplified to a slight degree (e.g., one extra copy per cell), and these new copies integrated into many different locations in the mouse genome. If the tumor mass proliferated heterogeneously, it would be difficult to detect these additional copies by either liquid hybridization or Southern blotting. For this reason, we consider the question of MuMTV amplification in nonvirally induced tumors to be open; indeed, the presence of additional copies of MuMTV DNA in some hormonally induced C57BL and C_3Hf mammary tumors has been reported (96).

MuMTV Amplification in Premalignant Mammary Tissue

Mouse mammary tumorigenesis is a multistep process involving the progression of normal, virus-infected tissue to premalignant tissue and finally to malignant tumors. As mentioned earlier, premalignant tissues are at higher risk of developing into tumors than normal gland, but they are not invasive and do not metastasize (89,90). Experiments designed to investigate the state of MuMTV proviral DNA in premalignant tissues have demonstrated the same phenomena as found in tumors. In fact, premalignant and tumor tissues from Balb/cfC_3H mice are indistinguishable in terms of tissue homogeneity and proviral amplification. Two types of benign premalignancies have been studied: hormone-dependent "tumors" of GR mice, and hyperplastic outgrowths (HOG) in GR and Balb/cfC_3H mice.

Hormone-dependent tumors (also known as pregnancy-dependent tumors, early mammary tumors, or plaques) are found in GR, DD, RIII, and CDf mice. These tumors arise during pregnancy or after stimulation with estrogen or synthetic androgens and regress upon parturition or removal of the hormones (128). Like hormone-independent tumors, these premalignant tumors in GR mice contain extra copies of the endogenous, highly oncogenic GR-MTV-2 provirus and appear to be more homogeneous than virus-infected gland tissue (54,98). Some hormone-dependent tumors progress to malignant hormone-independent tumors (98). The hormone-independent tumors contain the same MuMTV proviral restriction fragments as the progenitor hormone-dependent tumor and in many cases have additional restriction fragments as well (98).

Premalignant outgrowths derived by transplantation of HAN into gland-cleared fat pads have been studied in GR, Balb/c, and Balb/cfC_3H mice. Initial experiments involved the transplantation of Balb/c Dl premalignant outgrowth (89,90) into uninfected Balb/c and virus-infected Balb/cfC_3H mice (3). The Dl outgrowth, as well as tumors arising from the outgrowth, were then examined for acquired copies of MuMTV. In most cases, no amplification of MuMTV sequences could be detected (3), although scattered MuMTV-infected cells were detected by immunoperoxidase. Southern blot analysis of DNA from strong immunoperoxidase-positive tumors did show that acquired MuMTV(C_3H) proviruses were present, albeit at low copy number (3).

To examine the role of MuMTV in premalignancy in more detail, premalignant tissue lines were constructed by transplanting HAN from Balb/cfC_3H animals into virus-free Balb/c mice (4). Restriction mapping of the MuMTV proviruses in these lines demonstrated the presence of acquired MuMTV(C_3H) proviral DNA, that these proviruses had integrated into many different locations in the mouse genome, and that the HPOs were quite homogeneous (27). These results suggested that premalignant tissues were similar to tumors in terms of MuMTV amplification and in the proliferation of only a subset of viral-infected cells (Fig. 2).

Subsequent studies have directly compared the MuMTV proviral restriction patterns in Balb/cfC_3H premalignant tissues and tumors arising from them. In many cases, only minor changes in MuMTV proviral DNA occurred during the transformation of premalignant to malignant tissue. This finding suggested that the major role of MuMTV in mouse mammary tumorigenesis occurred during the progression of normal infected tissue to premalignant tissue (R. D. Cardiff, D. W. Morris, and L. J. Young, *unpublished observations*).

MuMTV proviral DNA amplification has also been investigated in premalignant tissues (HPO) of the GR mouse. Additional copies of the highly oncogenic GR-MTV-2 provirus were detected in these tissues, and the number of extra copies (two to three per cell) was identical to that found in GR mammary tumors (54). Restriction enzyme/Southern blot analyses indicated a large degree of heterogeneity in the cell population (54), suggesting that both premalignant tissues and hormone-dependent tumors of GR mice consist of many different cell populations, while the hormone-independent tumors are more homogeneous (54). This is in contrast to Balb/cfC_3H premalignant tissue and tumors arising from them; both tissues appear to be homogeneous by restriction enzyme/Southern blot criteria (27).

MuMTV Amplification in Normal Tissue

To our knowledge, the only normal tissue in which extra copies of MuMTV proviral DNA have been convincingly demonstrated is mammary gland tissue of virion-producing female mice (30,31,54).

METHYLATION

Introduction

The correlation between gene expression and DNA hypomethylation has been considered in detail by others (18,49,114,115) and can be summarized as follows: the rare base 5-methylcytosine is found predominantly in CG dinucleotides (60). Certain restriction enzymes, which have CG as part of their recognition sequence, are sensitive to the presence of 5-methylcytosine: when 5-methylcytosine is present in the recognition sequence, they will not cleave the DNA (12). These enzymes cleave eukaryotic genes poorly when DNA is extracted from cells in which the gene is not being expressed. Conversely, when DNA is extracted from cells in which the gene is actively transcribed, the enzymes cleave quite well.

The obvious conclusion from these observations is that methylated genes are inactive, while hypomethylated genes are transcribed. In several instances, this conclusion has been strengthened by the finding that hypomethylated genes are also DNAse I sensitive, DNAse I sensitivity being another indicator of gene activity (58,135). As convincing as this picture appears, it has not been observed in all cases (56,88,107,138). The following two examples are typical of a number of studies designed to test the "DNA hypomethylation-active gene" hypothesis.

DNA extracted from mouse tissues was cleaved with the methylation sensitive enzymes Hpa II (recognition sequence CCGG) and Hha I (GCGC), and the methylation patterns of the ribosomal DNA (rDNA) were examined after Southern blotting and hybridization with a labeled rDNA probe. In all tissues examined, the rDNA could be partitioned into a highly methylated fraction and a hypomethylated fraction. The hypomethylated fraction was also DNAse I sensitive, suggesting that this fraction contained the actively transcribed rDNA genes (13).

A more direct verification of the role of DNA methylation in gene activity was provided when cloned copies of a viral thymidine kinase (TK) gene were introduced into TK-deficient mouse cells. The active TK genes in these cells were hypomethylated. A derivative of this cell line, which did not express the TK genes, was shown to carry the genes in a highly methylated state. Revertants of this cell line, which again expressed TK, had hypomethylated copies of the gene. Finally, treatment of the cell line containing the methylated TK genes with 5-azacytidine, a potent methylation inhibitor, resulted in hypomethylation and expression of the TK gene (28).

MuMTV Hypomethylation in Mammary Tumors

The initial observation that MuMTV proviruses in mammary tumors were hypomethylated was made by Cohen (30). Mammary tumor DNA from Balb/cfC_3H, GR, and MuMTV(C_3Hf)-infected Balb/c mice were cleaved with the isoschizomer pair Msp I and Hpa II (both cleave at CCGG; Msp I is methylation insensitive, while Hpa II is sensitive to the presence of 5-methylcytosine) and the MuMTV restriction patterns compared with those of mouse liver DNA cleaved with the same enzymes. All tumors showed hypomethylation of MuMTV proviral DNA when compared to the highly methylated proviruses found in liver DNA. In several cases, the Msp I patterns of the acquired proviruses were found to be sufficiently different from those of the endogenous proviruses so that the acquired and endogenous DNAs could be distinguished. Since Hpa II fragments characteristic of the acquired provirus were detected in tumor DNA, it was concluded that many of the hypomethylated DNA fragments were derived from acquired proviruses (30). These initial observations of hypomethylated MuMTV proviral DNA in viral-induced mammary tumors have since been verified for both C_3Hf (51) and GR (54) tumors (Fig. 3).

In a subsequent study (16), hypomethylated MuMTV DNA was found in tumors arising from D1 premalignant tissue. Since the D1 tissue was derived from an uninfected Balb/c animal, it did not contain acquired proviruses. This finding suggested that spontaneous tumorigenesis (i.e., not dependent on infection by MuMTV) was correlated with the hypomethylation of endogenous MuMTV proviral DNA sequences (16). Interestingly, the hypomethylated MuMTV sequences in these D1 tumors were either contained within a subgenomic fragment or were part of a proviral 3′-LTR (16).

C_3H/Sm mice have lost the high oncogenic, milk-borne MuMTV(C_3H) virus normally found in the C_3H line and have a low (1 to 2%) mammary tumor incidence.

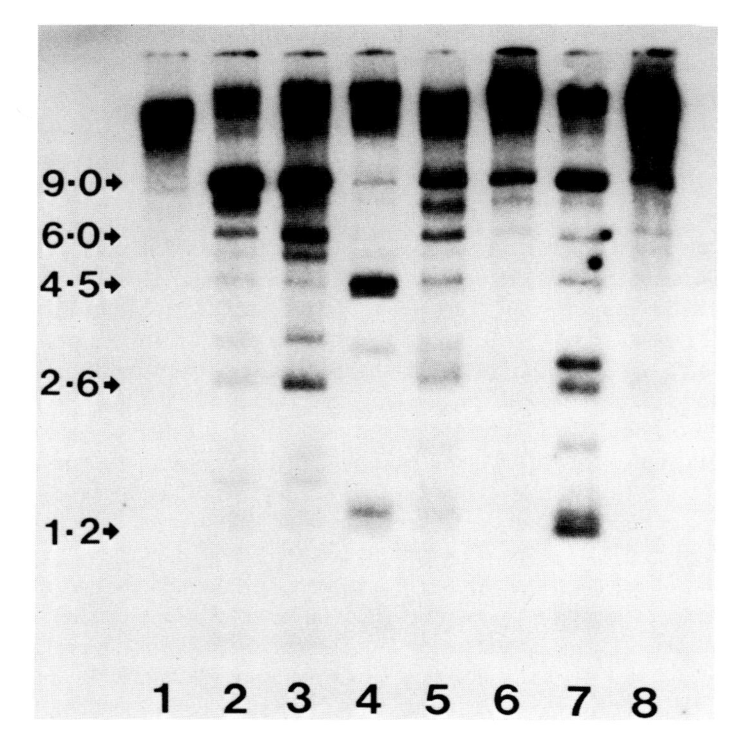

FIG. 3. Hypomethylated MuMTV proviral DNA in mouse tissues. GR mouse tissue DNAs were digested with the 5-methylcytosine-sensitive restriction enzyme Hha I (recognition sequence GCGC), Southern blotted, and probed with labeled MuMTV-specific DNA. The lanes contain DNA from *(1)* liver, *(2)* bone marrow, *(3)* spleen, *(4)* testes, *(5)* placenta, *(6)* fetuses, *(7)* mammary tumor, and *(8)* kidney. *Arrows*, fragment size in kb.

Viral-induced, spontaneous, and chemically induced mammary tumors in C_3H/Sm mice contained hypomethylated MuMTV proviral DNA (46). The hypomethylation of endogenous MuMTV DNA in spontaneous tumors of C_3H/Sm mice is analogous to that of the D1-derived tumors of Balb/c mice mentioned previously; as in D1 tumors, 3′-sequences of one or more endogenous proviruses appeared to be hypomethylated (46).

MuMTV Hypomethylation in Premalignant Mammary Tissue

Hypomethylation of MuMTV proviruses appears to be a constant feature of mouse mammary neoplasia, regardless of whether the tumors arose spontaneously (46) or were chemically (46) or virally (30,46,51,54) induced. However, is MuMTV proviral hypomethylation to be equated with malignant (e.g., invasive, metastatic) cells only?

As noted previously, hormone-dependent tumors arise during pregnancy in GR mice and regress upon parturition (128); this appearance/regression phenomenon is under hormonal control (128). Extensive hypomethylation of MuMTV proviral DNA has been detected in these tumors (54), demonstrating that even premalignant cells (noninvasive, nonmetastatic) are similar to tumor cells in terms of MuMTV hypomethylation.

Premalignant HPO, like hormone-dependent tumors, are noninvasive and nonmetastatic but have a higher risk of becoming malignant than normal mammary epithelium (89,90). HPO lines from GR mice contained hypomethylated MuMTV proviral DNA (54). In addition, the Z-series premalignant tissues (4) from Balb/cfC$_3$H mice also exhibited significant MuMTV hypomethylation (T. G. Fanning, *unpublished observations*). In fact, in both GR and Balb/cfC$_3$H HPO, the degree of MuMTV DNA hypomethylation was indistinguishable from that exhibited by tumors in the same mouse strains.

As we have seen previously, amplification of MuMTV proviral DNA is a characteristic of all MuMTV-induced mammary tumors; yet other mammary tumors (e.g., spontaneous, chemically induced) apparently exhibit no detectable increases in the amount of MuMTV DNA. If these results prove valid, they suggest alternative pathways of mammary tumorigenesis in mice (2). Similarly, the strong correlation seen between MuMTV DNA hypomethylation and malignancy suggested a cause and effect relationship (either tumorigenesis followed by hypomethylation or hypomethylation followed by tumorigenesis). However, the fact that premalignant tissues contain hypomethylated MuMTV DNA demonstrates that proviral hypomethylation per se is not unique to the malignant state. The finding of MuMTV hypomethylation in normal mouse tissues reinforces this viewpoint.

MuMTV Hypomethylation in Normal Tissue

MuMTV proviral methylation patterns in mammary tissue have routinely been compared with those in normal mouse organs. In most cases, the normal DNA has come from liver (30,46,51,54), although other organs (e.g., brain, kidney) have also been used (51,54). MuMTV proviruses in these normal, nonmammary tissues are highly methylated. On the other hand, normal mammary tissues of Balb/cfC$_3$H, C$_3$H/Sm, and GR mice display hypomethylated MuMTV proviral DNA (30,46,54). When compared to mammary tumors, however, the degree of hypomethylation is small. The slight degree of proviral hypomethylation in normal mammary tissue may be a consequence of either the heterogeneous nature of the tissue (only a few cells contain hypomethylated MuMTV DNA) or the fact that there are fewer MuMTV proviruses per cell than in tumors, or both (30,46,54).

The discovery that some normal, nonmammary tissues exhibited considerable hypomethylation of endogenous MuMTV proviral DNA was unexpected. Splenic DNAs from C$_3$Hf mice were digested with Msp I and Hpa II and shown to contain hypomethylated MuMTV proviruses; the degree of hypomethylation appeared to progress with the age of the animal (51). A comparison of the restriction profiles led to the conclusion that the hypomethylated provirus found in C$_3$Hf spleen DNA was the same provirus that was hypomethylated in C$_3$Hf tumors, i.e., unit V, the endogenous low oncogenic MuMTV of C_3Hf mice (51). Mammary gland from C_3Hf mice showed little or no hypomethylation of MuMTV (51), in contrast to the results obtained with Balb/cfC_3H, C_3H/Sm, and GR mice (30,46,54).

An evaluation of MuMTV methylation in normal mouse tissues detected extensive hypomethylation of endogenous proviral DNA in spleens and testes of GR, C_3H, Balb/c, and C57BL mice (Fig. 3) (W-S. Hu, T. G. Fanning, and R. D. Cardiff, *unpublished observations*). In addition, placental tissue and fetuses of GR mice also contained hypomethylated MuMTV DNA (Fig. 3). An analysis of RNA extracted from the nonmammary GR mouse organs that contained hypomethylated MuMTV DNA demonstrated that little if any MuMTV DNA was transcriptionally active (W-S. Hu, T. G. Fanning, and R. D. Cardiff, *unpublished observations*).

These results demonstrate that hypomethylated MuMTV proviral DNA exists in some normal nonmammary tissues, and that these hypomethylated proviruses are at best only sparingly transcribed. Thus MuMTV hypomethylation is not necessarily equatable with MuMTV transcription. The possibility remains, of course, that "critical" regions (129) of the provirus remain methylated in these cells, whereas the same regions are unmethylated in normal, premalignant, and malignant mammary cells.

MuMTV Hypomethylation and Gene Expression

MuMTV-induced mammary tumors contain hypomethylated acquired proviruses, which are actively transcribed (30,46,51,54). As we have seen in the preceding section, however, hypomethylation of MuMTV DNA is not sufficient per se to allow transcription. This leads to the proposal that the absence of 5-methylcytosine at specific sites on the provirus is important. Several additional lines of evidence support a correlation between MuMTV hypomethylation and transcription.

Cloned MuMTV proviruses are devoid of 5-methylcytosine by virture of being replicated in bacterial cells during the cloning procedure. These cloned MuMTV proviruses have been reintroduced into eukaryotic cells by the DNA-mediated transfection technique. Once inside the eukaryotic cell, the cloned proviruses usually remain unmethylated and are efficiently transcribed (20,42,72–74,108).

Hypomethylation of the 3′-LTR of one or more endogenous Balb/c MuMTV proviruses may be involved in the progression of D1 premalignant tissue to malignancy (16). A critical role for LTR hypomethylation in proviral transcription has recently been demonstrated (91). A cell line derived from a C57BL mouse thymoma was found to produce high levels of MuMTV RNA in the absence of glucocorticoids. Southern blot analysis of MuMTV DNA present in these cells demonstrated that the 5′-LTR region of at least one MuMTV provirus was hypomethylated, and that other proviral DNA sequences were highly methylated (91). Thus it is reasonable to assume that the lack of methyl groups at specific sites in the LTR may be involved in the ability of regulatory proteins to interact with this region of the provirus and initiate transcription (64,66,70,79,109,112,126). These putative critical sites may remain methylated in those normal organs containing hypomethylated MuMTV DNA and, if so, would account for the failure to find a correlation between hypomethylation and transcription of MuMTV in these tissues (W-S. Hu, T. G. Fanning, and R. D. Cardiff, *unpublished observations*).

DNA REARRANGEMENTS

The movement of mobile DNA sequences within the genome occasionally results in rearrangements of the mobile DNA and/or genomic DNA. The most thoroughly described examples come from the integration and excision of insertion elements (IS) in microorganisms (22). Inactivation of genes due to deletions occurring as a consequence of integration or excision of IS elements has been observed numerous times (22). In many instances, a portion of the element remains in the chromosome near the deleted segment. This phenomenon is not restricted to IS elements in microorganisms but also occurs during integration/excision of bacteriophages, e.g., phage lambda (23).

Mobile genetic elements in eukaryotes appear to cause rearrangements similar to those induced by IS elements in microorganisms. A number of unstable genes and mutations in *Drosophila* have been attributed to the action of transposable elements (59). The structural similarities between transposable elements and proviruses have been pointed out (83,124,131), and proviruses have been detected near large deletions in the host genome (82), although it is not known what role if any the provirus played in generating these deletions.

The limited data on hand indicate that most MuMTV proviruses integrate without rearrangements in proviral DNA or in host DNA; and once integrated into the genome, the proviruses remain stable. Most endogenous and acquired proviruses consist of approximately 7.5 kb of unique sequence DNA bracketed by two long terminal repeats (32,33,50,63). In the most extensively studied system, mouse DNAs flanking acquired MuMTV proviruses in 18 C_3H mouse mammary tumors were shown to have no rearrangements of host DNA (106), suggesting that in most cases, MuMTV proviral integration occurs by the simple insertion of the proviral DNA into the mouse genome.

Deletion/substitution of MuMTV DNA has been observed in at least one example. Keeping in mind the fact that cloned MuMTV proviruses are often deleted and rearranged due to the presence of a "poison" sequence near the 5′-end of the provirus (43,83), the results of Buetti and Diggelmann (20) are nevertheless intriguing. The authors cloned circular, unintegrated MuMTV(GR) proviruses from infected rat cells. Five of seven clones represented a 9 kb proviral form that had been described previously (20). The remaining two clones, however, were aberrant: one appeared to contain a duplication of proviral DNA (which could represent a cloning artifact), while the other contained a deletion/substitution in which a portion of proviral DNA had been replaced by rat sequences (20). This result could be due to a cloning artifact; however, the parallel between this case and that of transforming retroviruses which acquire host oncogenes (136) is noteworthy.

A truncated endogenous MuMTV provirus (unit I) has been described in Balb/c mice (32). Whether this partial MuMTV integrated as such (e.g., as a processed gene) or whether the complete proviral DNA suffered a deletion upon or subsequent to integration remains unknown. Nevertheless, this example demonstrates that aberrant MuMTV proviruses are present in mice.

Finally, intracellular 35S MuMTV RNA is the form that is normally packaged into virions (29). In a study involving reverse transcription of viral RNAs, Puma et al. (113) described several aberrant RNA species present in MuMTV(NIV) virions. Although the mechanism by which these altered RNAs were generated remains uncertain (some contained point mutations while others had deletions), infection of cells by virions containing the mutated RNAs could result in the integration of aberrant proviruses into the mouse genome.

MuMTV TRANSCRIPTION AND ONCOGENICITY

Two major changes occur in MuMTV proviral DNA during tumorigenesis: proviral DNA amplifies and concomitantly becomes hypomethylated. Amplification and enhanced expression of oncogenes have been described in other systems (36,37,84); it is conceivable that MuMTV amplification/hypomethylation is in some way intertwined with the expression of an oncogene. The putative oncogene may consist of viral or cellular sequences. MuMTVs do not contain sequences that are homologous to mouse sequences (21,33,35). Therefore, if MuMTV harbored an oncogene with no cellular counterpart, it would be similar to the spleen focus-forming virus in mice, whose envelope gene functions as an oncogene (80). On the other hand, integration of acquired MuMTVs might result in the expression of oncogenes near the site of integration, similar to the avian leukosis virus-induced expression c-*myc* in chickens (67,104,110).

The case for a MuMTV encoded oncogene has been described in detail by Bentvelzen (7). A brief summary of the argument is as follows: MuMTV proviruses contain exceptionally large LTRs (131). Sequencing studies have demonstrated that these LTRs contain long open reading frames, sufficient to code for a 35 to 36 kd protein (39,44,45,55,65,76,111). RNA transcribed from a cloned LTR, when translated *in vitro*, gave rise to several protein products, one of which was 35 to 36 kd in size (40). Antisera raised in mice against mammary tumor cells reacted with all MuMTV structural proteins and, in addition, reacted with a protein in tumor cells whose size was approximately 36 kd (7).

In more recent experiments, normal mammary gland and hormonally and chemically induced premalignant and malignant mammary tissues were shown to contain polyadenylated, spliced RNAs, which hybridized specifically with the U3 region (the open reading frame) of the MuMTV LTR (137). By inference, this RNA species is a mRNA encoding the protein specified by the MuMTV LTR. Surprisingly, the RNA was not found in MuMTV-induced tumors (137).

Evidence for the second model of MuMTV oncogenicity, transcriptional activation of a cellular oncogene due to MuMTV integration nearby, has also been presented recently (106). At least one newly integrated MuMTV(C_3H) provirus was found in a limited region of the mouse genome (int-1 locus) in about 70% of C_3H mammary tumors. This finding suggested that even though MuMTV proviruses integrate at many sites in the mouse genome, integration near the int-1 locus was especially tumorigenic (106). The idea of transcriptional activation was reinforced by the finding that in several tumors, but not in virus-infected mammary gland, RNA transcripts were detected which hybridized to a specific region of the int-1 locus (106). Whether these transcripts encode an oncogene is unknown. However, the possibility is certainly compelling, especially in view of the analogous ALV/c-*myc* case (67,104,110).

MuMTV proviral integration may result in the expression of different cellular oncogenes in different mouse strains or in different tissues of a particular mouse. In addition to the int-1 locus mentioned above, three other groups have isolated similar sequences: "int-1-like" sequences have been isolated from GR mammary tumor DNA (D. Morris, H. Bradshaw, and R. D. Cardiff, *unpublished*), GR T-cell lymphoma DNA (R. Michalides and J. Hilgers, *personal communication*) and Balb/cfC_3H mammary tumor DNA (G. Peters and C. Dickson, *personal communication*). In GR and Balb/cfC_3H mammary tumors, the integration sequences do not appear to be within the same area of the mouse genome that contains the int-1 locus. In fact, one region (int-2) appears to be located on a different chromosome than int-1 (G. Peters and C. Dickson, *personal communication*).

Similar sized restriction fragments from acquired proviruses in independently derived C_3Hf tumors have been interpreted as indicating region-specific integration of MuMTV proviruses (51). If this interpretation is correct, the region may be identical to one of those already identified, or it may represent a new locus entirely.

CONCLUSION

The detection of MuMTV amplification and hypomethylation in mouse tumors raised the hope that these features might have a direct causative role in mammary tumorigenesis. As reviewed here, however, both phenomena are found in premalignant and normal-infected mouse mammary epithelium. Therefore, amplification and hypomethylation are unlikely to play a direct role in tumorigenesis.

One major phenomenon that distinguishes premalignant and malignant neoplasms from normal mammary epithelium is the relative homogeneity of neoplastic tissue. This homogeneity suggests that only a few of the large number of host-virus interactions provide the selective growth advantage required for neoplastic transformation. The most important interaction may be related to the position in which the provirus integrates into the host DNA. Positional effects have been suggested by several investigators (130,133) and directly supported by the discovery of repeated integration of MuMTV into limited regions of the mouse genome (106).

A second biologically important conclusion concerns the similarity of MuMTV proviral restriction patterns in premalignant and tumor tissues (27). This suggests that the important host-virus interactions occurred early and affected the noduligenesis, rather than tumorigenesis, stage of tumor progression. Because of these similarities in restriction patterns, we predict that the integration sites already isolated will also be occupied in premalignant cells. If true, the major role of MuMTV is probably the induction of mammary hyperplasias, and research should now focus on the "real" mammary oncogenes, e.g., those genes affected by MuMTV insertion.

If the primary action of MuMTV is in nodulogenesis, the virus can be used as a convenient molecular probe to detect, identify, and characterize mammary genes related to nodule transformation. In addition, it is clear that additional genes are involved in malignant transformation. These mammary genes remain to be defined. However, the molecular cloning of often used integration regions, as well as the 3T3 transfection/transformation system (78,134), offer the tools necessary for this endeavor.

ACKNOWLEDGMENTS

We thank our colleagues for supplying numerous publications, preprints, and suggestions. Work done in this laboratory was supported by Public Health Service grant 5 RO1 CA21454 from the National Cancer Institute.

REFERENCES

1. Altrock, B. W., Cardiff, R. D., Puma, J. P., and Lund, J. K. (1982): Detection of acquired provirus sequences in mammary tumors from low-expressor, low-risk mice. *J. Natl. Cancer Inst.*, 68:1037–1041.
2. Arthur, L. O., Butel, J. S., Medina, D., Dusing-Swartz, S. K., Socher, S. H., and Smith G. H. (1982): Separate pathways for viral and chemical carcinogenesis in the mouse mammary gland. In: *Biological Carcinogenesis*, edited by M. A. Rich and P. Furmanski, pp. 169–182. Marcel Dekker, New York.
3. Ashley, R. L., Cardiff, R. D., and Fanning, T. G. (1980): Re-evaluation of the effect of mouse mammary tumor virus infection on the BALB/c mouse hyperplastic outgrowth. *J. Natl. Cancer Inst.*, 65:977–986.
4. Ashley, R. L., Cardiff, R. D., Mitchell, D. J., Faulkin, L. J., and Lund, J. K. (1980): Development and characterization of mouse hyperplastic mammary outgrowth lines from BALB/cfC$_3$H hyperplastic alveolar nodules. *Cancer Res.*, 40:4232–4242.
5. Bentvelzen, P. (1968): *Genetical Control of the Vertical Transmission of the Muhlbock Mammary Tumor Virus in the GR Mouse Strain*. Hollandia, Amsterdam.
6. Bentvelzen, P. (1974): Host-virus interactions in murine mammary carcinogenesis. *Biochem. Biophys. Acta*, 355:236–259.
7. Bentvelzen, P. (1982): The putative mam gene of the murine mammary tumor virus. In: *Advances in Viral Oncology, Vol. 1*, edited by G. Klein, pp. 141–152. Raven Press, New York.
8. Bentvelzen, P., and Hilgers, J. (1980): Murine mammary tumor virus. In: *Viral Oncology*, edited by G. Klein, pp. 311–355. Raven Press, New York.
9. Bern, H. A. (1960): Nature of the hormonal influence in mouse mammary cancer. *Science*, 131:1039–1040.
10. Bernhard, W. (1958): Electron microscopy of tumor cells and tumor viruses; a review. *Cancer Res.*, 18:491–509.
11. Bird, A. P. (1980): Gene reiteration and gene amplification. *Cell Biol.*, 3:61–111.
12. Bird, A. P., and Southern, E. M. (1978): Use of restriction enzymes to study eukaryotic DNA methylation: The methylation pattern in ribosomal DNA from Xenopus laevis. *J. Mol. Biol.*, 118:27–47.
13. Bird, A. P., Taggart, M. H., and Gehring, C. A. (1981): Methylated and unmethylated ribosomal RNA genes in the mouse. *J. Mol. Biol.*, 152:1–17.

14. Bishop, J. M., and Varmus, H. (1982): Functions and origins of retroviral transforming genes. In: *RNA Tumor Viruses*, edited by R. Weiss, N. Teich, H. Varmus, and J. Coffin, pp. 999–1108. Cold Spring Harbor Laboratory, New York.
15. Bittner, J. J. (1936): Some possible effects of nursing on the mammary tumor incidence in mice. *Science*, 84:162.
16. Breznik, T., and Cohen, J. C. (1982): Altered methylation of endogenous viral promoter sequences during mammary carcinogenesis. *Nature*, 295:255–257.
17. Britten, R. J., and Kohne, D. E. (1968): Repeated sequences in DNA. *Science*, 161:529–540.
18. Brown, D. D. (1981): Gene expression in eukaryotes. *Science*, 211:667–674.
19. Brutlag, D. L. (1980): Molecular arrangement and evolution of heterochromatic DNA. *Annu. Rev. Genet.*, 14:121–144.
20. Buetti, E., and Diggelmann, H. (1981): Cloned mouse mammary tumor virus DNA is biologically active in transfected mouse cells and its expression is stimulated by glucocorticoid hormones. *Cell*, 23:335–345.
21. Callahan, R., Drohan, W., Gallahan, D., D'Hoostelaere, L., and Potter, M. (1982): Novel class of mouse mammary tumor virus-related DNA sequences found in all species of Mus, including mice lacking the virus proviral genome. *Proc. Natl. Acad. Sci. USA*, 79:4113–4117.
22. Calos, M. P., and Miller, J. H. (1980): Transposable elements. *Cell*, 20:579–595.
23. Campbell, A. (1971): Genetic structure. In: *The Bacteriophage Lambda*, edited by A. D. Hershey, pp. 13–44. Cold Spring Harbor Laboratory, New York.
23a. Cardiff, R. D. (1984): Protoneoplasia: The molecular biology of murine mammary hyperplasia. *Adv. Cancer Res., (in press).*
24. Cardiff, R. D., Wellings, S. R., and Faulkin, L. J. (1977): Biology of breast preneoplasia. *Cancer*, 39:2734–2746.
25. Cardiff, R. D., and Altrock, B. W. (1978): Biology of mammary tumor viruses. In: *Origins of Inbred Mice*, edited by H. C. Morse III, pp. 321–342. Academic Press, New York.
26. Cardiff, R. D., and Young, L. J. T. (1980): Mouse mammary tumor biology: A new synthesis. In: *Viruses in Naturally Occurring Cancers*, edited by M. Essex, G. Todaro, and H. zur Hausen, pp. 1105–1114. Cold Spring Harbor Laboratory, New York.
27. Cardiff, R. D., Fanning, T. G., Morris, D. W., Ashley, R. L., and Faulkin, L. J. (1981): Restriction endonuclease studies of hyperplastic outgrowth lines from BALB/cfC_3H hyperplastic mammary nodules. *Cancer Res.*, 41:3024–3029.
28. Christy, B., and Scangos, G. (1982): Expression of transferred thymidine kinase genes is controlled by methylation. *Proc. Natl. Acad. Sci. USA*, 79:6299–6303.
29. Coffin, J. (1982): Structure of the retroviral genome. In: *RNA Tumor Viruses*, edited by R. Weiss, N. Teich, H. Varmus, and J. Coffin, pp. 261–368. Cold Spring Harbor Laboratory, New York.
30. Cohen, J. C. (1980): Methylation of milk-borne and genetically transmitted mouse mammary tumor virus proviral DNA. *Cell*, 19:653–662.
31. Cohen, J. C., Shank, P. R., Morris, V. L., Cardiff, R. D., and Varmus, H. E. (1979): Integration of the DNA of mouse mammary tumor virus in virus-infected normal and neoplastic tissues of the mouse. *Cell*, 16:333–345.
32. Cohen, J. C., Majors, J. E., and Varmus, H. E. (1979): Organization of mouse mammary tumor virus-specific DNA endogenous to BALB/c mice. *J. Virol.*, 32:483–496.
33. Cohen, J. C., and Varmus, H. E. (1979): Endogenous mammary tumor virus DNA varies among wild mice and segregates during inbreeding. *Nature*, 278:418–423.
34. Cohen, J. C., and Varmus, H. E. (1980): Proviruses of mouse mammary tumor virus in normal and neoplastic tissues from GR and C_3Hf mouse strains. *J. Virol.*, 35:298–308.
35. Cohen, J. C., Traina, V. L., Breznik, T., and Gardner, M. B. (1982): Development of a mouse mammary tumor virus-negative mouse strain: A new system for the study of mammary carcinogenesis. *J. Virol.*, 44:882–885.
36. Collins, S., and Groudine, M. (1982): Amplification of endogenous myc-related DNA sequences in a human myeloid leukemia cell line. *Nature*, 298:679–681.
37. Dalla Favera, R., Wong-Staal, F., and Gallo, R. C. (1982): Onc gene amplification in promyelocytic leukemia cell line HL-60 and primary leukemic cells of the same patient. *Nature*, 299:61–66.
38. DeOme, K. B., Faulkin, L. J., Bern, H. A., and Blair, P. B. (1959): Development of mammary tumors from hyperplastic alveolar nodules transplanted into gland-free mammary fat pads of female C_3H mice. *Cancer Res.*, 19:515–520.

39. Dickson, C., and Peters, G. (1981): Protein coding potential of mouse mammary tumor virus genome RNA as examined by in vitro translation. *J. Virol.*, 37:36–47.
40. Dickson, C., Smith, R., and Peters, G. (1981): In vitro synthesis of polypeptides encoded by the long terminal repeat region of mouse mammary tumor virus DNA. *Nature*, 291:511–513.
41. Dickson, C., Eisenman, R., Fan, H., Hunter, E., and Teich, N. (1982): Protein biosynthesis and assembly. In: *RNA Tumor Viruses*, edited by R. Weiss, N. Teich, H. Varmus, and J. Coffin, pp. 513–648. Cold Spring Harbor Laboratory, New York.
42. Diggelmann, H., Vessaz, A. L., and Buetti, E. (1982): Cloned endogenous mouse mammary tumor virus DNA is biologically active in transfected mouse cells and its expression is stimulated by glucocorticoid hormones. *Virology*, 122:332–341.
43. Donehower, L. A., Andre, J., Berard, D. S., Wolford, R. G., and Hager, G. L. (1980): Construction and characterization of molecular clones containing integrated mouse mammary tumor virus sequences. *Cold Spring Harbor Symp. Quant. Biol.*, 44:1153–1159.
44. Donehower, L. A., Huang, A. L., and Hager, G. L. (1981): Regulatory and coding potential of the mouse mammary tumor virus long terminal redundancy. *J. Virol.*, 37:226–238.
45. Donehower, L. A., Fleurdelys, B., and Hager, G. L. (1983): Further evidence for the protein coding potential of the mouse mammary tumor virus long terminal repeat: Nucleotide sequence of an endogenous proviral long terminal repeat. *J. Virol.*, 45:941–949.
46. Drohan, W. N., Benade, L. E., Graham, D. E., and Smith, G. H. (1982): Mouse mammary tumor virus proviral sequences congenital to C_3H/Sm mice are differentially hypomethylated in chemically induced, virus-induced, and spontaneous mammary tumors. *J. Virol.*, 43:876–884.
47. Duesberg, P. H., and Cardiff, R. D. (1968): Structural relationships between the RNA of mammary tumor viruses and those of other RNA tumor viruses. *Virology*, 36:696–700.
48. Dudley, J. P., and Varmus, H. E. (1981): Purification and translation of murine mammary tumor virus RNA's. *J. Virol.*, 39:207–218.
49. Ehrlich, M., and Wang, R. Y.-H. (1981): 5-Methylcytosine in eukaryotic DNA. *Science*, 212:1350–1357.
50. Etkind, P. R., Szabo, P., and Sarkar, N. H. (1982): Restriction endonuclease mapping of the proviral DNA of the endogenous RIII murine mammary tumor virus. *J. Virol.*, 41:855–867.
51. Etkind, P. R., and Sarkar, N. H. (1983): Integration of new endogenous mouse mammary tumor virus proviral DNA at common sites in the DNA of mammary tumors of C_3Hf mice and hypomethylation of the endogenous mouse mammary tumor virus proviral DNA in C_3Hf mammary tumors and spleens. *J. Virol.*, 45:114–123.
52. Fanning, T. G., Puma, J. P., and Cardiff, R. D. (1980): Selective amplification of mouse mammary tumor virus in mammary tumors of GR mice. *J. Virol.*, 36:109–114.
53. Fanning, T. G., Puma, J. P., and Cardiff, R. D. (1980): Identification and partial characterization of an endogenous form of mouse mammary tumor virus that is transcribed into the virion-associated RNA genome. *Nucleic Acids Res.*, 8:5715–5723.
54. Fanning, T. G., Vassos, A. B., and Cardiff, R. D. (1982): Methylation and amplification of mouse mammary tumor virus in normal, premalignant, and malignant cells of GR/A mice. *J. Virol.*, 41:1007–1013.
55. Fasel, N., Pearson, K., Buetti, E., and Diggelmann, H. (1982): The region of mouse mammary tumor virus DNA containing the long terminal repeat includes a long coding sequence and signals for hormonally regulated transcription. *EMBO J.*, 1:3–7.
56. Feinstein, S. C., Ross, S. R., and Yamamoto, K. R. (1982): Chromosomal position effects determine transcriptional potential of integrated mammary tumor virus DNA. *J. Mol. Biol.*, 156:549–565.
57. Foulds, L. (1958): The natural history of cancer. *J. Chronic Dis.*, 8:2–37.
58. Garel, A., and Axel, R. (1976): Selective digestion of transcriptionally active ovalbumin genes from oviduct nuclei. *Proc. Natl. Acad. Sci. USA*, 73:3966–3970.
59. Green, M. M. (1980): Transposable elements in Drosophila and other diptera. *Annu. Rev. Genet.*, 14:109–120.
60. Grippo, P., Iaccarino, M., Parisi, E., and Scarano, E. (1968): Methylation of DNA in developing sea urchin embryos. *J. Mol. Biol.*, 36:195–208.
61. Groner, B., Hynes, N. E., and Diggelmann, H. (1979): Identification of mouse mammary tumor virus-specific mRNA. *J. Virol.*, 30:417–420.
62. Groner, B., and Hynes, N. E. (1980): Number and location of mouse mammary tumor virus proviral DNA in mouse DNA of normal tissue and of mammary tumors. *J. Virol.*, 33:1013–1025.

63. Groner, B., Buetti, E., Diggelmann, H., and Hynes, N. E. (1980): Characterization of endogenous and exogenous mouse mammary tumor virus proviral DNA with site-specific molecular clones. *J. Virol.*, 36:734–745.
64. Groner, B., Kennedy, N., Rahmsdorf, U., Herrlich, P., van Ooyen, A., and Hynes, N. E. (1982): Introduction of proviral mouse mammary tumor virus gene and a chimeric MMTV-thymidine kinase gene into L cells results in their glucocorticoid responsive expression. In: *Hormones and Cell Regulation, Vol. 6*, edited by J. E. Dumont, J. Nunez, and G. Shultz, pp. 217–228. Elsevier, Amsterdam.
65. Hager, G. L., and Donehower, L. A. (1980): Observations on the DNA sequence of the extended terminal redundancy and adjacent host sequences for integrated mouse mammary tumor virus. In: *Animal Virus Genetics*, edited by B. N. Fields, R. Jaenisch, and C. F. Fox, pp. 255–263. Academic Press, New York.
66. Hager, G. L., Huang, A. L., Bassin, R. H., and Ostrowski, M. C. (1982): Analysis of glucocorticoid regulation by linkage of the mouse mammary tumor virus promoter to a viral oncogene. In: *Eukaryotic Viral Vectors*, edited by Y. Gluzman, pp. 165–169. Cold Spring Harbor Laboratory, New York.
67. Hayward, W. S., Neel, B. G., and Astrin, S. M. (1981): Activation of a cellular onc gene by promoter insertion in ALV-induced lymphoid leukosis. *Nature*, 290:475–480.
68. Herrlich, P., Hynes, N. E., Ponta, H., Rahmsdorf, U., Kennedy, N., and Groner, B. (1981): The endogenous proviral mouse mammary tumor virus genes of the GR mouse are not identical and only one corresponds to the exogenous virus. *Nucleic Acids Res.*, 9:4981–4995.
69. Hilkins, J., Hilgers, J., Michalides, R., van der Valk, M., van der Zeijst, B., Colombatti, A., Hynes, N., and Groner, B. (1980): Analysis of murine retroviral genes and genes for viral membrane receptors by somatic-cell genetics. In: *Viruses in Naturally Occurring Cancers*, edited by M. Essex, G. Todaro, and H. zur Hausen, pp. 1033–1048. Cold Spring Harbor Laboratory, New York.
70. Huang, A. L., Ostrowski, M. C., Bernard, D., and Hager, G. L. (1981): Glucocorticoid regulation of the Ha-MuSV p21 gene conferred by sequences from mouse mammary tumor virus. *Cell*, 27:245–255.
71. Hynes, N. E., Groner, B., Diggelmann, H., Van Nie, R., and Michalides, R. (1980): Genomic location of mouse mammary tumor virus proviral DNA in normal mouse tissue and in mammary tumors. *Cold Spring Harbor Symp. Quant. Biol.*, 44:1161–1168.
72. Hynes, N. E., Rahmsdorf, U., Kennedy, N., Fabiani, L., Michalides, R., Nusse, R., and Groner, B. (1981): Structure, stability, methylation, expression and glucocorticoid induction of endogenous and transfected proviral genes of mouse mammary tumor virus in mouse fibroblasts. *Gene*, 15:307–317.
73. Hynes, N. E., Kennedy, N., Rahmsdorf, U., and Groner, B. (1981): Hormone-responsive expression of an endogenous proviral gene of mouse mammary tumor virus after molecular cloning and gene transfer into cultured cells. *Proc. Natl. Acad. Sci. USA*, 78:2038–2042.
74. Hynes, N. E., Rahmsdorf, U., Kennedy, N., Herrlich, P., Hohn, B., and Groner, B. (1982): DNA mediated transfer and fate of a proviral gene of mouse mammary tumor virus in cultured cells. In: *Embryonic Development, Part A: Genetic Aspects*, pp. 159–171. Alan R. Liss, New York.
75. Jagadeeswaran, P., Forget, B. G., and Weissman, S. M. (1981): Short interspersed repetitive DNA elements in eucaryotes: Transposable DNA elements generated by reverse transcription of RNA Pol III transcripts? *Cell*, 26:141–142.
76. Kennedy, N., Knedlitschek, G., Groner, B., Hynes, N. E., Herrlich, P., Michalides, R., and van Ooyen, A. J. J. (1982): Long terminal repeats of endogenous mouse mammary tumor virus contain a long open reading frame which extends into adjacent sequences. *Nature*, 295:622–624.
77. Klein, G. (1981): The role of gene dosage and genetic transpositions in carcinogenesis. *Nature*, 294:313–318.
78. Lane, M.-A., Sainten, A., and Cooper, G. (1981): Activation of related transforming genes in mouse and human mammary carcinomas. *Proc. Natl. Acad. Sci. USA*, 78:5185–5189.
79. Lee, F., Mulligan, R., Berg, P., and Ringold, G. (1981): Glucocorticoids regulate expression of dihydrofolate reductase cDNA in mouse mammary tumor virus chimaeric plasmids. *Nature*, 294:228–232.
80. Linemeyer, D. L., Menke, J. G., Ruscetti, S. K., Evans, L. H., and Scolnick, E. M. (1982): Envelope gene sequences which encode the gp52 protein of spleen focus-forming virus are required for the induction of erythroid cell proliferation. *J. Virol.*, 43:223–233.

81. Macinnes, J. I., Chan, E. C. M. L., Percy, D. H., and Morris, V. L. (1981): Mammary tumors from GR mice contain more than one population of mouse mammary tumor virus-infected cells. *Virology*, 113:119–129.
82. Majors, J. E., Swanstrom, R., DeLorbe, W. J., Paynes, G. S., Hughes, S. H., Ortiz, S., Quintrell, N., Bishop, J. M., and Varmus, H. E. (1981): DNA intermediates in the replication of retroviruses are structurally (and perhaps functionally) related to transposable elements. *Cold Spring Harbor Symp. Quant. Biol.*, 45:731–738.
83. Majors, J. E., and Varmus, H. E. (1981): Nucleotide sequences at host-proviral junctions for mouse mammary tumor virus. *Nature*, 289:253–258.
84. McCoy, M. S., Toole, J. J., Cunningham, J. M., Chang, E. H., Lowy, D. R., and Weinberg, R. A. (1983): Characterization of a human colon/lung carcinoma oncogene. *Nature*, 302:79–81.
85. McGrath, C. M., Nandi, S., and Young, L. (1972): Relationship between organization of mammary tumors and the ability of tumor cells to replicate mammary tumor virus and to recognize growth-inhibitory contact signals in vitro. *J. Virol.*, 9:367–376.
86. McGrath, C. M., Marineau, E. J., and Voyles, B. A. (1978): Changes in MuMTV DNA and RNA levels in BALB/c mammary epithelial cells during malignant transformation by exogenous MuMTV and by hormones. *Virology*, 87:339–353.
87. McGrath, C. M., and Jones, R. F. (1978): Hormonal induction of mammary tumor virus and its implications for carcinogenesis. *Cancer Res.*, 38:4112–4125.
88. McKeon, C., Ohkubo, H., Pastan, I., and de Crombrugghe, B. (1982): Unusual methylation pattern of the alpha-2(I) collagen gene. *Cell*, 29:203–210.
89. Medina, D. (1973): Preneoplastic lesions in mouse mammary tumorigenesis. In: *Methods in Cancer Research, Vol. 7*, edited by H. Busch, pp. 3–52. Academic Press, New York.
90. Medina, D. (1978): Preneoplasia in breast cancer. In: *Breast Cancer, Vol. 2*, edited by W. L. McGuire, pp. 47–102. Plenum, New York.
91. Mermod, J.-J., Bourgeois, S., Defer, N., and Crepin, M. (1983): Demethylation and expression of murine mammary tumor proviruses in mouse thymoma cell lines. *Proc. Natl. Acad. Sci. USA*, 80:110–114.
92. Michalides, R., Vlahakis, G., and Schlom, J. (1976): A biochemical approach to the study of the transmission of mouse mammary tumor viruses in mouse strains RIII and C_3H. *Int. J. Cancer*, 18:105–115.
93. Michalides, R., Van Deemter, L., Nusse, R., and Van Nie, R. (1978): Identification of the Mtv-2 gene responsible for the early appearance of mammary tumors in the GR mouse by nucleic acid hybridization. *Proc. Natl. Acad. Sci. USA*, 75:2368–2372.
94. Michalides, R., Van Deemter, L., Nusse, R., Ropcke, G., and Boot, L. (1978): Involvement in mouse mammary tumor virus in spontaneous and hormone-induced mammary tumors in low-mammary-tumor mouse strains. *J. Virol.*, 27:551–559.
95. Michalides, R., Van Deemter, L., Nusse, R., and Hageman, P. (1979): Induction of mouse mammary tumor virus RNA in mammary tumors of BALB/c mice treated with urethane, X-irradiation, and hormones. *J. Virol.*, 31:63–72.
96. Michalides, R., Wagenaar, E., Groner, B., and Hynes, N. E. (1981): Mammary tumor virus proviral DNA in normal murine tissue and non-virally induced mammary tumors. *J. Virol.*, 39:367–376.
97. Michalides, R., Van Nie, R., Nusse, R., Hynes, N. E., and Groner, B. (1981): Mammary tumor induction loci in GR and DBAf mice contain one provirus of the mouse mammary tumor virus. *Cell*, 23:165–173.
98. Michalides, R., Wagenaar, E., and Sluyser, M. (1982): Mammary tumor virus DNA as a marker for genotypic variance within hormone-responsive GR mouse mammary tumors. *Cancer Res.*, 42:1154–1158.
99. Michalides, R., Wagenaar, E., Hilkens, J., Hilgers, J., Groner, B., and Hynes, N. E. (1982): Acquisition of proviral DNA of mouse mammary tumor virus in thymic leukemia cells from GR mice. *J. Virol.*, 43:819–829.
100. Morris, V. L., Mederios, E., Ringold, G. M., Bishop, J. M., and Varmus, H. E. (1977): Comparison of mouse mammary tumor virus-specific DNA in inbred, wild and asian mice, and in tumors and normal organs from inbred mice. *J. Mol. Biol.*, 114:73–91.
101. Morris, V. L., Gray, D. A., Jones, R. F., Chan, E. C. M. L., and McGrath, C. M. (1982): Mouse mammary tumor virus DNA sequences in tumorigenic and nontumorigeneic cells from a mammary adenocarcinoma. *Virology*, 118:117–127.

102. Muhlbock, O. (1972): Role of hormones in the etiology of breast cancer. *J. Natl. Cancer Inst.*, 48:1213–1216.
103. Nandi, S., and McGrath, C. M. (1973): Mammary neoplasia in mice. *Adv. Cancer Res.*, 17:353–414.
104. Neel, B. G., Hayward, W. S., Robinson, H. L., Fang, J., and Astrin, S. M. (1981): Avian leukosis virus-induced tumors have common proviral integration sites and synthesize discrete new RNAs: Oncogenesis by promoter insertion. *Cell*, 23:323–334.
105. Nowell, P. C. (1976): The clonal evolution of tumor cell populations. Science, 194:23–28.
106. Nusse, R., and Varmus, H. E. (1982): Many tumors induced by the mouse mammary tumor virus contain a provirus integrated in the same region of the host genome. *Cell*, 31:99–109.
107. Ott, M-O., Sperling, L., Cassio, D., Levilliers, J., Sala-Trepat, J., and Weiss, M. C. (1982): Undermethylation at the 5′ end of the albumin gene is necessary but not sufficient for albumin production by rat hepatoma cells in culture. *Cell*, 30:825–833.
108. Owen, D., and Diggelmann, H. (1983): Cloned mouse mammary tumor virus DNA exhibits glucocorticoid-dependent expression in simian virus 40-transformed mink cells. *J. Virol.*, 45:148–154.
109. Pavar, F., Wrange, O., Carlstedt-Duke, J., Okret, S., Gustafsson, J-A., and Yamamoto, K. (1981): Purified glucocorticoid receptors bind in vitro to a cloned DNA fragment whose transcription is regulated by glucocorticoids in vivo. *Proc. Natl. Acad. Sci. USA*, 78:6628–6632.
110. Payne, G. S., Bishop, J. M., and Varmus, H. E. (1982): Multiple arrangements of viral DNA and an activated host oncogene in bursal lymphomas. *Nature*, 295:209–213.
111. Peters, G., Smith, R., Brookes, S., and Dickson, C. (1982): Conservation of protein coding potential in the long terminal repeats of exogenous and endogenous mouse mammary tumor viruses. *J. Virol.*, 42:880–888.
112. Pfahl, M. (1982): Specific binding of the glucocorticoid-receptor complex to the mouse mammary tumor proviral promoter region. *Cell*, 31:475–482.
113. Puma, J. P., Fanning, T. G., Young, L. J. T., and Cardiff, R. D. (1982): Identification of a unique mouse mammary tumor virus in the BALB/cNIV mouse strain. *J. Virol.*, 43:158–165.
114. Razin, A., and Friedman, J. (1981): DNA methylation and its possible biological roles. *Prog. Nucleic Acid Res. Mol. Biol.*, 25:33–52.
115. Riggs, A. D. (1975): X inactivation, differentiation and DNA methylation. *Cytogenet. Cell Genet.*, 14:9–25.
116. Ringold, G. M., Yamamoto, K. R., Bishop, J. M., and Varmus, H. E. (1977): Glucocorticoid-stimulated accumulation of mouse mammary tumor virus RNA: Increased rate of synthesis of viral RNA. *Proc. Natl. Acad. Sci. USA*, 74:2879–2883.
117. Ringold, G. M., Cohen, J. C., Ring, J., Shank, P. R., Varmus, H. E., and Yamamoto, K. R. (1978): Mouse mammary tumor virus genes: Regulation of expression by glucocorticoids and structural analysis with restriction endonucleases. *J. Toxicol. Environ. Health*, 4:457–470.
118. Robertson, D. L., and Varmus, H. E. (1979): Structural analysis of the intracellular RNAs of murine mammary tumor virus. *J. Virol.*, 30:576–589.
119. Robertson, D. L., and Varmus, H. E. (1981): Dexamethasone induction of the intracellular RNAs of mouse mammary tumor virus. *J. Virol.*, 40:673–682.
120. Shank, P. R., Cohen, J. C., Varmus, H. E., Yamamoto, K. R., and Ringold, G. M. (1978): Mapping of linear and circular forms of mouse mammary tumor virus DNA with restriction endonucleases: Evidence for a large specific deletion occurring at high frequency during circularization. *Proc. Natl. Acad. Sci. USA*, 75:2112–2116.
121. Sharp, P. A. (1983): Conversion of RNA to DNA in mammals: alu-like elements and pseudogenes. *Nature*, 301:471–472.
122. Shyamala, G., and Dickson, C. (1976): Relationship between receptor and mammary tumor virus production after stimulation by glucocorticoid. *Nature*, 262:107–112.
123. Smith, G. P. (1976): Evolution of repeated DNA sequences by unequal crossovers. *Science*, 191:528–535.
124. Temin, H. M. (1980): Origin of retroviruses from cellular moveable genetic elements. *Cell*, 21:599–600.
125. Todaro, G. J. (1975): Evolution and modes of transmission of RNA tumor viruses. *Am. J. Pathol.*, 81:590–606.
126. Ucker, D. S., Ross, S., and Yamamoto, K. R. (1981): Mammary tumor virus DNA contains sequences required for its hormone-regulated transcription. *Cell*, 27:257–266.

127. Van Nie, R., and de Moes, J. D. (1977): Development of a congeneic line of the GR mouse strain without early mammary tumors. *Int. J. Cancer*, 20:588–594.
128. Van Nie, R., and Dux, A. (1971): Biological and morphological characteristics of mammary tumors in GR mice. *J. Natl. Cancer Inst.*, 46:885–897.
129. Vardimon, L., Kressmann, A., Cedar, H., Maechler, M., and Doerfler, W. (1982): Expression of a cloned adenovirus gene is inhibited by in vitro methylation. *Proc. Natl. Acad. Sci. USA*, 79:1073–1077.
130. Varmus, H. E. (1982): Recent evidence for oncogenesis by insertion mutagenesis and gene activation. *Cancer Surveys*, 1:309–319.
131. Varmus, H. E. (1982): Form and function of retroviral proviruses. *Science*, 216:812–820.
132. Varmus, H. E., and Swanstrom, R. (1982): Replication of retroviruses. In: *RNA Tumor Viruses*, edited by R. Weiss, N. Teich, H. Varmus, and J. Coffin, pp. 369–512. Cold Spring Harbor Laboratory, New York.
133. Varmus, H. E., Payne, G. S., Nusse, R., Luciw, P., Westaway, D., and Bishop, J. M. (1983): Insertional mechanisms in oncogenesis by retroviruses. In: *Perspectives on Genes and the Molecular Biology of Cancer*, edited by D. L. Robberson and G. F. Saunders, pp. 237–246. Raven Press, New York.
134. Weinberg, R. A. (1982): Transforming genes of nonvirus-induced tumors. In: *Advances in Viral Oncology, Vol. 1*, edited by G. Klein, pp. 235–241. Raven Press, New York.
135. Weintraub, H., and Groudine, H. (1976): Chromosomal subunits in active genes have an altered conformation. *Science*, 193:848–856.
136. Weiss, R., Teich, N., Varmus, H., and Coffin, J. (editors) (1982): *RNA Tumor Viruses*, Cold Spring Harbor Laboratory, New York.
137. Wheeler, D. A., Butel, J. S., Medina, D., Cardiff, R. D., and Hager, G. L. (1983): Transcription of mouse mammary tumor virus: Identification of a candidate mRNA for the long terminal repeat gene product. *J. Virol.*, 46:42–49.
138. Yamamoto, K. R., Chandler, V. L., Ross, S. R., Ucker, D. S., Ring, J. C., and Feinstein, S. C. (1981): Integration and activity of mammary tumor virus genes: Regulation by hormone receptors and chromosomal position. *Cold Spring Harbor Symp. Quant. Biol.*, 45:687–697.
139. Young, M. W. (1979): Middle repetitive DNA: A fluid component of the Drosophila genome. *Proc. Natl. Acad. Sci. USA*, 76:6274–6278.
140. Young, H. A., and Hager, G. (1979): Mouse mammary tumor virus and glucocorticoids. In: *Steroid Receptors and the Management of Cancer*, edited by E. B. Thompson and M. E. Lippman, pp. 45–57. CRC Press, Boca Raton, Florida.
141. Young, L. J. T., Cardiff, R. D., and Ashley, R. L. (1975): Long term primary culture of mouse mammary tumor cells: Production of virus. *J. Natl. Cancer Inst.*, 54:1215–1221.

Advances in Viral Oncology, Volume 4,
edited by George Klein. Raven Press,
New York © 1984.

Immunologic and Virologic Mechanisms in Retrovirus-Induced Murine Leukemogenesis

*James N. Ihle, †Alan Rein, and *Richard Mural

**Laboratory of Viral Immunobiology and †Laboratory of Molecular Virology and Carcinogenesis, National Cancer Institute, Frederick Cancer Research Facility, LBI, Basic Research Program, Frederick, Maryland 21701*

The etiologic relationship of retroviruses to leukemogenesis in mice has served as one of the most extensively studied model research systems over the past 30 years. The various types of murine leukemias are associated with two biologically distinct classes of retroviruses, including the acute leukemia viruses, which can directly transform hemopoietic cells *in vitro* and *in vivo*, and the nonacute viruses, which have no direct transforming activity. The acute retroviruses associated with leukemias in mice include: (a) the Abelson virus, which induces predominantly B cell lymphomas but can also transform erythroid cells *in vitro* (39,67,172,206); (b) the Harvey/Kirsten sarcoma viruses, which, although generally associated with sarcoma induction, have been shown to induce erythroid lineage transformation *in vitro* as well as splenomegaly and possible erythroleukemia *in vivo* (50,74); (c) a myeloproliferative syndrome-inducing virus, which transforms myeloid lineage cells *in vivo* and may have a transforming gene identical to that of Moloney sarcoma virus from which the variant was obtained (108,109,150); and (d) a virus isolated from chemically transformed fibroblasts, which induces splenic lymphomas (160,161). In addition, a virus possibly related to murine mammary tumor virus (MMTV) has been described, which induces T cell lymphomas in a manner suggestive of the presence of an acute transforming virus (6–9). In each of the cases, the viruses can transform hemopoietic cells *in vitro* and/or rapidly induce lymphomas *in vivo*. In addition to the above acute transforming viruses, the spleen focus forming virus (SFFV) can be classified as an acute leukemia virus (75,123–125,147). The virus rapidly induces erythroleukemia *in vivo* and contains a nonstructural protein, gp52, which when expressed in cells of the erythroid lineage alters their growth and/or differentiation properties. It has been argued that since the virus does not transform fibroblasts, and since the primary erythroleukemias are not transplantable, the disease should be termed a hyperproliferative disease rather than a true neoplasia.

The second group of leukemias that is the focus of this chapter include those etiologically associated with the wide variety of replication competent nonacute retroviruses. These types of leukemias are induced only after long latencies, during which an acute viremia is required. The viruses associated with this type of disease have not been shown to have a transforming activity *in vitro* and, where examined, have been shown to induce monoclonal tumors (22,192,201). Taken together, the above observations have demonstrated that this type of leukemia is the result of relatively rare events that occur as a consequence of extensive viral replication. Among the retrovirus-induced diseases in mice, this class of disease has been studied the most. Despite extensive research, however, a common underlying mechanism has not been identified. Three major areas of investigations, either alone or in combination, have been proposed to be important to leukemogenesis: (a) formation of more oncogenic viruses by recombination, (b) activation or alteration of oncogenes by virus integration, and (c) the role of immune responses against the virus in providing target cell populations for transformation. Perhaps the most virologic of these has been the hypothesis that the generation of a specific type of recombinant virus is required for leukemogenesis. The relationship of recombinant viruses to leukemia has evolved from a number of studies, which demonstrated the frequent association of recombinant viruses with lymphomas induced by a variety of ecotropic viruses. As discussed below, however, it is still unclear exactly how they contribute to leukemogenesis.

One of the most appealing models for a possible mechanism by which subacute viruses could induce or promote leukemia has come from studies of avian leukemias. The initial model stated that leukosis viruses occasionally might integrate near a potentially oncogenic normal host gene and activate (downstream promote) the transcription of the gene (11,12,41,101,170). The altered expression of such a gene out of context would cause transformation. This model is consistent with the requirements for viremia, long latencies and the monoclonality of the tumors. Evidence for this type of mechanism has come from a series of experiments by Hayward et al. (80–82), who demonstrated that in avian leukosis virus-induced B cell lymphomas in chickens, viral integration uniquely occurs near the cellular *myc*(c-*myc*) gene and causes increased transcription of c-*myc*, a transforming gene found in the acute transforming virus MC29. This type of mechanism has been demonstrated in other avian retrovirus-induced diseases (139). However, whether the mechanism involved in alteration of gene expression by viral integration is as simple as initially proposed is not clear (152). In addition, the ability to detect transforming genes unrelated to *myc*, by transfection experiments into fibroblasts has suggested that c-*myc* activation may be necessary but not sufficient for transformation (40).

In addition to the above possible factors contributing to leukemogenesis, a role for an immune response against the virus has been implicated (89,118,128,131). A possible role for immunologic events in leukemogenesis might be predicted since leukemia often involves transformation of components of the immune system. Second, the characteristics of viral leukemogenesis are such that it could be proposed that the virus simply creates an environment in which somatic events associated with transformation can occur. In this regard, the striking occurrence of trisomy 15 in approximately 50% of all murine T cell lymphomas is an intriguing observation and suggests that somatic events affecting gene expression may be important in leukemogenesis (47,106,107,208). In this case, the type of somatic mechanisms may be similar to those recently shown to be associated with the induction of plasmacytomas by mineral oil in which a unique translocation of the c-*myc* gene has been implicated in transformation (1,42,183). This type of somatic rearrangement appears to occur as a consequence of proliferation at the site of mineral oil injection without any apparent role for a retrovirus. In this chapter, the evidence for and the possible role of the above mechanisms in murine leukemias induced by the subacute, nontransforming group of retroviruses are evaluated.

RECOMBINANT MURINE LEUKEMIA VIRUSES AND THEIR ROLE IN LEUKEMOGENESIS

Work within the last decade has demonstrated that recombination between murine leukemia virus (MuLV) genomes is a frequent occurrence *in vivo*, and that recombinant viruses may play a significant role in many instances of virus-induced leukemogenesis. Various recombinant viruses have been isolated; one major thrust of current investigations is the characterization of these viruses and the search for correlations between molecular properties and biologic potential. After a brief introduction, we discuss the status of these investigations in two well-studied virus systems: MuLVs derived from *Akv* and those present in the Friend virus complex.

MuLVs can be subdivided according to many criteria; one method of classification that is particularly useful in yielding positive, unambiguous identification is by viral interference. Viruses interfere with each other only if they use the same receptor to penetrate the cell (168,188,189,205), so that identification of a virus by its interference properties is actually a determination of which receptor it uses. This is the basis of the grouping outlined in Table 1; with one exception (discussed below), all known replication-competent MuLVs fall into one of these four groups:

TABLE 1. *Classes of MuLV*

Virus group	Examples	Host range	Fusion of XC cells	Comments
Ecotropic	Friend, Moloney Rauscher, *Akv*	Infects only rodent cells	+	
Amphotropic	1504A, 4070A	Infects both rodent and many nonrodent	−	Isolated only from wild mice
MCF	MCF247, HIX	Relative infectivities for rodent and nonrodent cells vary over wide range	−	Some isolates virtually noninfectious for nonrodent cells; many isolates cytopathic for mink cells
Xenotropic	Balb-virus 2, ATS-124	Unable to infect mouse cells	−	

(a) ecotropic MuLVs, which infect only rodent cells and have the unique property of causing syncytium formation in XC cells; (b) amphotropic MuLVs, which have been isolated from wild mouse populations in California and which infect both rodent and nonrodent cells; (c) mink cell focus forming (MCF) MuLVs, which are discussed in detail below; and (d) xenotropic MuLVs, which cannot infect mouse cells.

The MCF class is the best studied group of recombinant MuLVs. The name of the class refers to the fact that many isolates in this group form cytopathic foci in cultures of mink lung cells (77). This property has been helpful in the detection and isolation of MCFs.

All MCFs arise in the somatic tissues of mice that are viremic for ecotropic MuLVs. They are formed by recombination between the ecotropic MuLV genome and other, endogenous sequences present in mouse DNA. The resulting MCFs differ from their respective ecotropic parents by a substitution which includes the region encoding the N-terminal portion of gp70 and which extends for varying distances in either direction in different isolates.

It is clear that the change in the N-terminal region of gp70 is responsible for those differences between MCFs and their ecotropic parents which are detectable in cell culture. Thus it is known from studies with purified gp70 (220) and from the properties of both natural mutants and recombinants constructed *in vitro* (165,167) that fusion of XC cells is a property of ecotropic gp70s. Furthermore, analysis of a recombinant derived by joining the 5′ portion of the amphotropic gp70 gene to the 3′ portion of an MCF gp70 gene indicates that receptor specificity is carried within the N-terminal ~200 amino acids of gp70 (A. Rein and A. Oliff, *unpublished data*). Finally, studies with *in vitro* recombinants between an ecotropic MuLV and a MCF indicate that the ability of the virus to replicate in and induce cytopathic foci on mink cells is a property of the MCF gp70 (N. Hopkins and J. W. Hartley, *personal communication*).

What are the nature and origin of the substituted sequences? By restriction mapping (26), heteroduplex analysis (14), biochemical and serologic analyses of gp70s (49,146), and direct DNA sequencing (16,169; C. Holland and N. Hopkins, *personal communication*), the sequences show a striking resemblance to the *env* genes of xenotropic MuLVs. It has been assumed, therefore, that the substituted sequences are from xenotropic MuLV genomes, which are present as endogenous proviruses in most if not all mouse cells. However, the MCF gp70s possess antigenic determinants which are not found in either ecotropic or in xenotropic MuLVs (36,177,193). MCFs also differ from both xenotropic and ecotropic MuLVs in their ability to infect mouse cells via the MCF-specific receptor (165).

As discussed previously, it is difficult to explain these distinctive biologic attributes as resulting from the new combination of ecotropic and xenotropic sequences in a chimeric *env* gene (165), particularly since some MCF isolates contain no ecotropic sequences detectable by oligonucleotide mapping of the gp70 coding region (126,154). In addition, detailed mapping of MCF genomes by RNAse T_1 oligonucleotide fingerprinting and by restriction enzyme analysis has shown that the substituted sequences are not, in fact, identical to those in any known xenotropic MuLV (27,68).

The simplest hypothesis that can explain these observations is that MCFs arise by substitution of endogenous *env* sequences, which are similar to those of xenotropic MuLVs but which are not present in mouse DNA as part of a replication-competent provirus. Indeed, *env* sequences possessing the expected restriction sites and hybridization properties have been detected in AKR embryo DNA by Chattopadhyay et al. (26) and have been molecularly cloned from C3H DNA by Roblin et al. (171) and from AKR DNA by Khan et al. (105). It is interesting that these endogenous sequences appear to be flanked on the 5′ side by long-terminal repeat (LTR), *gag*, and *pol* sequences and on the 3′ side by LTR sequences (26). It is likely that the nonecotropic sequences found in MCFs are inherited by mice as part of a proviral structure; since this provirus is not known to give rise to MuLV, it is probably defective for replication. The similarity between these sequences and those of xenotropic MuLVs might reflect a close evolutionary relationship.

One intriguing observation may offer a clue to the biologic significance of the substituted sequences: The cell surface antigen $G_{(AKSL2)}$ is expressed on thymus and bone marrow cells of young AKR mice and also is apparently encoded by many MCFs arising in older AKR mice. This raises the possibility that the substituted sequences are from a gene for a normal differentiation alloantigen of hematopoietic tissues (146,193).

VIRUSES DERIVED FROM AKV

The AKR strain of mouse has been a unique resource for analysis of virological events associated with thymic lymphomas. This analysis has been pioneered by Rowe and his colleagues. AKR mice all carry two or more proviral copies of the ecotropic virus (173), *Akv*, as chromosomal loci (24,28,29,173,174,191). This virus is spontaneously produced during late embryogenesis and is present at high titer in many tissues from birth onward (175). At about 5 to 6 months of age, MCFs begin to be produced in the thymus (77,104). Then, beginning at 6 to 7 months after birth, the mice develop thymic lymphomas. These lymphomas are monoclonal (22,191) and frequently are trisomic for chromosome 15 (47).

It has been clear for a number of years that replication of *Akv* is necessary but not sufficient for leukemogenesis in AKR mice. Thus inhibition of *Akv* spread by early passive immunization (86,181) or by the action of the cellular gene $Fv\text{-}1^b$ (122,140) prevents the later appearance of lymphomas; however, biologically pure preparations of *Akv* have never demonstrated leukemogenic activity in any test system. Significantly, development of lymphoma in AKR mice can be accelerated by injection of thymus extracts from old but not young AKR mice, despite the presence of comparable levels of *Akv* in both types of extract (79,100,138). This difference between young and old thymus can now be explained by the presence of MCFs in the latter: MCFs isolated and biologically purified from late preleukemic or leukemic AKR thymus do accelerate lymphoma development upon injection into young AKR mice (37,142,146).

Taken together, the above observations suggest the following hypothesis: AKR lymphomas are induced by infection of thymocytes by MCFs; *Akv* replication is required for leukemogenesis because *Akv* participates in the recombinational event that gives rise to MCFs. Many recombinant MuLVs have been isolated from AKR and from other mice infected with *Akv* or related viruses; we briefly review what is known about them before further considering this hypothesis.

Pathogenic MCFs

This class, which includes the prototype isolates AKR-247 and AKR-13 (77) and 69L1 (145), is probably the best characterized of the *Akv*-derived MCFs. These viruses accelerate lymphoma development upon either intraperitoneal injection of newborn or intrathymic injection of weanling AKR mice. Studies with genetically marked pathogenic MCFs show that the resulting lymphomas produce the inoculated MCF (144,154). AKR-247 is also somewhat leukemogenic in the low leukemia strain C3H/Bi; in general, however, these MCFs are negative in direct leukemogenicity tests in low leukemia strains (as opposed to acceleration in AKR mice) (37).

Class II MCFs

The class II MCFs have been isolated from congenic mouse strains, i.e., mice that carry the *Akv* locus from AKR in the genetic background of the low leukemia NFS strain (174). Like AKR mice, these mice become viremic early in life, but they then develop a variety of nonthymic, B-cell hematopoietic neoplasms (W. P. Rowe, J. Hartley, H. C. Morse, III, and T. Fredericksen, *unpublished*). These tumors produce class II MCFs. These MCFs have no detectable pathogenic activity and barely replicate upon injection into mice (35,37).

It is notable that pathogenic MCFs from late AKR thymus do induce thymic lymphoma (with only moderate efficiency) upon injection into congenic mice (37). This positive result has several possible implications. First, it shows that mice of the NFS genetic background are susceptible to thymoma induction by the same viruses that are pathogenic in AKR mice. Second, the fact that pathogenic MCFs induce thymic lymphomas in congenic mice but not in NFS mice themselves suggests that *Akv* plays another role in leukemogenesis in addition to its genetic contribution to the formation of pathogenic MCFs. One possibility is that genomic masking (encapsidation of MCF genomes in ecotropic coats) (57,72) protects the MCF *in vivo* from neutralization by antibodies or nonimmune factors in serum (58,121). Finally, the fact that a pathogenic MCF can replicate and induce thymoma can be taken as a reconstruction experiment, suggesting that if such a virus arose spontaneously in the congenic mice, it would be detectable. In turn, this implies that one difference between the low leukemic NFS strain and AKR may be that NFS lacks the nonecotropic endogenous sequences that can recombine with *Akv* to yield a pathogenic MCF.

A^+L^- MCFs

O'Donnell et al. (146) have described the isolation from late preleukemic AKR thymus of a group of MCFs with great potential importance. These viruses replicate well in the thymus after injection into AKR mice (as detected by antigen amplification; hence "A^+") but do not appreciably accelerate (or retard) leukemia development (hence "L^-"). It is interesting to note that when thymomas eventually appear in AKR mice injected with A^+L MCFs, they are found to produce pathogenic MCFs (P. V. O'Donnell, *personal communication*).

Gross Ecotropic MuLV

The existence of viruses that were leukemogenic in mice was first demonstrated by Gross (70), who injected extracts of AKR leukemias into newborn C3H/Bi mice. Serial passage of leukemic extracts in C3H mice resulted in the Gross Passage A (GPA) virus stock, which is capable of efficient leukemia induction in low leukemic mouse strains as well as acceleration in AKR mice (70). An ecotropic MuLV was subsequently isolated from the GPA stock by end point dilution passage in NIH mouse cells (J. W. Hartley, cited in refs. 19 and 56). This MuLV resembles *Akv* in many respects; unlike *Akv*, however, it is nearly as leukemogenic as the GPA virus complex from which it was derived (20,56,71).

GPA-V2

Analysis by polyacrylamide gel electrophoresis shows that cells infected with the GPA virus stock contain two distinct *env* protein precursors: one with the same mobility as that of Gross ecotropic MuLV or *Akv*, and the other migrating like the *env* precursors of *Akv*-derived MCFs (56). When the GPA stock was injected intrathymically into weanling C3Hf/Bi mice, it was found to induce thymomas that only contained the MCF-like *env* protein. It was then possible to isolate the MuLV that encodes this *env* protein. This virus, termed GPA-V2, grows poorly in cell cultures and has not been extensively characterized; however, it is a MCF, as defined by viral interference (A. Rein, *unpublished*).

The separation of Gross ecotropic MuLV and GPA-V2 has allowed Famulari et al. (56) to test the individual effects *in vivo* of these two components of the GPA virus complex. It was found, in agreement with earlier work (70,71), that Gross ecotropic MuLV induced thymomas when injected intraperitoneally into suckling C3Hf/Bi mice; the latent period was somewhat longer and the incidence lower than those of GPA itself. The thymomas induced by injection of biologically pure Gross ecotropic MuLV always contained a virus resembling GPA-V2 as well as the input ecotropic virus. Gross ecotropic MuLV was nonpathogenic if inoculated intrathymically into weanling mice. In contrast, GPA-V2 was pathogenic by either route of injection. Its pathogenicity was comparable to that of the GPA complex; and GPA-V2-induced thymomas contain only GPA-V2. Thus GPA-V2 replication appears to be sufficient for thymoma induction. The fact that it is invariably generated during pathogenesis by Gross ecotropic MuLV is consistent with (but does not prove) the idea that it is a necessary intermediate in this process.

SL Viruses

Yet another source of *Akv*-derived MuLVs is a series of cell lines established in culture from AKR thymomas by Nowinski and Hays and their colleagues (79,143). Several of these cell lines produce MuLVs that can accelerate AKR lymphoma development; two of them have been analyzed virologically in some detail.

One line, AKRSL2, produces a MuLV which, like GPA-V2, grows poorly in cell culture but is a MCF as judged by viral interference tests (165,166). Although this virus, termed AK45, has not been characterized further, its resemblance to the MCFs and GPA-V2 makes it likely that it is the agent in AKRSL2 culture fluids that can accelerate thymoma development in AKR mice. AKRSL2 cells also produce a replication-defective ecotropic MuLV that appears to have only minor genetic differences from *Akv* (167).

Two distinct MuLVs also have been isolated from a second tumor line, AKRSL3. Both grow well in culture and are highly leukemogenic in C3Hf/Bi as well as AKR mice (155). One of these, SL3-2, is a MCF with respect to viral interference, although its lack of infectivity for mink cells has sometimes led to its description as an ecotropic MuLV (A. Rein, *unpublished*). The other, SL3-3, is clearly ecotropic, since it forms XC plaques and is in the ecotropic interference group (155; A. Rein, *unpublished*). As in the case of Gross MuLV, leukemogenesis by SL3-3 is accompanied by the formation of additional, recombinant MuLVs; these are not yet well characterized (N. Famulari, *personal communication*). It should be noted that SL3-2 and SL3-3 were isolated only after propagation of AKRSL3 supernatant virus on mass cultures of NIH/3T3 cells (155). Since new viruses can be generated by mutation or recombination during passage of AKRSL3 virus in fibroblasts (143,166), it is possible that SL3-2 and SL3-3 are modified from the viruses actually released by the tumor cells, and that selection during growth in NIH/3T3 cells has increased their virulence.

The structure and properties of many of these viruses have been intensively studied in a number of laboratories in recent years. Some of the results of these studies are schematically summarized in Table 2, which reviews data obtained from oligonucleotide mapping of a number of MCF genomes and comparisons of the maps with that of *Akv*. This approach allows ready detection of large substitutions but cannot exclude the presence of small substitutions between the large oligonucleotides; also, the number of oligonucleotide markers in a small region (such as U3) is limited. As shown, the AKR pathogenic MCFs differ from *Akv* in the N-terminal region of the gp70 gene and also at the 3′ end of the genome, i.e., the 3′ end of the p15E gene and the U3 region of the LTR but not in the N-terminal part of the p15E gene. Since the pathogenic MCFs contain two substituted regions separated by an *Akv*-derived sequence (the N-terminal part of the p15E gene), their formation seems to require two substitution events. In contrast, the

TABLE 2. *Structure and properties of Akv-derived MuLVs*

Virus	Substituted[a] in					Interference group	Pathogenic in	
	N-term gp70	C-term gp70	N-term p15E	C-term p15E	U3[b]		AKR	C3H
Pathogenic AKR MCFs	+	+ or −	−	+ or −	+	MCF	+	−
Class II MCFs	+	+	+	+	−	MCF	−	−
A^+L^- MCFs	+	?	−	+[c]	?	MCF	−	?
Gross ecotropic MuLV	−	+	−	+	+	Eco	+	+
GPA-V2	?	?	?	?	?	MCF	+	+
SL3-2	+	+	−	±[d]	+	MCF	+	+
SL3-3	−	−	−	±[d]	+	Eco	++	+

[a]By fingerprinting and mapping; only AKR-247 has been sequenced.
[b]Only one oligonucleotide marker in U3.
[c]Nonecotropic substitution extends further into p15E from 3′ side than in pathogenic AKR MCFs.
[d]One oligonucleotide altered at one base.

class II, nonpathogenic MCFs derived from nonthymic tumors of congenic mice are apparently substituted throughout the gp70 and p15E coding regions but not in the U3 region (126).

The A^+L^- MCF genomes closely resemble those of the pathogenic MCFs. However, all the pathogenic MCFs studied by Lung et al. (126) share the RNAse T_1 oligonucleotide 18, located 797 nucleotides from the 3′ end of the provirus, with their *Akv* parent. In contrast, three of four A^+L^- isolates lack this oligonucleotide, although they contain the XbaI site of *Akv* at 854 nucleotides from the 3′ end of the genome. Thus the nonecotropic sequence at the 3′ end of the genomes of these A^+L^- isolates appears to begin between nucleotide -854 and -797, while that in the pathogenic MCFs begins to the 3′ side of residue -797 (M. L. Lung, P. V. O'Donnell, and N. K. Hopkins, *personal communication*).

The pathogenic viruses Gross ecotropic MuLV, SL3-3, and SL3-2 all contain a common alteration in the C-terminal region of the p15E gene and a small change in one oligonucleotide at the 5′ end of U3. They are all indistinguishable from *Akv* in the N-terminal part of the p15E gene; in addition, SL3-2 appears substituted throughout the gp70 gene, while Gross ecotropic MuLV contains some differences from *Akv* in the 3′ region of the gp70 gene (155). As noted, these MuLVs have probably been subjected to more complex selection prior to their isolation than have the pathogenic and class II MCFs.

Three additional studies on viruses in AKR mice should also be mentioned. The first is the analysis by Thomas and Coffin (197) of viruses obtained by cocultivation of thymocytes from young AKR mice with NIH/3T3 cells. Using RNAse T_1 mapping, the authors found that well before the emergence of MCFs, AKR thymocytes produce MuLVs that differ from *Akv* in the C-terminal portion of the p15(E) coding region and frequently in U3 as well, but not in the gp70 gene. The non-*Akv* oligonucleotides in these viruses are the same as those found in the corresponding region of pathogenic MCFs. The authors propose that these MuLVs are intermediates in the formation of pathogenic MCFs, having undergone one substitution (at the 3′ end of the genome) but not the other (at the 5′ end of the gp70 gene). They also point out that if these two substitution steps are separated in time, the two substituted sequences may come from different endogenous genomes.

The second investigation is the analysis by Herr and Gilbert (83) of somatically acquired proviruses in the DNA of AKR thymomas. They used Southern blotting with small subclones of the *Akv* genome as probes to map individual integrated proviruses in a series of AKR thymomas. Their results can be summarized as follows: (a) Between three and 20 somatically acquired proviruses (i.e., genomes found in the tumor but not in the germ line of the mice and containing at least some sequences from *Akv*) were found in the tumors. Most of the proviruses in a given tumor appear to be identical, as if each tumor is initially infected by a single MuLV, and then is repeatedly reinfected by the progeny of this MuLV before viral interference is established and further infection is prevented. [Similar results have been obtained in studies on productively infected fibroblasts in culture (192).] (b)

Ninety percent of the genomes found in the tumors are substituted in the N-terminal region of gp70; every tumor analyzed has several such (presumably MCF) genomes. (c) All tumors contain at least one genome with the C-terminal portion of gp70 from *Akv*. (d) In two of eight tumors, all the new proviruses were derived from *Akv* in the C-terminal region of p15E. (e) All the new genomes in all the tumors were from *Akv* in the N-terminal part of p15E.

Finally, Cloyd (35) has analyzed the target cell specificity of pathogenic and class II MCFs. He injected young AKR mice with these viruses and then determined the organ localization and phenotypic characteristics of MCF-producing cells in the infected animals. His results show that the pathogenic MCFs are produced by cells from the thymus, while no infectious centers were found among spleen, lymph node, bone marrow, brain, or pancreas cells. Thus these MuLVs exhibit a 10^3- to 10^4-fold selectivity for thymocytes *in vivo*. The properties of the infectious centers indicate that they are immature T-lymphocytes present in the thymic cortex. In contrast, the class II MCFs (which, as noted above, replicate poorly *in vivo*) were produced at a low level in the spleen, lymph nodes, and marrow but not in thymus or nonlymphoid organs. The infectious centers in these mice appeared to be B-lymphocytes, although some lacked surface Ig and may have been pre-B-cells. This report extends earlier findings suggesting that thymotropism is a prerequisite for pathogenicity in AKR mice (37,146). It is also interesting to note that some replication of class II MCFs can be detected in B-lymphocytes, since most of the lymphomas that eventually appear in congenic mice are B-cell tumors. Since all these MCFs were isolated and propagated in standard monolayer cultures, it would appear that the strict target organ specificity observed *in vivo* is not maintained *in vitro*.

It should be obvious from the above discussion that the mechanisms by which viruses cause AKR leukemias are not yet understood. However, several summary observations and speculations are possible:

1. The consistency of the changes in the pathogenic MCFs suggests that two alterations in *Akv* are required for leukemogenicity: one in the specific receptor binding region of gp70 and the other in the extreme 3′ portion of the genome (i.e., the U3 region of the LTR and/or the 3′ end of the p15E coding region). Between these domains, however, the 5′ end of the p15E gene apparently must come from *Akv*. This conclusion is substantiated by the observation of Herr and Gilbert (83) that most of the somatically acquired proviruses in thymomas conform to this pattern. If this is true, then the pathogenic ecotropic MuLVs (i.e., Gross ecotropic MuLV and SL3-3) might act by generating such MCFs with high efficiency, since their RNAse T_1 oligonucleotide maps suggest that they may have already undergone a recombinational event at the 3′ end of the genome. Perhaps they also contain sequence homologies that facilitate the second event. As noted above, MCFs are present in tumors induced by these viruses. Thus the viruses detected in young AKR mice by Thomas and Coffin (197) might be analogous to these pathogenic ecotropic MuLVs.

2. Why should the presence of these three regions (non-*Akv* in N-terminal gp70, *Akv* in N-terminal p15E, and non-*Akv* at the 3′ end of the genome) result in a virus that causes thymic lymphomas? It is clear that *Akv* cannot infect the majority of T-cells in the mouse, and it probably cannot infect progenitors of the tumor clone. Gisselbrecht et al. (64) showed that few if any AKR thymocytes or splenic T-cells are infectious centers in the XC test. Also, as noted above, AKR thymomas contain almost no unmodified *Akv* genomes among their somatically acquired proviruses (26,83,158,218). Thus one or both of the substitutions may be required in order to form a virus that can infect thymocytes and/or the target cells.

3. One relatively simple model would be that thymocytes (and/or the target cells) lack ecotropic MuLV receptors. Thus the altered gp70 would be required in order to infect the cells via their MCF receptors, while the altered 3′ end of the genome would be required for the leukemogenic transformation, perhaps by promoter insertion. This model has the appeal of analogy with the avian system, in which avian leukosis viruses apparently transform B-lymphocytes by integrating near the c-*myc* gene, so that the enhancer sequences in the U3 region of the LTR activate *myc* transcription (80,153). The presence of some ecotropic genomes in thymoma cells [the ecotropic SL viruses; 10% of the genomes noted by Herr and Gilbert (83)] then could be explained by entry of phenotypically mixed particles via the MCF receptor before viral interference is established. On the other hand, this model is not simply reconciled with the observation (197) that thymocytes of young AKR mice, i.e., several months before the appearance of MCFs, produce viruses with the *Akv* gp70 gene but with a substitution at their 3′ end. The model also provides no explanation for the requirement for a substitution at the 3′ end of the genome. One possibility is that the 3′ substitution, for unknown reasons, is a prerequisite for the substitution in gp70. Another is that the enhancers display cell type-specific (62,102,103) as well as species-specific (113) variation in activity, and that the *Akv* U3 region is incapable of activating cellular oncogenes in thymocytes. It would be of interest to compare the enhancer activities of MCF and *Akv* LTRs in both monolayer and lymphoid cells.

4. The proposal that the enhancer elements show cell type specificity raises the alternative possibility that it is the substituted U3 region that is required for successful infection of thymocytes and/or the target cells. In that case, it might be the altered gp70 that changes the physiology of the infected thymocyte or target cell to increase the risk of malignant transformation. The appeal of this model lies in its analogy with the spleen focus-forming virus (SFFV) component of Friend virus, as discussed below.

5. One potentially crucial observation is the existence of the A^+L^- MCFs, which replicate in the thymus but do not accelerate AKR leukemia (146). This finding suggests that pathogenic MCFs have a transforming function, which is distinct from their ability to infect thymocytes successfully. It is potentially consistent with either of the models discussed above. As noted above, however, A^+L^- MuLVs do not retard the development of thymomas, and thymomas appearing in A^+L^- MCF-infected mice produce pathogenic MCFs. It is difficult to understand why one

MCF, replicating the thymocyte population, does not block the spread of another by viral interference. Perhaps the target cell, i.e., the progenitor of the tumor cell, is a special, rare cell type that is not infected by A^+L^- MuLVs, unlike the majority of thymocytes. It would be important to know whether or not the thymoma cells in these mice are infected with the A^+L^- MCF as well as the pathogenic MCF.

6. All the pathogenic viruses listed in Table 2 contain *Akv* sequences in the N-terminal region of p15E; furthermore, the somatically acquired proviruses detected by Herr and Gilbert (83) retain this portion of the *Akv* genome. Both these findings strongly suggest that this region is required by leukemogenic viruses. The analysis of A^+L^- MuLV genomes also suggests that sequences in this region are major determinants of pathogenicity. Far too little is known about both structure-function relationships in transmembrane proteins and mechanisms of viral leukemogenesis for us to be able to imagine a connection between the two. One trivial explanation would be that AKR mice (unlike NFS mice, which give rise to class II viruses) lack endogenous genomes capable of providing a functional N-terminal portion of the p15E gene.

Several laboratories are now actively engaged in constructing recombinants between pathogenic and nonpathogenic MuLVs *in vitro*. Hopefully, analysis of these viruses should help identify the genes responsible for disease induction by the *Akv*-derived viruses. Recent results by Lenz and Haseltine (120) suggest that the *env* region is not a crucial determinant of pathogenicity.

FRIEND VIRUS

The role of recombinant viruses in leukemogenesis has also been studied intensively in the group of viruses known as the "Friend complex." Whereas *Akv* and its derivatives generally cause leukemias of the T-lymphocyte lineage, the Friend complex is principally (although not exclusively) associated with erythroid leukemias.

Friend complex contains three distinct types of MuLV: a nondefective ecotropic MuLV (F-MuLV), a nondefective MCF (Fr-MCF), and the replication-defective SFFV. Each of these is discussed in turn.

F-MuLV was initially isolated from the Friend complex by Steeves et al. (190) and subsequently was reisolated by Troxler and Scolnick (199) and MacDonald et al. (127). When newborn mice of susceptible strains (e.g., NIH Swiss or BALB/c) are inoculated with F-MuLV, their livers and spleens become infiltrated within 3 to 6 weeks by hyperbasophilic blast cells, resulting in hepatosplenomegaly and severe anemia. These blast cells, which also appear in peripheral blood, are in the erythroid lineage since they stain with antibodies to the erythroid protein spectrin and form colonies in methylcellulose if erythropoietin is added to the culture system (127). The size of these colonies, and the fact that cells in the colonies continue to divide for 7 to 8 days, indicates that they arise from immature erythroid precursors, possibly similar or identical to erythroid burst-forming units (137).

Although the disease caused by F-MuLV in susceptible newborn mice represents extraordinary proliferation of erythroid precursors, the question is frequently raised as to whether or not this is actual neoplastic growth. Cells from the diseased mice are not irrevocably committed to uncontrolled growth, since they do not continue their proliferation when transplanted into syngeneic mice or when placed in culture. Oliff et al. (149) reported that the disease can regress if the mice are hypertransfused with red blood cells from normal donors. It is likely, therefore, that the primary effect of F-MuLV in these mice is to block the normal course of erythroid differentiation. This block results in a severe anemia; the proliferation of the precursors then occurs as a normal response to overproduction of signals for erythroid growth and differentiation as the animal attempts to compensate for its anemia. However, the proliferation may also reflect a heightened responsiveness in the infected precursors.

In addition to the erythroid proliferation discussed above, the replication of F-MuLV in susceptible newborn mice is accompanied by the appearance of Fr-MCF (200). Indeed, MCF can be detected in these spleens as early as 1 week after inoculation of F-MuLV (99). Fr-MCF is a typical member of the MCF group of MuLVs with respect to its biologic properties *in vitro*, since it infects both mouse and mink cells, induces cytopathic foci on mink cells (200), and uses the MCF receptor on NIH/3T3 cells (165). In addition, it shares MCF-specific antigens with Moloney- and *Akv*-derived MCFs (36,177); also, like other MCFs, it encodes an *env* polyprotein that is slightly smaller than that of its ecotropic parent (176).

The rapid emergence of Fr-MCF in mice inoculated with F-MuLV naturally raises the question as to whether Fr-MCF could be the proximal pathogen in disease induction by F-MuLV. When newborn NIH Swiss or BALB/c mice are injected with Fr-MCF, they do not develop disease (176). However, if the Fr-MCF is prepared as a phenotypically mixed stock from cells coinfected with wild mouse amphotropic MuLV 4070A, then it does cause splenomegaly and anemia in some mice. (The requirement for coinfection with a second virus is not fully understood. Wild mouse amphotropic MuLV 4070A does not cause disease under these conditions. It may serve to protect Fr-MCF from inactivation by factors in mouse serum, as discussed above in connection with *Akv*-derived MCFs; it would be of interest to measure the replication of Fr-MCF in mice after solitary or mixed infections.) It should be noted that even in the presence of the amphotropic MuLV, the disease appears more slowly and with a lower incidence than that induced by F-MuLV (176).

This limited pathogenicity of Fr-MCF is consistent with the idea that this virus is the proximal pathogen in the early erythroproliferative disease induced by F-MuLV, but that F-MuLV is a more effective helper virus *in vivo* than is wild mouse amphotropic MuLV. On the other hand, the data are also compatible with the hypothesis that Fr-MCF is a weak pathogen that plays no significant role in disease induction by F-MuLV. An additional phenomenon that may be important in this regard is the resistance of DBA/2 mice to this disease. When F-MuLV is inoculated into newborn DBA/2 mice, it replicates as well as in NIH Swiss mice; however, this early replication is not accompanied by MCF production, and the mice do not develop splenomegaly or anemia. This resistance to both MCF replication and disease induction is dominant in crosses between DBA/2 and susceptible strains of mice (176). The lack of detectable Fr-MCF production in DBA/2 mice *in vivo* is reflected in the specific resistance of fibroblastic cells from DBA/2 embryos or tails to infection with MCFs *in vitro* (10,78). This resistance is not absolute but ranged from 25- to 250-fold in whole embryo or fibroblast cultures (78) and was high (≥ 1,000-fold) in a 3T3-like cell line established from DBA/2 embryo cells (10). In turn, this specific resistance to infection by MCFs is associated with the expression, in uninfected DBA/2 cells, of an endogenous MCF-related gp70. This *env* protein is synthesized constitutively in the apparent absence of other MuLV proteins and is found both in bone marrow cells and in the 3T3 cell line. Expression of this protein on bone marrow cells is also dominant in DBA/2 × BALB/c F1 mice (176).

Biologic studies with the cell line have shown that this protein is on the cell surface and blocks the MCF receptor by a mechanism directly analogous to standard viral interference (10). Hartley et al. (78) have found that the gene for resistance of DBA/2 fibroblasts to MCFs maps to chromosome 5. A similar resistance to the F-MuLV early erythroproliferative disease and to MCF infection in cell culture is also found in CBA/N mice (78,176) and probably is present in several other strains as well. One discrepancy between work in these two laboratories concerns C57BL/6 mice, which expressed the MCF-related gp70 in bone marrow and spleen and did not replicate MCF *in vivo* (176) but whose fibroblasts were permissive for MCF infection *in vitro* (78). As noted by Hartley et al. (78), this discordance may cast doubt on the connection between the *env* protein and the resistance; however, it may also reflect a different pattern of cell type-specific expression in C57BL/6 mouse cells. In conclusion, although further work is necessary to resolve these questions, it is possible that the association between resistance to MCFs and resistance to disease will finally show that Fr-MCF is the true pathogenic agent in the early disease induced by F-MuLV. This resistance may extend to other MCF-associated diseases (30).

F-MuLV is unique among the known ecotropic MuLVs in causing the rapid proliferation of erythroid precursors discussed above. What genes are responsible for this pathogenic phenotype? This question can now be systematically studied with recombinant DNA techniques. Using a "marker rescue" approach and exploiting spontaneous recombination between viral genes in NIH/3T3 cells, Oliff et al. (147) isolated an MuLV whose *gag* and *pol* genes are principally derived from amphotropic MuLV 1504A but whose gp70 is at least partly from F-MuLV. The origin of the LTR in this recombinant is unknown. This recombinant produced a disease in newborn NIH mice similar to that induced by F-MuLV, although the incidence was lower. The pathogenicity of this virus suggests that it is the *env* gene, and/or the LTR, of F-MuLV that carries its disease-inducing property.

In an extension of these studies, Oliff and Ruscetti (148) constructed a recombinant between F-MuLV and amphotropic MuLV 4070A by ligation of subgenomic fragments of cloned DNAs *in vitro*. The only F-MuLV sequences in this recombinant are 700 bp to the 5′ side of gp70, the entire gp70 gene, and the N-terminal 4/5 of p15E. This recombinant causes hepatosplenomegaly and anemia in newborn NIH Swiss mice, but the incidence of the disease is lower and the latency longer than with F-MuLV. The proliferating cells in the spleens are not transplantable (A. Oliff, *personal communication*). These results suggest that the pathogenicity of F-MuLV is determined in part by its *env* gene, but that other regions of the viral genome are also involved. Finally, it was noted that this ecotropic recombinant generates a second virus *in vivo*, which has several properties of MCFs; thus the *env* gene, or the 3′ end of the *pol* gene, of F-MuLV may also contain the information necessary for recombination with endogenous *env* sequences to form a MCF.

The foregoing discussion was limited to the early proliferation of erythroid precursors which is induced in newborn mice by F-MuLV. It was pointed out that this proliferation probably should not be considered malignant growth, since the cells are not transplantable *in vivo* or *in vitro* and the proliferation can be controlled by reversing the anemia with transfusions of mature red blood cells. In addition, it was noted that some mouse strains, such as DBA/2, are quite resistant to this disease. Thus there are some situations in which mice can survive for an extended period after infection with F-MuLV; a number of laboratories have now begun to analyze the events that occur only several months after infection.

Oliff et al. (149) prolonged the survival of F-MuLV-infected NIH Swiss mice by transfusions of red blood cells. They found that by 14 to 16 weeks after inoculation, the spleens of some of these mice contained cells that were transplantable to syngeneic adult mice. By serial passage of these cells, the authors were able to develop a cell line that is transplantable as a leukemia *in vivo* and grows well *in vitro* and thus appears to be malignant. The morphology of these cells and the fact that they stain with antispectrin antibody suggest that they are in the erythroid lineage. Similar results with BALB/c and ICFW mice were also reported by Shibuya and Mak (184) and by Wendling et al. (213), respectively, except that in the latter case, myeloid, lymphoid, and undifferentiated, as well as erythroid, malignancies were obtained.

Shibuya and Mak (184), Wendling et al. (213), and Chesebro et al. (32) reported that DBA/2 mice begin to develop leukemias several months or more after infection with F-MuLV as newborns. Significantly, these tumors were not restricted to the erythroid lineage: some mice had myelomonocytic or lymphoid, rather than erythroid, leukemias. It was possible to establish fully malignant tumor cell lines from such myelomonocytic leukemias.

Wendling et al. (213) and Chesebro et al. (32) also examined the latencies and types of disease induced in a number of other mouse strains by F-MuLV. They found that C57BL/6 or C57BL/10 mice showed an even longer latent period than DBA/2 but also ultimately developed malignant leukemias with myeloid or lymphoid phenotypes. Chesebro et al. (32) tested these leukemias for MCF production and detected MCF in three of six DBA/2 leukemias. Thus MCFs can arise in DBA/2 mice; once they arise, the block to their replication is not absolute. In addition, these authors found that C57BL/10 tumors never contained detectable MCF or MCF-specific antigens; interestingly, no erythroid tumors were found among the C57BL/10 leukemias.

In summary, the effects of F-MuLV can be divided into early and late phenomena. The early effect is a nonmalignant, specifically erythroid hyperplasia; it may be mediated by Fr-MCF rather than F-MuLV itself. Animals that do not undergo, or that survive the early hyperplasia invariably contract a truly malignant "late" disease. The late disease is not specifically erythroid but may be any one of a broad spectrum of leukemias. The findings of Chesebro et al. (32) suggest that Fr-MCF is not involved in the late induction of myeloid or lymphoid leukemias; its role in late erythroleukemia is uncertain at present.

The third member of the Friend complex, which is undoubtedly the most thoroughly investigated, is SFFV. SFFV has many of the properties of acute transforming viruses and thus is beyond the scope of this chapter; however, some aspects of its biology are noted because of its close sequence relationship to the MCFs.

SFFV induces the formation of macroscopic colonies of erythroid precursors on the spleens of susceptible mice, even when these mice are inoculated as adults. The cells in these colonies are actively engaged in functions characteristic of erythroid differentiation, including the synthesis of hemoglobin (194). The proliferation and differentiation of these cells are largely independent of the normal signals for erythrocyte production, since they continue in mice that have been hypertransfused (194). The cells in the foci do not appear to be malignant, however, since they cannot be transplanted *in vivo* or grown as cell lines *in vitro* (21,61,195).

The growth and differentiation that produce the foci are direct results of SFFV infection on its target cells, since they can be demonstrated *in vitro*: Infection of bone marrow cells with Friend complex causes the appearance of cells that can be grown into erythroid colonies in the absence of erythropoietin (33,73). This effect is due to the SFFV component, since it is also induced by SFFV rescued from fibroblasts with Moloney MuLV (75).

Recombinant DNA technology has now been applied to the analysis of the SFFV genome. The genome contains a partial *gag* gene, a partial *pol* gene, and a partial *env* gene between its LTRs (15,124,217). Several lines of evidence, including "marker rescue" experiments (125) and *in vitro* mutagenesis (123), indicate that the *env* gene is required for spleen focus formation, while the *gag* and *pol* genes, and at least the 3′ portion of the LTR, are not.

The sequence of the SFFV *env* gene has recently been determined (34,216). It can be considered a deleted MCF *env* gene. Thus it shows close homology with the Moloney MCF *env* gene (16), changing from a nonecotropic to an ecotropic sequence approximately 280 amino acids from the initiator methionine. However, a large internal deletion in SFFV removes approximately 109 amino acids from the C-terminal end of gp70 and approximately 87 amino acids from the N-terminal end of p15E. The product of this gene is the *env*-related protein gp52 (48,159,177–179), which is present in all SFFV-transformed cells and appears to be largely or entirely responsible for the transforming activity of SFFV.

In considering the mechanisms by which gp52 might induce proliferation and differentiation of erythroid precursors, it is crucial to remember that this protein is not released from the cell into virions (48,178). Biologic measurements also indicate that SFFV proteins are not functional *env* proteins, since they do not extend the host range of MuLVs synthesized in cells with SFFV (A. Rein, *unpublished*). Thus since gp52 apparently plays no role in the process of infection of cells by SFFV particles, it is likely that it contributes to spleen focus formation by means of a direct physiologic effect on the target cell.

OTHER VIRUSES

Moloney MuLV

Possible mechanisms of leukemogenesis by Moloney MuLV are considered in depth elsewhere in this chapter. It should be noted, however, that Moloney MuLV was the source of the first MCF to be described (60), and that this MCF causes lymphomas as efficiently as Moloney MuLV itself (59). Although tumors induced in mice by Moloney MuLV always contain substituted (presumably MCF) Moloney MuLV-derived proviruses (202), at least one Moloney MuLV-induced rat lymphoma appears to contain only the input, ecotropic Moloney MuLV genome (D. Steffen, *personal communication*).

Rauscher MuLV

Although the biology of the Rauscher virus complex has not been as intensively studied as that of Friend virus, there are many striking parallels between the two systems. Biologically purified Rauscher MCF induces hepatosplenomegaly, erythroblastosis, and, at a low frequency, thymic lymphomas (203,204). The N-terminal amino acid sequence of its gp70 has been determined and is almost identical with those deduced from nucleic acid sequences of Moloney- and *Akv*-derived MCFs and of Friend SFFV (180).

MuLVs from HRS/J Mice

HRS/J mice, which are homozygous for the *hr* gene, resemble AKR mice in a number of respects, including lifelong production of an ecotropic MuLV (84) and the appearance in mature mice of "polytropic viruses" (presumably MCFs) and thymic lymphomas. The polytropic viruses accelerate spontaneous leukemogenesis upon reinjection into newborn HRS/J mice and also induce thymic lymphomas in the low leukemia strain CBA/J. As noted above, oligonucleotide mapping shows that the substituted sequences in the *env* gene of the polytropic viruses are similar but not identical to *env* sequences of HRS/J xenotropic MuLVs (68).

Recombinants Derived from Amphotropic MuLV

The recombinants discussed thus far have arisen by somatic recombination between endogenous *env* sequences and ecotropic MuLV. However, Rasheed et al.

(162) have recently reported the appearance of a recombinant MuLV in NIH Swiss mice inoculated with wild mouse amphotropic (strain 1504A), rather than ecotropic, MuLV. This recombinant, designated 10A/1, is highly pathogenic, inducing splenic lymphomas within 1 to 2 months after injection into newborn NIH Swiss mice. The lymphomas lack both T and B cell markers and thus were termed "null cell" lymphomas (162).

Partial tryptic peptide mapping showed that the gp70 of 10A/1 shares some peptides with that of 1504A MuLV, its presumed parent, while other 1504A peptides have been replaced with novel peptides (162). Similarly, oligonucleotide mapping demonstrated that 10A/1 contains a substitution in the 5′ part of its *env* gene, relative to 1504A MuLV; it was also found that some of the oligonucleotides in the substitution are present in the xenotropic MuLV AT124 (112), which was originally isolated from NIH Swiss mice (198). No further molecular information on 10A/1 is available as yet.

Viral interference tests showed that 10A/1 MuLV enters NIH/3T3 cells by a unique receptor, not used by any other known MuLV. In addition, these tests indicated that 10A/1 MuLV still retains its affinity for the amphotropic MuLV receptor (A. Rein, *unpublished*). The fact that 10A/1 uses a unique receptor shows it to be clearly distinct from, but analogous to, the MCF MuLVs discussed in the bulk of this chapter. This finding also shows that there are at least two classes of endogenous *env* genes in mice. One class encodes the MCF and the other encodes the 10A/1 receptor specificities detected in recombinant viruses. Furthermore, since recombinants derived from ecotropic MuLVs appear to all use the MCF receptor, while a recombinant from amphotropic MuLV uses the 10A/1 receptor, there apparently is a specificity in the acquisition (or selection) of endogenous *env* genes *in vivo*. Finally, the pathogenicity of this distinct recombinant MuLV raises anew on the question of why such recombinant MuLVs cause disease.

GENERAL CONCLUSIONS

Our present state of knowledge of the role of recombinant MuLVs in murine leukemogenesis can be summarized as follows. First, certain MCFs can induce or accelerate lymphoid neoplasias. Since there is always a lag between infection and appearance of disease, it is likely that the viruses do not directly transform the cells but rather put the cells at risk for subsequent transforming events (e.g., chromosomal changes). MCFs appear not to be absolutely necessary for leukemogenesis, at least in rats, according to findings of Steffen (see above for discussion of Moloney MuLV). Data presently available indicate that the MCF gp70 and/or the MCF LTR, perhaps acting in concert with ecotropic sequences in p15E, are responsible for leukemogenesis.

Second, recombinant MuLVs derived from Friend MuLV affect the maturation of cells in the erythroid lineage. Thus the deleted MCF-type *env* protein of SFFV appears to alter late erythroid precursors in order to reduce their dependence on the normal physiologic signals for growth and differentiation. Friend MCF also may alter early erythroid precursors in order to block their maturation. In both these cases, a hyperplasia results; this early hyperplasia is followed by frank malignancy, which need not be erythroid. MCF apparently is not a requirement for the later malignant growth, since it is not found in C57BL10 leukemias (32).

MCFs may contribute to leukemogenesis by initiating hyperplastic growth in hematopoietic cells. The example of SFFV raises the possibility that alteration in growth control by MCFs is mediated by the MCF *env* gene product, but there is no direct evidence for this hypothesis.

One area that is not understood, and which should be amenable to experimentation, is the cell type specificity of virus-induced disease. Thus the pathogenic *Akv*-derived MCFs exhibit a high specificity for infection of thymocytes *in vivo* and are found to accelerate thymic lymphoma, while the immediate effects of F-MuLV + Fr-MCF and of SFFV are limited to erythroid cells. These specificities may play a crucial role in determining the pathogenic potentials of different MuLV isolates, but they have been difficult to analyze because they are not evident in standard monolayer cultures. Study of recombinants constructed *in vitro* likely will provide some insight into the mechanisms of specificity for infection of and pathogenesis in different cell types.

POSSIBLE IMMUNOLOGIC MECHANISMS IN MURINE LEUKEMOGENESIS

Initially, the concept was held that when mice were inoculated as newborns with various retroviruses, or when genetically acquired viruses were expressed early in life, such as in AKR mice, there would be immunologic tolerance to the virus. The lack of validity of this concept has been demonstrated in a variety of studies (90, 141). In situations of viremia and/or low grade expression of endogenous viruses, both cellular and humoral immune responses are readily detectable (52,88, 91,116,117). The immune responses have been shown to be directed against predominantly the envelope proteins gp70 and p15(E), although in certain situations, immune responses against the internal proteins p12, p15, and occasionally p30 have been demonstrated. Under conditions that do not result in the establishment of viremia, the immune responses are primarily associated with a readily detectable antibody response, which plays a major role in suppressing virus spread by the ability to neutralize virus infectivity. Under conditions that result in viremia, free circulating antibody is not detectable, although by standard techniques for the establishment of hybridoma cell lines, viral antigen reactive B cells can be isolated from splenic lymphocytes of viremic mice (J. N. Ihle and L. O. Arthur, *unpublished data*).

In addition to antibody responses, cellular immune responses against retroviruses occur which are potentially of more significance to leukemogenesis. Under conditions that do not lead to the establishment of viremia, a cellular immune response is transiently detectable and is demonstrable *in vitro* by the presence of antigen-specific T cells, which either mediate cytotoxicity or can mediate T cell blastogenesis. In both cases, the immunologic specificity appears similar to that observed in the B cell response. The predominant antigen involved in either cytotoxicity or T cell blastogenesis is gp70. The cellular response against p15(E) has not been studied in detail because of the unavailability of the purified protein. In addition, T cell blastogenic responses are also detectable against p12 with certain retroviruses. Under conditions of virus expression which lead to establishment of viremia and are associated with leukemia, a cytotoxic T cell population generally is not detectable *in vitro*. In contrast, in viremic mice, there generally is a T cell population which is detectable in T cell blastogenesis assays and which persists in such mice throughout most of the preleukemic period (116,117).

A role for a cellular immune response in leukemia was suggested by the observation that *in vivo*, the persistence of T cells reactive in *in vitro* blastogenesis assays was not typical of normal immune responses against retroviruses and was correlated with conditions leading to leukemia. In particular, such T cells only persisted *in vivo* under conditions in which viremia was maintained. This correlation with leukemia was found to occur with both AKR mice and BALB/c mice inoculated as newborns with Moloney leukemia virus (MoLV). The most compelling evidence came from a series of studies which compared the oncogenicity of MoLV in a variety of strains of mice with the ability to induce a cellular immune response. Among a variety of strains, there was an absolute correlation among the ability to establish viremia, the presence of T cells reactive in a blastogenesis assay, and the development of leukemia. The one exception was the CBA/N strain in which MoLV-inoculated mice developed an acute viremia comparable to that seen in other strains but in which T cells detectable in blastogenesis assays were not detectable (118). Moreover, MoLV-inoculated CBA/N mice rarely developed lymphomas. The lack of leukemias was not due to the absence of recombinant viruses which were in the initial virus stock used and which were detectable *in vivo* at levels comparable to that observed in BALB/c mice (J. Ihle and P. Fischinger, *unpublished data*). These results demonstrated that in MoLV-induced leukemia, neither extensive virus replication nor the presence of recombinant virus was sufficient for induction of leukemia and strongly implicated that an immune response was also required.

The potential role of a cellular immune response in leukemia became more apparent as fundamental information developed concerning the events generally associated with immune responses. In particular, the available *in vitro* assays can detect two unique subpopulations of T cells. Cytotoxic T cells are characterized by the expression of the cell surface markers Thy-1 and Lyt-2. Their regulation appears to involve a sequence of events which is initiated by antigen stimulation of a precursor lymphocyte to yield an activated T cell. The subsequent proliferation of this activated T cell is dependent on a specific lymphokine termed T cell growth factor (TCGF) or interleukin 2 (IL 2) (18,63,132,133,207). This expansion of functional cytotoxic cells does not require antigen and cannot be indefinitely sustained with IL 2; in the absence of IL 2, the activated cytotoxic T cells quickly lose viability.

The second antigen-specific T cell population detectable *in vitro* is that mediating the T cell blastogenesis reaction. This lymphocyte, as with cytotoxic cells, is antigen-specific but cannot mediate a cytolytic function and is distinguished by expressing the cell surface marker Lyt-1 rather than Lyt-2. Initially, it was assumed that in a standard blastogenesis assay, the antigen-specific Thy-1^+, Lyt-1^+,2^- lymphocyte bound antigen and was subsequently induced to proliferate. This simplistic interpretation, however, is not correct. In particular, Thy-1^+, Ly 1^+,2^- lymphocytes produce a variety of factors, including IL 2, as a consequence of binding antigen (51,53,63,66,95,136,182). The production of these lymphokines does not require proliferation.

The majority of the proliferation observed in blastogenesis assays is largely a nonantigen-specific subpopulation of cells responding to the lymphokines. Whether the antigen-specific Thy-1^+, Lyt 1^+,2^- lymphocyte actively proliferates in response to the antigen in addition to producing lymphokines is not clear. It has been shown that for the proliferation of cloned lymphokine-producing T cells *in vitro*, antigen is required, unlike cloned cytotoxic T cells, which require only IL 2. Nevertheless, the results distinguish two functional subpopulations of antigen-specific T cells. One type mediates cytotoxicity, is dependent on antigen for activation, and is largely if not exclusively dependent on IL 2 for proliferation. The second population specifically produces a variety of lymphokines in response to antigen and is uniquely detected in blastogenesis assays *in vitro*. The factors required for the differentiation and/or proliferation of this class of T cell are not known, although roles for antigen and perhaps IL 2 have been suggested.

Imperative in understanding the physiologic significance of the antigen-specific T cells detected in T cell blastogenesis assays is an understanding of the mechanisms of action of some of the lymphokines involved. The biochemical and biologic properties of most of the factors produced by gp70-stimulated lymphocytes are not known. Using DEAE-cellulose columns, multiple peaks of activity are detected which induce proliferation of splenic lymphocytes from MoLV-inoculated preleukemic mice and induce proliferation only to a lesser extent with normal splenic lymphocytes (J. N. Ihle and J. Keller, *unpublished data*). In the response to gp70, three distinct factors have been identified among the multiple proliferation-inducing activities detected following DEAE cellulose chromatography: IL 2, interleukin 3 (IL 3), and a unique T cell colony stimulating activity (CSF-2).

As noted above, IL 2 specifically causes the proliferation or amplification of antigen-activated, differentiated cytotoxic T cells. In addition, it may cause the proliferation of antigen-specific T cells capable of producing lymphokines. In both cases, IL 2 predominantly causes proliferation without any evidence that differentiation might also occur. Also, the cells responding to IL 2 are functionally differentiated and represent terminal stages of T cell differentiation; conversely, no evidence exists to suggest that IL 2 regulates early phases of T cell differentiation. In contrast to IL 2, the physiologic significance of CSF-2 is not known. This factor induces bone marrow cells to proliferate *in vitro* and, under cloning conditions in soft agar cultures, gives rise to mixed colonies containing myeloid and macrophage-like cell types.

Among the lymphokines produced by antigen-specific lymphocytes, IL 3 has been of particular interest. It was initially identified as a lymphokine which induces the expression of an enzyme, 20-alpha-hydroxysteroid dehydrogenase (20αSDH), characteristically expressed in mature T cells (96,98,156,209–211). IL 3 has been purified to homogeneity and shown to be a glycoprotein with an apparent molecular weight of 28,000 daltons on SDS-PAGE (93). In addition to inducing 20αSDH in appropriate cultures of stem cells, IL 3 mediates a number of immunologic phenomena, including the following (87,92,94,97): (a) it induces the proliferation of a variety of cell lines derived from long-term bone marrow cultures; (b) it induces formation of colonies in soft agar cultures of bone marrow cells; (c) it promotes the differentiation of mast-like cells in long-term cultures of bone marrow or spleen cells; (d) it maintains *in vitro* a cell that, when given to lethally irradiated mice, can give rise to colonies in the spleen; and (e) it induces Thy-1 in cultures of bone marrow cells.

A detailed analysis of the events occurring in cultures of lymphocytes responding to IL 3 demonstrates that 20αSDH induction represents only the first step in a differentiation sequence. In particular, the first step in the sequence involves induction of 20αSDH in an appropriate stem cell population, which normally is found in the bone marrow. Following induction, the next event is the acquisition of the cell surface markers Ly 5 and H-11. From this population is also generated a Thy-1$^+$ prothymocyte-like population, which has been speculated to be a precursor for differentiation of more mature T cell populations (87). In addition, Thy-1$^-$, Ly 5$^+$, H-11$^+$, Ia$^+$ cells are generated, which subsequently give rise to either an adherent monocyte-like population or mast-like cells. In both cases, the proliferation and continued differentiation require IL 3; in its absence, cell viability is rapidly lost.

Much of the information concerned with the regulation of proliferation of lymphoid cells by IL 3 has come from studies utilizing a series of factor-dependent cell lines derived from long-term bone marrow cultures. These cell lines have been derived in two laboratories using long-term bone marrow cultures and were obtained by subculturing nonadherent cells from such cultures into conditioned media from WEHI-3 cells (46,69). Once established, these cell lines were found to be absolutely dependent on WEHI-3-conditioned media for growth *in vitro*. In subsequent studies, it was demonstrated that among the lines examined, all uniquely required IL 3 (92) for growth from among a variety of potential growth factors, including CSF-2 and IL 2 found in either WEHI-3-conditioned media or conditioned media from activated T cells. All the cell lines were characterized by the expression of the enzyme 20αSDH, although the cell surface phenotypes of the cell lines varied considerably. The phenotypes, however, were those observed in the sequence of differentiation promoted by IL 3 using normal lymphocyte populations. Based on these observations, it was proposed that the cell lines were transformed, in that they represent intermediates in a pathway of differentiation which are unable to terminally differentiate, as occurs with normal lymphocytes.

A number of the characteristics of the IL 3-dependent cell lines are of interest in the understanding of the physiologic effects associated with IL 3 production *in vivo*. As might be expected, all the cell lines examined have readily demonstrable cell surface receptors for IL 3, with K_ds of approximately 1×10^{-11} M (151). The IL 3 receptors are found on IL 3-dependent lymphoid cells and not on a variety of other hemopoietic cell types. They constitute a specific marker for identifying IL 3 lineage cells.

The requirements for IL 3 for proliferation of all the cell lines examined have been quite dramatic. In the absence of IL 3, the cells begin to lose viability after about 4 to 6 hr with a first order decay rate such that half the population is lost every 1 to 2 hr. Within the cell cycle, IL 3 is apparently required throughout the first portion of the G1 phase but is not required in S, G2, or M (J. N. Ihle and A. Scott, *unpublished data*). In none of the cell lines does there appear to exist a state in which viability is maintained in the absence of a commitment to proliferate. Therefore, *in vivo* the proliferation and continued differentiation of IL 3 lineage cells can be speculated to be absolutely dependent on a continual source of IL 3. More specifically, in the absence of IL 3, viability of any cells in the sequence would be rapidly lost. As noted above, since IL 3 is a product of antigen-activated Thy-1^+, Lyt-1^+,2^- helper T cells *in vivo*, the *in vivo* concentration of IL 3 is dependent on the presence of an appropriate helper T cell and sufficient concentrations of antigen to induce its production.

The chronic presence of a Thy-1^+, Lyt-1^+,2^- antigen-specific lymphocyte in retrovirus-infected mice has been shown to have a profound effect on normal lymphocyte proliferation and differentiation. In particular, *in vivo* the concentrations of viral proteins, such as gp70, are comparable to those that are required *in vitro* to induce the production of lymphokines by Thy-1^+, Lyt-1^+,2^- lymphocytes. Consequently, it could be proposed that *in vivo* one of the consequences of the extensive viremia and the presence of lymphocytes capable of producing lymphokines is a chronic production of lymphokines. In this regard, it should be noted that just as there exists a dose response curve for lymphokine production *in vitro*, subtle differences of the extent of viremia *in vivo*, which are known to influence leukemogenesis, could be expected to affect the extent of production of lymphokines (J. N. Ihle and A. Scott, *unpublished data*). Experiments to directly measure the levels of various lymphokines *in vivo* in preleukemic mice have consistently demonstrated that levels are not achieved that are sufficient to allow direct measurements. However, the function of several of the lymphokines produced is to promote the differentiation and/or proliferation of a variety of lineages of cells. This effect causes *in vitro* and *in vivo* an expansion of lymphokine-responsive populations. In MoLV viremic, preleukemic BALB/c mice, there is a readily detectable expanded population of lymphokine-responsive cells (119). This expansion constitutes a 50- to 200-fold increase throughout the preleukemic phase of lymphocytes capable of proliferating in response to the lymphokines produced by gp70-stimulated Thy-1^+,

Lyt-1^+,2^- lymphocytes. This expansion does not occur in MoLV-inoculated, viremic CBA/N mice, demonstrating that the expansion is dependent on the presence of an antigen-specific Lyt-1^+,2^- lymphocyte and is not due simply to viremia. As above, the resistance of CBA/N mice to leukemia suggests that the expansion observed in mice capable of responding immunologically to the virus may play a role in virus-induced leukemogenesis. The data suggest that as a consequence of the establishment of viremia and the continual production of lymphokines by antigen-specific lymphocytes, there is an increase in the number of lymphocytes of various lineages which are proliferating *in vivo* in response to specific lymphokines.

Based on the above observations, a model for virus-induced leukemia has been proposed which emphasizes the possible role of a chronic immune response. In particular, the results strongly suggest that the presence of a detectable population of antigen-specific lymphocytes capable of producing lymphokines in response to antigenic stimulation plays a major role in the events that ultimately result in leukemia. Several possibilities have been proposed. One is that the increased proliferation and/or differentiation induced by the production of lymphokines is important in providing a "target" cell population for virus infection and thus increasing the probability of virus replication-related events. In particular, a large target cell population may be required to allow a sufficient number of proviral integrations to occur to result in one with the proper configuration for the activation of potentially oncogenic sequences. Second, the proliferation may be required to increase the probability of a somatic event associated with DNA replication which might be associated with transformation. In particular, trisomy 15 may be the critical event or a translocation type of event comparable to that observed in plasmacytomas. In the latter case, it is interesting to note that the translocation and concomitant activation of *myc* have been postulated to occur as a consequence of a specific type of immunologic stimulation which occurs in response in injection of mineral oil.

A second distinct immunologically related mechanism for viral leukemogenesis is termed the "receptor-mediated" model of leukemia, which has been proposed by McGrath et al. (128–131,212). This hypothesis states that the transformed lymphocyte is a viral antigen-specific lymphocyte; following infection, it is continually induced to proliferate by the production of the antigen by the normally physiologic regulation mediated by antigen-specific cell surface receptors. The initial evidence for this model was the observation that fluorescenated viruses, derived from several lymphoma lines, bind to the surface of the lymphomas, suggesting the presence of an antigen-specific receptor. The specificity of binding of various viruses was such that this receptor does not appear to be the normal receptor used for virus infection of a variety of cell types but rather has a specificity more consistent with an immune specific receptor. In one case, the BCL_1 B cell lymphoma line, a surface Ig, was detected, which specifically binds the retrovirus produced by the lymphoma. Furthermore, monoclonal antibodies recognize idiotypic determinants on the isolated Igs and could inhibit the binding of virus to

BCL_1 cells. Seven of eight other spontaneous B cell lymphomas in this study were also found to express virus receptors. A few of these receptors were found to share a cross-reactive idiotype with the BCL_1 Ig. *In vivo* virus-binding thymocytes are not detectable by the assay throughout the preleukemic phase but appear with onset of disease, suggesting a strong correlation between the phenotype and leukemia. Unlike the above model of chronic immune stimulation, the receptor-mediated model involves only the antigen-specific subpopulations of lymphocytes. The prediction of the model is that primary lymphomas would only require the appropriate viral antigens for continued proliferation *in vivo* as well as presumably *in vitro*. Last, the phenotype of the lymphomas should be either B cell or a functionally mature helper T cell phenotype. Conversely, more immature lymphocytes, such as terminal transferase (TdT) positive lymphocytes or mature, IL 2-dependent T cell phenotypes, should not be included, since in neither case does there exist evidence for an antigen requirement for proliferation.

LINEAGE CHARACTERISTICS AND LYMPHOKINE RESPONSIVENESS OF PRIMARY RETROVIRUS-INDUCED LYMPHOMAS

The lineage characteristics of murine lymphomas induced by several murine retroviruses have been studied. Perhaps the first and most relevant point that has emerged from these studies is that a variety of lymphoid types are observed, including thymic lymphosarcomas, B cell tumors, splenic T cell lymphomas, and splenic null cell lymphomas (3–5,17,64,156,164,187). In AKR mice, the predominant lymphoma type involves a thymic tumor in which TdT positive, hydrocortisone-sensitive cells constitute the major cell phenotype. This type of lymphoma is also the predominant type in radiation-induced leukemia but is a relatively minor type in most other leukemias, including Moloney or Rauscher virus-induced leukemias and the tumors observed with a wild mouse ecotropic virus in NFS mice (115) (S. Morse and P. Hoffman, *personal communication*).

The possible relationships of TdT positive lymphomas to normal aspects of lymphocyte differentiation and regulation are largely unknown. Under normal conditions, TdT positive thymocytes constitute the major thymic cortical population. Although it was initially proposed (23) that they represented an immature population which differentiated to the immunologically more mature medullary population, several lines of evidence from recent studies have suggested that TdT positive thymocytes may be terminally differentiated (31,185,186). As a consequence of these studies, it has been proposed that this population, perhaps because of the expression of TdT, is undergoing clonal deletion due to the expression of potentially autoreactive lymphocyte receptors (13,97). Nevertheless, few data exist concerning the normal regulation of differentiation of this subpopulation of lymphocytes relative to the understanding of the types of alterations that may have occurred to give a transformed phenotype. One intriguing possibility is that transformation in some way alters the ability of TdT to be lethal, either by lowering the rate of transcription or by physically altering the enzyme to change its catalytic properties. Under these conditions, the autoreactivity might be speculated to provide the proliferative signal. In a highly specialized case in which a viral antigen is involved, this type of situation might also conform to the receptor-mediated hypothesis of McGrath and Weissman (130,131).

Several murine retroviruses, including AKR, MoLV, and RLV, induce B cell lymphomas. Interestingly, it has been reported that the predominant lymphoma induced in certain strains of mice by RLV are B cell lymphomas (115). Perhaps the most intriguing but as yet unaddressed question is whether there is a mechanistic relationship between murine retrovirus-induced B cell lymphomas and mineral oil-induced plasmacytomas, the B cell lymphomas induced by the Abelson acute transforming virus, or the B cell lymphomas induced by avian leukosis virus. In particular, this unique subset of lymphomas might be characterized by downstream promotion of c-*abl*. Alternatively, as noted above, the chronic inflammatory response induced by injection of mineral oil has been implicated in promoting B cell lymphomas by a mechanism involving a translocation of c-*myc*. As a consequence of the immune response against retroviruses, factors associated with this type of inflammatory response may be produced and occasionally allow the emergence of a comparable transformed phenotype.

A third group of retrovirus-induced lymphomas involve splenic T cell or null cell lymphomas. Unlike the thymic T cell lymphomas, these tumors involve a unique subpopulation of lymphocytes characterized by the lack of TdT and the presence of the T cell lineage-related enzyme 20αSDH (5,156). With some types of retroviruses, including MoLV and a wild mouse ecotropic virus (Cas-B-M), 60 to 80% of the lymphomas may be of this type. In contrast, among the lymphomas in AKR mice, only approximately 5% are of this type. The characteristic expression of 20αSDH in these lymphomas and the relationship of IL 3 to a lineage of cells characterized by the expression 20αSDH prompted studies to further examine this type of lymphoma. Typical results are illustrated in Table 3. Among this group of splenic lymphomas, four were highly positive for Thy-1 expression; another was positive, although the degree of fluorescence was lower; and one lymphoma was negative or only weakly positive. Among the tumors, three had extremely low levels of 20αSDH, whereas three were positive, suggesting that the latter might constitute IL 3 lineage-related lymphomas.

To further explore this possibility, the primary lymphomas were assayed for the ability to bind ^{125}I-labeled IL 3 as a general measure of the abundance of potentially IL 3-regulated lymphocytes. As indicated, a low level of ^{125}I-IL 3 binding was detectable with normal splenic lymphocytes. Binding activity was significantly increased in preparations of splenic lymphocytes from preleukemic mice. This increase correlates directly with the increases that occur in the lymphocytes capable of proliferating in response to IL 3. Among the lymphomas examined, the three lymphomas having the lowest levels of 20αSDH also had the lowest levels of binding activity. Among the other lymphomas, the null cell lymphoma had a high level of

TABLE 3. *Phenotypic and growth characteristics of MoLV-induced lymphomas*

Lymphoma no.	Tissue[a]	Thy-1[b]	pmoles/hr/10^8 cells 20αSDH	Percent ^{125}I-IL 3 binding/10^7 cells[c]	Growth *in vitro*[d]	
					Media	+IL 3
118	Spleen	++	46	0.65	—	—
119	Spleen	++	29	0.43	—	—
120	Spleen	++	62	0.95	—	—
121	Spleen	±	404	3.93	—	Cell line
122	Spleen	+	483	1.25	—	Cell line
123	Spleen	++	2736	1.30	—	Cell line
Normal	Spleen	ND	120	0.2	—	Limited growth
Preleukemic	Spleen	ND	370	0.72	—	Limited growth

[a]All lymphomas examined involved a primarily splenic localized tumor. BALB/c mice were inoculated as newborns with MoLV, and lymphomas arose after approximately 3 to 4 months. Preleukemic mice were approximately 2 months of age and were not overtly lymphomatous. Normal mice were 1- to 2-month-old BALB/c mice.

[b]Thy-1 expression determined by immunofluorescence.

[c]^{125}I-labeled IL 3 was used in a receptor binding assay to assess the general level of IL 3 receptor positive cells. Percentage binding is proportionally related to the frequency of receptor positive cells, although factors including receptor density could significantly distort the relative relationship among tumors.

[d]Growth *in vitro* was assessed by culturing lymphoma cells at a density of 2.0×10^6 cells/ml in RPMI-1640 containing 10% FCS with or without IL 3. Negative indicates no apparent growth. In the presence of IL 3, splenic lymphocytes from either normal or preleukemic mice will proliferate for approximately 4 to 6 weeks in culture but rarely give rise to continuous cell lines.

binding activity, and the two Thy-1^+ lymphomas had levels of binding significantly higher that that observed with preleukemic splenic lymphocytes.

To further assess the possible relationship of 20αSDH-positive ^{125}I-IL 3 binding lymphomas to an IL 3 lineage, the *in vitro* growth characteristics of the lymphomas in the presence or absence of IL 3 were also examined. As described in detail elsewhere (87,97), both normal and preleukemic splenic lymphocytes show significant *in vitro* growth in response to IL 3. This proliferation is associated with a sequence of differentiation which results in 3 to 6 weeks in cultures of cells having the properties of a mast-like cell with limited proliferative capabilities. In contrast to this type of response, the 20αSDH-positive lymphomas proliferated extensively *in vitro* in response to IL 3 and gave rise to continuous IL 3-dependent cell lines. No growth of the other lymphomas was detectable in either the presence or absence of IL 3. The results demonstrated that a significant number of splenic lymphomas are phenotypically related to an IL 3 lineage, and that this type can be uniquely expanded *in vitro* in IL 3 and can give rise to continuous cell lines requiring IL 3 for continued growth.

The properties of several IL 3-dependent lymphoma cell lines established from MoLV-induced lymphomas are summarized in Table 4 and compared to a MoLV-induced B cell lymphoma (5F4-B) and a radiation-induced, TdT positive T cell

TABLE 4. *Phenotypic characteristics of IL 3-dependent lymphoma cell lines*

Cell line[a]	20αSDH pmoles/hr/10^8 cells	IL 3 receptors[b]	Growth dependence[c]	Cell surface phenotype[d]
DA-1	100	+	IL 3	Thy-$1^{\pm}$, Ly-5^{+}, H-11^{+}, Ia^{-}
DA-3	3,500	+	IL 3	Thy-1^{-}, Ly-5^{+}, H-11^{+}, Ia^{+}
DA-4	50	+	IL 3	Thy-1^{-}, Ly-5^{+}, H-11^{+}, Ia^{+}
DA-5	300	+	IL 3	Thy-$1^{\pm}$, Ly-5^{+}, H-11^{+}, Ia^{-}
DA-7	200	+	IL 3	Thy-1^{-}, Ly-5^{+}, H-11^{+}, Ia^{+}
DA-11	ND	+	IL 3	Thy-1^{-}, Ly-5^{+}, H-11^{+}, Ia^{+}
5F4-B	<10	−	None	Thy-1^{-}, Ly-5^{+}, H-11^{-}
RL-12	<10	−	None	Thy-1^{+}, Ly-5^{+}, H-11^{+}

[a]DA lymphoma cell lines were derived from lymphomas induced in BALB/c mice or in (CBA/N × BALB/c) × CBA/N backcrossed mice (DA-4, DA-7) by MoLV.

[b]Presence of IL 3 receptors was determined using ^{125}I-IL 3 in a receptor binding assay (93,151).

[c]Growth dependence for IL 3 was assessed by a standard proliferation assay which examines the dose response of proliferation for purified IL 3 (92).

[d]Cell surface phenotypes were determined by a fluorescence cell sorter and were kindly performed by Herbert Morse, III (National Institutes of Health). ±, populations of both positive and negative populations were present.

lymphoma (RL-12). As expected, the IL 3-dependent lymphomas all express 20αSDH and have detectable receptors for IL 3. In contrast, the B cell and TdT positive lymphomas have no detectable 20αSDH. None of the control lines had detectable receptors for IL 3. The cell surface phenotypes of the individual IL 3-dependent lymphomas were found to vary. Some lymphoma lines contain Thy-1^{+} and Thy-1^{-} cells. Some of the lines express Ia, whereas all the lines shown expressed Lyt-5 and H-11. These phenotypes are expressed at different stages in the differentiation of IL 3 lineage cells. The lymphoma lines vary from normal IL 3-dependent lymphocytes only in their inability to terminally differentiate; conversely, the lymphoma cells are unique in their ability to continue to proliferate *in vitro* without differentiation. Nevertheless, the data suggest that transformation of IL 3 lineage cells, sufficient to be manifested as a lymphoma *in vivo*, does not require (and in fact in the vast majority of cases does not involve) a loss of dependency on IL 3 for growth.

In addition to MoLV-inoculated BALB/c mice, the splenic lymphomas induced by Cas-Br-M MuLV in NFS/N mice have also been examined (J. N. Ihle and H. Morse, *unpublished data*). This virus is a biologically cloned ecotropic MuLV-isolated from wild mice trapped in the Lake Casitas region of Southern California, which in addition to lymphomas induces hind limb paralysis (76,85). Among the lymphomas studied to date, approximately 75% are of the IL 3 lineage and can be readily grown *in vitro* in the presence of IL 3. Whether other viruses induce comparable levels of IL-3 lineage lymphomas is not known. Similarly, it is unclear whether the host may contribute to the pathologic properties. In particular, the IL 3 lineage may be more often involved in BALB/c or NFS/N strains.

POSSIBLE MOLECULAR MECHANISMS OF TRANSFORMATION IN MURINE LEUKEMOGENESIS

The mechanisms that ultimately lead to transformation in the leukemias induced by murine retroviruses are largely unknown. In general, from other studies, it can be proposed that most leukemias involve the activation or alteration of cellular genes (c-*onc*), which contribute to a pathologic phenotype. Alternatively, a mechanism comparable to that found in erythroleukemias, which involves a specific viral gene product that acts as or alters the response to a proliferation/differentiation-inducing factor, may be postulated. This is unlikely since, unlike retroviral-induced leukemias, erythroleukemias are polyclonal diseases, and the pathology can be shown to be dependent on the expression of a specific viral product. Similarly, the receptor-mediated model proposed by McGrath and Weissman (130,131) does not require a c-*onc* gene activation. It is not clear whether such lymphomas would appear monoclonal perhaps due to a limited immunologic repertoire or the ability of a single high affinity receptor to predominate in outgrowth. Nor is it clear how such a mechanism would explain the high incidence of trisomy 15 found among murine lymphomas.

In a number of experimental systems, activation of specific cellular genes has been demonstrated to occur in transformed cells, which may be responsible for transformation. In particular, the expression of c-*myc* has been implicated in a variety of systems, including avian bursal lymphomas (80,81,139), murine and human B cell lymphomas (1,42,43,55,135,183,196), colon carcinoma-derived cell lines (2), and the promyelocytic leukemia HL-60 (38,44). Whether c-*myc* expression occurs in a significant number of murine retrovirus-induced lymphomas is not known.

The possible role of c-*myc* in murine retrovirus-induced lymphomas is of particular interest, since c-*myc* is located on chromosome 15 (42), which has been shown to be trisomic in a number of lymphomas. Interestingly, in a study of c-*myc* expression in B cell lymphomas, it was demonstrated that at least one T cell line had *myc* RNA at levels comparable to that observed in the B cell lymphomas (135). The significance of this expression was not examined. In examining several MoLV-induced lymphomas from BALB/c mice for structural alterations of the c-*myc* gene by restriction analysis, no evidence has been obtained for rearrangements (R. Mural and J. N. Ihle, *unpublished data*). The ability of Rauscher MuLV to uniquely induce B cell lymphomas suggested that mechanisms comparable to avian B cell lymphomas may be involved, although preliminary experiments have not detected alterations of c-*myc* in this system (S. Aaronson, *personal communication*). In contrast, using a slightly different approach, proviral integrations near c-*myc* in three of 11 lymphomas induced by retroviruses in rats have been observed (D. Steffen, *personal communication*). While the results do not rule out a role for c-*myc* in murine leukemias, the data suggest that a high frequency of the types of events associated with either mineral oil-induced B cell lymphomas in mice or avian leukosis virus-induced lymphomas in chickens are not observed in murine lymphomas.

Although a major emphasis has involved screening for changes in c-*myc*, various potential *onc* genes exist, which may be relevant to murine leukemogenesis. In particular, v-*abl* can cause transformation of cells in the B cell and erythroid lineage (39,67,173,206,215). In addition, translocations involving c-*abl* have been observed in chronic myelocytic leukemia (45). An examination of the potential downstream promotion of c-*abl* in the B cell lymphomas induced by RLV as well as by other retroviruses will be of considerable interest. Mechanisms involving either provirus insertion and downstream promotion or translocation and activation may be possible. To date, however, no evidence exists that c-*abl* activation is associated with any murine lymphoma.

The rearrangement of c-*mos* in a pristine-induced mouse myeloma has been demonstrated (163). In this case, neither proviral insertion nor translocation was involved; rather, the c-*mos* sequences were substituted with a novel insertion sequence-like element immediately 5′ to c-*mos*. By sequence analysis, this segment of DNA has close homology with the LTR of a known intracisternal A-particle gene (111). We have examined a number of MoLV-induced lymphomas for rearrangements involving c-*mos* (R. Mural and J. N. Ihle, *unpublished data*). In none of the lymphomas examined to date have rearrangements been detected. The *onc* genes related to Ha-*ras* and Ki-*ras* are also of interest. The altered expression of Ha-*ras* related c-*ras* has been implicated in the transformation of a human bladder carcinoma cell line (25,152). Among the lymphoma cell lines induced by MoLV that have been examined, one was found to have a Ha-*ras* related c-*ras* allele which contained a 5′ alteration consistent with but not necessarily due to proviral insertion. This lymphoma line also expressed p21, which was slightly smaller than the Ha-*ras* p21. More extensive studies will be required to determine how frequently the c-*ras* gene is altered in primary lymphomas.

Implicit in studying possible alterations of c-*onc* genes is the assumption that any or all of the *onc* genes have the potential to mediate the type of transformation observed in primary retrovirus-induced lymphomas. In this regard, data from transfection experiments suggest that unique *onc* genes may be associated with particular cell lineages (40,41,110,114,134,157). This concept has been supported by the observation that within a particular system, a predominant *onc* gene may be involved (e.g., the situation observed in avian leukosis virus-induced B cell lymphomas or murine plasmacytomas). With respect to other lymphomas of either viral or nonviral etiology, however, a common *onc* gene or particular mechanism may not exist. In particular, as noted above, a significant number of MoLV-induced lymphomas are of an IL 3-regulated lineage. These lymphomas can be readily established as cell lines and retain an absolute requirement for IL 3 for growth *in vitro*. Among these lymphomas, transformation appears to involve a phenotype in which, although the dependency for a normal growth factor is maintained, the cells fail to terminally differentiate. This knowledge, coupled with the availability of IL 3 and *in vitro* systems that allow the proliferation and differentiation of IL 3 lineage cells, has allowed studies to determine whether murine transforming virus can induce a phenotype comparable to that observed with MoLV-induced primary lymphomas (J. N. Ihle, D. Blair, and A. Rein, *unpublished data*).

In these studies, we examined the ability of v-*mos* or v-*ras* either by transfection of cloned DNAs or by infectious virus, to allow the establishment of continuous factor-dependent cell lines from cultures of normal lymphocytes responding to IL 3. In all cases, continuous cell lines were obtained which phenotypically were identical to some of the MoLV-induced primary lymphomas. Consequently, the initial data suggest that there is no specificity within this group of lymphomas with respect to potentially oncogenic genes.

In the absence of demonstrable activation or alteration of known *onc* genes, more generalized approaches have been taken to determine whether specific *onc* genes exist. Based on the avian leukosis virus model, attempts have been made to identify viral-cellular RNAs of the type predicted by downstream promotion. Alternatively, integrated provirus-flanking cellular junction sequences have been cloned to determine whether a set of flanking sequences are found near an integrated provirus in multiple lymphomas. This approach assumes that a mechanism exists involving proviral insertion and promotion, and that the number of *onc* genes involved is sufficiently limited to detect a recurring event. In a series of AKR thymic lymphomas, suggestive evidence was obtained for possible common integration sites (219). In studies that examined MoLV-induced lymphomas in rats (201), two mRNAs were found which were likely to contain cellular sequences. Among the 20 tumors studied, eight had a 2.4 kb mRNA which hybridized to only a LTR probe; seven tumors had a 7 kb mRNA that hybridized only to a representative probe; and nine had no detectable altered viral mRNAs. A proviral junction fragment was cloned from one of the lymphomas. The cellular sequences of this one 3′ junction fragment showed rearrangements in five of the 16 lymphomas examined, suggesting that viral integrations may occur frequently at this site. However, there was no correlation between the presence of a possible integrated provirus within this sequence and the presence of hybrid virus-host mRNAs. The evaluation of the significance of these results will require more data.

In contrast to the rapidity with which documentation of possible mechanisms has occurred in models such as avian leukosis virus-induced bursal lymphomas or mineral oil-induced plasmacytomas in mice, progress has been slow in murine retrovirus-induced leukemias. This has been due in part to the background of genetically acquired proviruses and the multiplicity of newly acquired proviruses in most primary lymphomas. A second consideration is the frequency with which a particular type of event is associated with transformation. Both avian bursal lymphomas and mineral oil-induced plasmacytomas are relatively homogeneous tumor models. In contrast, the types of lymphomas induced by murine retroviruses tend to be more varied. There is no reason to assume that either a common *onc* gene or mechanism exists. Nevertheless, the approaches developed in other model systems will be rapidly applied to murine leukemias and hopefully shed light on the mechanisms of transformation over the next few years.

ACKNOWLEDGMENTS

The research reported herein was sponsored by the National Cancer Institute, DHHS, under contract no. NO1-CO-23909 with Litton Bionetics, Inc. The contents of this chapter do not necessarily reflect the views or policies of the Department of Health and Human Services, nor does mention of trade names, commercial products, or organizations imply endorsement by the United States Government.

REFERENCES

1. Adams, J. M., Gerondakis, S., Webb, E., Corcoran, L. M., and Cory, S. (1983): Cellular *myc* oncogene is altered by chromosome translocation to an immunoglobulin locus in murine plasmacytomas and is rearranged similarly in human Burkitt lymphomas. *Proc. Natl. Acad. Sci. USA*, 80:1982–1986.
2. Alitalo, K., Schwab, M., Lin, C. C., Varmus, H. E., and Bishop, J. M. (1983): Homogeneously staining chromosomal regions contain amplified copies of an abundantly expressed cellular oncogene (c-*myc*) in malignant neuroendocrine cells from a human colon carcinoma. *Proc. Natl. Acad. Sci. USA*, 80:1707–1711.
3. Asjo, B., Buetti, E., Fenyo, E. M., Diggelmann, H., and Klein, G. (1981): Moloney murine leukemia virus variants with distinct p30 peptide maps are associated with different clinical types of leukemia. *Eur. J. Cancer*, 17:187–192.
4. Asjo, B., Fenyo, E. M., Spria, J., and Klein, G. (1980): Appearance and distribution of virally determined antigens in lymphoid organs of mice during leukemogenesis by Moloney leukemia virus. *Leuk. Res.*, 4:89–103.
5. Asjo, B., Skoog, L., Fenyo, E. M., and Klein, G. (1982): Different T-cell subtypes are associated with pathologically distinct forms of Moloney leukemia virus (M-MuLV)-induced lymphoma. *Int. J. Cancer*, 29:163–167.
6. Ball, J. K. (1979): Leukaemogenesis by an endogenous virus isolated from the CFW mouse. II. Early effects of virus on thymus gland and bone marrow cell populations. *J. Natl. Cancer Inst.*, 62:1517–1522.
7. Ball, J. K., Dekaban, G. A., McCarter, J. A., and Loosemore, S. M. (1983): Molecular biological characterization of a highly leukaemogenic virus isolated from the mouse. III. Identity with mouse mammary tumor virus. *J. Gen. Virol. (in press)*.
8. Ball, J. K., and McCarter, J. A. (1971): Repeated demonstration of a mouse leukaemia virus after treatment with chemical carcinogen. *J. Natl. Cancer Inst.*, 42:575–591.
9. Ball, J. K., and McCarter, J. A. (1979): Biological characterization of leukaemogenic virus isolated from the CFW mouse. *Cancer Res.*, 39:3080–3088.
10. Bassin, R. H., Ruscetti, S., Ali, I., Haapala, D. K., and Rein, A. (1982): Normal DBA/2 mouse cells synthesize a glycoprotein which interferes with MCF virus infection. *Virology*, 123:139–151.
11. Bishop, J. M., Courtneidge, S. A., Levinson, A. D., Opperman, H., Quintrell, W., Sheiness, D. K., Weiss, W., and Varmus, H. E. (1979): Origin and function of avian retrovirus transforming genes. *Cold Spring Harbor Symp. Quant. Biol.*, XLIV:919–930.
12. Blair, D. G., Oskarsson, M., Wood, T. G., McClements, W. L., Fischinger, P. J., and Vande Woude, G. F. (1981): Activation of the transforming potential of a normal cell sequence: A molecular model for oncogenesis. *Science*, 212:941–942.
13. Bollum, F. J. (1981): Terminal transferase: Experienced biochemical reagent seeks biological assignment. In: *Trends in Biochemical Sciences*, p. 41. Elsevier/North-Holland, Amsterdam.
14. Bosselman, R. A., van Griensven, L. J. L. D., Vogt, M., and Verma, I. M. (1979): Genome organization of retroviruses. VI. Heteroduplex analysis of ecotropic and xenotropic sequences of Moloney mink cell focus-inducing viral RNA obtained from either a cloned isolate or a thymoma cell line. *J. Virol.*, 32:968–978.
15. Bosselman, R. A., van Griensven, L. J. L. D., Vogt, M., and Verma, I. M. (1980): Genome organization of retroviruses. IX. Analysis of the genomes of Friend spleen focus-forming (F-SFFV) and helper murine leukemia viruses by heteroduplex formation. *Virology*, 102:234–239.
16. Bosselman, R. A., van Straaten, F., Van Beveren, C., Verma, I. M., and Vogt, M. (1982): Analysis

of the *env* gene of a molecularly cloned and biologically active Moloney mink cell focus-forming proviral DNA. *J. Virol.*, 44:19–31.
17. Boyer, B., Gisselbrecht, S., Debre, P., McKenzie, I., and Levy, J. P. (1980): Genetic control of sensitivity to Moloney leukemia virus in mice. IV. Phenotypic heterogeneity of the leukemic mice. *J. Immunol.*, 125:1415–1420.
18. Brown, R. L., Griffith, R. L., Neubauer, R. H., and Rabin, H. (1982): The effect of T cell growth factor on the cell cycle of primate T cells. *J. Immunol.*, 129:1849.
19. Buchhagen, D. L., Pedersen, F. S., Crowther, R. L., and Haseltine, W. A. (1980): Most sequence differences between the genomes of the AKV virus and a leukemogenic Gross A virus passaged *in vitro* are located near the 3′ terminus. *Proc. Natl. Acad. Sci. USA*, 77:4359–4363.
20. Buchhagen, D. L., Pincus, T., Stutman, O., and Fleissner, E. (1976): Leukemogenic activity of murine type C virus after long-term passage *in vitro*. *Int. J. Cancer*, 18:835–842.
21. Buffett, R. F., and Furth, J. A. (1959): A transplantable reticulum-cell sarcoma variant of Friend's viral leukemia. *Cancer Res.*, 19:1063–1069.
22. Canaani, E., and Aaronson, S. A. (1979): Restriction enzyme analysis of mouse cellular type C viral DNA: Emergence of new viral sequences in spontaneous AKR/J lymphomas. *Proc. Natl. Acad. Sci. USA*, 76:1677–1681.
23. Cantor, H., and Weissman, I. (1976): Development and function of subpopulations of thymocytes and T lymphocytes. *Prog. Allergy*, 20:1.
24. Chan, H. W., Bryan, T., Moore, J. L., Staal, S. P., Rowe, W. P., and Martin, M. A. (1980): Identification of ecotropic proviral sequences in inbred mouse strains with a cloned subgenomic DNA fragment. *Proc. Natl. Acad. Sci. USA*, 77:5779–5783.
25. Chang, E. H., Furth, M. E., Scolnick, E. M., and Lowy, D. R. (1982): Tumorigenic transformation of mammalian cells induced by a normal human gene homologous to the oncogene of Harvey murine sarcoma virus. *Nature*, 297:479–483.
26. Chattopadhyay, S. K., Cloyd, M. W., Linemeyer, D. L., Lander, M. R., Rands, E., and Lowy, D. R. (1982): Cellular origin and role of mink cell focus-forming viruses in murine thymic lymphomas. *Nature*, 295:25–31.
27. Chattopadhyay, S. K., Lander, M. R., Gupta, S., Rands, E., and Lowy, D. R. (1981): Origin of mink cytopathic focus-forming (MCF) viruses: Comparison with ecotropic and xenotropic murine leukemia virus genomes. *Virology*, 113:465–483.
28. Chattopadhyay, S. K., Lander, M. R., Rands, E. R., and Lowy, D. R. (1980): The structure of murine leukemia virus DNA in mouse genomes. *Proc. Natl. Acad. Sci. USA*, 77:5774–5778.
29. Chattopadhyay, S. K., Rowe, W. P., Teich, N. M., and Lowy, D. R. (1975): Definitive evidence that the murine C-type virus inducing locus AKV-1 is viral genetic material. *Proc. Natl. Acad. Sci. USA*, 72:906–910.
30. Chen, S., and Lilly, F. (1982): Suppression of spontaneous lymphoma by previously undiscovered dominant genes in crosses of high- and low-incidence mouse strains. *Virology*, 118:76–85.
31. Chen, W.-F., Scollay, R., and Shortman, K. (1982): The functional capacity of thymus subpopulations: Limit-dilution analysis of all precursors of cytotoxic lymphocytes and of all T cells capable of proliferation in subpopulations separated by the use of peanut agglutinin. *J. Immunol.*, 129:18.
32. Chesebro, B., Portis, J. L., Wehrly, K., and Nishio, J. (1983): Effect of murine host genotype of MCF virus expression, latency and leukemia cell type of leukemias induced by Friend murine leukemia helper virus. *Virology*, 128:221–233.
33. Clarke, B. J., Axelrad, A. A., Shreeve, M. M., and McLeod, D. L. (1975): Erythroid colony induction without erythropoietin by Friend leukemia virus *in vitro*. *Proc. Natl. Acad. Sci. USA*, 72:3556–3560.
34. Clark, S. P., and Mak, T. W. (1983): Complete nucleotide sequence of an infectious clone of Friend spleen focus-forming provirus: gp55 is an envelope fusion glycoprotein. *Proc. Natl. Acad. Sci. USA*, 80:5037–5041.
35. Cloyd, M. W. (1983): Characterization of target cells for MCF viruses in AKR mice. *Cell*, 32:217–225.
36. Cloyd, M. W., Hartley, J. W., and Rowe, W. P. (1979): Cell-surface antigens associated with recombinant mink cell focus inducing murine leukemia viruses. *J. Exp. Med.*, 149:702–712.
37. Cloyd, M. W., Hartley, J. W., and Rowe, W. P. (1980): Lymphomagenicity of recombinant mink cell focus-inducing murine leukemia viruses. *J. Exp. Med.*, 151:542–552.
38. Collins, S., and Groudine, M. (1982): Amplification of endogenous *myc*-related DNA sequences in a human myeloid leukaemia cell line. *Nature*, 298:679–681.

39. Cook, W. (1982): Rapid thymomas induced by Abelson murine leukemia virus. *Proc. Natl. Acad. Sci. USA*, 79:2917–2921.
40. Cooper, G. M., and Neiman, P. E. (1980): Transforming genes of neoplasms induced by avian lymphoid leukosis viruses. *Nature*, 287:656–659.
41. Cooper, G. M., Okenquist, S., and Silverman, L. (1980): Transforming activity of DNA of chemically transformed and normal cells. *Nature*, 284:418–421.
42. Crews, S., Barth, R., Hood, L., Prehn, J., and Calame, K. (1982): Mouse c-myc oncogene is located on chromosome 15 and translocated to chromosome 12 in plasmacytomas. *Science*, 218:1319–1321.
43. Dalla-Vavera, R., Bregni, M., Erikson, J., Patterson, D., Gallo, R. C., and Croce, C. M. (1982): Human c-*myc onc* gene is located on the region of chromosome 8 that is translocated in Burkitt lymphoma cells. *Proc. Natl. Acad. Sci. USA* 79:7824–7827.
44. Dalla-Favera, R., Wong-Staal, F., Gallo, R. C. (1982): *onc* Gene amplification in promyelocytic leukaemia cell line HL-60 and primary leukaemic cells of the same patient. *Nature*, 299:61–63.
45. de Klein, A., van Kessel, A. G., Grosveld, G., Bartram, C. R., Hagemeijer, A., Bootsma, D., Spurr, N. K., Heisterkamp, N., Groffen, J., and Stephenson, J. R. (1982): A cellular oncogene is translocated to the Philadelphia chromosome in chronic myelocytic leukaemia. *Nature*, 300:765–767.
46. Dexter, T. M., Garland, J., Scott, D., Scolnick, E., and Metcalf, D. (1980): Growth of factor-dependent hemopoietic precursor cell lines. *J. Exp. Med.*, 152:1036.
47. Dofuku, R., Biedler, J. L., Spengler, B. A., and Old, L. J. (1975): Trisomy of chromosome 15 in spontaneous leukemia of AKR mice. *Proc. Natl. Acad. Sci. USA*, 72:1515–1517.
48. Dresler, S., Ruta, M., Murray, M. J., and Kabat, D. (1979): Glycoprotein encoded by the Friend spleen focus-forming virus. *J. Virol.*, 30:564–575.
49. Elder, J. H., Gautsch, J. W., Jensen, F. C., Lerner, R. A., Hartley, J. W., and Rowe, W. P. (1977): Biochemical evidence that MCF murine leukemia viruses are envelope gene recombinants. *Proc. Natl. Acad. Sci. USA*, 74:4676–4680.
50. Ellis, R. W., Lowy, D. R., and Scolnick, E. M. (1982): The viral and cellular p21 *ras* gene family. In: *Advances in Viral Oncology, Vol. 1*, edited by G. Klein, pp. 107–126. Raven Press, New York.
51. Ely, J. M., Prystowsky, M. B., Eisenberg, L., Quintans, J., Goldwasser, E., Glasebrook, A. L., and Fitch, F. W. (1981): Alloreactive cloned T cell lines. V. Differential kinetics of IL 2, LCSF, and BCSF release by a cloned T amplifier cell and its variant. *J. Immunol.*, 127:2345.
52. Enjuanes, L., Lee, J. C., and Ihle, J. N. (1979): Antigenic specificities of the cellular immune response of C57BL/6 mice to the Moloney leukemia/sarcoma virus complex. *J. Immunol.*, 122:665–674.
53. Enjuanes, L., Lee, J., and Ihle, J. N. (1981): T-cell recognition of Moloney sarcoma virus proteins during tumor regression. I. Lack of a requirement for macrophages and the role of blastogenic factor(s) in T cell proliferation. *J. Immunol.*, 126:1478–1484.
54. Erikson, J., Ar-Rushdi, A., Drwinga, H. L., Nowell, P. C., and Croce, C. M. (1983): Transcriptional activation of the translocated c-*myc* oncogene in Burkitt lymphoma. *Proc. Natl. Acad. Sci. USA*, 80:820–824.
55. Eva, A., Robbins, K. C., Andersen, P. R., Srinivasan, A., Tronick, S. R., Reddy, E. P., Ellmore, N. W., Galen, A. T., Lautenberger, J. A., Papas, T. S., Westin, E. H., Wong-Staal, F., Gallo, R. C., and Aaronson, S. A. (1982): Cellular genes analogous to retroviral *onc* genes are transcribed in human tumor cells. *Nature*, 195:116–119.
56. Famulari, N. G., Koehne, C. F., and O'Donnell, P. V. (1982): Leukemogenesis by Gross passage A murine leukemia virus: Expression of viruses with recombinant *env* genes in transformed cells. *Proc. Natl. Acad. Sci. USA*, 79:3872–3876.
57. Fischinger, P. J., Blevins, C. S., and Dunlop, N. M. (1978): Genomic masking of nondefective recombinant murine leukemia virus in Moloney virus stocks. *Science*, 201:457–459.
58. Fischinger, P. J., Ihle, J. N., Bolognesi, D. P., and Schafer, W. (1976): Inactivation of murine xenotropic oncornavirus by normal mouse sera is not immunoglobulin-mediated. *Virology*, 71:346–351.
59. Fischinger, P. J., Ihle, J. N., deNoronha, F., and Bolognesi, D. P. (1977): Oncogenic and immunogenic potential of cloned HIX virus in mice and cats. *Med. Microbiol. Immunol.*, 164:119–129.
60. Fischinger, P. J., Nomura, S., and Bolognesi, D. P. (1975): A novel murine oncornavirus with dual eco- and xenotropic properties. *Proc. Natl. Acad. Sci. USA*, 72:5150–5155.

61. Friend, C., and Haddad, J. R. (1960): Tumor formation with transplants of spleen or liver from mice with virus-induced leukemia. *J. Natl. Cancer Inst.*, 25:1279–1289.
62. Fujimura, F. K., Deininger, P. L., Friedmann, T., and Linney, E. (1981): Mutation near the polyoma DNA replication origin permits productive infection of F9 embryonal carcinoma cells. *Cell*, 23:809–814.
63. Gillis, S., and Smith, K. A. (1977): Long-term culture of tumor-specific cytotoxic T cells. *Nature*, 268:154–155.
64. Gisselbrecht, S., Blaineau, C., Hurot, M.-A., Pozo, F., and Levy, J. P. (1978): Prevalence of non-T-cells in the replication of the N-tropic, type C virus of young AKR mice. *Cancer Res.*, 38:939–941.
65. Glasebrook, A. L., and Fitch, F. W. (1980): Allo-reactive cloned T cell lines. I. Interactions between cloned amplifier and cytolytic T cell lines. *J. Exp. Med.*, 151:876.
66. Glasebrook, A. L., Quintans, J., Eisenberg, L., and Fitch, F. W. (1981): Alloreactive cloned T cell lines. II. Polyclonal stimulation of B cells by a cloned helper T cell lines. *J. Immunol.*, 126:240–244.
67. Goff, S. P., and Baltimore, D. (1982): The cellular oncogene of the Abelson murine leukemia virus genome. In: *Advances in Viral Oncology, Vol. 1*, edited by G. Klein, pp. 127–139. Raven Press, New York.
68. Green, N., Hiai, H., Elder, J. H., Schwartz, R. S., Khiroya, R. H., Thomas, C. Y., Tsichlis, P. N., and Coffin, J. M. (1980): Expression of leukemogenic recombinant viruses associated with a recessive gene in HRS/J mice. *J. Exp. Med.*, 152:249–264.
69. Greenberger, J. S., Gans, P. J., Davisson, P. B., and Moloney, W. C. (1979): In vitro induction of continuous acute promyelocytic cell lines in long-term bone marrow cultures by Friend or Abelson leukemia virus. *Blood*, 53:987.
70. Gross, L. (1957): Development and serial cell-free passage of a highly potent strain of mouse leukemia virus. *Proc. Soc. Exp. Biol. Med.*, 94:767–771.
71. Gross, L., and Dreyfuss, Y. (1978): Relative loss of oncogenic potency of mouse leukemia virus (Gross) after prolonged propagation in tissue culture. *Proc. Natl. Acad. Sci. USA*, 75:3989–3992.
72. Haas, M., and Patch, V. (1980): Genomic masking and rescue of dual-tropic murine leukemia viruses: Role of pseudotype virions in viral leukemogenesis. *J. Virol.*, 35:583–591.
73. Hankins, W. D., Kost, T. A., Koury, M. J., and Krantz, S. B. (1978): Erythroid bursts produced by Friend leukaemia virus *in vitro*. *Nature*, 276:506–508.
74. Hankins, W. D., and Scolnick, E. M. (1981): Harvey and Kirsten sarcoma viruses promote the growth and differentiation of erythroid precursor cells *in vitro*. *Cell*, 26:91–97.
75. Hankins, W. D., and Troxler, D. (1980): Polycythemia- and anemia-inducing erythroleukemia viruses exhibit differential erythroid transforming effects *in vitro*. *Cell*, 22:693–699.
76. Hartley, J. W., and Rowe, W. P. (1976): Naturally occurring murine leukemia viruses in wild mice: Characterization of a new "amphotropic" class. *J. Virol.*, 19:19–25.
77. Hartley, J. W., Wolford, N. K., Old, L. J., and Rowe, W. P. (1977): A new class of murine leukemia virus associated with the development of spontaneous lymphomas. *Proc. Natl. Acad. Sci. USA*, 74:789–792.
78. Hartley, J. W., Yetter, R. A., and Morse, H. C., III. (1983): A mouse gene on chromosome 5 that restricts infectivity of MCF-type recombinant murine leukemia viruses. *J. Exp. Med.*, 158:16–24.
79. Hays, E. F., and Vredevoe, D. L. (1977): A discrepancy in XC and oncogenicity assays for murine leukemia virus in AKR mice. *Cancer Res.*, 37:726–730.
80. Hayward, W. A., Neel, B. G., and Astrin, S. M. (1981): Activation of a cellular *onc* gene by promoter insertion in ALV-induced lymphoid leukosis. *Nature*, 290:475–480.
81. Hayward, W. S., and Neel, B. G. (1981): Retroviral gene expression. *Curr. Top. Microbiol. Immunol.*, 91:217–276.
82. Hayward, W., Neel, B. G., and Astrin, S. (1981): Activation of a cellular *onc* gene by promoter insertion of ALV-induced lymphoid leukosis. *Nature*, 290:475–480.
83. Herr, W., and Gilbert, W. (1983): Somatically acquired recombinant murine leukemia proviruses in thymic leukemias of AKR/J mice. *J. Virol.*, 46:70–82.
84. Hiai, H., Morrissey, P., Khiroya, R., and Schwartz, R. S. (1977): Selective expression of xenotropic virus in congenic HRS/J (hairless) mice. *Nature*, 270:247–249.
85. Hoffman, P. M., Davidson, W. F., Ruscetti, S. K., Chused, T. M., and Morse, H. C., III. (1981): Wild mouse ecotropic murine leukemia virus infection of inbred mice: Dual-tropic virus expression precedes the onset of paralysis and lymphoma. *J. Virol.*, 39:597–602.

86. Huebner, R. J., Gilden, R. V., Toni, R., Hill, R. W., Trimmer, R. W., Fish, D. C., and Sass, B. (1976): Prevention of spontaneous leukemia in AKR mice by type-specific immunosuppression of endogenous ecotropic virogenes. *Proc. Natl. Acad. Sci. USA*, 73:4633–4635.
87. Ihle, J. N. (1983): Biochemical and biological properties of interleukin 3: A lymphokine mediating the differentiation of a lineage of cells which includes prothymocytes and mast-like cells. In: *Contemporary Topics in Molecular Immunology*, edited by F. P. Inman. Plenum, New York *(in press)*.
88. Ihle, J. N., Domotor, J. J., Jr., and Bengali, K. M. (1976): Characterization of the type and group specificities of the immune response in mice to murine leukemia viruses. *J. Virol.*, 18:124–131.
89. Ihle, J. N., Enjuanes, L., Lee, J. C., and Keller, J. (1982): The immune response to C-type viruses and its potential role in leukemogenesis. In: *Current Topics in Microbiology and Immunology, Vol. 101*, pp. 31–49. Springer-Verlag, Berlin.
90. Ihle, J. N., Hanna, M. G., Jr., Roberson, L. E., and Kenney, F. T. (1974): Autogenous immunity to endogenous RNA tumor virus: Identification of antibody reactivity in select viral antigens. *J. Exp. Med.*, 139:1568–1581.
91. Ihle, J. N., and Joseph, D. R. (1978): Genetic analysis of the endogenous C3H murine leukemia virus genome: Evidence of one locus unlinked to the endogenous murine leukemia virus genome of C57BL/6 mice. *Virology*, 87:298–306.
92. Ihle, J. N., Keller, J., Greenberger, J. S., Henderson, L., Yetter, R. A., and Morse, H. C., III. (1982): Phenotype characteristics of cell lines requiring interleukin 3 for growth. *J. Immunol.*, 129:1377–1383.
93. Ihle, J. N., Keller, J., Henderson, L., Klein, F., and Palaszynski, E. W. (1982): Procedures for the purification of interleukin 3 to homogeneity. *J. Immunol.*, 129:2431.
94. Ihle, J. N., Keller, J., Oroszlan, S., Henderson, L., Copeland, T., Fitch, F., Prystowsky, M. B., Goldwasser, E., Schrader, J. W., Palaszynski, E., Ky, M., and Lebel, B. (1983): Biological properties of homogeneous interleukin 3: I. Demonstration of WEHI-3 growth factor activity, colony stimulating factor activity and histamine producing cell stimulating factor activity. *J. Immunol.* 131:282–287.
95. Ihle, J. N., Lee, J. C., and Rebar, L. (1981): T cell recognition of Moloney leukemia virus proteins. III. T cell proliferative responses against gp70 are associated with the production of a lymphokine inducing 20 alpha hydroxysteroid dehydrogenase in splenic lymphocytes. *J. Immunol.*, 127:2565–2570.
96. Ihle, J. N., Pepersack, L., and Rebar, L. (1981): Regulation of T cell differentiation: *In vitro* induction of 20 alpha hydroxysteroid dehydrogenase in splenic lymphocytes is mediated by a unique lymphokine. *J. Immunol.*, 126:2184–2189.
97. Ihle, J. N., Rebar, L., Keller, J., Lee, J. C., and Hapel, A. (1981): Interleukin 3: Possible roles in the regulation of lymphocyte differentiation and growth. *Immunol. Rev.*, 63:101–128.
98. Ihle, J. N., and Weinstein, Y. (1983): Interleukin 3: Regulation of a lineage of lymphoid cells characterized by the expression of 20 α hydroxysteroid dehydrogenase. In: *Recognition and Regulation in Cell-Mediated Immunity*, edited by J. D. Watson and J. Marbrook. Marcel Dekker, New York *(in press)*.
99. Ishimoto, S., Adachi, A., Sakai, K., Yorifuji, T., and Tsuruta, S. (1981): Rapid emergence of mink cell focus-forming (MCF) virus in various mice infected with NB-tropic Friend virus. *Virology*, 113:644–655.
100. Kaplan, H. S. (1967): On the natural history of the murine leukemias: Presidential address. *Cancer Res.*, 27:1325–1340.
101. Karess, R. E., Hayward, W. S., and Hanafusa, H. (1979): Cellular information in the genome of recovered avian sarcoma virus directs the synthesis of transforming protein. *Proc. Natl. Acad. Sci. USA*, 76:3154–3158.
102. Katinka, M., Vasseur, M., Montreau, N., Yaniv, M., and Blangy, D. (1981): Polyoma DNA sequences involved in control of viral gene expression in murine embryonal carcinoma cells. *Nature*, 290:720–722.
103. Katinka, M., Yaniv, M., Vasseur, M., and Blangy, D. (1980): Expression of polyoma early functions in mouse embryonal carcinoma cells depends on sequence rearrangements in the beginning of the late region. *Cell*, 20:393–399.
104. Kawashima, K., Ikeda, H., Hartley, J. W., Stockert, E., Rowe, W. P., and Old, L. J. (1976): Changes in expression of murine leukemia virus antigens and production of xenotropic virus in the late preleukemic period of AKR mice. *Proc. Natl. Acad. Sci. USA*, 73:4680–4684.

105. Khan, A. S., Rowe, W. P., and Martin, M. A. (1982): Cloning of endogenous murine leukemia virus-related sequences from chromosomal DNA of BALB/c and AKR/J mice: Identification of an *env* progenitor of AKR-247 mink cell focus-forming proviral DNA. *J. Virol.*, 44:625–636.
106. Klein, G. (1981): The role of gene dosage and genetic transpositions in carcinogenesis. *Nature*, 294:313–318.
107. Klein, G. (1983): Specific chromosomal translocations and the genesis of B-cell-derived tumors in mice and men. *Cell*, 32:311–315.
108. Klein, B., Le Bousse, C., Fagg, B., Smajda-Joffe, F., Vehmeyer, K., Mori, K. J., Jasmin, C., and Ostertag, W. (1981): Effects of myeloproliferative sarcoma virus on the pluripotential stem cell and granulocyte precursor cell populations of DBA/2 mice. *J. Natl. Cancer Inst.*, 66:935–940.
109. Klein, B., Le Bousse, C., Smadja-Joffe, F., Pragnell, I., Ostertag, W., and Jasmin, C. (1982): A study of added GM-CSF independent granulocyte and macrophage precursors in mouse spleen infected with myeloproliferative sarcoma virus (MPSV). *Exp. Hematol.*, 10:373–382.
110. Krontiris, T. G., and Cooper, G. M. (1981): Transforming activity of human tumor DNAs. *Proc. Natl. Acad. Sci. USA*, 78:1181–1184.
111. Kuff, E. L., Feenstra, A., Lueders, K., Rechavi, G., Givol, D., and Canaani, E. (1983): Homology between an endogenous viral LTR and sequences inserted in an activated cellular oncogene. *Nature*, 302:547–548.
112. Lai, M. M. C., Rasheed, S., Shimizu, C. S., and Gardner, M. B. (1982): Genomic characterization of a highly oncogenic *env* gene recombinant between amphotropic retrovirus of wild mouse and endogenous xenotropic virus of NIH Swiss mouse. *Virology*, 117:262–266.
113. Laimins, L. A., Khoury, G., Gorman, C., Howard, B., and Gruss, P. (1982): Host-specific activation of transcription by tandem repeats from simian virus 40 and Moloney murine sarcoma virus. *Proc. Natl. Acad. Sci. USA*, 79:6453–6457.
114. Lane, M. A., Sainten, A., and Cooper, G. M. (1981): Activation of related transforming genes in mouse and human mammary carcinomas. *Proc. Natl. Acad. Sci. USA*, 78:5185–5189.
115. Langdon, W. Y., Hoffman, P. M., Silver, J. E., Buckler, C. E., Hartley, J. W., Ruscetti, S. K., and Morse, H. C., III. (1983): Identification of a spleen focus-forming virus in erythroleukemic mice infected with a wild-mouse ecotropic murine leukemia virus. *J. Virol.*, 46:230–238.
116. Lee, J. C., Horak, I., and Ihle, J. N. (1981): Mechanisms in T cell leukemogenesis. II. T cell responses of preleukemic BALB/c mice to Moloney leukemia virus antigens. *J. Immunol.*, 126:715–722.
117. Lee, J. C., and Ihle, J. N. (1979): Mechanisms of C-type viral leukemogenesis. I. Correlation of *in vitro* lymphocyte blastogenesis to viremia and leukemia. *J. Immunol.*, 123:2351–2358.
118. Lee, J. C., and Ihle, J. N. (1981): Chronic immune stimulation is required for Moloney leukemia virus-induced lymphomas. *Nature*, 209:407–409.
119. Lee, J. C., and Ihle, J. N. (1981): Increased responses to lymphokines are correlated with preleukemia in Moloney virus inoculated mice. *Proc. Natl. Acad. Sci. USA*, 78:7712–7716.
120. Lenz, J., and Haseltine, W. S. (1983): Localization of the leukemogenic determinants of SL3-3, an ecotropic, XC-positive, murine leukemia virus of AKR origin. *J. Virol.*, 47:317–328.
121. Levy, J. A., Ihle, J. N., Oleszko, D., and Barnes, R. D. (1975): Virus-specific neutralization by a soluble nonimmunoglobulin factor found naturally in normal mouse sera. *Proc. Natl. Acad. Sci. USA*, 72:5071–5075.
122. Lilly, F., Duran-Reynals, M. L., and Rowe, W. P. (1975): Correlation of early murine leukemia virus titer and H-2 type with spontaneous leukemia in mice of the BALB/c ×AKR cross: A genetic analysis. *J. Exp. Med.*, 141:882–889.
123. Linemeyer, D. L., Menke, J. G., Ruscetti, S. K., Evans, L. H., and Scolnick, E. M. (1982): Envelope gene sequences which encode the gp52 protein of spleen focus-forming virus are required for the induction of erythroid cell proliferation. *J. Virol.*, 43:223–233.
124. Linemeyer, D. L., Ruscetti, S. K., Menke, J. G., and Scolnick, E. M. (1980): Recovery of biologically active spleen focus-forming virus from molecularly cloned spleen focus-forming virus-pBR322 circular DNA by cotransfection with infectious type C retroviral DNA. *J. Virol.*, 35:710–721.
125. Linemeyer, D. L., Ruscetti, S. K., Scolnick, E. M., Evans, L. H., and Duesberg, P. H. (1981): Biological activity of the spleen focus-forming virus is encoded by a molecularly cloned subgenomic fragment of spleen focus-forming virus DNA. *Proc. Natl. Acad. Sci. USA*, 78:1401–1405.
126. Lung, M. L., Hartley, J. W., Rowe, W. P., and Hopkins, N. H. (1983): Large RNase T_1-resistant oligonucleotides encoding p15E and the U3 region of the long terminal repeat distinguish two

biological classes of mink cell focus-forming type C viruses in inbred mice. *J. Virol.*, 45:275–290.

127. MacDonald, M. E., Mak, T. W., and Berstein, A. (1980): Erythroleukemia induction by replication-competent type C viruses cloned from the anemia- and polycythemia-inducing isolates of Friend leukemia virus. *J. Exp. Med.*, 151:1493–1503.
128. McGrath, M. S., Pillemer, E., and Weissman, I. L. (1980): Murine leukaemogenesis: Monoclonal antibodies to T-cell determinants arrest T-lymphoma cell proliferation. *Nature*, 285:259–261.
129. McGrath, M. S., Pillemer, E., Kooistra, D., and Weissman, I. L. (1980): The role of MuLV receptors on T lymphoma cells in lymphoma cell proliferation. In: *Contemporary Topics of Immunobiology, Vol. 11*, edited by N. L. Warner, pp. 157–184. Plenum, New York.
130. McGrath, M. S., and Weissman, I. L. (1978): A receptor mediated model of viral leukemogenesis: Hypothesis and experiments. Cold Spring Harbor Meeting on Differentiation of Normal and Neoplastic Hematopoietic Cells, pp. 577–589. Cold Spring Harbor Laboratory, Cold Spring Harbor, New York.
131. McGrath, M. S., and Weissman, I. L. (1979): AKR leukemogenesis: Identification and biological significance of thymic lymphoma receptors for AKR retroviruses. *Cell*, 17:65–75.
132. Meier, J. W., and Gallo, R. C. (1982): Human T cell growth factor (TCGF): Biochemical properties and interaction with production by normal and neoplastic human T cells. In: *Lymphokines, Vol. 6*, edited by S. B. Mizel, pp. 137–163. Academic Press, New York.
133. Morgan, D. A., Ruscetti, F. W., and Gallo, R. C. (1976): Selective *in vitro* growth of T lymphocytes from normal human bone marrows. *Science*, 193:1007–1008.
134. Murray, M. J., Shilo, B.-Z., Shih, C., Cowing, D., Hsu, H. W., and Weinberg, R. A. (1981): Three different human tumor cell lines contain different oncogenes. *Cell*, 25:355–361.
135. Mushinski, J. F., Bauer, S. R., Potter, M., and Reddy, E. P. (1983): Increased expression of *myc*-related oncogene mRNA characterizes most BALB/c plasmacytomas induced by pristane or Abelson murine leukemia virus. *Proc. Natl. Acad. Sci. USA*, 80:1073–1077.
136. Nabel, G., Greenberger, J. S., Sakakeeny, M. A., and Cantor, H. (1981): Multiple biologic activities of a cloned inducer T-cell population. *Proc. Natl. Acad. Sci. USA*, 78:1157–1161.
137. Niho, Y., Shibuya, T., and Mak, T. W. (1982): Modulation of erythropoiesis by the helper-independent Friend leukemia virus F-MuLV. *J. Exp. Med.*, 156:146–158.
138. Nishizuka, Y., and Nakakuki, K. (1968): Acceleration of leukemogenesis in AKR mice by grafts, cell suspensions, and cell-free centrifugates of thymuses from preleukemic AKR donors. *Int. J. Cancer*, 3:203–210.
139. Noori-Daloii, M. R., Swift, R. A., Kung, H.-J., Crittenden, L. B., and Witter, R. L. (1981): Specific integration of REV proviruses in avian bursal lymphomas. *Nature*, 294:574–576.
140. Nowinski, R. C., Brown, M., Doyle, T., and Prentice, R. L. (1979): Genetic and viral factors influencing the development of spontaneous leukemia in AKR mice. *Virology*, 96:186–204.
141. Nowinski, R. C., and Kachler, S. L. (1974): Antibody to leukemia virus: Widespread occurrence in inbred mice. *Science*, 185:869–871.
142. Nowinski, R. C., and Hays, E. F. (1978): Oncogenicity of AKR endogenous leukemia viruses. *J. Virol.*, 27:13–18.
143. Nowinski, R. C., Hays, E. F., Doyle, T., Linkhart, S., Medeiros, E., and Pickering, R. (1977): Oncornaviruses produced by murine leukemia cells in culture. *Virology*, 81:363–370.
144. O'Donnell, P. V., Nowinski, R. C., and Stockert, E. (1982): Amplified expression of murine leukemia virus (MuLV)-coded antigens on thymocytes and leukemia cells after infection by dualtropic (MCF) MuLV. *Virology*, 119:450–464.
145. O'Donnell, P. V., Stockert, E., Obata, Y., DeLeo, A. B., and Old, L. J. (1980): Murine leukemia virus-related cell surface antigens as serological markers of AKR ecotropic, xenotropic, and dualtropic MuLV. *Cold Spring Harbor Symp. Quant. Biol.*, 44:1255–1264.
146. O'Donnell, P. V., Stockert, E., Obata, Y., and Old, L. J. (1981): Leukemogenic properties of AKR dualtropic (MCF) viruses: Amplification of murine leukemia virus-related antigens on thymocytes and acceleration of leukemia development in AKR mice. *Virology*, 112:548–563.
147. Oliff, A., Linemeyer, D., Ruscetti, S., Lowe, R., Lowy, D. R., and Scolnick, E. (1980): Subgenomic fragment of molecularly cloned Friend murine leukemia virus DNA contains the gene(s) responsible for Friend murine leukemia virus-induced disease. *J. Virol.*, 35:924–936.
148. Oliff, A., and Ruscetti, S. (1983): A 2.4 kbp fragment of the F-MuLV genome contains the sequences responsible for F-MuLV-induced leukemia. *J. Virol.*, 46:718–725.
149. Oliff, A., Ruscetti, S., Douglass, E. C., and Scolnick, E. (1981): Isolation of transplantable

erythroleukemia cells from mice infected with helper-independent Friend murine leukemia virus. *Blood*, 58:244–254.
150. Ostertag, W., Vehmeyer, K., Fagg, B., Pragnell, I. B., Paetz, W., Le Bousse, M. C., Smadja-Joffe, F., Klein, B., Jasmin, C., and Eisen, H. (1980): Myeloproliferative virus, a cloned murine sarcoma virus with spleen focus-forming properties in adult mice. *J. Virol.*, 33:573–582.
151. Palaszynski, E. W., and Ihle, J. N. (1983): Evidence for specific receptors for interleukin 3 on lymphokine dependent cell lines established from long-term bone marrow cultures. *J. Immunol. (in press)*.
152. Parada, L. F., Tabin, C. J., Shih, C., and Weinberg, R. A. (1982): Human EJ bladder carcinoma oncogene is homologue of Harvey sarcoma virus *ras* gene. *Nature*, 297:474–478.
153. Payne, G. S., Bishop, J. M., and Varmus, H. E. (1982): Multiple arrangements of viral DNA and an activated host oncogene in bursal lymphomas. *Nature*, 295:209–214.
154. Pedersen, F. S., Buchhagen, D. L., Chen, C. Y., Hays, E. F., and Haseltine, W. A. (1980): Characterization of virus produced by a lymphoma induced by inoculation of AKR MCF-247 virus. *J. Virol.*, 35:211–218.
155. Pedersen, F. S., Crowther, R. L., Tenney, D. Y., Reimold, A. M., and Haseltine, W. A. (1981): Novel leukaemogenic retroviruses isolated from cell line derived from spontaneous AKR tumour. *Nature*, 292:167–170.
156. Pepersack, L., Lee, J. C., McEwan, R., and Ihle, J. N. (1980): Phenotypic heterogeneity of Moloney leukemia virus-induced T cell lymphomas. *J. Immunol.*, 124:279–285.
157. Perucho, M., Goldfarb, M., Shimizu, K., Lama, C., Fogh, J., and Wigler, M. (1981): Human-tumor-derived cell lines contain common and different transforming genes. *Cell*, 27:467–476.
158. Quint, W., Quax, W., van der Putten, H., and Berns, A. (1981): Characterization of AKR murine leukemia virus sequences in AKR mouse substrains and structure of integrated recombinant genomes in tumor tissues. *J. Virol.*, 39:1–10.
159. Racevkis, J., and Koch, G. (1977): Viral protein synthesis in Friend erythroleukemia cell lines. *J. Virol.*, 21:328–337.
160. Rapp, U. R., Goldsborough, M. D., Mark, G. E., Bonner, T. I., Greifen, J., Reynolds, F. H., Jr., and Stephenson, J. R. (1983): Structure and biological activity of r-*raf*: A new oncogene transduced by a retrovirus. *Proc. Natl. Acad. Sci. USA*, 80:4218–4222.
161. Rapp, U. R., Reynolds, F. H., Jr., and Stephenson, J. R. (1983): New mammalian transforming retrovirus: Demonstration of a polyprotein gene product. *J. Virol.*, 45:914–924.
162. Rasheed, S., Pal, B. K., and Gardner, M. B. (1982): Characterization of a highly oncogenic murine leukemia virus from wild mice. *Int. J. Cancer*, 29:345–350.
163. Rechavi, G., Givol, D., and Canaani, E. (1982): Activation of a cellular oncogene by DNA rearrangement: Possible involvement of an IS-like element. *Nature*, 300:607–611.
164. Reedy, E. P., Dunn, C. Y., and Aaronson, S. A. (1980): Different lymphoid cell targets for transformation by replication-competent Moloney and Rauscher mouse leukemia viruses. *Cell*, 19:663–669.
165. Rein, A. (1982): Interference grouping of murine leukemia viruses: A distinct receptor for the MCF-recombinant viruses on mouse cells. *Virology*, 120:251–257.
166. Rein, A., Athan, E., Benjers, B. M., Bassin, R. H., Gerwin, B. I., and Slocum, D. R. (1979): Isolation of a replication-defective murine leukaemia virus from cultured AKR leukaemia cells. *Nature*, 282:753–754.
167. Rein, A., Lowy, D. R., Gerwin, B. I., Ruscetti, S. K., and Bassin, R. H. (1982): Molecular properties of a *gag*$^-$ *pol*$^-$ *env*$^+$ murine leukemia virus from cultured AKR leukemia cells. *J. Virol.*, 41:626–634.
168. Rein, A., Schultz, A. M., Bader, J. P., and Bassin, R. H. (1982): Inhibitors of glycosylation reverse retroviral interference. *Virology*, 119:185–192.
169. Repaske, R., O'Neill, R. R., Khan, A. S., and Martin, M. A. (1983): Nucleotide sequence of the *env*-specific segment of NFS-Th-1 xenotropic murine leukemia virus. *J. Virol.*, 46:204–211.
170. Robinson, H. L., Parson, M. N., DeSimone, D. W., Tsichlis, P. N., and Coffin, J. M. (1980): Subgroup E avian leukosis virus-associated disease in chickens. *Cold Spring Harbor Symp. Quant. Biol.*, 44:1133–1142.
171. Roblin, R., Young, J. M., Mural, R. J., Bell, T. E., and Ihle, J. N. (1982): Molecular cloning and characterization of murine leukemia virus-related sequences from C3H/Hen mouse DNA. *J. Virol.*, 43:113–126.

172. Rosenberg, N., and Baltimore, D. (1980): Abelson virus. In: *Viral Oncology*, edited by G. Klein, pp. 187–203. Raven Press, New York.
173. Rowe, W. P. (1973): Genetic factors in the natural history of murine leukemia virus infection. *Cancer Res.*, 33:3061–3068.
174. Rowe, W. P., and Kozak, C. (1980): Germ-line reinsertions of AKR murine leukemia virus genomes in *Akv*-1 congenic mice. *Proc. Natl. Acad. Sci. USA*, 77:4871–4874.
175. Rowe, W. P., and Pincus, T. (1972): Quantitative studies of naturally occurring murine leukemia virus infection of AKR mice. *J. Exp. Med.*, 135:429–436.
176. Ruscetti, S., Davis, L., Feild, J., and Oliff, A. (1981): Friend murine leukemia virus-induced leukemia is associated with the formation of mink cell focus-inducing viruses and is blocked in mice expressing endogenous mink cell focus-inducing xenotropic viral envelope genes. *J. Exp. Med.*, 154:907–920.
177. Ruscetti, S., Linemeyer, D., Feild, J., Troxler, D., and Scolnick, E. (1978): Type-specific radioimmunoassays for the gp70s of mink cell focus-inducing murine leukemia viruses: Expression of a cross-reacting antigen in cells infected with the Friend strain of the spleen focus-forming virus. *J. Exp. Med.*, 148:654–663.
178. Ruscetti, S. K., Linemeyer, D., Feild, J., Troxler, D., and Scolnick, E. M. (1979): Characterization of a protein found in cells infected with the spleen focus-forming virus that shares immunological cross-reactivity with the gp70 found in mink cell focus-inducing virus particles. *J. Virol.*, 30:787–798.
179. Ruscetti, S. K., Troxler, D., Linemeyer, D., and Scolnick, E. (1980): Three distinct strains of spleen focus-forming virus: Comparison of their genomes and translational products. *J. Virol.*, 33:140–151.
180. Schultz, A. M., Rein, A., Henderson, L., and Oroszlan, S. (1983): Biological, chemical and immunological studies of Rauscher ecotropic and mink cell focus forming viruses from JLS-V9 cells. *J. Virol.*, 45:995–1003.
181. Schwarz, H., Fischinger, P. J., Ihle, J. N., Thiel, H.-J. Weiland, F., Bolognesi, D. P., and Schafer, W. (1979): Properties of mouse leukemia viruses. XVI. Suppression of spontaneous fatal leukemias in AKR mice by treatment with broadly reacting antibody against the viral glycoprotein gp71. *Virology*, 93:159–174.
182. Shaw, J., Caplan, B., Paetkau, V., Pilarski, L. M., Delovitch, T. L., and McKenzie, I. F. C. (1980): Cellular origins of costimulator (IL 2) and its activity in cytotoxic T lymphocyte responses. *J. Immunol.*, 124:231.
183. Shen-Ong, G. L. C., Keath, E. J., Piccoli, S. P., and Cole, M. D. (1982): Novel *myc* oncogene RNA from abortive immunoglobulin-gene recombination in mouse plasmacytomas. *Cell*, 31:443–452.
184. Shibuya, T., and Mak, T. W. (1982): Host control of susceptibility to erythroleukemia and to the types of leukemia induced by Friend murine leukemia virus: Initial and late stages. *Cell*, 31:483–493.
185. Shortman, K. (1977): The pathway of T-cell development within the thymus. In: *Progress in Immunology, Vol. III.*, edited by T. E. Mandel. pp. 197–205, North Holland, Amsterdam.
186. Shortman, K., and Jackson, H. (1974): The differentiation of T lymphocytes. I. Proliferation kinetics and interrelationships of subpopulations of mouse thymus cells. *Cell. Immunol.*, 12:230.
187. Spira, J., Asjo, B., Cochran, A., Shen, F. W., Wiener, F., and Klein, G. (1981): Chromosomal, histopathological and cell surface marker studies on Moloney virus induced lymphomas. *Leuk. Res.*, 5:113–121.
188. Steck, F. T., and Rubin, H. (1966): The mechanism of interference between an avian leukosis virus and Rous sarcoma virus. I. Establishment of interference. *Virology*, 29:628–641.
189. Steck, F. T., and Rubin, H. (1966): The mechanism of interference between an avian leukosis virus and Rous sarcoma virus. II. Early steps of infection by RSV of cells under conditions of interference. *Virology*, 29:642–653.
190. Steeves, R. A., Eckner, R. J., Bennett, M., Mirand, E. A., and Trudel, P. J. (1971): Isolation and characterization of a lymphatic leukemia virus in the Friend virus complex. *J. Natl. Cancer Inst.*, 46:1209–1217.
191. Steffen, D., Bird, S., Rowe, W. P., and Weinberg, R. A. (1979): Identification of DNA fragments carrying ecotropic proviruses of AKR mice. *Proc. Natl. Acad. Sci. USA*, 76:4554–4558.
192. Steffen, D., and Weinberg, R. A. (1978): The integrated genome of murine leukemia virus. *Cell*, 15:1003–1010.

193. Stockert, E., DeLeo, A. B., O'Donnell, P. V., Obata, Y., and Old, L. J. (1979): $G_{(AKSL2)}$: A new cell surface antigen of the mouse related to the dualtropic mink cell focus-inducing class of murine leukemia virus detected by naturally occurring antibody. *J. Exp. Med.*, 149:200–215.
194. Tambourin, P., and Wendling, F. (1971): Malignant transformation and erythroid differentiation by polycythaemia-inducing Friend virus. *Nature*, 234:230–233.
195. Tambourin, P. E., Wendling, F., Jasmin, C., and Smadja-Joffe, F. (1979): The physiopathology of Friend leukemia. *Leuk. Res.*, 3:117–129.
196. Taub, R., Kirsch, I., Morton, C., Lenoir, G., Swan, D., Tronick, S., Aaronson, S., and Leder, P. (1982): Translocation of the c-*myc* gene into the immunoglobulin heavy chain locus in human Burkitt lymphoma and murine plasmacytoma cells. *Proc. Natl. Acad. Sci. USA*, 79:7837–7841.
197. Thomas, C. Y., and Coffin, J. M. (1982): Genetic alterations of RNA leukemia viruses associated with the development of spontaneous thymic leukemia in AKR/J mice. *J. Virol.*, 43:416–426.
198. Todaro, G. J., Arnstein, P., Parks, W. P., Lennette, E. H., and Huebner, R. J. (1973): A type-C virus in human rhabdomyosarcoma cells after inoculation into NIH Swiss mice treated with antithymocyte serum. *Proc. Natl. Acad. Sci. USA*, 70:859–862.
199. Troxler, D. H., and Scolnick, E. M. (1978): Rapid leukemia induced by cloned Friend strain of replicating murine type C virus. *Virology*, 85:17–27.
200. Troxler, D. H., Yuan, E., Linemeyer, D., Ruscetti, S., and Scolnick, E. M. (1978): Helper-independent mink cell focus-inducing strains of Friend murine type-C virus: Potential relationship to the origin of replication-defective spleen focus-forming virus. *J. Exp. Med.*, 148:639–653.
201. Tsichlis, P. N., Strauss, P. G., and Hu, L. F. (1983): A common region for proviral DNA integration in MoMuLV induced rat thymic lymphomas. *Nature*, 302:445–449.
202. van der Putten, H., Quint, W., van Raaij, J., Maandag, E. R., Verma, I. M., and Berns, A. (1981): M-MuLV-induced leukemogenesis: Integration and structure of recombinant proviruses in tumors. *Cell*, 24:729–739.
203. van Griensven, L. J. L. D., and Vogt, M. (1980): Rauscher "mink cell focus-inducing" (MCF) virus causes erythroleukemia in mice: Its isolation and properties. *Virology*, 101:376–388.
204. Vogt, M. (1982): Virus cloned from the Rauscher virus complex induces erythroblastosis and thymic lymphoma. *Virology*, 118:225–228.
205. Vogt, P. K., and Ishizaki, R. (1966): Patterns of viral interference in the avian leukosis and sarcoma complex. *Virology*, 30:368–374.
206. Waneck, G. L., and Rosenberg, N. (1981): Abelson leukemia virus induces lymphoid and erythroid colonies in infected fetal cell cultures. *Cell*, 26:79–89.
207. Watson, J., Frank, M. B., Mochizuki, P., and Gillis, S. (1982): The biochemistry and biology of interleukin 2. In: *Lymphokines in Antibody and Cytotoxic Responses Vol. 6.*, edited by S. B. Mizel, pp. 95–116. Academic Press, New York.
208. Weiner, F., Ohno, S., Spira, J., Haran-Ghera, W., and Klein, G. (1978): Chromosomal changes (trisomy 15 and 12) associated with tumor progression in leukemias induced by radiation leukemia virus (RadLV). *J. Natl. Cancer Inst.*, 61:227–233.
209. Weinstein, Y. (1977): 20αHydroxysteroid dehydrogenase: A T lymphocyte associated enzyme. *J. Immunol.*, 119:1223.
210. Weinstein, Y. (1981): Expression of 20 alpha hydroxysteroid dehydrogenase in the mouse marrow cells: Strain differences, thymic effect on enzymatic activity, and possible localization in pre T lymphocytes. *Thymus*, 2:305–320.
211. Weinstein, Y., Linder, H. R., and Eckstein, B. (1977): Thymus metabolizes progesterone, a possible enzymatic marker for T lymphocytes. *Nature*, 266:632–633.
212. Weissman, I. L., and Baird, S. (1977): Oncornavirus leukemogenesis as a model for selective neoplastic transformation. In: *Life Sciences Research Report 7. Neoplastic Transformation: Mechanisms and Consequences*, edited by H. Koprowski, pp. 135–152. Dahlem Konferenzen, Berlin.
213. Wendling, F., Heard, J.-M., and Tambourin, P. (1983): Phenotypic heterogeneity of leukemia induced by ecotropic replication-competent virus isolated from Friend virus complex. *Blood (submitted)*.
214. Westin, E. H., Wong-Staal, F., Gelmann, E. P., Dalla-Favera, R., Papas, T. S., Lautenberger, J. A., Eva, A., Reddy, E. P., Tronick, S. R., Aaronson, S. A., and Gallo, R. C. (1982): Expression of cellular homologues of retroviral *onc* genes in human hematopoietic cells. *Proc. Natl. Acad. Sci. USA*, 79:2490–2494.
215. Whitlock, C. A., Ziegler, S. F., Treiman, L. J., Stafford, J. I., and Witte, O. N. (1983): Differ-

entiation of cloned populations of immature B cells after transformation with Abelson murine leukemia virus. *Cell*, 32:903–911.

216. Wolff, L., Scolnick, E., and Ruscetti, S. (1983): Envelope gene of the Friend spleen focus forming virus: Deletion and insertions in 3′ gp70/p15E coding region have resulted in unique features in the primary structure of its protein product. *Proc. Natl. Acad. Sci. USA*, 80:4718–4722.
217. Yamamoto, Y., Gamble, C. L., Clark, S. P., Joyner, A., Shibuya, T., MacDonald, M. E., Mager, D., Bernstein, A., and Mak, T. W. (1981): Clonal analysis of early and late stages of erythroleukemia induced by molecular clones of integrated spleen focus-forming virus. *Proc. Natl. Acad. Sci. USA*, 78:6893–6897.
218. Yoshimura, F., and Breda, M. (1981): Lack of AKR ecotropic provirus amplification in AKR leukemic thymuses. *J. Virol.*, 39:808–815.
219. Yoshimura, F. K., and Levine, K. L. (1983): AKR thymic lymphomas involving mink cell focus-inducing murine leukemia viruses have a common region of provirus integration. *J. Virol.*, 45:576–584.
220. Zarling, D. A., and Keshet, I. (1979): Fusion acitivity of virions of murine leukemia virus. *Virology*, 95:185–196.

Advances in Viral Oncology, Volume 4,
edited by George Klein. Raven Press,
New York © 1984.

Recent Developments in Plasmacytomagenesis in Mice

Michael Potter, *Francis Wiener,
and J. Frederic Mushinski

*Laboratory of Genetics, National Cancer Institute, National Institutes of Health, Bethesda, Maryland 20205; and *Department of Tumor Biology, Karolinska Institutet, S-104 01, Stockholm, Sweden*

In the final steps of B-lymphocyte development, lymphocytes become immunoglobulin (Ig)-secreting cells called plasma cells. Mitoses in normal plasma cells are rarely seen, and it is thought that the acquisition of Ig secretory capability and the extensive protein-secreting cytoplasm characteristic of plasma cell histology signal the emergence of a permanent postmitotic state and imminent cellular elimination.

The development of Ig-secreting tumors (plasmacytomas or immunocytomas) presents an intriguing model system of oncogenesis, as these tumor cells have retained the potential to both divide and secrete Ig. There are several well-known forms of plasmacytomagenesis: multiple myeloma in man, the spontaneous immunocytomas in the Lou/Wsl rats (11,12), and the plasmacytomas that are induced in the peritoneal tissues of the mouse by different types of nondigestible mild irritants, such as mineral oils and plastics. Mouse plasmacytomagenesis has been reviewed in articles replete with details of this experimental tumor system (54,68,97).

In this chapter, we briefly review the biology of plasmacytomagenesis in the mouse and discuss the possible role of endogenous retroviruses in the pathogenesis, as well as evidence that nonrandom chromosomal translocations and specific oncogenes are involved.

PLASMACYTOMA INDUCTION

Agents

The most extensively studied model of plasmacytomagenesis in mice is the induction of plasmacytomas by intraperitoneal mineral oil injections in the inbred BALB/c (8,9,67–70,73) and NZB (56,97) strains. All medicinal preparations of mineral oils so far tested and several available branched alkanes, such as pristane (2,6,10,14-tetramethylpentadecane), phytane (2,6,10,14-tetramethylhexadecane), or 7-N-hexyloctadecane, induce plasmacytomas in BALB/c mice (8,9). Equally effective agents are solid plastic materials, such as Lucite (plexiglas) discs or shavings (48), that are implanted in the peritoneal cavity. Both types of agents, although different in physical form and chemical composition, are not metabolized and remain in the peritoneal space indefinitely.

When mineral oils and related substances are injected into the peritoneal cavity, they induce the formation of an oil granuloma, a form of a chronic granulomatous tissue. Although the granulomatous tissue that develops in association with implanted solid plastic materials has not been studied in detail, it also appears to be the product of a chronic peritoneal inflammation. A few relevant facts are available about the process of plasmacytoma induction by pieces of solid plastic. First, the yield of plasmacytomas increases with the size of the implanted disc, i.e., discs that are 21 mm in diameter are more effective than smaller (17.5 mm) ones (48). Second, the greatest incidence of plasmacytomas occurred when the implanted discs had rough edges, which probably continuously injured peritoneal surfaces. Intraperitoneally implanted plastic discs induce both fibrosarcomas and plasmacytomas. The fibrosarcomas develop in the fibrous capsules that form around the discs, while the plasmacytomas appear to arise on peritoneal surfaces (48).

While the available evidence suggests that the physical properties of mineral oils and their resistance to metabolic catabolism are responsible for their plasmacytomagenic activity, small amounts of genotoxic (mutagenic) contaminants could be present in mineral oils, plastics, or even commercial pristane preparations. Since pristane is chemically the best defined of the known plasmacytomagenic agents, it is the easiest to study in detail. Commercial pristane is of biogenic origin and is obtained from whale livers. Filter feeding whales acquire pristane from ingested marine zooplankton (15). These organisms convert the plant product phytane (2,6,10,14-tetramethylhexadecane) into pristane (15). In this process, other intermediates, including olefinic compounds that are potentially metabolically active, are generated (14,16). Commercial pristane is known to be contaminated with ultraviolet-absorbing materials, perhaps olefins, which can be removed by passage of pristane over alumina columns. It is unlikely, however, that such reactive compounds are the active carcinogens in pristane; it has been shown that pristane after alumina absorption is as effective as crude pristane for inducing plasma cell tumors (77). The evidence favors the notion that the primary biologic action of pristane and other agents is to induce the formation of a chronic granulomatous tissue (see below).

Latent Period of Plasmacytoma Development

Before discussing the biologic properties of the oil granuloma, it is relevant to describe the long latent period that intervenes between the first injection of oil and the appearance of plasma cell tumors.

Peritoneal plasmacytomas in mineral oil- (or pristane-) treated BALB/c mice appear after latent periods of 6 to 12 months. This places the process of plasmacytoma development in the same category as many other experimental forms of induced and spontaneous tumor development, e.g., spontaneous thymic leukemogenesis in AKR mice, spontaneous immunocytoma development in Lou/Wsl rats, retrovirus-induced bursal lymphomas in the chicken, and radiation leukemogenesis in mice.

Several pristane dose regimens for the induction of plasmacytomas have been used (77). A high yield of plasmacytomas results from the injection of 0.5 ml pristane three times separated by 2-month intervals. The incidence of plasmacytomas in BALB/c mice ranges from 50 to 70% in the first year after the first injection (77). As shown in Fig. 1, plasmacytoma cells may be detected in stained smears of ascites fluid as early as 120 days after the first injection of pristane, but most tumors begin to appear after 180 days and can continue to develop during the next 200 or more days.

Two other regimens, consisting of single injections of 0.5 or 1.0 ml pristane, have also been studied (77). The yield of plasmacytomas following injection of 1.0 ml is approximately twice that obtained with 0.5 ml, but the mean latent periods for plasmacytoma induction by pristane are similar in all three regimens. Thus far, no method has been found for decreasing the latent period by altering the dose of pristane.

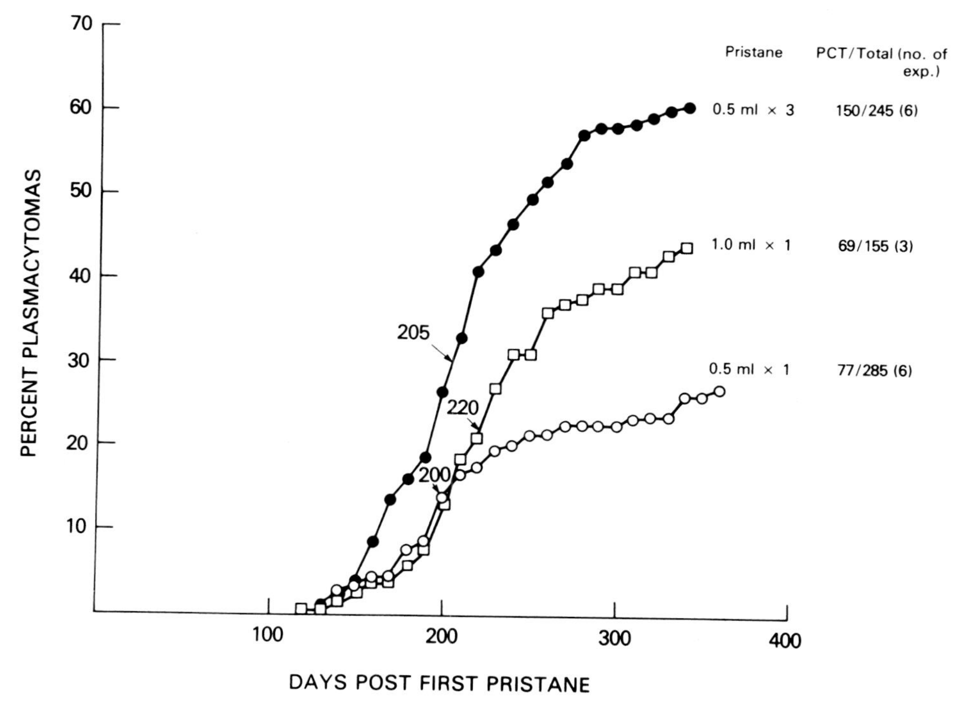

FIG. 1. Summary of data reported in ref. 77 on the incidence of plasmacytomas in BALB/c π (of BALB/c An origin) mice injected intraperitoneally with different doses of pristane. The chart shows the percentage of mice that developed plasmacytomas within 365 days after the first or only injection of pristane. The numbers followed by the arrows are the day at which 50% of the mice in each experiment developed a plasmacytoma (the mean latent period).

Oil Granuloma

It appears that the intraperitoneal oils act as mild irritants and induce the influx of blood monocytes and other inflammatory cells into the peritoneal cavity (21). The monocytes develop into peritoneal macrophages which phagocytize oil droplets. These phagocytic cells adhere to peritoneal surfaces (mesentery, diaphragm, abdominal wall, posterior peritoneum, and pelvic tissues) and form an organized pathologic tissue, called an oil granuloma (71). Such oil granulomas can be seen histologically as early as a few days after injection of oil. This tissue soon becomes vascularized and its surface covered with mesothelium. Much of the oil incorporated into the granuloma appears in individual macrophages, but some larger globules can be seen surrounded by several macrophages. The early oil granuloma is highly cellular and contains numerous nucleated cells, many of which are not macrophages (71). Later, granulocytes, lymphocytes, and a few plasma cells are found widely dispersed among the oil-laden macrophages.

The importance of the oil granuloma and other chronic peritoneal granulomatous tissues (i.e., the granuloma induced by solid plastics) in plasmacytomagenesis is indicated by the fact that the plasmacytomas arise in this tissue (71) and that primary plasmacytomas appear to require cells and factors from the oil granuloma for growth (21,74). There is only one study on the histogenesis of oil-induced plasmacytomas (71). Mice that had not developed ascites were autopsied during the time when plasmacytomas first began appearing. In these mice, and in others that were autopsied for different reasons during the course of other plasmacytoma induction studies, focal plasma cell hyperplasias and microscopic developing plasmacytomas were described. These appeared in the oil granulomatous tissue, usually close to a mesothelial surface. Developing plasmacytomas apparently shed tumor cells into the peritoneal space; thus even early plasmacytomas are associated with the appearance of plasmacytoma cells in the ascites and multiple sites of plasma cell tumor growth on peritoneal surfaces.

Additional evidence that the oil granuloma (or its equivalent in the plastic system) plays a basic role in plasmacytoma development comes from the requirements for transplantation of primary plasmacytomas (74). Plasmacytoma cells from primary hosts usually do not grow if they are introduced into the normal peritoneal cavities of syngeneic hosts (74). In contrast, primary plasmacytomas grow rapidly when introduced into a syngeneic mouse that has been conditioned by an intraperitoneal injection of pristane, given on the day of transplantation or several days before. It can be shown that this effect is related to the influx of macrophages into the peritoneum, for if this process is blocked by the administration of hydrocortisone (initiated before pristane injection), a highly reduced incidence of primary growths is obtained (21). If hydrocortisone treatment is initiated after pristane has been given and then primary plasmacytoma cells are injected, the tumor cells grow just as if the mice had received only pristane. Once sufficient macrophages enter the oil-conditioned peritoneum, they are able to replicate and form an oil granuloma. In contrast, repeated thioglycollate injections do not induce an adherent granuloma and do not condition the peritoneum for the growth of primary plasmacytoma cells

(21). The evidence suggests that a dividing population of macrophages and the ability of the macrophages to adhere to tissues are factors necessary for the conditioning of the peritoneum.

Chronic inflammatory tissues, such as the oil granuloma, are probably sources of growth factors that play a role in B-lymphocyte behavior. This matter has not been systematically studied, but there is indirect evidence that factors that stimulate growth of established plasmacytomas are generated by the oil granuloma (49,59,60,67).

Genetic Basis of Susceptibility

Pristane, mineral oils, or plastics are not universal plasmacytomagens in mice. In fact, most of the common inbred strains, C57BL/6, C57BL/Ka, C3H/He, A/J, DBA/2, and CBAT6T6, are resistant to the induction of intraperitoneal plasmacytomas by these agents. The incidence of plasmacytomas is 5% or less in these strains, and many develop no tumors (see ref. 54). In contrast, approximately 60% of BALB/c An (and related sublines) consistently develop plasmacytomas within 1 year after the first injection of pristane (77). Plasmacytomas develop in 35% (97) or 15% (56) of NZB mice. The differences in incidence of plasmacytomas in NZB mice in these two reports may be due to susceptibility differences in the different sublines used. The latent periods of plasmacytoma development in the two strains differ; the mean latent periods range from 180 to 240 days for BALB/c (77) and about 380 days for NZB (56).

The striking genetic susceptibility of the BALB/c mouse provides a good model system for identifying genes that influence the process of plasma cell neoplastic transformation. Susceptibility (S)-genes hypothetically determine the increased probability that cells in the BALB/c B-lymphocytic lineage will undergo neoplastic transformation, while resistance (R)-genes decrease these probabilities. R-genes are defined in a relative context, i.e., by comparing the susceptible BALB/c AnN mouse with a specific resistant strain. Probably most resistant strains differ from BALB/c by more than one gene.

Another approach for identifying S- and R-genes that determine susceptibility to developing plasmacytomas is to examine sublines of BALB/c mice for variations in plasmacytoma incidence. An important subline difference in susceptibility exists between BALB/c AnN (Andervont, NIH substrains) and BALB/c Jax. The BALB/c Jax subline was separated from BALB/c An in 1937; in contrast to BALB/c AnN where the incidence ranges from 50 to 70%, less than 20% of BALB/c Jax develop plasmacytomas after three 0.5 ml i.p. injections of pristane (76). Few genetic differences are known to exist between BALB/c An and BALB/c Jax, and most of the genes that are commonly polymorphic in the mouse are known to be the same in these two sublines (76). One striking difference is known: the regulator of α-fetoprotein (AFP) (RAF-1) gene, which controls the expression of high neonatal levels of AFP throughout adult life (64). The linkage and relationship of RAF-1 to plasmacytomagenesis resistance has not been conclusively determined. It could be argued that in 1937, when these two sublines were separated, the BALB/c mice were still segregating those genes that determine resistance and susceptibility to plasmacytomagenesis. Alternatively, it is possible that mutations affecting S-genes occurred in BALB/c Jax.

Immunologic Aspects of Plasmacytomagenesis

A characteristic of the Ig-producing tumors in inbred mice or rats is the predilection for different inbred strains to develop tumors expressing characteristic types of heavy chains. Plasmacytomas induced in BALB/c mice by injections of mineral oils express predominantly (~60%) IgA heavy chains (55,68). Lou/Ws1 rats spontaneously develop Ig-producing tumors (immunocytomas), of which a high proportion (~30%) secrete IgE monoclonal Igs (11,12). Two studies have been reported on NZB mice: one by Warner (97), who reported predominantly IgA-secreting tumors, and the other by Morse et al. (56), who found a predominance of IgG secretors. These differences may be due to subline differences. Plasmacytoma induction experiments utilize conventionally raised mice that carry a normal gut and respiratory microbial flora and which are exposed to common murine viral and microbial infections. Germ-free BALB/c mice injected with mineral oil are resistant to developing plasma cell tumors, and those few that occur in these mice do so after long latent periods (47). The normal microbial flora, dietary antigens, and exposures to mild infectious agents probably are essential for the stimulation and development of lymphoid tissues which contain the cells at risk for plasmacytomagenesis.

The reason why a high proportion of plasmacytomas induced by mineral oil in BALB/c mice produce IgA is not established. IgA-type antibodies are normally found in mucosal secretions, such as in the gut. Approximately 5% of the myeloma proteins from BALB/c mice bind environmental antigens that can be found in the gut microbial flora or diet (69). Furthermore, many of these antigen-binding myeloma proteins share antigenic (idiotypic) specificities with natural antibodies (69). There is substantial experimental evidence that many IgA-producing cells originate from germinal center lymphocytes in Peyer's patches (see ref. 18). It is also generally accepted that these cells are stimulated by antigens that enter via the gut, even though the precise mechanism of antigen presentation is not yet understood. Peyer's patch germinal center lymphocytes express IgA on their plasma membranes (mIgA). In addition, $mIgA^+$ lymphocytes migrate from Peyer's patches through the mesenteric lymph nodes, circulate, and home into the lamina propria of the gut and respiratory tracts. Circulating $mIgA^+$ cells may be diverted into the oil granuloma induced by the intraperitoneal injection of pristane. Here they may undergo further development and mature into IgA-secreting plasma cells. Thus a possible cellular precursor of the neoplastic plasma cell may be an $mIgA^+$ lymphocyte. Some data support this possibility. The MOPC315 plasmacytoma has been shown to contain two neoplastic cell types: an $mIgA^+$ cell and an IgA-secreting cell type. The $mIgA^+$ cells are lymphoid and develop into the secretory types (50,85,86) as the tumor grows. Thus some plasmacytomas may also contain neoplastic lymphoid precursors of plasma cells.

As described below, the nonrandom chromosomal translocation T(12;15) occurs in high frequency in BALB/c plasmacytomas. The locus most frequently found at the junction site in chromosome (chr) 12 is Sα, the switch region sequence just 5′ of the first exon of Cα (IgA heavy chain constant region gene). This association could be due to the relatively high risk of neoplastic transformation in cells differentiating to the IgA heavy chain class.

NZB mice, in contrast to BALB/c, produce a high proportion of IgG-secreting myelomas. These mice also spontaneously develop autoimmunity, which is associated with hyperactivity of IgG-producing cells. Thus with NZB, plasmacytomagenesis may preferentially involve a B-lymphocyte population different from that transformed in BALB/c mice.

RETROVIRUSES

Endogenous retroviruses are potentially important components of any process of lymphoma development in mice because they are potential mutagenic agents. There are three large families of endogenous retroviruses: (a) the type C leukemia viruses, including xenotropic and ecotropic forms, (b) mammary tumor viruses (MTV), and (c) intracisternal A particles (IAP).

Type C Leukemia Viruses

There are two major classes of endogenous type C viruses in mice: the xenotropic and ecotropic forms. Both are transmitted vertically by proviral loci. Inbred strains of mice carry multiple xenotropic proviral loci (37), but as a rule these viruses do not infect mouse cells. Any role that they might play in plasmacytomagenesis is probably not as an infectious agent but as a potential mutagen.

Strains of mice differ in their number of loci for ecotropic proviruses (39). BALB/c has a single ecotropic locus, env-1, on chr 5, while NZB has none (53). The env-1 locus in BALB/c codes for an N-tropic form of type C virus; this form is inefficient in establishing infections in other BALB/c cells because of the restrictions of the Fv-1^b allele carried by BALB/c. The Fv-1^b gene interferes with N-tropic virus propagation by impeding successful integration of the virus (see ref. 38); thus even the endogenous ecotropic virus may not be a factor in plasmacytomagenesis.

New forms of type C viruses can arise in cells through recombinations of ecotropic and xenotropic proviral loci or possibly between two activated endogenous viruses. One such recombinant form is the B-ecotropic type C virus, which is a recombinant form that contains sequences in its *gag* genes that are derived from both ecotropic and xenotropic viruses. These recombinants can produce infections in cells of Fv-1^{bb} hosts (i.e., BALB/c) (13).

A second type of recombinant ecotropic retrovirus are the mink cell focus forming (MCF) viruses. The *env* genes of these viruses contain sequences of ecotropic and xenotropic origin (28,62). MCF viruses can spread efficiently from cell to cell. Furthermore, cells infected with these viruses express the novel products of the *env* genes on their surfaces. These viral genes may be biologically active and alter the physiology of the cells in which they are expressed. MCF viruses appear to have specific cellular tropisms that are associated with tumor formation

(23). They are associated with AKR thymic leukemogenesis (24,34) and other kinds of lymphomas, including B-cell neoplasms in old mice (65).

Endogenous type C viral particles are actively produced in all primary and transplantable plasmacytomas so far studied (7,10,54). The injection of pristane i.p. in BALB/c mice activates xenotropic and ecotropic retroviruses (J. Hartley and M. Potter, *unpublished observations*). This is not surprising as B cell mitogens are also known to activate endogenous retroviruses (4,52). The xenotropic forms are most commonly expressed, but N- and B-ecotropic viruses have also been isolated (54). A search for MCF-like viruses in BALB/c plasmacytomas has produced only one such isolate among 26 so far studied (54).

In NZB mice, which lack ecotropic proviral loci, xenotropic viruses are produced continuously throughout life. These viruses are not infectious for other NZB cells. Thus a spreading somatic cell infection probably is not a factor in NZB plasmacytomagenesis; nor are actions attributable to recombinant forms of N- and B-ecotropic virus applicable to this model.

MTV

A few preliminary studies of MTV activity in plasmacytomas are available (20). The BALB/c mouse genome carries multiple MTV-proviral loci. In plasmacytomas, several of these appear to be activated. Some plasmacytomas transcribe novel mRNA from DNA found in flanking sequences around MTV loci. The significance of these findings awaits additional studies.

IAP

The IAP represent a special and distinct class of retroviruses that have diverged extensively from other retroviruses in the mouse (41,45). During the evolution of the *Mus musculus* species, a dramatic amplification of IAP sequences occurred in the germ cells. It is estimated that approximately 1,000 copies of IAP proviruses are integrated and distributed on all the chromosomes of the mouse (44). All inbred strains so far studied appear to have the same complement of IAP genes (41).

IAP bud into the cisternae of the rough endoplasmic reticulum. Since they appear to remain within cells, they are not considered to be able to spread from cell to cell. They have been of great interest in plasmacytomagenesis because virtually every plasmacytoma so far examined contains an abundance of these particles, while none are found in normal plasma cells (27). IAP are not specific for plasmacytomas but have been found in many other tumors (101), as well as in mouse embryos.

The mechanism of IAP activation is not known. Apparently, in plasmacytomas, a large number of different sized mRNAs are transcribed. While IAP are not infectious for other cells, they are capable of reintegrating into host cell chromosomes. Recently, two examples of IAP-associated reintegrations have been described (35,40,42,82). Both were found in BALB/c plasmacytomas. In one, an IAP genome is transposed to an intervening sequence between two Ig kappa exons (35). In a second case, Rechavi et al. (82) and Kuff et al. (42) discovered an IAP insertion within the c-*mos* oncogene in the XRPC24 plasmacytoma. The IAP insert was found in the opposite transcriptional direction to *mos* gene elements; thus it cannot act as a promotor for *mos* gene elements. However, the sequence does provide a potential open reading frame that can permit transcription of c-*mos*. In addition, the long terminal repeat of the virus may enhance transcription of the usually quiescent gene of c-*mos*. It will be important to screen primary plasma cell tumors for other rc-*mos* sequences to determine if this event is frequently associated with plasmacytomagenesis and is not an artifact or secondary change that occurred during transplantation.

Abelson Virus

The Abelson virus has been reviewed recently (83,87). Here we summarize its association with plasmacytomagenesis. The Abelson transforming viral element (A-MuLV) with a Moloney leukemia virus helper induces lymphoid tumors (Abelson lymphosarcomas, ABLS) with great rapidity in neonatal mice of most inbred strains (1). These tumors arise in bone marrow and lymph nodes. Large tumors often protrude from the skull, vertebrae, and long bones. Lymphosarcomas induced *in vitro* have been shown by DNA hybridization studies (6,93) to resemble early pre-B stages of development by usually having only their heavy chain gene loci rearranged. In many of the *in vitro* transformed cell lines derived from fetal liver, one of the heavy chain loci is productively rearranged, and small amounts of μ-chains are demonstrable in the cytoplasm (93). In other lines of similar origin, only L chain loci are rearranged. The long-term culture of normal B-cell lines has provided a more defined source of target cells for transformation by A-MuLV. Whitlock et al. (98) have shown that all stages of early B-lymphocyte development are susceptible to transformation by p160 type Abelson virus.

The undifferentiated lymphosarcoma derived from the B-cell lineage is also the predominant neoplasm induced *in vivo* by A-MuLV (83,84). In some of these tumors, small amounts of cell surface Ig may be detected by immune precipitation (72,78), but expression of membrane Ig usually is an unstable property and often is lost as the tumors are successively passaged (79).

In contrast to the susceptibility of most neonates to induction of tumors by A-MuLV, the adults of only a few inbred strains are susceptible. The most susceptible are BALB/c and SEC, while DBA/2, SEA, and SWR are partially susceptible (79). Two genes, AV-1 and AV-2, determine susceptibility, and these genes appear to be associated with expression of differentiation antigens on target cells (83). Adult C57BL strain mice are resistant as adults to developing tumors, although their cells can be infected with Moloney helper virus (79). Why tumors can be produced in neonates and not in adults is not yet understood.

A-MuLV infection alone does not induce plasmacytoma development in BALB/c mice; however, the susceptibility of these mice has made it possible to study the effect of A-MuLV infection on plasmacytoma development in pristane-conditioned mice (75). This study was prompted by a chance finding. A-MuLV-induced lymphosarcomas were being transplanted in pristane-conditioned mice to determine if these cells required this microenvironment for growth. Smears of the ascites from one of these lines revealed the presence of typical plasmacytoma cells admixed with the lymphosarcoma cells. During further serial passages, the tumor became a pure plasmacytoma (ABPC4). This curious phenomenon has been seen since. We then began to infect pristane-conditioned BALB/c mice with A-MuLV. In these experiments, the mice were conditioned with a single 0.5 ml i.p. injection of pristane for short periods of time (10 to 40 days) before A-MuLV infection. Lymphosarcomas appeared 21 to 80 days after virus infection in a large number of the mice; however, 5 to 25% of the mice in different experiments also developed plasmacytomas. These tumors secreted myeloma proteins predominantly of the IgA class, like the plasmacytomas induced by mineral oil or pristane in this strain (75).

All the A-MuLV-associated plasmacytomas so far studied contain integrated Abelson proviral genomes (57,58), and most produce Moloney helper virus (and hence infectious A-MuLV) as well. When the latent periods of A-MuLV-induced plasmactyomas are compared with latent periods of plasmacytomas induced with 0.5 ml pristane alone (or by other dose regimens) (Fig. 2), it can be seen that the mean latent periods (when assessed from the time of either pristane injection or virus infection) are dramatically shortened by the presence of A-MuLV. There is essentially no overlap with the earliest tumors induced by pristane only. This result indicates that A-MuLV completes the transformation steps of preplasmacytoma cells, or possibly even in a single step transforms appropriate target cells to become

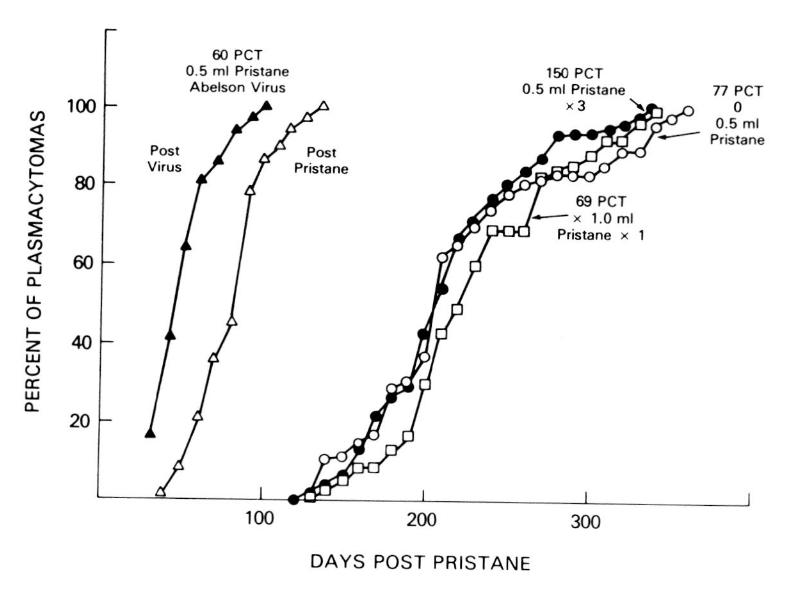

FIG. 2. Comparison of the latent periods of plasmacytomas (PCT) induced by pristane alone or by Abelson virus infection in pristane-conditioned mice. The latent period or time of detection of plasmacytomas induced by Abelson virus and pristane is present in two forms: *Open triangles*, latent periods measured from the day when 0.5 ml pristane was first injected; *filled triangles*, latent periods for individual tumors when measured from the day Abelson virus (A-MuLV) was injected.

plasmacytomas. In BALB/c mice injected only with pristane, the plasmacytomagenic process probably occurs in several steps over a much longer period of time. We hypothesize that the pristane conditioning produces preplasmacytoma cells that are vulnerable to direct transformation by A-MuLV.

A third type of B-cell tumor is also seen in pristane-conditioned mice infected with A-MuLV (72). The cells of these tumors have slightly more cytoplasm than lymphosarcomas, and the cytoplasm stains intensely like that in plasmacytomas (72). Electron microscopic studies indicated that these tumors also had a better developed rough endoplasmic reticulum than the ABLS cells (72). We call these tumors plasmacytoid lymphosarcomas (ABPL). We recently demonstrated that both ABLS and ABPL tumors, but not ABPCs, express abundant amounts of a 3.8 kb RNA transcript of the c-*myb* gene (58). ABPLs have three unusual characteristics that set them apart biochemically from the ABLSs and ABPCs: (a) The ABPLs do not contain A-MuLV integrated as a provirus in their genomes and do not express abundant *abl* RNA transcripts from such a provirus; (b) ABPL contain an unusually large *myb* RNA (4.5 to 5.2 kb) that differs in size in different tumors; and (c) a DNA rearrangement in or near one end of the c-*myb* locus is seen in Southern blots of restriction endonuclease digests of ABPL tumor DNA (58). We have speculated that the alterations in *myb* DNA and RNA may be due to the unstable integration of A-MuLV near the *myb* locus followed by an imprecise excision of the provirus. This "hit and run" phenomenon may enable the ABPL cells to remain neoplastic in the absence of transforming virus. It has been shown with several ABLS lines of C57BL origin that A-MuLV proviruses can be deleted and that the cell types that have lost provirus are still neoplastic (31). It remains to be demonstrated whether the hit and run transformation in these cells utilizes a mechanism similar to that operating in ABPL cells.

NONRANDOM CHROMOSOMAL TRANSLOCATIONS

Yosida et al. (103) studied 16 transplantable mineral oil-induced plasmacytomas in the BALB/c mouse and found that all were highly aneuploid with modal chromosome numbers in the subtetraploid range. In a separate study of 14 different primary or early transfer generation tumors, Yosida et al. (102) again found three diploid and one hyperdiploid tumors, while the rest were in the near tetraploid range. During serial transplantation, all the tumors changed and acquired near tetraploid chromosome numbers. Although tetraploidy is not uncommon in neoplasms, the reason why BALB/c plasmacytomas so regularly have subtetraploid chromosome numbers is not known.

With the availability of G-banding techniques that permit the identification of the individual chromosomes in the mouse, specific chromosome translocations were identified. Shepard et al. (89–91) identified reciprocal translocations of chr 6 and 15 [rcpT(6;15)] in MOPC31C and rcpT(12;15) in MOPC21, MOPC315, and X5563. The break point in chr 15 was at the same site for both kinds of translocations (92). Yosida et al. (104,105), using Q-banding, found deleted (del) chromosome 15s and T(12;15) in two plasmactyomas. Ohno et al. (63) studied seven primary pristane-induced plasmacytomas in CBB-22 mice [a BALB/c congenic strain carrying Igh, (chr 12), C (chr 7), and B (chr 4) genes from C57BL/Ka]. All seven tumors had one of two characteristic translocations, either rcpT(6;15) or T(12;15) + (del) 15 (63). The break sites on chr 15 were the same as previously observed between the D3 and E bands.

In subsequent karyotypic studies of early transfer generation pristane-induced plasmacytomas in BALB/c and BALB/c congenic mice, 85% were found to have T(12;15) + del 15 and (11%) rcpT(6;15) (63; F. Wiener, S. Ohno, G. Klein, and M. Potter, *unpublished observations*). This result dramatically illustrates the association of nonrandom translocations involving chr 15 and plasmacytomagenesis.

A characteristic feature of the chromosomal translocations in mouse plasmacytomas is that they involve only one chr 15 in diploid cells or two in the near tetraploid cells. Furthermore, rcpT(6;15) and T(12;15) have not been seen in the same cell. If both types of translocation would have occurred in the same diploid cell, that cell would still have one normal chr 6 and 12 and thus potentially be able to synthesize Ig, provided productive rearrangements occurred on both these chromosomes. However, both chr 15 in a diploid cell would be disrupted by a double translocation process; if the break sites in both chr 15 occurred within the *myc* gene, then those cells would be able to express *myc* gene sequences only from a rearranged form of the gene.

After a productive DNA rearrangement process in Ig genes has taken place on one chromosome (e.g., chr 6 or 12), the other Ig-bearing chromosomes probably do not continue to rearrange. Similarly, in tetraploid plasmacytomas, other copies of intact chr 6, 12, and 15 remain in the nontranslocated state. Thus the translocations do not appear to be an inevitable phenomenon related to hypersusceptible break sites on these chromosomes associated with the neoplastic state. Rather, they are more likely to be related to functional changes in DNA that occur during development, such as those known for Ig gene rearrangements and, perhaps, for actively transcribed genes, such as *myc*.

In many cases, the translocations occur on the nonproductively rearranged Ig gene-bearing chromosome; e.g., in IgA-producing tumors, the translocation usually involves the switch α (Sα) region or, in rare cases, the Sμ 5′ of the Cμ gene. It is worth noting that a translocation to chr 12 at Sα would still permit transcription of complete heavy chain messages from Cμ or other proximal genes.

The exact time of the translocation processes in relation to B-lymphocyte differentiation has not been determined. They may be temporally associated with one of the gene rearrangement processes in Ig gene-bearing chromosomes. Break sites occurring near the joining (J) genes could be related to V-J joining, which occurs early in B-lymphocyte development. Break sites occurring in switch regions could be considered late events that take place in more mature B-lymphocytes. Homologous types of translocations involving human Ig gene-bearing chr 2 (κ) 22(λ) and 14 (h) with 8, the *myc* gene chromosome, occur in the B-lymphocytic neoplastic disease Burkitt's lymphoma (43a). Thus translocations of this type are not limited to Ig-secreting mouse plasma cells. It is possible, then, that the translocations in the mouse occur in B-lymphocytes prior to the plasma cell stage. In general, the evidence favors the hypothesis that the translocations occur prior to the development of the plasma cell stage (Ig-secreting). Two exceptions that cannot be easily explained are the tumors CBBTEPC 1 and CBBTEPC8 (63) in which both T(12;15) and rcpT(6;15) were seen during the early transplantation history. These cases may represent biclonal tumors.

Tumors induced by pristane alone show a great preponderance (85%) of T(12;15) translocations (Table 1), while only 52% of those induced by pristane and A-MuLV have T(12;15) (63a). There were three tumors in the Abelson series (Table 1) that had no translocations, and one in the series with pristane alone. Subsequently, it has been found that three of these tumors have an interstitial deletion in chr 15 in the vicinity of the *myc* gene (100a). Thus, only one tumor (ABPC22) has no discernible chr 15 translocation or internal deletion.

The limits of the conventional G-banding did not permit an accurate definition of the breakpoints on chromosomes involved in the plasmacytoma associated translocations. It was also unclear whether the T(12;15) rearrangement was reciprocal or of deletion type. By using the high resolution banding (HRB) technique (65a), it was possible to define the breakpoints on chromosomes 15, 12 and 6 with more accuracy. The HRB revealed two main facts. First, it proved the reciprocal nature of the T(12;15) translocation. On chromosome 15, the breakpoint was located

TABLE 1. *Karyotypes of plasmacytomas determined by Ohno and Wiener*

Translocation	Pristane only plasmacytomas		Abelson virus + pristane plasmacytomas
rcpT(12;15)	CBBTEPC1[a]	PI-4 19-93(1°)[d]	ABPC18
	CBBTEPC6	PI-4 15-72(1°)[d]	ABPC24
	CBBTEPC8(10)[b]	PI-4 1193(1°)[d]	ABPC33
	CBBTEPC9	PI-4 20-102(1°)[d]	ABPC47
	CBBPC49(λ)	MCF 16-78(1°)[d]	ABPC48
	TEPC1072(λ)	MCF TEPC2021	ABPC52
	TEPC1081(λ)	MCF TEPC2027	ABPC60
	TEPC1082(λ)	MCF TEPC2030	ABPC65
	TEPC1094(λ)	TEPC1165	ABPC72
	XRPC24[c]		ABPC89
	TEPC1017(IgD)		
	TEPC1033(IgD)		
rcpT(6;15)	CBB22TEPC1[a]		ABPC4
	CBB22TEPC8[b]		ABPC17
	CBBTEPC3		ABPC105
Internal deletion: del 15(qD1qE)	CBPC112(IgM)		ABPC26
			ABPC45
None			ABPC22(IgM)

[a]In the primary and transfer generation-2, T(12;15) was seen; in transfer generation-1, rcpT(6;15) was seen.
[b]In transfer generation-1, both T(12;15) and rcpT(6;15) were seen.
[c]Tissue culture line is trisomic for T(12;15).
[d]1°, Tumor studied only as a primary tumor that was not transplanted.

TABLE 2. *Relationship of chromosomal translocations to rearrangement of the* myc *oncogene and transcription of* myc *gene mRNA*

Translocation	*myc* gene DNA	Tumors	*myc* gene transcription
rcpT(6;15)	Rearranged	ABPC17 (IgA κ)	c-*myc*
	Germline	ABPC4 (IgA κ)	c-*myc*
		ABPC103 (IgG_1 κ)	c-*myc*
		ABPC105 (IgA κ)	c-*myc*
		CBBPC3 (IgA κ)	c-*myc*
T(12;15) + deleted 15	Rearranged	ABPC18 (IgG_{2b} κ)	rc-*myc*
		ABPC24 (IgA κ)	rc-*myc*
		ABPC33	c-*myc*
		ABPC48 (IgA κ)	c-*myc*
		ABPC72 (IgA)	rc-*myc*
		TEPC1165(IgA κ)	rc-*myc*[a]
		TEPC1017 (IgD κ)	rc-*myc*
		TEPC1033 (IgD κ)	c-*myc*
	Germline	ABPC47(IgA κ)	c-*myc*
		ABPC52(IgA κ)	c-*myc*
		ABPC60(IgG_1)	c-*myc*
		ABPC89	c-*myc*
None	Rearranged	ABPC45(IgA κ)	c-*myc*
	Germline	ABPC22(IgM κ)	c-*myc*
		ABPC26(IgA)	c-*myc*

[a]c-*myc* indicates the synthesis of a 2.4 kb *myc* RNA; rc-*myc* indicates the synthesis of a 1.8 kb *myc* RNA; but TEPC 1165 uniquely contains 1.8, 2.4 and 4.5 kb *myc* RNAs.

between bands D2 and D3 (D2/3) and on chromosome 12 within band F1. The F2 band of chromosome 12 was translocated at D2/3 on chromosome 15. The reciprocal nature of the rearrangement was also confirmed by molecular evidence by Cory et al. (24a). Second, it showed unequivocally that the breakpoint on chromosome 15 in rcpT (6;15) is identical with that identified in rcpT (12;15). This strongly suggested the involvement of identical 15 chromosome segment in both translocations. On chromosome 6, the breakpoint was located on band C2 in agreement with that previously defined by G-banding.

The identification of specific genes at the break sites on chr 6 and 12 has been possible through DNA hybridization studies (see below). The break site on chr 6 is roughly compatible with the locations of the Igkappa complex locus. The break site in chr 12 is near its distal end and has been shown to involve sites within the IgC_H gene complex by DNA sequencing at the junction site. The karyotypes of NZB plasmacytomas have not been systematically studied. This would be of considerable interest because trisomy of chr 15 has been reported to occur spontaneously in this strain (81).

The high susceptibility of the BALB/c mouse strain to plasmacytomagenesis raised the question whether this feature is determined by some peculiarity of the c-*myc* carrying segment of the BALB/c-derived chromosome 15. This possibility was examined in plasmacytomas induced in (BALB/c × AKR) × BALB/c backcross mice carrying cytogenetically distinguishable chromosomes 15 (100b). The results showed that there was no significant preference for the BALB/c-derived 15 chromosome. Thus the genes determining susceptibility to the development of plasmacytomas lie outside of the chromosomes involved in plasmacytoma associated chromosomal rearrangements.

Role of the *myc* Oncogene

Investigations into the molecular nature of the chromosome translocations found so frequently in plasmacytomas have focused on the *myc* oncogene, which has been mapped close to the part of chr 15 that breaks and recombines with one of the Ig-encoding chromosomes (2,19,25,32,33,88). It was not immediately recognized that this oncogene locus was intimately involved with these chromosome translocations, so this locus was initially called NIARD (nonimmunoglobulin associated rearranging DNA) (32,33), NIRD (standing for roughly the same concept) (19,25), or LyR (for lymphocyte rearranging DNA) (3). The identification of these rearranging portions of chr 15 as the *myc* locus was made simultaneously by many groups, but it was Shen-Ong and colleagues (88) who published the initial, comprehensive description of (a) how the *myc* gene was always found near the break point junction of chr 15; (b) how transcription of the *myc* gene was usually altered in cells containing translocations of chr 15 such that a *myc* RNA of 1.8 kb was synthesized instead of the normal 2.4 kb *myc* RNA; and (c) how *myc* gene rearrangements usually took the form of translocating the *myc* portion of chr 15 into the switch region of IgA α chain constant region gene. Shen-Ong et al. (88) did not think that this *myc* rearrangement resulted in any increase in *myc* RNA transcription, but other groups did present data that indicated that plasmacytomas contained greater amounts of *myc* RNA than other lymphoid tumors, or than normal spleen, liver, or thymus (46,57). The most appropriate control with which the levels of *myc* RNA in plasmacytomas should be compared (namely, normal plasma cells) has never been used; thus this point remains controversial.

As can be seen in Tables 1–3, the ABPC tumors have much the same changes in *myc* gene rearrangement and expression as do the plasmacytomas induced by pristane alone. *myc* DNA rearrangements are seen in *Eco*RI endonuclease digests of seven ABPCs: five from the tumors with T(12;15), one from the rcpT(6;15), and one from the tumors with no chromosome translocation. All the ABPCs express abundant *myc* RNA, mostly in the form of the c-*myc* transcript of 2.4 kb, which is also seen in normal cells. In three cases, the *myc* RNA takes the form of a tumor-specific 1.8 kb transcript. The 1.8 kb rc *myc* RNA is found only in tumors with T(12;15) and a rearrangement in the *Eco*RI *myc* restriction fragment. It is

INTER- AND INTRACHROMOSOMAL REARRANGEMENTS IN PLASMACYTOMAS AND BURKITT'S LYMPHOMAS

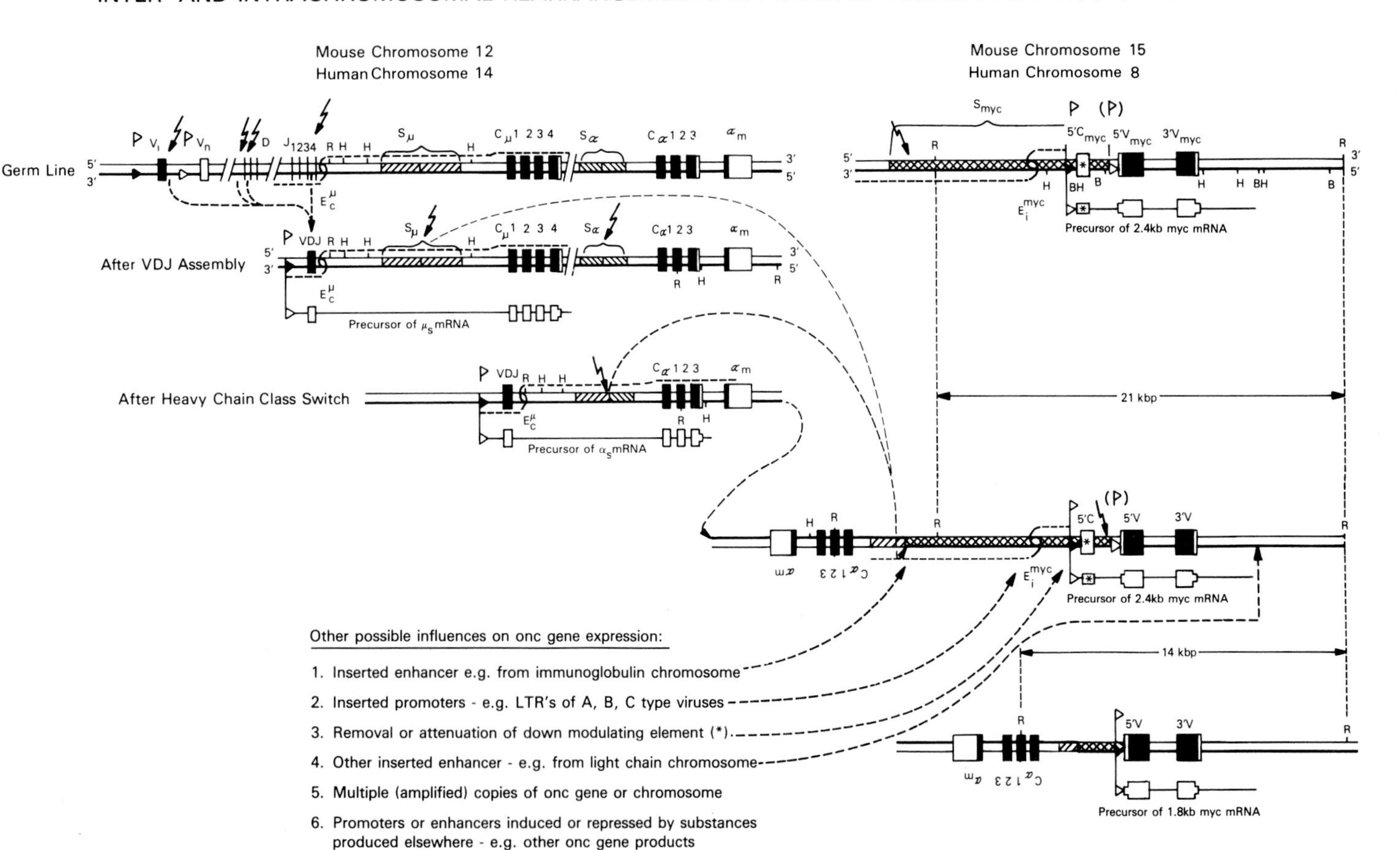

Other possible influences on onc gene expression:

1. Inserted enhancer e.g. from immunoglobulin chromosome
2. Inserted promoters - e.g. LTR's of A, B, C type viruses
3. Removal or attenuation of down modulating element (*).
4. Other inserted enhancer - e.g. from light chain chromosome
5. Multiple (amplified) copies of onc gene or chromosome
6. Promoters or enhancers induced or repressed by substances produced elsewhere - e.g. other onc gene products

important to emphasize that not all ABPCs with *myc* genomic rearrangements synthesize the 1.8 kb *myc* RNA, and that *myc* genomic rearrangements can be seen in ABPCs with rcpT(6;15) and in some with no translocation.

It is important to consider mechanisms that control *myc* transcription. Some of these are presented diagrammatically in Fig. 3, which depicts the most common and best studied situation wherein the *myc*-containing portion of chr 15 is joined to chr 12 at a point in the Sα region. If the break point in chr 15 occurs to the left of the *Eco*RI site 5′ to the *myc* gene segments, then no change is seen in the size of *myc Eco*RI hybridizing band, and no change is seen in the size of the *myc* RNA transcript. If the break point occurs within the *myc* gene itself (shown in the lowest right diagram), then a smaller *Eco*RI *myc* fragment would be created and a smaller

FIG. 3. Diagrammatic summary of the molecular events known to occur when T(12;15) chromosomal rearrangements bring the *myc* gene complex close to the Cα gene complex. On the top left are depicted the germ line arrangements of the Ig heavy chain locus. The arrows pointing left indicate points in the DNA where breaks and recombination occur during VDJ assembly of the heavy chain variable region gene that is first expressed attached to the IgM μ-chain constant region genes and expressed as the variable part of an IgM heavy chain. After VDJ assembly, more breaks and recombinations (at locations indicated by arrows) occur in the switch regions (Sμ and Sα), permitting the same VDJ to be expressed on an IgA α chain mRNA. Still another DNA rearrangement involving breakage and recombination in the Sα region is required for the most common type of Sα-*myc* recombination seen in T(12;15) chromosome translocations. The arrow pointing right points to a site in the Sα region where the breakage in chr 12 may occur in a nonproductive heavy chain rearrangement. Whether this breakage and recombination occurs after heavy chain switch or in the germ line, unrearranged configuration is not known; to simplify the diagram, it is depicted after class switch. At the top right is shown a possible germ line configuration of the c-*myc* gene. It consists of two exons, 5′ V_{myc} and 3′ V_{myc}, which are homologous to parts of the viral v-*myc*. There are additional sequences transcribed from the region 5′ to those two exons, but this exon is thought to contain only untranslated sequences. These sequences are present in 2.4 kb *myc* mRNAs, but not in 1.8 kb *myc* mRNAs. The arrow points out one possible site for breakage and recombination with DNA from chr 12. This recombination, in opposite order of RNA transcription, is depicted in the middle right diagram. The recombination event depicted here occurs 5′ of the *Eco*RI site (R) and, therefore, does not alter the germ line 21 kbp *Eco*RI *myc*-containing fragment. Similarly, transcription occurs at the usual c-*myc* promotor [▷], perhaps influenced by nearby but weak enhancer sequences [§, E_i^{myc}]. Such enhancers may act on promotors in either direction or on either strand, as long as they are not too distant (within the interrupted lines). The lowest right diagram shows the configuration that probably ensues if the break and recombination point is within the *myc* gene complex, effectively deleting the *myc* enhancer and the usual promotor. In this case a weaker promotor is used to initiate RNA transcription at a potential but usually unused promotor [(▷)]. The *Eco*RI site in Cα2 now defines one end of the *myc Eco*RI fragment at around 14 kbp, and the *myc* transcript is smaller, e.g., 1.8 kb. These diagrams are freely adapted from the data presented in refs. 2, 88, and 94. The symbols in the diagram include ▷ for promotor, and ▷ for site of initiation of RNA transcription. *Thicker line*, DNA strand used as template for RNA transcription; *V*, heavy chain variable region gene segments; *D*, diversity gene segments; *J*, heavy chain joining gene segments; Sμ and Sα, switch regions associated with Cμ and Cα (constant region gene segments for μ and α heavy chains); §, enhancer sequences with influence extending as far as the interrupted line indicates; $E_c\mu$, constitutive enhancer associated with the 4 μ constant region gene segments (Cμ, 1,2,3,4); E_i^{myc}, inducible enhancer associated with the *myc* gene); α_m, membrane binding portion of the α chain; 5′V_{myc} and 3′ V_{myc}, two gene segments homologous to V_{myc}; 5′ C_{myc}, c-*myc* gene segment with sequences not found in v-*myc*; restriction endonuclease sites R = *Eco*RI, H = *Hind* III, B = *Bam* HI. The six influences shown at lower left are possible mechanisms that could operate on the *myc* gene in addition to or instead of the rearrangements depicted here in detail for 12;15 chromosome translocations.

myc RNA synthesized. In both cases, the *myc* and Sα are in opposite transcriptional orientations, as shown by the arrows and the upside-down labels.

Since the Sα sequence does not have an active promotor, the interrupted *myc* probably activates a cryptic or weaker promotor within the *myc* gene (▹), as has been proposed by Stanton et al. (94) and Adams et al. (2). It is clear that high levels of 2.4 kb *myc* RNA are commonly seen in plasmacytomas, even when 1.8 kb *myc* RNA does not appear (Tables 2 and 3) (57). Furthermore, high levels of 2.4 kb *myc* RNA are seen in some of the plasmacytomas that have no translocation of chr 15 (Tables 2 and 3). Therefore, this translocation is not a *sine qua non* for *myc* activation, and there must be additional regulatory mechanisms that can result in *myc* RNA transcription. These possibilities include (a) amplification of *myc* genes either in tandem (5,26) or due to polysomy of chr 15 (29,100); (b) enhancement of *myc* RNA transcription due to the approximation of strong promotor or enhancer sequences to the *myc* locus (66); and (c) insertion of active viral promoters in the form of long terminal repeats of leukemia virus (36,61).

An enhancer sequence is probably associated with the Sμ gene segment which activates the promotor associated with each V_H gene when V_H and C_H are close together in functional VDJ joining, as illustrated to the left in Fig. 3. The enhancer is not included in the portion of chr 12 that is recombined with the *myc*-containing portion of chr 15, so it is not thought to play a role in *myc* expression in plasmacytomas. Whereas the Sμ enhancer may be constitutive (E_c^{μ}), it is possible that *myc* has an enhancer normally associated with it which may be inducible (E_i^{myc}). Thus it may be activated by events or factors unrelated to chromosome translocation or in normal cells at different stages of B-lymphocyte maturation. Whether any or only some of these are operative in plasmacytomas is currently under investigation.

TABLE 3. *Characteristics of A-MuLV + pristane-induced plasmacytomas*

Tumor	Transplant generation	v-*abl* RNA + DNA	2.4 kb c-*myc* RNA	1.8 kb c-*myc* RNA	Rearr. c-*myc* DNA	Chr translocation
ABPC 4	14	+	+++	−	−	6;15
17	20	+	+++	−	+	6;15
18	31;32	+	++	++	+	12;15
20	9	+	++	−	−	ND
22	7	+	+	−	−	0
24	9	+	−	++	+	12;15
26	7,8	+	+++	−	−	0
33	24	+	+++	−	+	12;15
45	23,24	+	+++	−	+	0
47	8,9	+	+++	−	−	12;15
48	8	+	++	−	+	12;15
52	17;18	+	+++	−	−	12;15
60	3	+	+++	−	−	12;15
65	4,5	+	+++	−	ND	12;15
72	6,7	+	−	++	+	12;15
89	3,4,5	+	+++	−	−	12;15
103	10,11	+	++	−	−	6;15
105	8,11	+	+++	−	−	6;15

Another aspect of the *myc* findings that have been associated with mouse plasmacytomas should be considered here. Although a quantitative increase in amount of *myc* RNA appears to be characteristic of plasmacytomas (46,57), it is equally plausible that a qualitative change in *myc* RNA structure, i.e., an alteration in the sequence encoding the *myc* gene protein product could render the *myc* gene effective in the oncogenic process.

A general picture is emerging about the role of the *myc* gene in oncogenic transformation. Transfection or transduction of *myc* gene sequences (of varied origins) alone do not by themselves bring about neoplastic transformation of cells *in vitro* (42a). However, it has been recently shown that tumorigenic transformation of rat embryo fibroblasts *in vitro* can be made by the introduction of two oncogenes, e.g., *myc* plus *ras* (42a), whereas neither one alone is effective. There are interesting parallels with *in vivo* plasmacytomagenesis. First, the activation and dysregulation of *c-myc* by chromosomal alterations *in vivo* may constitute one essential step in transformation which is insufficient by itself. A second oncogene in mineral oil plasmacytomagenesis has not been identified as yet, but may exist (25,43). The *c-mos* gene may in some cases act as a second gene (82). The *abl* gene in the form of *v-abl* also appears to act as a second oncogene (63a).

Land et al. (42a) has proposed that *myc* belongs to a group of oncogenes (e.g., large T, E1a, *myc*) whose function *in vitro* relates to "establishment/immortalization." Normally *myc* is associated with active cellular proliferation (30). It is possible that the dysregulation of *c-myc* by chromosomal alterations compromises the ability of cells to mature to a post-mitotic state. Nonetheless, these cells can go on to develop the cellular organelles needed for Ig secretion. The second oncogene acting in this vulnerable cell type completes the oncogenic transformation process producing a neoplastic cell type that continuously secretes Ig but still acts as an actively replicating B-lymphocyte.

SUMMARY

BALB/c plasmacytomas arise in association with intraperitoneal granulomatous tissue induced by nonmetabolizable oils or plastic fragments. Differences in susceptibility to plasmacytomagenesis between the closely related mouse strains BALB/c and BALB/Jax indicate that genetic susceptibility and resistance are probably determined by only a few genes, but the mode of action of these genes is not yet known.

All but one pristane-induced plasmacytomas so far studied exhibit nonrandom chromosome alterations, either rcpT (12;15), rcpT (6;15) or an interstitial chr 15 deletion. These alterations all occur in the vicinity of the *myc* gene which is located in chr 15 near the D-band, and they are associated with high levels of *myc* gene transcription. It is suspected the *myc* gene activation is a critical but not the only step in mineral-oil plasmacytomagenesis. The reaction of a second oncogene is probably involved. Abelson virus infection of mineral-oil (pristane) treated BALB/c mice can induce plasmacytomas with short latent periods.

REFERENCES

1. Abelson, H. T., and Rabstein, L. S. (1970): Lymphosarcoma: Virus induced thymic-independent disease in mice. *Cancer Res.*, 30:2213–2222.
2. Adams, J. M., Gerondakis, S., Webb, E., Corcoran, L. M., and Cory, S. (1983): Cellular *myc* oncogene is altered by chromosome translocation to an immunoglobulin locus in murine plasmacytomas and rearranged similarly in Burkitt lymphomas. *Proc. Natl. Acad. Sci. USA*, 80:1982–1986.
3. Adams, J. M., Gerondakis, S., Webb, E., Mitchell, J., Bernard, O., and Cory, S. (1982): Transcriptionally active DNA region that rearranges frequently in murine lymphoid tumors. *Proc. Natl. Acad. Sci. USA*, 79:6966–6970.
4. Alberto, B. P., Callahan, L. F., and Pincus, T. (1982): Evidence that retrovirus expression in mouse spleen cells results from B-cell differentiation. *J. Immunol.*, 129:2768.
5. Alitalo, K., Schwab, M., Lin, C. C., Varmus, H. E., and Bishop, J. M. (1983): Homogeneously staining chromosomal regions contain amplified copies of an abundantly expressed cellular oncogene (c-myc) in malignant neuroendocrine cells from a human colon carcinoma. *Proc. Natl. Acad. Sci. USA*, 80:1701–1711.
6. Alt, F., Rosenberg, N., Lewis, S., Thomas, E., and Baltimore, D. (1981): Organization and reorganization of immunoglobulin genes in A-MuLV transformed cells. Rearrangement of heavy but not light chain genes. *Cell*, 27:381–390.
7. Aoki, T., Potter, M., and Sturm, M. M. (1973): Analysis by immunoelectron microscopy of type C viruses associated with primary and short term transplanted mouse plasma cell (H) tumors. *J. Natl. Cancer Inst.*, 51:1609–1617.
8. Anderson, P. N. (1970): Plasma cell tumor induction in BALB/c mice. *Proc. Am. Assoc. Cancer Res.*, 11:3 (Abstr.).
9. Anderson, P. N., and Potter, M. (1969): Induction of plasma cell tumors in BALB/c mice with 2,6,10,14-tetramethylpentadecane (pristane). *Nature*, 222:994–995.
10. Armstrong, M. Y. K., Ebenstein, P., Konigsberg, W. H., and Richards, F. F. (1978): Endogenous RNA tumor viruses are activated during chemical induction of murine plasmacytomas. *Proc. Natl. Acad. Sci. USA*, 75:4549–4552.
11. Bazin, H., Deckers, C., Beckers, A., and Heremans, J. F. (1972): Transplantable immunoglobulin secreting tumors in rats. I. General features of Lou/wsl strain rat immunocytomas and their monoclonal proteins. *Int. J. Cancer*, 10:568.
12. Bazin, H., Beckers, A., Deckers, C., and Moriame, M. (1973): Transplantable immunoglobulin secreting tumors in rats. V. Monoclonal immunoglobulins secreted by 250 ileocecal immunocytomas in Lou/wsl rats. *J. Natl. Cancer Inst.*, 51:1359.
13. Benade, L. E., Ihle, J. N., and Decleve, A. (1978): Serological characterization of B-tropic viruses of C57BL mice: Possible origin by recombination of endogenous N-tropic and xenotropic viruses. *Proc. Natl. Acad. Sci. USA*, 75:4553–4557.
14. Blumer, M., and Thomas, D. (1965): Phytadienes in zooplankton. *Science*, 147:1148–1149.
15. Blumer, M., Mullin, M. M., and Thomas, D. W. (1963): Pristane in zooplankton. *Science*, 140:974.
16. Blumer, M., Robertson, J. C., Gordon, J. E., and Sass, J. (1969): Phytol derived C_{19} di- and triolefinic hydrocarbons in marine zooplankton and fishes. *Biochemistry*, 8:4067–4074.
17. Breindl, M., Bacheler, L., Fan, H., and Jaenisch, R. (1980): Chromatin conformation of integrated Moloney leukemia virus DNA sequences in tissues of BALB/c Mo mice and in virus infected cell lines. *J. Virol.*, 34:373–382.
18. Butcher, E. C., Rouse, R. V., Coffman, R. L., Nottenburg, C. N., Hardy, R. G., and Weissman, I. L. (1982): Surface phenotype of Peyer's patches germinal center cells. Implications for the role of germinal centers in B-cell differentiation. *J. Immunol.*, 129:2698.
19. Calame, K., Kim, S., Lalley, P., Hill, R., Davis, M., and Hood, L. (1982): Molecular cloning of translocations involving chromosome 15 and the immunoglobulin Cα gene from chromosome 12 in two murine plasmacytomas. *Proc. Natl. Acad. Sci. USA*, 79:6994.
20. Callahan, R. (1983): In: *Mechanisms of B Cell Neoplasia*, pp. 189–193. Roche, Basel, Switzerland.
21. Cancro, M., and Potter, M. (1976): The requirement of an adherent substratum for the growth of developing plasmacytoma cells *in vivo*. *J. Exp. Med.*, 144:1554–1566.
22. Chattopadhyay, S. K., Cloyd, M. W., Linemeyer, D. L., Lander, M. E., Rands, E., and Lowy, D. R. (1982): Cellular origin and role of mink cell focus-forming viruses in murine thymic lymphomas. *Nature*, 295:25–31.
23. Cloyd, M. W. (1983): Characterization of target cells for MCF viruses in AKR mice. *Cell*, 32:217–225.

24. Cloyd, M. W., Hartley, J. W., and Rowe, W. P. (1980): Lymphomagenicity of recombinant mink cell focus-inducing murine leukemia viruses. *J. Exp. Med.*, 151:542–552.
24a. Cory, S., Gerondakis, S., and Adams, J. M. (1983): Interchromosomal recombination of the cellular oncogene c-*myc* with the immunoglobulin heavy chain locus in murine plasmacytomas is a reciprocal exchange. *EMBO J.*,2:697–703 .
25. Crews, S., Barth, R., Hood, L., Prehn, J., and Calame, K. (1982): Mouse c-myc oncogene is located on chromosome 15 and translocated to chromosome 12 in plasmacytomas. *Science*, 218:1319.
26. Dalla-Favera, R., Wong-Staal, F., and Gallo, R. C. (1982): Onc gene amplification in promyelocytic leukemia cell line HL-60 and primary leukemia cells of the same patient. *Nature*, 299:61–63.
27. Dalton, A. J., Potter, M., and Merwin, R. M. (1961): Some ultrastructural characteristics of a series of primary and transplanted plasma cell tumors of the mouse. *J. Natl. Cancer Inst.*, 26:1221–1267.
28. Elder, J. H., Gautsch, J. W., Jensen, F. C., Lerner, R. A., Hartley, J. W., and Rowe, W. P. (1977): Biochemical evidence that MCF murine leukemia viruses are envelope (env) gene recombinants. *Proc. Natl. Acad. Sci. USA*, 74:4676–4680.
29. Fialkow, P. J., Reddy, A. L., and Briandt, J. I. (1980): Clonal origin and trisomy of chromosome 15 in murine B cell malignancies. *Int. J. Cancer*, 26:603–608.
30. Gorette, M., Petropoulous, C. J., Shank, P. R., and Fausto, N. (1983): Expression of a cellular oncogene during liver regeneration. *Science*, 219:510.
31. Grunwald, D. J., Dale, B., Dudley, J., Lamph, W., Sugden, B., Ozanne, B., and Risser, R. (1982): Loss of viral gene expression and retention of tumorigenicity by Abelson lymphoma cells. *J. Virol.*, 43:92–103.
32. Harris, L. J., Lang, R. B., and Marcu, K. B. (1982): Non-immunoglobulin associated DNA rearrangements in mouse plasmacytomas. *Proc. Natl. Acad. Sci. USA*, 79:4175.
33. Harris, L. J., D'Eustachio, P., Ruddle, F. H., and Marcu, K. B. (1982): DNA sequence associated with chromosome translocations in mouse plasmacytomas. *Proc. Natl. Acad. Sci. USA*, 79:6622.
34. Hartley, J. W., Wolford, N. K., Old, L. J., and Rowe, W. P. (1977): A new class of murine leukemia virus associated with the development of spontaneous lymphomas. *Proc. Natl. Acad. Sci. USA*, 74:789–792.
35. Hawley, R. G., Shulman, M. J., Murialdo, H., Gibson, D. M., and Hozumi, N. (1982): Mutant immunoglobulin genes have repetitive DNA elements inserted into their intervening sequences. *Proc. Natl. Acad. Sci. USA*, 79:7425.
36. Hayward, W. S., Neel, B. G., and Astrin, S. M. (1981): Activation of a cellular onc gene by promoter insertion in ALV-induced lymphoid leukosis. *Nature*, 290:475.
37. Hoggan, M. D., Buckler, C. E., Sears, J. F., Chan, H. W., Rowe, W. P., and Martin, M. A. (1982): Internal organization of endogenous proviral DNAs of xenotropic murine leukemia viruses. *J. Virol.*, 43:8–17.
38. Jolicoeur, P. (1979): The Fv-1 gene of the mouse audits control of murine leukemia virus replication. *Curr. Top. Microbiol. Immunol.*, 86:67–122.
39. Kozak, C., and Rowe, W. P. (1980): Genetic mapping of xenotropic murine leukemia virus inducing loci in five mouse strains. *J. Exp. Med.*, 152:219–228.
40. Kuff, E. L., Feenstra, A., Lueders, K., Smith, L., Hawley, R., Hozumi, N., and Shulman, M. (1983): Intracisternal A particle genes as movable elements in the mouse genome. *Proc. Natl. Acad. Sci. USA*, 80:1992–1996.
41. Kuff, E. L., Smith, L. A., and Lueders, K. K. (1981): Intracisternal A particle genes in Mus musculus. A conserved family of retrovirus-like elements. *Mol. Cell. Biol.*, 1:216–227.
42. Kuff, E. L., Feenstra, A., Lueders, K., Rechavi, E., Givol, D., and Canaani, E. (1983): Sequences inserted in a rearranged and activated cellular oncogene; rc-mos are homologous to an endogenous retroviral (intracisternal A particle) LTR. *Nature*, 302:547–548.
42a. Land, H., Parada, L. F., and Weinberg, R. A. (1983): Tumorigenic conversion of primary embryo fibroblasts requires at least two cooperating oncogenes. *Nature*, 304:596–602.
43. Lane, M. A., Sainten, A., and Cooper, G. M. (1982): Stage-specific transforming genes of human and mouse B- and T-lymphocyte neoplasms. *Cell*, 28:873–880.
43a. Lenoir, G. M., Preud'homme, J. L., Berheim, A., and Berger, R. (1982): Correlation between immunoglobulin light chain expression and variant translocation in Burkitt's lymphoma. *Nature*, 298:474–476.
44. Lueders, K. K., and Kuff, E. L. (1977): Sequences associated with intracisternal A particles are reiterated in the mouse genome. *Cell*, 12:963–972.

45. Lueders, K. K., and Kuff, E. L. (1979): Genetic individuality of intracisternal A particles of Mus musculus. *J. Virol.*, 30:225–231.
46. Marcu, K. B., Harris, L. J., Stanton, L. W., Erikson, J., Watt, R., and Croce, C. (1983): Transcriptionally active c-myc oncogene is contained within NIARD. A DNA sequence associated with chromosome translocation in B cell neoplasia. *Proc. Natl. Acad. Sci. USA*, 80:519.
47. McIntire, K. R., and Princler, G. L. (1969): Prolonged adjuvant stimulation in germ-free BALB/c mice. Development of plasma cell neoplasia. *Immunology*, 17:481–487.
48. Merwin, R. M., and Redmon, L. W. (1963): Induction of plasma cell tumors and sarcomas in mice by diffusion chambers placed in the peritoneal cavity. *J. Natl. Cancer Inst.*, 31:998–1007.
49. Metcalf, D. (1974): The serum factor stimulating colony formation *in vitro* by murine plasmacytoma cells: Response to antigens and mineral oil. *J. Immunol.*, 113:235.
50. Milburn, G. L., Hoover, R. G., and Lynch, R. G. (1982): Immunoregulatory cell interactions that govern the growth and differentiation of murine myeloma cells. In: *B and T Cell Tumors*, pp. 335–347. Academic Press, New York.
51. Mladenov, Z., Heine, U., Beard, D., and Beard, J. W. (1967): Strain MC-29 avian leukosis virus myelocytoma endothelioma and renal growths. *J. Natl. Cancer Inst.*, 38:251.
52. Moroni, C, and Schumann, G. (1976): Mitogen induction of murine type C viruses. II. Effect of B-lymphocyte mitogens. *Virology*, 73:17–22.
53. Morse, H. C., III, and Hartley, J. W. (1982): Expression of xenotropic murine leukemia viruses. *Curr. Top. Microbiol. Immunol.*, 98:17–26.
54. Morse, H. C., III, Hartley, J. W., and Potter, M. (1980): Genetic considerations in plasmacytomas of BALB/c, NZB and (BALB/c × NZB)F_1 mice. In: *Progress in Myeloma. Biology of Myeloma*, edited by M. Potter, pp. 263–279. Elsevier/North Holland, New York.
55. Morse, H. C., III, Pumphrey, J. G., Potter, M., and Asofsky, R. (1976): Murine plasma cells secreting more than the class of immunoglobulin heavy chain. I. Frequency of two or more M-components in ascitic fluids from 788 primary plasmacytomas. *J. Immunol.*, 117:541–547.
56. Morse, H. C., III, Riblet, R., Asofsky, R., and Weigert, M. (1978): Plasmacytomas of the NZB mouse. *J. Immunol.*, 121:1969–1972.
57. Mushinski, J. F., Bauer, S. R., Potter, M., and Reddy, E. P. (1983): Increased expression of myc related oncogene mRNA characterizes most BALB/c plasmacytomas induced by pristane or Abelson virus. *Proc. Natl. Acad. Sci. USA*, 80:1073–1077.
58. Mushinski, J. F., Potter, M., Bauer, S. R., and Reddy, E. P. (1983): DNA rearrangement and altered RNA expression of the c-myc oncogene in mouse plasmacytoid lymphosarcomas. *Science*, 220:795.
59. Namba, Y., and Hanaoka, M. (1972): Immunocytology of cultured IgM-forming cells of mouse. I. Requirement of phagocytic cell factor for the growth of IgM-forming tumor cells in tissue culture. *J. Immunol.*, 109:1193–1200.
60. Namba, Y., and Hanoaka, M. (1974): Immunocytology of cultured IgM forming cells of mouse. II. Purification of phagocytic cell factor and its role in antibody formation. *Cell. Immunol.*, 12:74–84.
61. Neel, B. G., Hayward, W. S., Robinson, H. L., Fang, J., and Astrin, S. M. (1981): Avian leukosis virus induced tumors have common proviral integration sites and synthesize discrete new RNAs: Oncogenesis by promoter insertion. *Cell*, 23:323.
62. O'Donnell, P. V., and Nowinski, R. C. (1980): Serological analysis of antigenic determinants on the env gene products of AKR dual tropic (MCF) murine leukemia viruses. *Virology*, 107:81–88.
63. Ohno, S., Babonits, M., Wiener, F., et al. (1979): Non-random chromosome changes involving the Ig gene-carrying chromosomes 12 and 6 in pristane-induced mouse plasmacytomas. *Cell*, 18:1001–1007.
63a. Ohno, S., Migita, S., Wiener, F., Babonits, M., Klein, G., Mushinski, J. F., and Potter, M. (1984): Cytogenetic studies of plasmacytomas induced by Abelson virus in pristane conditioned mice. *(Submitted.)*
64. Olsson, M., Lindahl, G., and Ruoslahti, E. (1977): Genetic control of alpha fetoprotein synthesis in the mouse. *J. Exp. Med.*, 145:819–827.
65. Pattengale, P. K., Taylor, C. R., Twomey, P., et al. (1982): Immunopathology of B-cell lymphomas induced in C57BL/6 mice by dual-tropic murine leukemia virus (MuLV). *Am. J. Pathol.*, 107:362–376.
65a. Rybak, J., Tharadel, P., Robinett, S., Garcia, M., Makinen, G., and Freeman, M. (1982): A simple reproducible method for prometaphase chromosome analysis. *Hum. Genet.*, 60:328–333.

66. Payne, G. S., Bishop, J. M., and Varmus, H. E. (1982): Multiple arrangements of viral DNA and an activated host oncogene in bursal lymphomas. *Nature*, 295:209.
67. Platica, M., Bojko, C., Clincea, R., and Hollander, V. (1982): Lipopolysaccharide and pristane induced peritoneal exudate requirements for plasma cell tumor cell colony formation. *J. Natl. Cancer Inst.*, 68:325.
68. Potter, M. (1972): Immunoglobulin producing tumors and myeloma proteins in mice. *Physiol. Rev.*, 52:631–719.
69. Potter, M. (1977): Antigen binding myeloma proteins of mice. *Adv. Immunol.*, 25:141–210.
70. Potter, M., and Boyce, C. (1962): Induction of plasma cell neoplasms in strain BALB/c mice with mineral oil and mineral oil adjuvants. *Nature*, 193:1086–1087.
71. Potter, M., and MacCardle, R. C. (1964): Histology of developing plasma cell neoplasia induced by mineral oil in BALB/c mice. *J. Natl. Cancer Inst.*, 33:497–515.
72. Potter, M., Premkumar-Reddy, E., and Wivel, N. A. (1978): Immunoglobulin production by lymphosarcomas induced by Abelson virus in mice. In: *Gene Expression and Regulation in Cultured Cells*, edited by K. K. Sanford, pp. 311–320. NCI Monograph 48. Bethesda, Maryland.
73. Potter, M., Pumphrey, J. G., and Bailey, D. W. (1975): Genetics of susceptibility of plasmacytoma induction. I. BALB/cAnN(C), C57BL/6N(B6) C57BL/Ka(BK), $(C \times B6)F_1$, $(C \times BK)F_1$ and $C \times B$ recombinant inbred strains. *J. Natl. Cancer Inst.*, 54:1413–1417.
74. Potter, M., Pumphrey, J. G., and Walters, J. L. (1972): Growth of primary plasmacytomas in the mineral oil-conditioned peritoneal environment. *J. Natl. Cancer Inst.*, 49:305–308.
75. Potter, M., Sklar, M. D., and Rowe, W. P. (1973): Rapid viral induction of plasmacytomas in pristane primed BALB/c mice. *Science*, 182:592–594.
76. Potter, M., and Wax, J. S. (1981): Genetics of susceptibility to pristane induced plasmacytomas in BALB/cAn: Reduced susceptibility in BALB/cJ with a brief description of pristane-induced arthritis. *J. Immunol.*, 127:1591.
77. Potter, M., and Wax, J. S. (1983): Peritoneal plasmacytomagenesis in mice. A comparison of three pristane dose regimens. *J. Natl. Cancer Inst.*, 71:391–395.
78. Premkumar, E., Potter, M., Singer, P. A., and Sklar, M. D. (1975): Synthesis surface deposition and secretion of immunoglobulins by Abelson virus-transformed lymphosarcoma cell lines. *Cell*, 6:149–159.
79. Raschke, W. C. (1977): Transformation by Abelson murine leukemia virus. Properties of the transformed cells. *Cold Spring Harbor Symp. Quant. Biol.*, 1187–1194.
80. Raschke, W. C., Baird, S., Ralph, P., and Nakoinz, I. (1978): Functional macrophage cell lines transformed by Abelson leukemia virus. *Cell*, 15:261.
81. Raveche, E. S., Tjio, J. H., and Steinberg, A. D. (1979): Genetic studies in NZB mice. II. Hyperdiploidy in the spleen of NZB mice and their hybrids. *Cytogenet. Cell Genet.*, 23:182–193.
82. Rechavi, G., Givol, D., and Canaani, E. (1982): Activation of a cellular oncogene by DNA rearrangement: Possible involvement of an IS-like element. *Nature*, 300:607–611.
83. Risser, R. (1982): The pathogenesis of Abelson virus lymphomas of the mouse. *Biochim. Biophys. Acta*, 651:213–244.
84. Risser, R., Potter, M., and Rowe, W. P. (1978): Abelson virus induced lymphomagenesis in mice. *J. Exp. Med.*, 148:714–729.
85. Rohrer, J. W., and Lynch, R. G. (1977): Specific immunological regulation of differentiation of immunoglobulin expression in MOPC315 cells during *in vivo* growth in diffusion chambers. *J. Immunol.*, 119:2045–2053.
86. Rohrer, J. W., Vasa, K., and Lynch, R. G. (1977): Myeloma cell immunoglobulin: Expression during *in vivo* growth in diffusion chambers: Evidence for repetitive cycles of differentiation. *J. Immunol.*, 119:861–866.
87. Rosenberg, N. (1982): Abelson leukemia virus. *Curr. Top. Microbiol. Immunol.*, 101:95–126.
88. Shen-Ong, G. L. C., Keath, E. J., Piccoli, S. P., and Cole, M. D. (1982): Novel myc oncogene RNA from abortive immunoglobulin gene recombination in mouse plasmacytomas. *Cell*, 31:443–452.
89. Shepard, J. S., Pettengill, O. S., Wurster-Hill, D. H., and Sorenson, G. D. (1974): Alterations of karyotype and oncogenicity in mouse myeloma MOPC315 and 5-bromodeoxyuridine-resistant cell line. *Cancer Res.*, 34:2852–2858.
90. Shepard, J. S., Wurster-Hill, D. H., Pettengill, O. S., and Sorenson, G. D. (1974): Giemsa banded chromosomes of mouse myeloma in relationship to oncogenicity. *Cytogenet. Cell Genet.*, 13:279–304.
91. Shepard, J. S., Pettengill, O. S., Wurster-Hill, D. H., and Sorenson, G. D. (1974): Karyotype marker formation and oncogenicity in mouse plasmacytomas. *J. Natl. Cancer Inst.*, 56:1003–1011.

92. Shepard, J. S., Pettengill, O. S., Wurster-Hill, D. H., and Sorenson, G. D. (1978): A specific chromosome breakpoint associated with mouse plasmacytomas. *J. Natl. Cancer Inst.*, 61:225–258.
93. Siden, E. J., Baltimore, D., Clark, D., and Rosenberg, N. E. (1979): Immunoglobulin synthesis by lymphoid cells transformed *in vitro* by Abelson murine leukemia virus. *Cell*, 16:389–396.
94. Stanton, L. W., Watt, R., and Marcu, K. B. (1983): Translocation breakage and truncated transcripts of c-myc oncogene in murine plasmacytomas. *Nature*, 303:401–406.
95. VanNess, B. G., Shapiro, M., Kelley, D. E., Perry, R. P., Weigert, M., D'Eustachio, P., and Ruddle, F. (1983): Aberrant rearrangement of the K-light chain locus involving the heavy chain locus and chromosome 15 in a mouse plasmacytoma. *Nature*, 301:425–427.
96. Varmus, H. E., Quintrell, N., and Ortiz, S. (1981): Retroviruses as mutagens: Insertion and excision of a non-transforming provirus alter expression of a resident transforming provirus. *Cell*, 25:23–26.
97. Warner, N. L. (1975): Review. Autoimmunity and the pathogenesis of plasma cell tumor induction in NZB and hybrid mice. *Immunogenetics*, 2:1–20.
98. Whitlock, C. A., Ziegler, S. F., Treiman, L. J., Stafford, J. I., and Witte, O. N. (1983): Differentiation of cloned populations of immature B-cells after transformation with Abelson murine leukemia virus. *Cell*, 32:903–911.
99. Wiener, F., Babonits, M., Spira, J., Klein, G., and Potter, M. (1980): Cytogenetic studies on IgA/Lambda producing murine plasmacytomas: Regular occurrence of a T(12;15) translocation. *Somatic Cell Genet.*, 6:731–738.
100. Wiener, F., Babonits, M., Spira, J., Bregula, U., Klein, G., Merwin, R. M., Asofsky, R. H., Lynes, M., and Haughton, M. (1981): Chromosome 15 trisomy in spontaneous and carcinogen induced murine lymphomas of B-cell origin. *Int. J. Cancer*, 27:51–58.
100a. Wiener, F., Ohno, S., Babonits, S., Sumegi, J., Wirschubsky, Z., Klein, G., Mushinski, J. F., and Potter, M. (1984): Hemizygous interstitial deletion of chromosome 15 (band D) in three translocation negative plasmacytomas. *Proc. Natl. Acad. Sci. (in press).*
100b. Wiener, F., Babonits, M., Bregula, U., Klein, G., Leonard, A., Wax, J., and Potter, M. (1984): High resolution banding analysis of the involvement of strain BALB/c and AKR derived chromosomes no. 15 in plasmacytoma specific translocations. *J. Exp. Med. (in press).*
101. Wivel, N. A., and Smith, G. H. (1971): Distribution of intracisternal A particles in a variety of normal and neoplastic mouse tissues. *Int. J. Cancer*, 7:167–175.
102. Yosida, T. H., Imai, H. T., and Moriwaki, K. (1970): Chromosomal alteration and development of tumors. XXI. Cytogenetic studies of primary plasma cell neoplasms induced in BALB/c mice. *J. Natl. Cancer Inst.*, 45:411–418.
103. Yosida, T. H., Imai, H. T., and Potter, M. (1968): Chromosomal alteration and development of tumors. XIX. Chromosomal constitution of tumor cells in 16 plasma cell neoplasms of BALB/c mice. *J. Natl. Cancer Inst.*, 41:1083–1087.
104. Yosida, M. C., and Moriwaki, K. (1975): Specific marker chromosomes involving a translocation (12;15) in a mouse myeloma. *Proc. Jpn. Acad.*, 51:588–592.
105. Yosida, M. C., Moriwaki, K., and Migita, S. (1978): Specificity of the deletion of chromosome no. 15 in mouse plasmacytoma. *J. Natl. Cancer Inst.*, 60:235–238.

Advances in Viral Oncology, Volume 4,
edited by George Klein. Raven Press,
New York © 1984.

Transformation Parameters Expressed by Tumor-Virus Transformed Cells

Bernard Perbal

Institut Curie, Section de Biologie, Centre Universitaire, Orsay, France

For many years, attempts have been made to define model systems that allow a study of the changes in cellular regulation associated with oncogenic transformation. Since we generally think of a cancer cell as a normal cell having lost the ability to respond normally to signals associated with proliferation control, it is worth considering what is commonly called "normal cells" before examining the characteristic modifications, known as transformation parameters, expressed in transformed cells.

In the adult animal, continued proliferation of cells is limited primarily to epithelial cells (e.g., of the skin, gut, or bronchii) and to stem cells (e.g., thymus, bone marrow, and lymph nodes). Proliferation is controlled in balance with natural cell death. Other cells are capable of limited proliferation only in response to a specific mitogenic stimulus. This is the case for liver cells following abnormal tissue damage and for epithelial cells of mammary glands. In the absence of these stimuli, tissues are generally devoid of mitotic structures (274). Other cells, such as nerve cells and striated muscle cells, are irreversibly arrested in the Go phase of the cell cycle (61).

Normal cells used *in vitro* usually derive from two different origins: primary cultures or established cell lines. Primary cultures are composed of cells obtained after dissociation of mammalian tissues by mechanical and enzymatic procedures (for a review, see ref. 139). Such cells may be grown on glass or plastic in the presence of growth medium, generally supplemented with 5 to 10% serum. A fundamental characteristic of these cells is that they form monolayers of parallel cells on the support as a result of a firm adhesion to the substratum and of contact inhibition of movement (1,2). Under such conditions, cells are generally spread and flattened; following a collision, they turn aside and move in another direction (99). The cell population ceases to grow after a limited number of divisions, as most of the cells rapidly die (89,96) for unknown reasons. The limited lifespan of these normal primary cells has led investigators to search for cell lines that are able to grow indefinitely without being tumorigenic when injected into animals.

Established cell lines cannot be defined simply as cell populations that are able to grow for several months under laboratory conditions (24,43,63), since cells can be grown for many generations without becoming established (96,190). The early studies of Todaro and Green (243) showed that cultured mouse fibroblasts cease to divide after a period of time corresponding to 10 to 20 generations from the beginning of the culture, and that a small proportion of cells could escape from the "crisis" period and continue to proliferate with a rate comparable to that of initial cells. The growth rate of such cells was never found to decrease, and they could be grown for a considerable number of generations. Such established cell lines have been obtained from several different animal species, but not from chicken or human cells.

Primary cultures and established cell lines offer complementary properties. On one hand, primary cultures probably are more likely to represent the physiologic state of mammalian cells *in vivo* than are established cells which have been selected for a nonphysiologic property, that is, continuous growth ability. On the other hand, established cells can be cloned, thereby providing genetically homogeneous cellular populations, whereas primary cultures have been shown to be a heterogeneous mixture of different cell types with variable growth abilities. Alterations of the cellular karyotype may occur during the selection of established cells (109,144,209). Rodent cells, and especially murine cells, have been reported to be genetically unstable and tend to give rise to heterogeneous cell populations (108,243). Slight karyotypic changes have also been reported in the case of established cell lines from hamster and pig (150,212,297). This demonstrates the need for a careful definition of the systems used in the study of cellular transformation and expression of related parameters.

It is possible to isolate normal cells that cease to divide after they have reached a particular density corresponding to the formation of a confluent monolayer of flat cells. This saturation density is independent of the initial seeding density but is dependent on the cell type and the growth medium used. The limitation of cell proliferation *in vitro* has been attributed to contact-mediated growth control (234), the accumulation of cell-produced inhibitory substances, or depletion of essential substances in the medium (104,105).

Among normal cells that stop dividing at confluence, many agents can induce the formation of transformed cells which are not susceptible to the control processes involved in the regulation of cell proliferation. These will continue to grow, pile up, and form dense foci of cells arranged in irregular crisscross patterns on top of the monolayer of normal cells. In most cases, such *in vitro* transformed cells are able to induce the formation of tumors when inoculated into susceptible animals and, therefore, are considered to be comparable to tumor cells arising *in vivo*.

Oncogenic transformation has been reported to occur readily after treatment of normal cells with chemical carcinogens or radiations and after infection with RNA or DNA tumor viruses. Spontaneous transformation has also been described. In most cases, the procedure used to isolate transformed cells was based on the expression of one particular feature of these cells which is not shared by normal cells.

We summarize below the different phenotypic properties (transformation parameters) usually displayed by transformed cells induced by tumor viruses.

TRANSFORMATION PARAMETERS RELATED TO CELLULAR MORPHOLOGY

Morphologic Changes

Transformed cells are characterized by a great number of changes at the level of their external morphology and by a high pleomorphism (numerous different shapes) in culture (Fig. 1). Comparison of normal and transformed cell morphology has been performed in several different systems, including cells transformed by DNA tumor viruses, such as adenovirus (81), polyoma virus (185), simian virus 40 (SV40) (30,91,177,185), and herpes viruses (36), and RNA tumor viruses, such as Rous sarcoma virus (RSV) (13,17,21,22), avian myelocytomatosis virus (MC29) (210), Fujinami, PRC II, PRC IIp, and Y73 strains of avian sarcoma viruses (ASV) (88), and kirsten murine sarcoma virus (Ki-MSV) (52). In all cases, the trans-

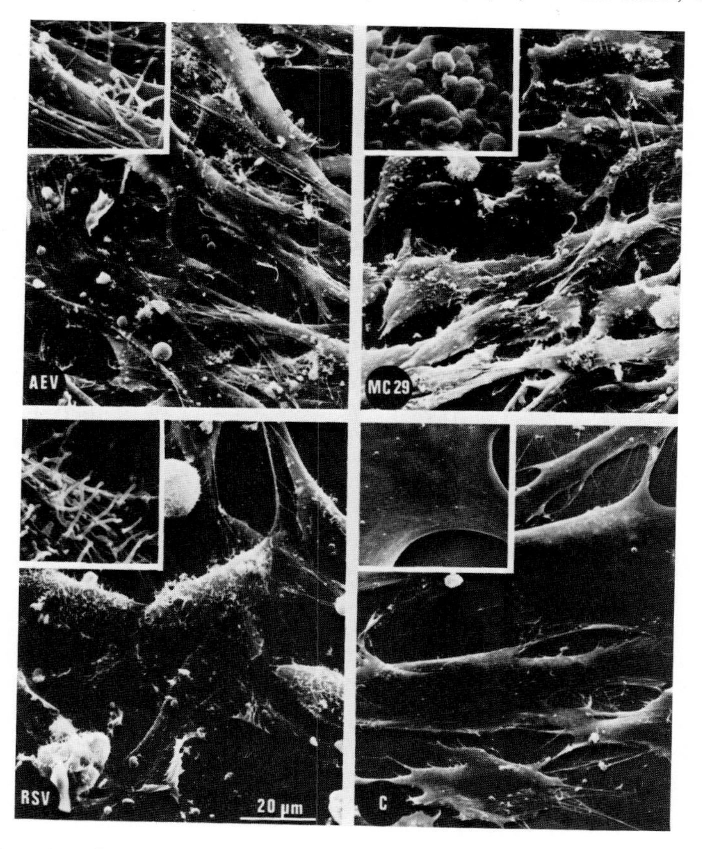

FIG. 1. Scanning electron micrographs of clone 77 chicken fibroblasts transformed by AEV, MC29, and RSV. Control culture (C) was infected with RAV-1. All cells were passaged twice after infection. Cells were seeded on glass coverslips 1 day prior to fixation in 2.4% glutaraldehyde. [Reproduced from ref. 210, with permission of T. Graf (MIT copyright).]

formed cells were described as thicker, more convex over their center, sometimes spherical, and refringent. The normal cells, by comparison, are usually thin, flat, and smooth when they form monolayers and become spherical as they begin mitosis, whereas transformed cells have been described as hemispherical or spherical throughout their life cycle, either in suspension or in monolayer cultures (60,81,211). An increased number of cytoplasmic strands protruding from the cell surface (12, 83,152,177,184,185) accompanied by formation of flower-like ruffles and spherical blebs have been visualized by scanning electron microscopy on most transformed cells. An increased number of microvilli have been found in many transformed cells (13,27,30,81,152,177,185), although this is not always the case (46). Variable distributions, shapes, and sizes were reported (12).

Reorganizations of the cytoskeleton are thought to play a prominent role in the appearance of the morphologic modifications described above. The cytoskeleton is composed of at least three different networks formed by microtubules, microfilaments, and intermediate filaments. Microtubules are long, fibrous elements of 22 to 24 nm diameter which extend throughout the cell cytoplasm, parallel to the longitudinal axis of bipolar cells and which disappear when the cells round up at mitosis (31,64,183,286).

Disorganization and loss of microtubules have been reported to occur in RSV transformed chicken cells (58,265), whereas normal rat kidney cells infected by RSV do not display gross reorganization of the microtubule pattern (220,265). A comparison of the microtubule patterns in 3T3 cells transformed by polyoma or SV40 reveals that the modifications of these structures induced upon cell transformation (Fig. 2) may be expressed more or less. SV40 transformed cells show a residual organization of microtubules (166,278), whereas in polyoma transformed cells, microtubules are completely disorganized (A. M. Hill, R. Seif, and D. Pantaloni, *personal communication*). An increased stability of tubulin has been correlated with the expression of early SV40 region in transformed cells (279).

In normal cells, microfilaments are located in the cortex of the dorsal and ventral cell surfaces (82,149,286). They are composed of polymerized actin, myosin, tropomyosin, filamin, and α-actinin (15,78,80,93,101,122,141,142,148,149,157,176, 182,216,217,263,270,277) organized in bundles (also called stress fibers or cables) and involved in cell motility (7,77,79,82,99,149,277) and cell-to-cell and cell-to-substratum interactions (91,148,149,262).

Two different kinds of specialized sites, namely, focal adhesion and close contacts, seem to be involved in cell-to-substratum adhesion (123). Focal adhesion represents sites for strong adhesion (3,197), whereas close contacts are more likely to be involved in cell adhesion (98). Upon transformation, a decrease or disappearance of the focal adhesion plaque has been observed (51), together with striking changes in the organization of vinculin and α-actinin. A majority of RSV transformed cells exhibit a cluster of small patches (rosette) (Fig. 3) that are immunolabeled for both vinculin and α-actinin (51). It has been proposed that the redistribution of vinculin and α-actinin observed in RSV transformed cells might reflect intramolecular events that lead to the weak adhesive contact of transformed cells (51).

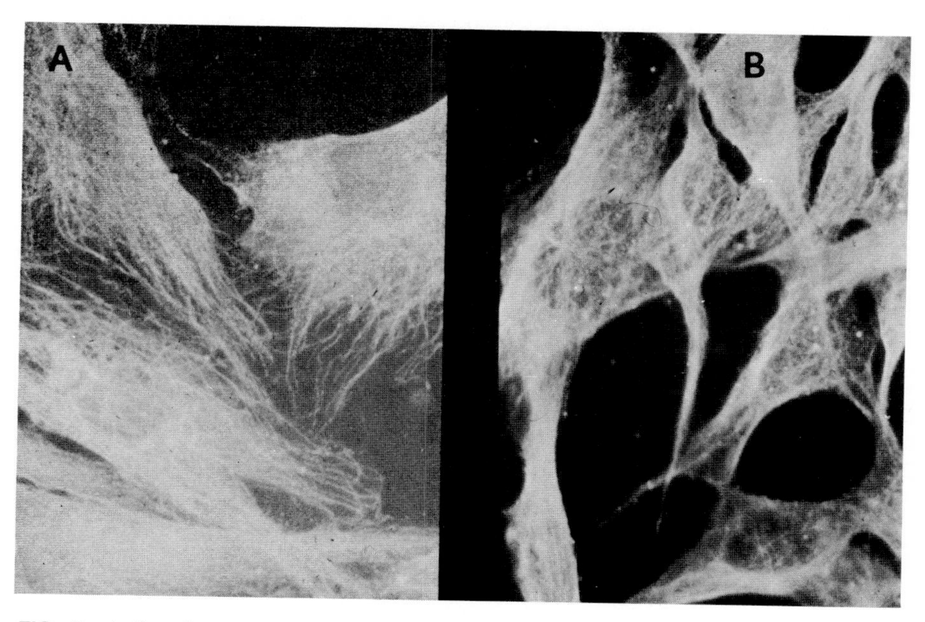

FIG. 2. Indirect fluorescence staining of cells with antitubulin antibody to visualize microtubules. **A:** Untransformed rat 3T3 cells (221). **B:** SV40 transformed rat 3T3 cells. (A. M. Hill, R. Seif, D. Pantaloni, *personal communication.*)

The patches described would be associated with sites of close contact with the substratum at the ventral surface of the cell. In the case of SV40 transformed cells, disorganization of microfilaments has been shown to require the expression of small t protein (84,219). While most transformed cells exhibit alterations of microfilament bundles, it does not seem to be a strict rule (82), indicating that these structures might be related only to adhesiveness and might not be needed for the loss of growth control characteristic of transformed cells.

Much less information is available concerning the possible modifications of the intermediate filament network upon transformation. These fibrous structures, whose function is not clearly established, are composed of 10 nm fibers (167) characterized by a high degree of polymorphism (168). Disorganization of the intermediate filaments has been described in the case of herpes simplex virus transformed NIL8 cells (119).

Cell Surface Organization and Composition

A consequence of the cell surface changes that occur upon cellular transformation is their ability to be preferentially agglutinated by lectins (35,162). These proteins, essentially isolated from plants, are able to combine specifically with the sugar moiety of glycolipids and glycoproteins of the cell surface and, hence, lead to the binding together of different cells. For example, the lectin agglutinability of avian

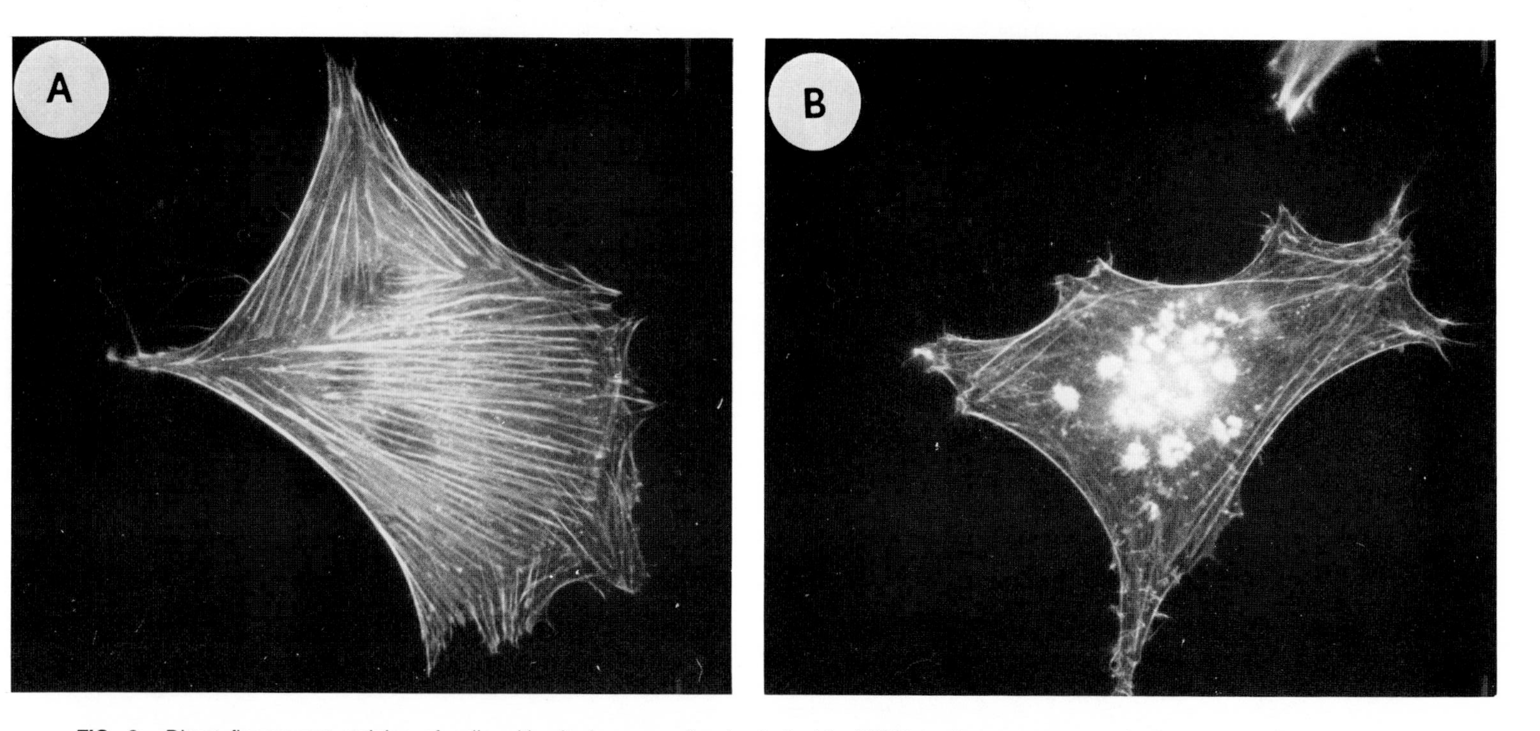

FIG. 3. Direct fluorescent staining of cells with nitrobenzooxadiazole-phallacidin (NBD) to demonstrate organization of microfilament network. **A:** Untransformed chick embryo fibroblasts. **B:** RSV *ts* mutant (NY68) transformed chick embryo fibroblasts showing rosettes (see text). (Photograph kindly provided by C. Krycève-Martinerie.)

leukemia virus tranformed cells was obtained at lectin concentrations 10 to 30 times lower than required for the agglutination of normal cells (210). Studies performed with HSV transformed rat embyro cells revealed that the higher level of agglutinability by lectins may not be regarded as a good indicator of transformation and tumorigenicity by itself, since HSV1 transformed cells required approximately five times the concentration of lectin necessary for half-maximal agglutination of normal cells (36), although they were found to be tumorigenic. On the other hand, the HSV2 transformed and tumor cell lines required approximately half the concentration needed for half-maximal agglutination (36).

The increased susceptibility of transformed cells to lectin is not due to an elevated number of binding sites on the cell surface (44,172). The lectin receptors appear to be randomly distributed on both normal and transformed cells (14,207,226,251). Rather, it is thought that addition of lectin would aggregate the receptors in patches leading to the formation of caps at several sites on the cell surface. The redistribution of sites following the binding of lectins would be inhibited in normal cells (16,161,207,251). It is unlikely that lipids play a controlling role in cell agglutinability (67,68,191,193,296), whereas mild treatment of normal cells by proteases can increase their agglutinability (34,120,163), suggesting that modifications of cell surface proteins are involved in this phenomenon.

Cell Surface Proteins in Transformed Cells

The identification and comparison of the normal and transformed cell surface proteins has led to the distinction of two types of proteins: those whose synthesis is increased after transformation and those whose synthesis is decreased in transformed cells.

Membrane proteins with molecular weights of 90,000 to 95,000 and 73,000 to 79,000 daltons have been found to be present in increased amounts in transformed cells (39,121,237). A correlation between the accumulation of proteins with similar molecular weights (i.e., 75,000 and 95,000 daltons) and glucose starvation has been postulated (187,229). Furthermore, the synthesis of these proteins has been induced by inhibition of glycosylation (187), whereas feeding of transformed cells with higher amounts of glucose may prevent the induction of synthesis. Many more proteins have been reported to be present in reduced amounts in transformed cells. We focus here on the best studied, namely, collagen and fibronectin.

Collagen is synthesized as a precursor, procollagen, which is not entirely cleaved by cells (28). Both collagen and procollagen may be found incorporated in the extracellular matrix of the cells. The synthesis of collagen has long been known to be depressed in transformed cells (87,129,145,178,250), probably because of a 5- to 10-fold decrease in the amount of procollagen mRNA synthesized, at least in chicken fibroblasts (4).

The properties and possible role of fibronectin in cell transformation have already been extensively reviewed over the past few years [see, for example, reviews by Hynes (112), Yamada and Olden (293), and Vaheri and Mosher (258)]. We present below a brief summary of the salient features of these properties.

Fibronectin (59,131) is a large, external glycoprotein with an apparent molecular weight of 230,000 daltons; it is also known as LETS protein (large, external, transformation sensitive) (114) or cell surface protein (CSP) (287). It is located on the surface of normal cells and arranged in fibrillar arrays (10,25,29,37,40,41,100, 124,156,213,214,232,257,258,267,288,292). The external location of this protein has allowed its detection by surface radioactive and enzymatic labeling (48,49,71, 72,102,110,125,159) and has suggested that it is involved in cell-to-cell interactions, cell-to-substratum interactions, cell motility, and cell morphology (10,11,289,290,293). Fibronectin binds specifically to biologic macromolecules, such as collagen, fibrin, and glycosaminoglycans (59,90) and therefore could play an important role in the organization of the extracellular matrix (29,41,259). Fibronectin is also found as a circulating protein in the plasma, with a slightly lower molecular weight. Both cellular and plasma forms of fibronectin are found as dimers or polymers involving the formation of disulfide bonds (38,117,132,147,158,291). Purification of chymotrypic fragments from cellular fibronectin led to the identification of a 40,000 dalton fragment involved in binding with collagen and a 160,000 dalton fragment found to contain the cell surface binding sites of cellular fibronectin (259). These observations are consistent with the possibility that fibronectin polypeptidic chains contain different structural domains responsible for its biologic properties (8,9,38).

Considerable reduction or loss of cell surface fibronectin has been shown to occur in the case of adenovirus (37), polyoma (69,71,110,173), SV40 (48,69, 102,256,257), RSV (49,70,115,121,125,154,164,204,237,267), avian erythroblastosis virus (AEV) (210), MSV (102), and herpes simplex virus (HSV) (36) transformed cells.

It is interesting to note that MC29 transformed fibroblasts expressed normal levels of fibronectin (210), and that normal human or mouse mammary epithelial cells in primary culture do not express fibronectin (295). Also, primary culture cells of normal pig periodontal ligament (138) and mouse salivary gland and bladder (281) lack the typical distribution of surface fibronectin, suggesting that in some systems, there might be no direct correlation between neoplastic transformation and loss of fibronectin.

Several morphologic aspects of the transformed cells can be reverted by the addition of purified fibronectin to the culture medium (10,41,118,282,288,290), whereas several other transformation parameters are still expressed in the presence of added fibronectin (10,41,282,288,290). The reduction of fibronectin biosynthesis observed in some transformed cells (113,164) has been shown to result from a reduced synthesis of the corresponding mRNA species (4), while an increased turnover of fibronectin in transformed cells has also been reported (164).

The possible role for fibronectin and other matrix components in the process of oncogenic transformation has been discussed in several reviews (113,258,293). We stress here only that it is far from being clarified and unequivocally established, since the study of several transformed cell lines obtained after either viral infection or mutagenesis, or spontaneously, revealed that loss of fibronectin is neither specifically associated with the ability to grow without anchorage (see below) nor required for tumorigenicity *in vivo* (128).

PARAMETERS RELATED TO CELL PROLIFERATION

Transformed cells, by definition, are not sensitive to the control processes involved in the regulation of normal cell proliferation and therefore exhibit several properties that have been valuable for their isolation. Thus reduced serum requirements, loss of contact inhibition, and ability to grow without anchorage are properties used to select transformed cells.

Reduced Serum Requirements

Whereas normal cells in culture have generally high serum requirements (55, 103,104,234,236), transformed cells are able to grow in the presence of reduced serum concentrations, sometimes in serum-free medium (53,230). For example, the saturation densities obtained for SV40 and polyoma transformed rat fibroblasts grown in the presence of 2% calf serum were found to be 4 and 7×10^6 cells per 6 cm plates, respectively (175), whereas parental FR3T3 rat cells grown under similar conditions reached 10^6 cells per plate (221). Plating of SV40-infected BALB/c 3T3 mouse cells in the presence of low serum concentrations allowed the isolation of transformed cell lines (230).

Since serum added to culture medium is a complex mixture of components that might be involved in normal cell growth, efforts continue to be directed toward the identification of indispensable and dispensable serum components for cell growth and to the formulation of semisynthetic culture media. Among the indispensable factors identified so far are hormones [e.g., insulin, epidermal growth factor (EGF), fibroblast growth factor (FGF), platelet derived growth factor (PDGF), and prostaglandins], binding proteins (e.g., transferrin and bovine serum albumin), and factors involved in spreading and adhesiveness (e.g., fibronectin, fetuine, or collagen).

Some of these factors have been purified and their properties established: EGF has been shown to bind on specific cell receptors whose number is decreased after RSV transformation but unchanged after DNA tumor virus transformation (246). PDGF is known to be required but not sufficient for the transition from Go to S phase (208,218). It is probable that the interaction of all these different factors with the cells results in controlled growth and that reduced serum requirement of transformed cells may be the consequence of several modifications at the level of these interactions. The production of transforming growth factors (TGF) by transformed cells (for review, see ref. 233) also might be involved in these processes. The reduced serum requirement does not seem to be a universal characteristic for transformed cells, however, since three independently isolated HSV type 1 (HSV1) transformed rat embryo cell lines did not show increased saturation densities at low serum concentrations but rather needed serum for their growth (36).

Generation Time

Most of the transformed cells were shown to grow more rapidly than their untransformed counterparts, due to a markedly decreased generation time. This is true for (a) Ki-MSV transformed rat kidney cells, whose doubling time in culture is 15 hr, compared to 80 hr for normal rat kidney cells (52), and (b) polyoma and SV40 transformed established rat cells, whose generation time is 12 and 15 to 20 hr, respectively, compared to 25 hr for the untransformed parental cells (175). Interestingly, the SV40 transformed cell lines having a generation time of 20 hr display an intermediate transformation phenotype (175). A reduction of doubling time was also observed in low serum concentrations. Thus SV40 transformed primary rat embryo cells grown in 1% fetal calf serum had generation times in the range of 20 to 35 hr, whereas normal primary cultures showed doubling times of 100 to 130 hr under similar conditions (203). Herpes simplex transformed rat embryo cells did not behave as described above. Their generation times (32 to 41 and 34 hr) were higher than that of control parental cells, which have a doubling time of 22 to 29 hr (36).

Anarchic Proliferation of Transformed Cells

As mentioned above, normal cells that form monolayers in culture and exhibit a contact inhibition of movement (1,2) can be isolated. Their transformed derivatives appear to be less sensitive to contact inhibition of movement and therefore move over or under each other. It has been shown that transformed cells underlap (20,92), probably because they are stuck to a lesser degree to the substratum (reduced adhesiveness described above). Consequently, transformed cells will form multilayers and, in most cases, the cultures will show a crisscross aspect. The isolation of transformants among infected cells has been possible because of the anarchic proliferation of transformed cells, which leads to the formation of dense foci on the monolayer of untransformed cells (153,235,242,244). In many cases, the frequency of transformation induced by tumor viruses after infection of normal cells is low. For instance, transformation frequencies of FR3T3 rat cells by wild type strain and thermosensitive mutants of polyoma virus were about 0.15% (134, 174,189,221). Transformation frequencies obtained under similar conditions with wild type strain and thermosensitive mutants of SV40 were in the range of 0.4 to 0.8% (194). Foci can be trypsinized and the transformed cells replated at low densities in fresh medium. Under these conditions, they will form isolated dense colonies, which in turn can be picked, thereby allowing transformed cell lines to be cloned.

Anchorage-Independent Growth

Except for some specialized cells, such as blood cell precursors or chondrocytes, most normal cells must be attached to a solid substratum in order to multiply and will not proliferate when suspended in liquid or semisolid media. In contrast, transformed cells often acquire the ability to grow *in vitro* without being attached and therefore can be selected directly from mixed populations by seeding cultures in agar (0.35%) or methocel (1.2%). Under these conditions, normal cells will not divide, whereas transformed cells will proliferate and form colonies which can be easily isolated and replated, either in semisolid or liquid media. This procedure was first described by Macpherson and Montagnier (151) for the isolation of polyoma virus transformed cells.

Loss of anchorage dependence has long been considered an absolute criterion for cell transformation and has been successfully used for the isolation of transformed cells induced by tumor viruses. However, some indications suggest that growth of transformed cells in agar may require profound modifications of cell physiology. Thus transformed cells selected in this way could represent an extreme state of cellular transformation and might not be suitable systems for the study of mechanisms involved in the establishment of the transformed state.

TRANSFORMATION PARAMETERS DETECTED BY BIOCHEMICAL ASSAYS

Increased Hexose Uptake

Increased rate of hexose uptake has been a consistently described transformation parameter in RNA tumor virus transformed cells (for review, see ref. 94). Comparison of the relative rates for different nutrients in normal and transformed cells led to the conclusion that the elevation of hexose uptake is not the result of a generalized increase of transport systems, which could be due to an elevated membrane activity, since only hexose uptake was significantly enhanced upon transformation (273). The increased rate of hexose uptake is usually measured by using nonmetabolizable glucose analogs, such as 2-deoxyglucose, which is transported and phosphorylated, or 3-O-methylglucose, which is a nonphosphorylatable compound.

Kinetic studies performed with these analogs revealed that only the V_{max} for hexose uptake is increased in transformed cells, and that no change in K_m for either one of the analogs used could be detected (210,215,268). A three- to eight-fold increase in the rate of hexose uptake has been reported in the case of chicken, mouse, or rat fibroblasts transformed by RSV (95,155), AEV (210), SV40 (175,194), and polyoma virus (57,175), compared to that of normal growing fibroblasts. In chick embryo fibroblasts transformed by Fujinami, PRCII, and Y73 avian sarcoma viruses, a three- to four-fold increase of hexose transport rate was reported (88), and Ki-MSV transformed rat kidney cells showed an approximately two-fold increase in glucose uptake compared to normal cells (52). None of the HSV transformed rat embryo cells or MC29 transformed chicken embryo fibroblasts tested expressed increased rate of hexose uptake (36,210).

The possible ways by which transformation could lead to an increased rate of hexose uptake have been examined by Weber et al. (273). They concluded that the increased rate of hexose uptake results from an increased synthesis of the transporter and a redistribution of the transporter between membrane fractions during transformation.

It has been suggested that the increased rate of glycolysis frequently observed in malignant cells (266) may be a consequence of an increased rate of glucose transport across the cell membrane (26,268). Since both RSV mutants defective in their ability to induce increased glucose transport (273) and a transformed cell mutant defective in glucose uptake (188) form slowly growing tumors in nude mice, enhanced ability to utilize glucose may not be necessary for loss of growth control but may provide a selective advantage for growth of tumors.

Proteases and Plasminogen Activator Production

Proteases in general are known to induce cell surface modifications similar to those observed in transformed cells (for review, see ref. 111). Strikingly, most malignant cells derived from tumors or obtained *in vitro*, regardless of the transforming agent, were shown to produce elevated levels of proteases. One of them, a protease able to function as a plasminogen activator, has been extensively studied over the past few years because of its potential importance in neoplastic transformation (42,76,116,181,200,271).

Production of plasminogen activator (PA) by transformed cells has been detected by qualitative and quantitative tests. The qualitative tests rely on the fact that upon incubation in the presence of PA, plasminogen becomes activated into plasmin, a trypsin-like enzyme. The first assay introduced by Goldberg (75) measures plasmin-mediated caseinolysis, whereas the second qualitative assay developed by Jones et al. (126) detects plasmin-mediated fibrinolytic activity. In both cases, clear halos of lysis are scored on a turbid background where PA is produced (Fig. 4). Both these tests have also been used directly on top of polyacrylamide gels (85) to detect the presence of PA, thereby allowing estimation of the molecular weight of the activator produced by different transformed cell lines. Quantitative tests include those that measure plasmin activity and those that measure directly the activation of plasminogen or the hydrolytic activity of PA.

Plasmin activity induced by PA from transformed cells has been quantitatively detected with the use of ^{125}I-labeled fibrin-coated plates (169,254) or ^{3}H-labeled casein (175) as substrates. Fluorometric assays and assays using radiolabeled N-toluene sulfonyl-L-arginine methyl ester (TAME) assay have also been described (45). A direct fluorescent assay using a synthetic fluorogenic peptide substrate [7 (N-Cbz, glycylglycyl argininamido)-4 methylcoumarin trifluoroacetate] has allowed the study of kinetic properties of PA produced by RSV transformed cells (298). Other approaches included measuring the amount of plasmin generated in the presence of PA after either electrophoretic separation of ^{125}I-labeled plasminogen and plasmin (50) or chromatographic separation of ^{3}H-labeled plasmin heavy and light chains (B. Perbal, *unpublished*). Enhanced production of PA has been reported

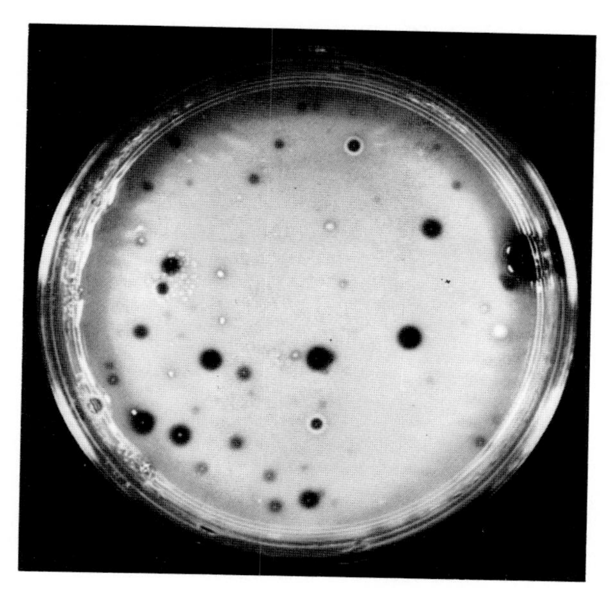

FIG. 4. PA production by polyoma transformed rat cells. Clones were overlaid with 0.7% agarose containing 2.5% nonfat dried milk and 20 μg purified plasminogen. Note that extent of casein lysis (halos) is not correlated with the size of the colony.

to occur in hamster, mouse, and rat embyro fibroblasts transformed by either DNA or RNA tumor viruses (42,75,107,169,174,175,179,192,198,254,294). Chick embryo fibroblasts transformed by RSV exhibit a 40- to 100-fold increase in fibrinolytic activity, whereas no increase in PA production is detected after infection with nontransforming avian leukosis viruses or nontransforming cytocidal RNA and DNA viruses (169,254).

Different levels of PA production characterize transformed cell lines from different origins. For example, MC29 transformed fibroblasts, which exhibit some transformation parameters, do not release high levels of PA (210), whereas chick embryo fibroblasts transformed by AEV, Fujinami, PRCII, and Y73 produce elevated levels (52). Similarly, SV40 transformed FR3T3 rat cells do not produce high levels of PA when compared to normal cells, whereas polyoma virus transformed FR3T3 do excrete high levels (175).

PAs produced by different transformed cells have been purified and characterized for their physical properties, substrate specificity, and location in the cell (42). They appear as a serine protease of molecular weight 40,000 to 60,000 daltons, depending on the cell type used to isolate transformants and not on the oncogenic virus (192). They are dimers of two polypeptidic chains held by disulfide bonds. The light chain carries the active site and binds diisopropylfluorophosphate (DFP). PA is an arginine-specific protease which does not activate zymogens, such as trypsinogen, chymotrypsinogen, or pepsinogen (42,75), but cleaves plasminogen into two polypeptides which are identical to those produced after digestion of plasminogen by urokinase.

It has been suggested that the increased level of proteolytic activity and PA production observed in transformed cells could be involved in the modifications of the cell surface proteins reported above. Removal of plasminogen from the serum used in transformed cell cultivation is accompanied by the disappearance of some transformation parameters. Thus the morphology of transformed cells grown in dog serum was shown to be plasminogen dependent (170,171), while morphologic and adhesiveness alterations reported for RSV transformed cells were not expressed in plasminogen-free media (269). It has also been reported that PA production is often related to the ability to grow in agar (179) and correlates with disorganization of actin cables (180); but removal of plasminogen from serum had no effect on the increased rate of hexose uptake and loss of fibronectin associated with the transformed state.

Furthermore, the use of several proteases inhibitors, such as soybean trypsin inhibitor, trasylol, antipain, leupeptin, N-α-tosyl-L-phenylalanyl-chloromethane (TPCK), N-α-tosyl-L-lysyl-chloromethane (TLCK), phenyl methyl sulfonyl fluoride (PMSF), TAME, or nitrophenyl guanido benzoate (NPGB), did not always provide evidence for an involvement of proteolytic enzymes in the modification of cell surface proteins (116). The failure of leupeptin and soybean trypsin inhibitor to block the transformation suggests that proteolytic enzyme activity is not required for the establishment of the transformed state (116).

Elevated membrane-bound proteolytic activities, different from PA, have also been observed in polyoma virus transformed rat fibroblasts, whereas total protease levels measured in SV40 transformed cells were similar to those of untransformed parental cells (175). A considerable variation in the production of PA, measured by fibrinolytic activity, has been reported to occur among clones of SV40 transformed mouse cells (202). Also, the frequency of PA-producing transformants obtained after polyoma virus and SV40 infection of FR3T3 rat cells clearly showed that these two viruses induce different transformation patterns in rat cells. After polyoma virus infection, 96% of the transformants selected in agarose were PA producers, whereas only 17% of the colonies obtained under the same conditions with SV40 were producers. The secretion of PA inhibitors by transformed cells has been reported in several cases (146,205). No such production of inhibitor has been detected with SV40 transformed rat cells, confirming that the difference of PA production levels corresponds to different levels of transformation (175).

Conflicting results have been reported concerning PA production by herpes viruses transformed cells. Several hamster cells transformed by HSV2, HSV1, and cytomegalovirus (MCV) were found to produce high levels of PA (5,107,294), whereas HSV transformed rat embryo cells did not (36), suggesting that PA production is not a common property of HSV transformed cells. None of the four cell lines tested (three HSV1 and one HSV2 transformed lines) was found positive with the fibrinolysis assay, although one was positive for PA production when measured by caseinolysis. On the other hand, cells derived from tumors induced by HSV2 transformed cells (one line tested) were found positive for PA production as measured by both tests (36).

There is an increasing body of evidence suggesting that the great variability observed in the expression of the various transformation parameters described above is the result of different cellular responses to both apparently unrelated oncogene products and different selection procedures applied for the isolation and propagation of transformed cells.

REGULATION OF TRANSFORMATION PARAMETER EXPRESSION

The involvement of viral transforming gene functions in the establishment and maintenance of the transformed phenotype has been clearly established using different types of mutants (247,275). However, such evidence has never been obtained for adenoviruses or herpes viruses. It has been proposed, therefore, that transformation induced by herpes viruses might proceed through the activation of cellular *onc* sequences (73), as described for ALV-induced lymphoid leukosis (97,160). Most of the studies conducted to determine whether a single viral function could be involved in the expression of the entire transformation phenotype were performed with RSV transformed chicken cells and papovavirus (SV40 and polyoma) transformed mouse and rat cells.

Early studies performed with thermosensitive (*ts*) RSV mutants altered in their capacity to transform fibroblasts at high temperatures (41 to 42° C) led to the conclusion that the expression of the different transformation parameters was under the control of the *src* gene product. As visualized by immunofluorescence, chick embryo fibroblasts transformed with *ts* mutants of RSV display a normal microfilament bundle organization when maintained at the restrictive temperature, whereas they rapidly lose their organized actin filament cables when incubated back at the permissive temperature (16,58,264). Similarly, the reduction of cell-associated fibronectin takes place only at the permissive temperature (115,125,154,164,204, 237,267).

Temperature shift experiments conducted with *ts* mutants allow temporal studies of transformation parameter expression. In *ts* (RSV) transformed cells kept at 41.5° C, the rate of hexose uptake was found to be the same as in uninfected cells. When shifted from the high temperature to 36° C, a stimulation of hexose uptake occurred by 6 hr, while morphologic alterations were apparent only by 12 to 18 hr after the shift (155). In shift-up experiments (from 36° to 41.5° C), *ts*-infected cells exhibited a normal phenotype after 4 hr incubation (155). Since increased rate of transport occurs in the presence of the DNA synthesis inhibitor actinomycin D but is blocked by cycloheximide (130,137), and since the increase of hexose uptake induced by shift to the permissive temperature does not proceed in the presence of cordycepin, an inhibitor of mRNA polyadenylation (135,136), it is probable that the regulation of hexose transport is dependent on posttranscriptional events in the nucleus. This would be consistent with the observation that enucleated transformed cells do not exhibit increased rates of hexose uptake (23).

PA production was also found to be expressed in a temperature-dependent way in chicken fibroblasts transformed with *ts* RSV (200,254,272). In shift-down experiments (from restrictive to permissive temperatures), PA production was detected as early as 2 hr after the shift, indicating that PA production is one of the earliest events associated with transformation (42). Furthermore, the tumorigenicity of the *ts* RSV transformed cells was found to be lost or reduced because of the high body temperature of the birds used (18,74).

Later studies indicated that some *ts* mutants of RSV were able to induce growth-related transformation parameters in the absence of morphologic transformation (18), and that the cell surface neoantigen known as tumor-specific surface antigen (TSSA) could be detected in transformed cells without expression of other transformation parameters (74). It has also been reported that chick embryo fibroblasts and rat kidney fibroblasts transformed by *ts* mutants of RSV were able to induce tumors on the chorioallantoic membrane of chicken eggs at both permissive and restrictive temperatures (186). These observations suggest that different cellular targets of the viral *src* gene product ($pp60^{src}$) might be involved in the establishment and maintenance of the transformed state.

In order to obtain information about the cellular targets of the transforming *src* product of RSV, Beug et al. (23) studied whether the presence of the cell nucleus is necessary for the expression of different transformation parameters. The *ts* mutant transformed cells were grown at both permissive and restrictive temperatures and enucleated with cytochalasin B. The resulting cytoplasts were shifted to either restrictive or permissive temperatures. Different transformation parameters, such as morphologic modifications, disorganization of actin cables, appearance of microvilli, loss of fibronectin, and increased agglutinability were expressed in a temperature-dependent way in cytoplasts, whereas increased uptake of hexoses by transformed cells required the presence of the nucleus. These observations are consistent with the presence in the cell of different targets being involved in the expression of the transformed phenotype.

Similar conclusions were drawn from studies performed with SV40 and polyoma transformed cells. Several different mutants altering the transforming properties of these two viruses have been isolated and mapped in a region corresponding to about one-half the genome, expressed early after infection and in transformed cells. SV40 early region codes for two proteins known as large T (94,000 daltons) and small t (17,000 daltons) (62,196), whereas polyoma early region codes for three proteins, designated large T (100,000 daltons), middle t (55,000 daltons), and small t (22,000 daltons) (66,231).

To determine whether the SV40 early region is involved in the expression of the transformed phenotype, Brugge and Butel (33) transformed normal cells of different origins (mouse, hamster, and human) with SV40 mutants of complementation group A (corresponding to the early functions) and examined the phenotype of the transformants at both permissive and restrictive temperatures. All transformation parameters tested (morphology; saturation density; colony formation on plastic, on cell monolayers, and in soft agar; and hexose uptake) were expressed at low temperature (permissive for the expression of the A gene functions) but not at the restrictive temperature. These observations and others (32,133,134,165,241) led to the conclusion that the expression of the transformation phenotype was directly dependent on viral A function.

Studies performed with polyoma and SV40 transformed rat cells revealed that, depending on the conditions to which the infected cells are submitted for several days after infection, different kinds of transformants can be obtained. Infection of actively growing cells by early *ts* mutants gave rise to transformants whose *in vitro* growth properties were expressed in a temperature-dependent way, whereas infection of resting cells led to the isolation of transformants whose growth properties were expressed at both permissive and restrictive temperatures (221). Subsequent studies indicated that large T and small t proteins were not required for the maintenance of the transformed state in SV40 transformants isolated from resting cells (195,222,223). The expression of several transformation parameters was studied in these two kinds of transformants (175).

First, striking differences were observed between SV40 and polyoma transformants isolated under similar conditions (175), allowing an operational distinction between a maximally transformed state and several intermediate phenotypes, as previously reported for polyoma virus-infected hamster cells (261) and for SV40-infected mouse cells (201). Thus all polyoma virus transformed rat cells on one side exhibited a well-characterized transformed phenotype for any property that was assayed (growth characteristics on solid substrate with high or low serum concentration, ability to grow without anchorage, rate of hexose uptake, and production of PA), whereas all the SV40 transformants studied expressed intermediate levels of transformation.

Although all SV40 transformants exhibited an increased rate of hexose uptake as compared to normal parental cells, none produced PA and grew well in the presence of low serum concentrations (175). This dual control of transformation parameter expression is similar to that observed in the case of RSV transformed cells (272,284,285).

In order to determine whether the expression of the different transformation parameters in transformed cells was dependent on the expression of functional early viral products, the transformation phenotype of different cell lines was examined at both permissive and restrictive temperatures. All cell lines isolated from actively growing infected cells displayed a temperature-sensitive phenotype, whereas the expression of transformation parameters in transformants derived from resting cells was expressed in a temperature-independent manner (175). These observations indicated that the expression of different transformation parameters could be controlled by an early gene function in some polyoma transformed cells. However, an uncoupling of the temperature dependence of transformation parameters was observed after high numbers of passages, indicating that different cellular targets might be responsible for the expression of the transformed state, as already suggested for RSV transformed cells (see above).

Analysis of a greater number of polyoma transformed rat cells obtained with *tsa* and *ts25* mutants (174) and with SV40 revealed that the expression of several transformation parameters, such as ability to grow in agar, PA production, and deoxyglucose uptake, could be under separate controls in transformed cells; it has been possible to isolate clones expressing either one of these parameters in a temperature-independent fashion. In baby hamster kidney cells (BHK21) transformed by polyoma virus mutant *ts*-3 (located in the late region), two transformation parameters (topoinhibition and lectin agglutinability) reverted to normal at the nonpermissive temperature, whereas two other properties (growth without anchorage and wound serum requirement) were still expressed (54,56).

Cellular factors involved in the expression of transformation parameters include the origin and the physiologic state of the cells. Studies performed by Risser et al. (202) demonstrated that SV40 could induce different transformation patterns in established or primary embryonic cells, and that some of the transformants obtained expressed only some of the changes of *in vitro* growth properties generally associated with transformation. Three distinct classes of SV40 transformed mouse 3T3 cells could be distinguished. One corresponded to standard maximally SV40 transformed cells described before (245). The two other classes, namely, minimal and intermediate transformants, corresponded to cells exhibiting the parameters tested at different levels. Great variations in PA production and plating efficiency in methyl cellulose (<0.001 to 2%) were also observed among SV40 transformed primary rat cells.

Since PA production varied significantly from one transformed cell line to the other, it was interesting to examine if the expression of this transformation parameter could be influenced by the selective conditions used to isolate the transformed cells. A study performed with SV40 and polyoma virus-infected FR3T3 rat cells showed that PA production by transformants was dependent on the growth state of the infected cells during a period of several days after infection (174). In other words, when infected cells are kept in active growth after infection, PA-producing transformants are counterselected, whereas a high proportion of PA producers is obtained among transformants derived from infected cells seeded directly in agar (174).

Considerable variations in levels of PA activity have been found in RSV transformed chick embryo fibroblasts isolated in agar (284,285). Several clones showed little if any PA activity; yet they did not differ from clones with elevated levels of proteases in their morphology, hexose uptake, and efficiency of cloning in agar. The fraction of PA nonproducer cells also varies with the selection procedure used. One to three percent of foci produced by infection of fibroblasts with RSV did not show increased levels of proteases, whereas as many as 13 to 32% of clones selected in agar after infection with RSV showed no detectable or low PA activity.

The levels of PA production by normal cells were also found to be subject to the physiologic and developmental state of the cells (19,199,239,255) and modulated by several different compounds, such as steroid and polypeptidic hormones, lectins, cyclic nucleotides, glucocorticoids, retinoids, lymphoid products, and growth factors (19,86,143,225,238,240,260,276,280,283). These observations could explain why some mammalian cells do not always produce increased levels of PA after transformation. In addition, a decline of both cell-associated and excreted PA has been shown to occur with increased passaging of normal cells (206) and of some polyoma transformants (B. Perbal, *unpublished data*).

CONCLUSIONS

The expression of different transformation parameters by transformed cells is not a single and unique response to oncogenic stimuli, but rather a complex interplay between cellular and viral factors. No universal transformation phenotype can be defined at present.

In contrast, a wide range of transformation levels can be induced in cells infected by the same virus. This is the case for RSV tranformed fibroblasts and SV40 and HSV transformed rat embryo fibroblasts (see above). Different transformation phenotypes, i.e., cellular responses, are also expressed after infection of the same cells by different viruses, probably because of the distinct nature of the viral oncogene products and their corresponding cellular targets. Thus SV40 transformed rat fibroblasts express low levels of transformation as compared to polyoma transformants obtained under similar conditions (175). Similarly, MC29 transformed chick embryo fibrobalsts do not display the typical transformation phenotype described for AEV (another acute leukemia virus) transformed fibroblasts; yet they express normal levels of LETS protein, rate of deoxyglucose uptake, and proteolytic activity, exhibit some modifications of their cytoskeleton organization, and are able to grow without anchorage (210).

A third factor that has been shown to play an important role in the expression of the different transformation parameters and their control by viral functions is the physiologic state of the cells up to several days after infection (174,222). The expression of several transformation parameters has been found to vary along with the maintenance of cells in culture, indicating that the transformation phenotype expressed by cells may vary according to controlled environmental selection pressures. This emphasizes the need for carefully defined growth conditions in future studies of transformation parameter expression.

How are the transformed cell phenotypes described here related to tumor cell phenotypes? How is the expression of the different transformation parameters studied *in vitro* related to tumorigenicity? Several different studies have been performed to obtain answers to these questions; it is not the aim of this chapter to discuss them in detail. It is interesting to note that most of the transformation parameters induced by viral infection have been found to be expressed in cells explanted directly from tumors. Whether the maximally transformed states defined *in vitro* correspond to more tumorigenic states remains to be established. It is clear, however, that elevated production of PA and ability to grow without anchorage—although not necessary related—are two parameters expressed by highly tumorigenic cells. The possible effects of protease inhibitors on tumors and tumorigenesis have been addressed in different systems. Troll et al. (248,249) showed that several synthetic inhibitors could be used to inhibit tumor promotion but not tumor initiation. Similarly, fibrinolytic activity and tumorigenesis of mouse fibroblasts transformed by Ki-MSV and of HSV transformed cells can be inhibited by proteinase inhibitors (6,47). Inhibition of both tumor growth and metastasis have been reported to occur in the presence of leupeptin, bestatin (252,253), and aprotin (140).

Freedman and Shin (65) and Shin et al. (227) proposed that anchorage independence is the *in vitro* transformation parameter that is best correlated with tumorigenicity in nude mice. Similar conclusions were obtained in other systems (106, 127). There are a few exceptions to the association between ability to grow without anchorage and tumorigenic potential; it appears that loss of anchorage requirement *in vitro* is a necessary but not sufficient correlate of cellular malignancy (65,228).

It has been proposed that the different levels of transformation observed *in vitro* may correspond to different stages toward the tumoral state. The recent finding that tumor cells derived from SV40 transformed cells do not express the viral proteins found in the original transformants used to induce the tumors (224) illustrates the considerable selection pressures exerted on transformed cell propagation *in vivo*. This observation might help to explain some of the discrepancies reported in the study of the transformation phenotypes expressed by transformed cells *in vitro* and related tumor cells obtained *in vivo*. The relationships between expression of particular transformation parameters by transformed cells and their tumorigenicity *in vivo* should be reconsidered.

Finally, the recent developments brought about by cloning and engineering of viral oncogenes (see Volume 1 of this series) should lead to a better understanding of the processes involved, at a molecular level, in the expression of the transformed cell phenotypes after viral infection.

ACKNOWLEDGMENTS

I thank Dr. Ethel Moustacchi and Dr. Roland Seif for their critical reading of the manuscript, and Cécile Kryceve-Martinerie for helpful suggestions. Thanks also are due to Françoise Arnouilh for her skillful assistance in typing this chapter.

REFERENCES

1. Abercrombie, M. (1970): Contact inhibition in tissue culture. *In Vitro*, 6:128–142.
2. Abercrombie, M., and Heaysman, J. E. M. (1954): Observations on the social behavior of cells in tissue culture. II. Monolayering of fibroblasts. *Exp. Cell Res.*, 6:293–306.
3. Abercrombie, M., and Dunn, G. A. (1975): Adhesions of fibroblasts to substratum during contact inhibition observed by interference reflection microscopy. *Exp. Cell Res.*, 92:57–62.
4. Adams, S. L., Sobel, M. E., Howard, B. H., Olden, K., Yamada, K. M., de Combrugghe, B., and Pastan, I. (1977): Levels of translatable mRNAs for cell surface protein, collagen precursors and two membrane proteins are altered in Rous sarcoma virus transformed chick embryo fibroblasts. *Proc. Natl. Acad. Sci. USA*, 74:3399–3403.
5. Adelman, S. F., Howett, M. K., and Rapp, F. (1980): Quantification of plasminogen activator activity associated with herpes virus transformed cells. *J. Gen. Virol.*, 50:101–110.
6. Adelman, S. F., Howett, M. K., and Rapp, F. (1982): Protease inhibitors suppress fibrinolytic activity of herpes virus-transformed cells. *J. Gen. Virol.*, 60:15–24.

7. Albrecht-Buehler, G. (1977): Daughter 3T3 cells, are they mirror images of each other? *J. Cell Biol.*, 72:595–603.
8. Alexander, S. S., Jr., Colonna, G., Yamada, K. M., Pastan, I., and Edelhoch, H. (1978): Molecular properties of a major cell surface protein from chick embryo fibroblasts. *J. Biol. Chem.*, 253:5820–5824.
9. Alexander, S. S., Jr., Colonna, G., and Edelhoch, H. (1979): The structure and stability of human plasma cold-insoluble globulin. *J. Biol. Chem.*, 254:1501–1505.
10. Ali, I. U., Mautner, V. M., Lanza, R. P., and Hynes, R. O. (1977): Restoration of normal morphology, adhesion and cytoskeleton in transformed cells by addition of a transformation-sensitive surface protein. *Cell*, 11:115–126.
11. Ali, I. U., and Hynes, R. O. (1978): Effects of LETS glycoprotein on cell motility. *Cell*, 14:439–446.
12. Allred, L. E., and Porter, K. R. (1979): Morphology of normal and transformed cells. In: *Surface of Normal and Malignant Cells*, edited by R. O. Hynes, pp. 21–63. Wiley, New York.
13. Ambros, V. R., Chen, L. B., and Buchanan, J. M. (1975): Surface ruffles as markers for studies of cell transformation by Rous sarcoma virus. *Proc. Natl. Acad. Sci. USA*, 72:3144–3148.
14. Arndt-Jovin, D. J., and Berg, P. (1971): Quantitative binding of ^{125}I concanavalin A to normal and transformed cells. *J. Virol.*, 8:716–721.
15. Ash, J. F., and Singer, S. J. (1976): Concanavalin A-induced transmembrane linkage of concanavalin A surface receptors to intracellular myosin-containing filaments. *Proc. Natl. Acad. Sci. USA*, 73:4575–4579.
16. Ash, J. F., Vogt, P. K., and Singer, S. J. (1976): Reversion from transformed to normal phenotype by inhibition of protein synthesis in rat kidney cells infected with a temperature-sensitive mutant of Rous sarcoma virus. *Proc. Natl. Acad. Sci. USA*, 73:3603–3607.
17. Bader, A. V., and Bader, J. P. (1976): Transformation of cells by Rous sarcoma virus: Cytoplasmic vacuolization. *J. Cell. Physiol.*, 87:33–46.
18. Becker, D., Kurth, R., Critchley, D., Friis, R., and Bauer, H. (1977): Distinguishable transformation defective phenotypes among temperature sensitive mutants of Rous sarcoma virus. *J. Virol.*, 21:1042–1055.
19. Beers, W. H., Strickland, S., and Reich, E. (1975): Ovarian plasminogen activator: Relationship to ovulation and hormonal regulation. *Cell*, 6:387–394.
20. Bell, P. B. (1977): Locomotory behavior, contact inhibition, and pattern formation of 3T3 and polyoma virus-transformed 3T3 cells in culture. *J. Cell Biol.*, 74:963–982.
21. Beug, H., Peters, J. M., and Graf, T. (1976): Expression of virus specific morphological cell transformation induced in enucleated cells. *Z. Naturforsch.*, 31c:766–768.
22. Beug, H., and Graf, T. (1977): The role of the cell nucleus in the expression of Rous sarcoma virus (RSV)-induced cell transformation. In: *Avian RNA Tumor Viruses*, edited by S. Barlati and C. de Giuli, pp. 280–295. Piccin Editore, Padova, Italy.
23. Beug, H., Claviez, M., Jockush, B. M., and Graf, T. (1978): Differential expression of Rous sarcoma virus-specific transformation parameters in enucleated cells. *Cell*, 14:843–856.
24. Billen, D., and Debrunner, G. A. (1960): Continuously propagating cells derived from the normal mouse bone marrow. *J. Natl. Cancer Inst.*, 25:1127–1134.
25. Birdwell, C. R., Gospodarowicz, D., and Nicolson, G. L. (1978): Identification, localization and role of fibronectin in cultured bovine endothelial cells. *Proc. Natl. Acad. Sci. USA*, 75:3273–3277.
26. Bissell, M. (1976): Transport as a rate limiting step in glucose metabolism in virus-transformed cells: Studies with cytochalasin B. *J. Cell. Physiol.*, 89:701–710.
27. Borek, C., and Fenoglio, M. (1976): Scanning electron microscopy of surface features of hamster embryo cells transformed in vitro by X-irradiation. *Cancer Res.*, 36:1325–1334.
28. Bornstein, P. (1974): Biosynthesis of collagen. *Ann. Rev. Biochem.*, 43:567–603.
29. Bronstein, P., and Ash, J. F. (1977): Cell surface associated structural proteins in connective tissue cells. *Proc. Natl. Acad. Sci. USA*, 74:2480–2484.
30. Bowman, P. D., and Daniel, C. W. (1975): Aging of human fibroblasts in vitro: Surface features and behavior of aging WI-38 cells. *Mech. Ageing Dev.*, 4:147–158.
31. Brinkley, B. R., Fuller, G. M., and Highfield, D. P. (1975): Cytoplasmic microtubules in normal and transformed cells in culture: Analysis by tubulin antibody immunofluorescence. *Proc. Natl. Acad. Sci. USA*, 72:4981–4985.
32. Brockman, W. W. (1978): Transformation of BALB/c 3T3 cells by tsA mutants of simian virus

40: Temperature sensitivity of the transformed phenotype and retransformation by wild type virus. *J. Virol.*, 25:860–870.
33. Brugge, J. S., and Butel, J. S. (1975): Role of simian virus 40 gene A function in maintenance of transformation. *J. Virol.*, 15:619–635.
34. Burger, M. M. (1969): A difference in the architecture of the surface membrane of normal and virally transformed cells. *Proc. Natl. Acad. Sci. USA*, 62:994–1001.
35. Burger, M. M. (1973): Surface changes in transformed cells detected by lectins. *Fed. Proc.*, 32:91–101.
36. Büültjens, T. E. J., and Macnab, J. C. M. (1981): Characterization of rat embryo cells transformed by *ts* mutants and sheared DNA of Herpes Simplex Virus types 1 and 2 and a derived tumor cell line. *Cancer Res.*, 41:2540–2547.
37. Chen, L. B., Gallimore, P. H., and McDougall, J. K. (1976): Correlation between tumor induction and the large external transformation sensitive protein on the cell surface. *Proc. Natl. Acad. Sci. USA*, 73:3570–3574.
38. Chen, A. B., Amrami, D. L., and Mosesson, M. W. (1977): Heterogeneity of the cold-insoluble globulin of human plasma (CIg) a circulating cell surface protein. *Biochim. Biophys. Acta*, 493:310–322.
39. Chen, Y., Hayman, M., and Vogt, P. K. (1977): Properties of mammalian cells transformed by temperature sensitive mutants of avian sarcoma virus. *Cell*, 11:513–521.
40. Chen, L. B., Maitland, N., Ballimore, P. H., and McDougall, J. K. (1977): Detection of the large external transformation sensitive protein on some epithelial cells. *Exp. Cell Res.*, 103:39–46.
41. Chen, L. B., Murray, A., Segal, R. A., Bushnell, A., and Walsh, M. L. (1978): Studies on intracellular LETS glycoprotein matrices. *Cell*, 14:377–391.
42. Christman, J. K., Acs, G., Silagi, S., and Silverstein, S. C. (1975): Plasminogen activator: Biochemical characterization and correlation with tumorigenicity. In: *Proteases and Biological Control*, edited by E. Reich, D. B. Rifkin, and E. Shaw, pp. 827–839. Cold Spring Harbor Laboratory, Cold Spring Harbor, New York.
43. Clausen, J. J., and Syverton, J. T. (1962): Comparative chromosomal study of 31 cultured mammalian cell lines. *J. Natl. Cancer Inst.*, 28:117–129.
44. Cline, M. J., and Livingston, D. C. (1971): Binding of (^{3}H) concanavalin A by normal and transformed cells. *Nature*, 232:155–156.
45. Coleman, P. L., Latham, H. G., Jr., and Shaw, E. N. (1976): Some sensitive methods for the assay of trypsin like enzyme. In: *Methods in Enzymology, Vol. 45*, edited by S. Colowick and N. O. Kaplan, pp. 12–26. Academic Press, New York.
46. Collard, J. G., and Temmink, J. H. M. (1976): Surface morphology and agglutinability with concanavalin A in normal and transformed murine fibroblasts. *J. Cell Biol.*, 68:101–112.
47. Collen, D., Billiau, A., Edy, J., and De Somer, P. (1977): Identification of the human plasma proteins which inhibit fibrinolysis associated with malignant cells. *Biochim. Biophys. Acta*, 499:194–201.
48. Critchley, D. R. (1974): Cell surface proteins of NIL 1 hamster fibroblasts labelled by a galactose oxidase, tritiated borohydride method. *Cell*, 3:121–125.
49. Critchley, D. R., Wyke, J. A., and Hynes, R. O. (1976): Cell surface and metabolic labelling of the proteins of normal and transformed chicken cells. *Biochim. Biophys. Acta*, 436:335–352.
50. Danø, K., and Reich, E. (1979): Plasminogen activator from cells transformed by an oncogenic virus. Inhibitors of the activation reaction. *Biochim. Biophys. Acta*, 566:138–151.
51. David-Pfeuty, T., and Singer, S. J. (1980): Altered distributions of the cytoskeletal proteins vinculin and α-actinin in cultured fibroblasts transformed by Rous sarcoma virus. *Proc. Natl. Acad. Sci. USA*, 77:6687–6691.
52. Devouge, M. W., Mukherjeen, B. B., and Pena, S. D. J. (1982): Kirsten murine sarcoma virus-coded $p21^{ras}$ may act on multiple targets to effect pleiotropic changes in transformed cells. *Virology*, 121:327–344.
53. Dulbecco, R. (1970): Topoinhibition and serum requirement of transformed and untransformed cells. *Nature*, 227:802–806.
54. Dulbecco, R., and Eckhart, W. R. (1970): Temperature dependent properties of cells transformed by a thermosensitive mutant of polyoma virus. *Proc. Natl. Acad. Sci. USA*, 67:1775–1781.
55. Dulbecco, R., and Elkington, J. (1973): Conditions limiting multiplication of fibroblastic and epithelial cells in dense cultures. *Nature*, 246:197–199.
56. Eckhart, W. R., Dulbecco, R., and Burger, M. (1971): Temperature dependent surface changes

in cells infected or transformed by a thermosensitive mutant of polyoma virus. *Proc. Natl. Acad. Sci. USA*, 68:283–286.
57. Eckhart, W., and Weber, M. J. (1974): Uptake of 2-deoxyglucose by Balb/3T3 cells: Changes after polyoma infection. *Virology*, 61:223–228.
58. Edelman, G. M., and Yahara, I. (1976): Temperature sensitive changes in surface modulating assemblies of fibroblasts transformed by mutants of Rous sarcoma virus. *Proc. Natl. Acad. Sci. USA*, 73:2047–2051.
59. Engvall, E., and Ruoslahti, E. (1977): Binding of soluble form of fibroblast surface protein, fibronectin, to collagen. *Int. J. Cancer*, 20:1–5.
60. Enlander, D., Scott, T., and Tobey, R. A. (1974): Observations of the surfaces of synchronized chinese hamster ovary cells in suspension culture. In: *Scanning Electron Microscopy, Part III*, pp. 573–580. Research Institute, Chicago.
61. Epifanova, O. I., and Terskikh, V. V. (1969): On the resting periods in the cell life cycle. *Cell Tissue Kinet.*, 2:75–80.
62. Fiers, W., Contreras, R., Haegeman, G., Rogiers, R., Vande-Voorde, A., Van Heuverswyn, H., Van Herreweghe, J., Volckaert, G., and Ysebaert, M. (1978): Complete nucleotide sequence of SV40 DNA. *Nature*, 273:113–120.
63. Foley, G. E., Drolet, B. P., McCarthy, R. E., Goulet, K. A., Dokos, J. M., and Filler, D. A. (1960): Isolation and serial propagation of malignant and normal cells in semi-defined media. *Cancer Res.*, 20:930–936.
64. Fonte, V., and Porter, K. R. (1974): Topographical changes associated with the viral transformation of normal cells to tumorigenicity. In: *Abstracts of the Eighth International Congress on Electron Microscopy, Vol. II.*, Canberra.
65. Freedman, V. H., and Shin, S. (1978): Use of nude mice for studies on the tumorigenicity of animal cells. In: *The Nude Mouse in Experimental and Clinical Research*, edited by J. Fogh and B. C. Giovanella, pp. 353–384. Academic Press, New York.
66. Friedman, T., Esty, A., Laporte, P., and Deininger, P. (1979): The nucleotide sequence and genome organization of the polyoma early region: Extensive nucleotide and amino-acid homology with SV40. *Cell*, 17:715–724.
67. Gaffney, B. J. (1975): Fatty acid chain flexibility in the membranes of normal and transformed fibroblasts. *Proc. Natl. Acad. Sci. USA*, 72:664–668.
68. Gaffney, B. J., Branton, P. E., Wickus, G. G., and Hirschberg, C. B. (1974): Fluid lipid regions in normal and Rous sarcoma virus transformed chick embryo fibroblasts. In: *Viral Transformation and Endogenous Viruses*, edited by A. S. Kaplan, pp. 97–115. Academic Press, New York.
69. Gahmberg, C. G., and Hakomori, S. I. (1973): Altered growth behavior of malignant cells associated with changes in externally labelled glycoprotein and glycolipid. *Proc. Natl. Acad. Sci. USA*, 70:3329–3333.
70. Gahmberg, C. G., and Hakomori, S. I. (1974): Organization of glycolipids and glycoproteins in surface membranes: Dependency on cell cycle and on transformation. *Biochem. Biophys. Res. Commun.*, 59:283–291.
71. Gahmberg, C. G., Kiehn, D., and Hakomori, S. I. (1974): Changes in a surface-labelled galactoprotein and in glycolipid concentrations in cells transformed by a temperature-sensitive polyoma virus mutant. *Nature*, 248:413–415.
72. Gahmberg, C. G., and Hakomori, S. I. (1975): Surface carbohydrates of hamster fibroblasts. *J. Biol. Chem.*, 250:2447–2451.
73. Galloway, D. A., and McDougall, J. K. (1983): The oncogenic potential of herpes simplex viruses: Evidence for a "hit-and-run" mechanism. *Nature*, 302:21–24.
74. Gionti, E., Kryceve-Martinerie, C., Aupoix, M. C., and Calothy, G. (1980): Phenotypic heterogeneity among temperature-sensitive mutants of Rous sarcoma virus. Studies with inhibitors of protein synthesis. *Virology*, 100:219–228.
75. Goldberg, A. R. (1974): Increased protease levels in transformed cells. A casein overlay assay for the detection of plasminogen activator production. *Cell*, 2:95–102.
76. Goldberg, A. R., Wolf, B. A., and Lefebvre, P. A. (1975): Plasminogen activators of transformed and normal cells. In: *Proteases and Biological Control*, edited by E. Reich, D. B. Rifkin, and E. Shaw, pp. 841–847. Cold Spring Harbor Laboratory, Cold Spring Harbor, New York.
77. Goldman, R. D., and Follet, A. C. (1969): The structure of the major cell processes of isolated BHK-21 fibroblasts. *Exp. Cell Res.*, 57:263–276.

78. Goldman, R. D., and Knipe, D. M. (1972): Functions of cytoplasmic fibers in non-muscle cell motility. *Cold Spring Harbor Symp. Quant. Biol.*, 37:523–538.
79. Goldman, R. D., Pollack, R., and Hopkins, N. H. (1973): Preservation of normal behavior by enucleated cells in culture. *Proc. Natl. Acad. Sci. USA*, 70:750–754.
80. Goldman, R. D., Lazarides, E., Pollack, R., and Weber, K. (1975): The distribution of actin in non-muscle cells: The use of actin antibody in the localization of actin within the microfilament bundles of mouse 3T3 cells. *Exp. Cell Res.*, 90:333–344.
81. Goldman, R. D., Pollack, R., Chang, C. M., and Bushnell, A. (1975): Properties of enucleated cells. III. Changes in cytoplasmic architecture of enucleated BHK21 cells following trypsinization and replating. *Exp. Cell Res.*, 93:175–183.
82. Goldman, R. D., Yerna, M. J., and Schloss, J. A. (1976): Localization and organization of microfilaments and related proteins in normal and virus-transformed cells. *J. Supramol. Struct.*, 5:155–183.
83. Gonda, M. A., Aaronson, S. A., Ellmore, N., Zeve, V. H., and Nagashima, K. (1976): Ultrastructural studies of surface features of human normal and tumor cells in tissue culture by scanning and transmission electron microscopy. *J. Natl. Cancer Inst.*, 56:245–263.
84. Graessman, A., Graessman, M., Tijian, R., and Topp, W. C. (1980): Simian virus 40 small t protein is required for loss of actin cable networks in rat cells. *J. Virol.*, 33:1182–1191.
85. Granelli-Piperno, A., and Reich, E. (1978): A study of proteases and protease inhibitor complexes in biological fluids. *J. Exp. Med.*, 147:223–234.
86. Granelli-Piperno, A., Vassali, J. D., and Reich, E. (1977): Secretion of plasminogen activator by human polymorphonuclear leukocytes. Modulation by glucocorticoids and other effectors. *J. Exp. Med.*, 146:1693–1706.
87. Green, M., Todaro, G. J., and Goldberg, B. (1966): Collagen synthesis in fibroblasts transformed by oncogenic viruses. *Nature*, 209:916–917.
88. Guyden, J. C., and Martin, S. (1982): Transformation parameters of chick embryo fibroblasts transformed by Fujinami, PRC II, PRC II-p, and Y73 avian sarcoma viruses. *Virology*, 122:71–83.
89. Haff, R. F., and Swim, H. E. (1956): Serial propagation of 3 strains of rabbit fibroblasts. *Proc. Soc. Exp. Biol. Med.*, 93:200–208.
90. Hahn, L. H. E., and Yamada, K. M. (1979): Isolation and biological characterization of active fragments of the adhesive glycoprotein fibronectin. *Cell*, 18:1043–1051.
91. Hanafusa, H. (1977): Cell transformation by RNA tumor viruses. In: *Comprehensive Virology, Vol. 10*, edited by H. Fraenkel-Conrat and R. R. Wagner, pp. 401–483. Plenum, New York.
92. Harris, A. K. (1973): Location of cellular adhesions to solid substrate. *Dev. Biol.*, 35:97–114.
93. Hartwig, J. H., and Stossel, T. P. (1975): Isolation and properties of actin, myosin and a new actin binding protein in rabbit alveolar macrophages. *J. Biol. Chem.*, 250:5696–5705.
94. Hatanaka, M. (1974): Transport of sugars in tumor cell membranes. *Biochim. Biophys. Acta*, 355:77–104.
95. Hatanaka, M., and Hanafusa, H. (1970): Analysis of a functional change in membrane in the process of cell transformation by Rous sarcoma virus: Alteration in the characteristics of sugar transport. *Virology*, 41:647–652.
96. Hayflick, L., and Moorhead, P. S. (1961): The serial cultivation of human diploid cell strains. *Exp. Cell Res.*, 25:585–621.
97. Hayward, W. S., Neel, B. G., and Astrin, S. (1981): Activation of a cellular *onc* gene by promoter insertion in ALV-induced lymphoid leukosis. *Nature*, 290:475–480.
98. Heath, J. P., and Dunn, G. A. (1978): Cell to substratum contacts of chick fibroblasts and their relation to the microfilament system. A correlated interference-reflexion and high-voltage electron microscope study. *J. Cell Sci.*, 29:197–211.
99. Heaysman, J. E. M., and Pegrum, S. M. (1973): Early contacts between fibroblasts, an ultrastructural study. *Exp. Cell Res.*, 78:71–78.
100. Hedman, K., Vaheri, A., and Wartiovaara (1978): External fibronectin of cultured human fibroblasts is predominantly a matrix protein. *J. Cell Biol.*, 76:748–760.
101. Heggeness, M. H., Wang, K., and Singer, S. J. (1977): Intracellular distribution of mechanochemical proteins in cultured fibroblasts. *Proc. Natl. Acad. Sci. USA*, 74:3883–3887.
102. Hogg, N. M. (1974): A comparison of membrane proteins of normal and transformed cells by lactoperoxidase labelling. *Proc. Natl. Acad. Sci. USA*, 71:489–492.

103. Holley, R. W., and Kiernan, J. A. (1968): "Contact inhibition" of cell division in 3T3 cells. *Proc. Natl. Acad. Sci. USA*, 60:300–304.
104. Holley, R. W., and Kiernan, J. A. (1974): Control of the initiation of DNA synthesis in 3T3 cells: Serum factors. *Proc. Natl. Acad. Sci. USA*, 71:2908–2911.
105. Holley, R. W., Armour, R., and Baldwin, J. H. (1978): Density dependent regulation of growth of BSC-1 cells in cell culture: Growth inhibitors formed by these cells. *Proc. Natl. Acad. Sci. USA*, 75:1864–1866.
106. Howell, A. N., and Sager, R. (1978): Tumorigenicity and its suppression in hybrids of mouse and chinese hamster cell lines. *Proc. Natl. Acad. Sci. USA*, 75:2358–2362.
107. Howett, M. K., High, C. S., and Rapp, F. (1978): Production of plasminogen activator by cells transformed by herpes viruses. *Cancer Res.*, 38:1075–1078.
108. Hsu, T. C. (1961): Chromosomal evolution in cell populations. *Int. Rev. Cytol.*, 12:69–75.
109. Hsu, T. C., Billen, D., and Levan, A. (1961): Mammalian chromosomes in vitro. XV. Patterns of transformation. *J. Natl. Cancer Inst.*, 27:515–527.
110. Hynes, R. O. (1973): Alteration of cell-surface proteins by viral transformation and by proteolysis. *Proc. Natl. Acad. Sci. USA*, 70:3170–3174.
111. Hynes, R. O. (1974): Role of surface alterations in cell transformation: The importance of proteases and surface proteins. *Cell*, 1:147–156.
112. Hynes, R. O. (1976): Cell surface proteins and malignant transformation. *Biochim. Biophys. Acta*, 458:73–107.
113. Hynes, R. O. (1979): Proteins and glycoproteins. In: *Surface of Normal and Malignant Cells*, edited by R. O. Hynes, pp. 103–148. Wiley, New York.
114. Hynes, R. O., and Bye, J. M. (1974): Density and cell cycle dependence of cell surface proteins in hamster fibroblasts. *Cell*, 3:113–120.
115. Hynes, R. O., and Wyke, J. A. (1975): Alterations in surface proteins in chicken cells transformed by temperature-sensitive mutants of Rous sarcoma virus. *Virology*, 64:492–504.
116. Hynes, R. O., Wyke, J. A., Bye, J. M., Humphryes, K. C., and Pearlstein, E. S. (1975): Are proteases involved in altering surface proteins during viral transformation? In: *Proteases and Biological Control*, edited by E. Reich, D. B. Rifkin, and E. Shaw, pp. 931–944. Cold Spring Harbor Laboratory, Cold Spring Harbor, New York.
117. Hynes, R. O., and Destree, A. T. (1977): Extensive disulfide bonding at the mammalian cell surface. *Proc. Natl. Acad. Sci. USA*, 74:2855–2859.
118. Hynes, R. O., Destree, A. T., Mautner, V. M., and Ali, I. U. (1977): Synthesis, secretion, and attachment of LETS glycoprotein in normal and transformed cells. *J. Supramol. Struct.*, 7:397–408.
119. Hynes, R. O., and Destree, A. T. (1978): 10 nm filaments in normal and transformed cells. *Cell*, 13:151–163.
120. Inbar, M., and Sachs, L. (1969): Interaction of the carbohydrate-binding protein concanavalin A with normal and transformed cells. *Proc. Natl. Acad. Sci. USA*, 63:1418–1425.
121. Isaka, T., Yoshida, M., Owada, M., and Toyoshima, K. (1975): Alterations in membrane polypeptides of chick embryo fibroblasts induced by transformation with avian sarcoma viruses. *Virology*, 65:226–237.
122. Ishikawa, H., Bischoff, R., and Holtzer, H. (1969): Formation of arrowhead complexes with heavy meromyosin in a variety of cell types. *J. Cell Biol.*, 43:312–328.
123. Izzard, C. S., and Lochner, L. R. (1976): Cell to substrate contacts in living fibroblasts. An interference reflexion study with an evaluation of the technique. *J. Cell Sci.*, 21:129–159.
124. Jaffe, E. A., and Mosher, D. (1978): Synthesis of fibronectin by cultured human endothelial cells. In: *Fibroblasts Surface Protein*, edited by A. Vaheri, E. Ruoslahti, and D. Mosher, pp. 122–131. Academic Press, New York.
125. Jone, C., and Hager, L. P. (1976): Iodination of zeta protein by lactoperoxidase, chloroperoxidase and chloramine T. *Biochem. Biophys. Res. Commun.*, 68:16–20.
126. Jones, P. A., Benedict, W. F., Strickland, S., and Reich, E. (1975): Fibrin overlay methods for the detection of single transformed cells and colonies of transformed cells. *Cell*, 5:323–329.
127. Jones, P. A., Laug, W. E., Gardner, A., Nye, C. A., Fink, L. M., and Benedict, W. F. (1976): *In vitro* correlates of transformation in C3H/10 T 1/2 clone 8 mouse cells. *Cancer Res.*, 36:2863–2867.
128. Kahn, P., and Shin, S. (1979): Cellular tumorigenicity in nude mice. *J. Cell Biol.*, 82:1–16.
129. Kamine, J., and Rubin, H. (1977): Coordinate control of collagen synthesis and cell growth in

chick embryo fibroblasts and the effect of viral transformation on collagen synthesis. *J. Cell. Physiol.*, 92:1–12.

130. Kawai, S., and Hanafusa, H. (1971): The effects of reciprocal changes of temperature on the transformed state of cells infected with a temperature-sensitive Rous sarcoma virus mutant. *Virology*, 46:470–479.
131. Keski-Oja, J., Mosher, D. F., and Vaheri, A. (1976): Cross linking of a major fibroblast surface-associated glycoprotein (fibronectin) catalysed by blood coagulation factor XIII. *Cell*, 9:29–35.
132. Keski-Oja, J., Mosher, D. F., and Vaheri, A. (1977): Dimeric character of fibronectin, a major cell surface associated glycoprotein. *Biochem. Biophys. Res. Commun.*, 74:699–706.
133. Kimura, G., and Itagaki, A. (1975): Initiation and maintenance of cell transformation by simian virus 40: A viral genetic property. *Proc. Natl. Acad. Sci. USA*, 72:673–677.
134. Kimura, G., Itagaki, A., and Summers, J. (1975): Rat cell line 3Y1 and its virogenic polyoma and SV40 transformed derivatives. *Int. J. Cancer*, 15:694–706.
135. Kletzien, R. F., and Perdue, J. F. (1974): Alterations in sugar transport following transformation by a temperature sensitive mutant of the Rous sarcoma virus. *J. Biol. Chem.*, 249:3375–3382.
136. Kletzien, R. F., and Perdue, J. F. (1975): Regulation of sugar transport in chick embryo fibroblasts infected with a temperature-sensitive mutant of RSV. *Cell*, 6:513–520.
137. Kletzien, R. F., and Perdue, J. F. (1976): Regulation of sugar transport in chick embryo fibroblasts and in fibroblasts transformed by a temperature-sensitive mutant of the Rous sarcoma virus. *J. Cell. Physiol.*, 89:723–728.
138. Konoza, R. J. J., Brunette, D. M., Purdon, A. D., and Sodek, J. (1978): Isolation and identification of epithelial like cells in culture by a collagenase-separation technique. *In Vitro*, 14:746–753.
139. Kruse, P. F., and Patterson, M. K., Jr. (editors) (1973): *Tissue Culture: Methods and Applications*. Academic Press, New York.
140. Lage, A., Diaz, J. W., and Gonzalez, P. (1978): Effects of proteinase inhibitors in experimental tumors. *Neoplasma*, 25:257–259..
141. Lazarides, E., and Weber, K. (1974): Actin antibody: The specific visualization of actin filaments in non-muscle cells. *Proc. Natl. Acad. Sci. USA*, 71:2268–2272.
142. Lazarides, E., and Burridge, K. (1975): α-Actinin: Immunofluorescent localization of a muscle structural protein in non-muscle cells. *Cell*, 6:289–298.
143. Lee, L. S., and Weinstein, I. B. (1978): Epidermial growth factor, like phorbol esters, induces plasminogen activator in HeLa cells. *Nature*, 274:696–697.
144. Levan, A., and Biesele, J. J. (1958): Role of chromosomes in cancerogenesis as studied in serial tissue culture of mammalian cells. *Ann. NY Acad. Sci.*, 71:1022–1026.
145. Levinson, W., Bhatnagar, R. S., and Liu, T. Z. (1975): Loss of ability to synthesize collagen in fibroblasts transformed by Rous sarcoma virus. *J. Natl. Cancer Inst.*, 55:807–810.
146. Loskutoff, D., and Edgington, T. S. (1977): Synthesis of a fibrinolytic activator and inhibitor by endothelial cells. *Proc. Natl. Acad. Sci. USA*, 74:3903–3907.
147. McConnell, M. R., Blumberg, P. M., and Rossow, P. W. (1978): Dimeric and high molecular weight forms of the large external transformation-sensitive protein on the surface of chick embryo fibroblasts. *J. Biol. Chem.*, 253:7522–7530.
148. McNutt, N. S., Culp, L. A., and Black, P. H. (1971): Contact inhibited revertant cell lines isolated from SV40 transformed cells. II. Ultrastructure study. *J. Cell Biol.*, 50:691–708.
149. McNutt, N. S., Culp, L. A., and Black, P. H. (1973): Contact inhibited revertant cell lines isolated from SV40 transformed cells. IV. Microfilament distribution and cell shape in untransformed, transformed and revertant Balb/c 3T3 cells. *J. Cell Biol.*, 56:412–428.
150. Macpherson, I., and Stocker, M. (1962): Polyoma transformation of hamster cell clones. An investigation of genetic factors affecting cell competence. *Virology*, 16:147–151.
151. Macpherson, I., and Montagnier, L. (1964): Agar suspension culture for the selective assay of cells transformed by polyoma virus. *Virology*, 23:291–294.
152. Malick, L. E., and Langenbach, R. (1976): Scanning electron microscopy of *in vitro* chemically transformed mouse embryo cells. *J. Cell Biol.*, 68:654–664.
153. Manaker, R. A., and Groupe, V. (1956): Discrete foci of altered chicken embryo cells associated with Rous sarcoma virus in tissue culture. *Virology*, 2:838–840.
154. Marciani, D. J., and Bader, J. P. (1975): Polypeptide composition of cell membranes from chick embryo fibroblasts transformed by Rous sarcoma virus. *Biochim. Biophys. Acta*, 401:386–398.
155. Martin, G. S., Venuta, S., Weber, M., and Rubin, H. (1971): Temperature dependent alterations

in sugar transport in cells infected by a temperature sensitive mutant of Rous sarcoma virus. *Proc. Natl. Acad. Sci. USA*, 68:2739–2741.
156. Mautner, V. M., and Hynes, R. O. (1977): Surface distribution of LETS protein in relation to the cytoskeleton of normal and transformed cells. *J. Cell Biol.*, 75:743–758.
157. Mooseker, M. S., and Tilney, L. G. (1975): Organization of an actin filament membrane complex. Filament polarity and membrane attachment in the microvilli of intestinal epithelial cells. *J. Cell Biol.*, 67:725–743.
158. Mosesson, M. W., Chen, A. B., and Huseby, R. M. (1975): The cold-insoluble globulin of human plasma: Studies of its essential structural features. *Biochim. Biophys. Acta*, 386:509–524.
159. Mosher, D. F. (1977): Labeling of a major fibroblast surface protein (fibronectin) catalyzed by blood coagulation factor XIII_a. *Biochim. Biophys. Acta*, 491:205–210.
160. Neel, B. G., Hayward, W. S., Robinson, H., Fang, J., and Astrin, S. (1981): Avian leukosis virus-induced tumors have common proviral integration sites and synthesize discrete new RNAs: Oncogenesis by promoter insertion. *Cell*, 23:323–334.
161. Nicolson, G. L. (1973): Temperature-dependent mobility of concanavalin A sites on tumor cell surfaces. *Nature*, 243:218–220.
162. Nicolson, G. L. (1974): The interactions of lectins with animal cell surfaces. *Int. Rev. Cytol.*, 39:89–100.
163. Nicolson, G. L., and Blaustein, J. (1972): The interaction of *Ricinus communis* agglutinin with normal and tumor cell surfaces. *Biochim. Biophys. Acta*, 266:543–547.
164. Olden, K., and Yamada, K. M. (1977): Mechanism of the decrease in the major cell surface protein of chick embryo fibroblasts after transformation. *Cell*, 11:957–969.
165. Osborn, M., and Weber, K. (1975): Simian virus 40 gene A function and maintenance of transformation. *J. Virol.*, 15:636–644.
166. Osborn, M., and Weber, K. (1977): The display of microtubules in transformed cells. *Cell*, 12:561–571.
167. Osborn, M., Franke, W., and Weber, K. (1977): Visualization of a system of filaments 7-10 nm thick in cultured cells of an epithelioid line (PtK2) by immunofluorescence microscopy. *Proc. Natl. Acad. Sci. USA*, 74:2490–2494.
168. Osborn, M., Franke, W., and Weber, K. (1980): Direct demonstration of the presence of two immunologically distinct intermediate-sized filament systems in the same cell by double immunofluorescence microscopy. *Exp. Cell Res.*, 125:37–46.
169. Ossowski, L., Unkeless, J. C., Tobia, A., Quigley, J. P., Rifkin, D. B., and Reich, E. (1973): An enzymatic function associated with transformation of fibroblasts by oncogenic viruses. II. Mammalian fibroblast cultures transformed by DNA and RNA tumor viruses. *J. Exp. Med.*, 137:112–126.
170. Ossowski, L., Quigley, J. P., and Reich, E. (1974): Fibrinolysis associated with oncogenic transformation: Morphological correlates. *J. Biol. Chem.*, 249:4312–4320.
171. Ossowski, L., Quigley, J. P., and Reich, E. (1975): The relationship between cell migration and plasminogen activation. In: *Proteases and Biological Control*, edited by E. Reich, D. B. Rifkin, and E. Shaw, pp. 901–913. Cold Spring Harbor Laboratory, Cold Spring Harbor, New York.
172. Ozanne, B., and Sambrook, J. (1971): Binding of radioactively labelled concanavalin A and wheat germ agglutinin to normal and virus transformed cells. *Nature*, 232:156–160.
173. Pearlstein, R., and Waterfield, M. D. (1974): Metabolic study on ^{125}I labeled baby hamster kidney cell plasma membrane. *Biochim. Biophys. Acta*, 362:1–12.
174. Perbal, B. (1980): Transformation phenotype of polyoma virus transformed rat fibroblasts: Plasminogen activator production is modulated by the growth state of the cells and regulated by the expression of an early viral gene function. *J. Virol.*, 35:420–427.
175. Perbal, B., and Rassoulzadegan, M. (1980): Distinct transformation phenotypes induced by polyoma virus and Simian virus 40 in rat fibroblasts and their control by an early viral gene function. *J. Virol.*, 33:697–707.
176. Perdue, J. F. (1973): The distribution, ultrastructure and chemistry of microfilaments in cultured chick embryo fibroblasts. *J. Cell Biol.*, 58:265–283.
177. Perecko, J. P., Berezesky, I. K., and Grimley, P. M. (1973): Surface features of some murine cell lines under various conditions of oncogenic virus infection. In: *Scanning Electron Microscopy, Part III*, pp. 521–528. Research Institute, Chicago.
178. Peterkofsky, B., and Prather, W. B. (1974): Increased collagen synthesis in Kirsten sarcoma virus-transformed BALB 3T3 cells grown in the presence of dibutyryl cyclic AMP. *Cell*, 3:291–299.

179. Pollack, R., Risser, R., Conlon, S., and Rifkin, D. (1974): Plasminogen activator production accompanies loss of anchorage regulation in transformation of primary rat embryo cells by simian virus 40. *Proc. Natl. Acad. Sci. USA*, 71:4792–4796.
180. Pollack, R., and Rifkin, D. (1975): Actin-containing cables within anchorage-dependent rat embryo cells are dissociated by plasmin and trypsin. *Cell*, 6:495–506.
181. Pollack, R., Risser, R., Conlon, S., Freedman, V., Shin, S. I., and Rifkin, D. B. (1975): Production of plasminogen activator and colonial growth in semisolid medium are *in vitro* correlates of tumorigenicity in the immune-deficient nude mouse. In: *Proteases and Biological Control*, edited by E. Reich, D. B. Rifkin, and E. Shaw, pp. 885–899. Cold Spring Harbor Laboratory, Cold Spring Harbor, New York.
182. Pollard, T. D., and Weihing, R. R. (1974): Actin and myosin and cell movement. *CRC Crit. Rev. Biochem.*, 2:1–65.
183. Porter, K. R. (1966): In: *Cytoplasmic Microtubules and their Functions, Ciba Foundation Symposium on Principles of Biomolecular Organizations*, pp. 308–345. Churchill, London.
184. Porter, K. R., and Fonte, V. G. (1973): Observations of the topography of normal and cancer cells. In: *Scanning Electron Microscopy, Part III*, pp. 683–688. Research Institute, Chicago.
185. Porter, K. R., Todaro, G. J., and Fonte, V. G. (1973): A scanning electron microscope study of surface features of viral and spontaneous transformants of mouse Balb/c 3T3 cells. *J. Cell Biol.*, 59:633–642.
186. Poste, G., and Flood, M. K. (1979): Cells transformed by temperature sensitive mutants of avian sarcoma virus cause tumors *in vivo* at the permissive and nonpermissive temperatures. *Cell*, 17:789–800.
187. Pouyssegur, J., Shiu, R. P. C., and Pastan, I. (1977): Induction of two transformation-sensitive membrane polypeptides in normal fibroblasts by a block in glycoprotein synthesis or glucose deprivation. *Cell*, 11:941–947.
188. Pouyssegur, J., Franchi, A., Salomon, J. C., and Silvestre, P. (1980): Isolation of a Chinese hamster fibroblast mutant defective in hexose transport and aerobic glycolysis. Its use to dissect the malignant phenotype. *Proc. Natl. Acad. Sci. USA*, 77:2698–2701.
189. Prasad, I., Zouzias, D., and Basilico, C. (1976): State of the viral DNA in rat cells transformed by polyoma virus. *J. Virol.*, 18:436–444.
190. Puck, T. T., Cieciura, S. J., and Robinson, A. (1958): Genetics of somatic mammalian cells. III. Long term cultivation of euploid cells from human and animal subjects. *J. Exp. Med.*, 108:945–955.
191. Quigley, J. P., Rifkin, D. B., and Reich, E. (1971): Phospholipid composition of Rous sarcoma virus, host cell membranes, and other enveloped viruses. *Virology*, 46:106–116.
192. Quigley, J. P. (1979): Proteolytic enzymes. In: *Surfaces of Normal and Malignant Cells*, edited by R. O. Hynes, pp. 247–285. Wiley, New York.
193. Quigley, J. P., Rifkin, D. B., and Reich, E. (1972): Lipid studies of Rous sarcoma virus and host cell membranes. *Virology*, 50:550–557.
194. Rassoulzadegan, M., Perbal, B., and Cuzin, F. (1978): Growth control in Simian virus 40-transformed rat cells: Temperature independent expression of the transformed phenotype in tsA transformants derived by agar selection. *J. Virol.*, 28:1–5.
195. Rassoulzadegan, M., Seif, R., and Cuzin, F. (1978): Conditions leading to the establishment of the N (*a* gene dependent) and A (*a* gene independent) transformed states after polyoma virus infection in rat fibroblasts. *J. Virol.*, 28:421–426.
196. Reddy, V. B., Thimmapaya, B., Dhar, R., Subramanian, K. N., Zain, B. S., Pan, J., Celma, C. L., and Weissman, M. (1978): The genome of simian virus 40. *Science*, 200:494–502.
197. Rees, D. A., Lloyd, C. W., and Thom, D. (1977): Control of grip and stick in cell adhesion through lateral relationships of membrane glycoproteins. *Nature*, 267:124–128.
198. Reich, E. (1973): Tumor associated fibrinolysis. *Fed. Proc.*, 32:2174–2175.
199. Rifkin, D. B. (1978): Plasminogen activator synthesis by cultured human embryonic long cells: Characterization of the suppressive effects of corticosteroids. *J. Cell. Physiol.*, 97:421–428.
200. Rifkin, D. B., Beal, L. P., and Reich, E. (1975): Macromolecular determinants of plasminogen activator synthesis. In: *Proteases and Biological Control*, edited by E. Reich, D. B. Rifkin, and E. Shaw, pp. 841–847. Cold Spring Harbor Laboratory, Cold Spring Harbor, New York.
201. Risser, R., and Pollack, R. (1974): A non-selective analysis of SV40 transformation of mouse 3T3 cells. *Virology*, 59:477–489.
202. Risser, R., Rifkin, D. B., and Pollack, R. (1974): The stable classes of transformed cells induced

by SV40 infection of established 3T3 cells and primary rat embryonic cells. *Cold Spring Harbor Symp. Quant. Biol.*, 39:317–324.

203. Risser, R., and Pollack, R. (1979): Factors affecting the frequency of transformation of rat embryo cells by simian virus 40. *Virology*, 92:82–90.
204. Robbins, P. W., Wickus, G. G., Branton, P. E., Gaffney, B. J., Hirschberg, C. B., Fuchs, P., and Blumberg, P. M. (1974): The chick fibroblast cell surface after transformation by Rous sarcoma virus. *Cold Spring Harbor Symp. Quant. Biol.*, 39:1173–1180.
205. Roblin, R. O., Young, P. L., and Bell, T. E. (1978): Concomitant secretion by transformed SVWI38-VA-13-2RA cells of plasminogen activator(s) and substance(s) which prevent their detection. *Biochem. Biophys. Res. Commun.*, 82:165–172.
206. Rohrlich, S. T., and Rifkin, D. B. (1977): Patterns of plasminogen activator production in cultured normal embryonic cells. *J. Cell Biol.*, 75:31–42.
207. Rosenblith, J. Z., Ukena, T. E., Yin, H. H., Berlin, R. D., and Karnousky, M. J. (1973): A comparative evaluation of the distribution of concanavalin A-binding sites on the surfaces of normal, virally-transformed and protease-treated fibroblasts. *Proc. Natl. Acad. Sci. USA*, 70:1625–1629.
208. Ross, R., and Vogel, A. (1978): The platelet-derived growth factor. *Cell*, 14:203–210.
209. Rothfels, K., and Parker, R. C. (1959): The karyotypes of cell lines recently established from normal mouse tissue. *J. Exp. Zool.*, 142:507–514.
210. Royer-Pokora, B., Beug, H., Claviez, M., Winkhardt, M. J., Friis, R. R., and Graf, T. (1978): Transformation parameters in chicken fibroblasts transformed by AEV and MC29 avian leukemia viruses. *Cell*, 13:751–760.
211. Rubin, R. W., and Everhart, L. P. (1973): The effect of cell to cell contact on the surface morphology of chinese hamster ovary cells. *J. Cell Biol.*, 57:837–844.
212. Ruddle, F. H. (1961): Chromosome variation in cell populations derived from pig kidney. *Cancer Res.*, 21:885–891.
213. Ruoslahti, E., Vaheri, A., Kuusela, P., and Linder, E. (1973): Fibroblast surface antigen: A new serum protein. *Biochim. Biophys. Acta*, 322:352–358.
214. Ruoslahti, E., and Vaheri, A. (1974): Novel human serum protein from fibroblast plasma membrane. *Nature*, 248:789–791.
215. Salter, D. W., and Weber, M. J. (1979): Glucose specific cytochalasin B binding is increased in chicken embryo fibroblasts transformed by Rous sarcoma virus. *J. Biol. Chem.*, 254:3554–3561.
216. Sanger, J. W. (1975): Presence of actin during chromosomal movement. *Proc. Natl. Acad. Sci. USA*, 72:2451–2455.
217. Sanger, J. W. (1975): Changing patterns of actin localization during cell division. *Proc. Natl. Acad. Sci. USA*, 72:1913–1916.
218. Scher, C. D., Shepard, R. C., Antoniades, H. N., and Stiles, C. D. (1979): Platelet-derived growth factor and the regulation of the mammalian fibroblast cell cycle. *Biochim. Biophys. Acta*, 560:217–241.
219. Schlegel, R., and Benjamin, T. L. (1978): Cellular alterations dependent upon the polyoma virus hrt function: Separation of the mitogenic from transforming capacities. *Cell*, 14:587–599.
220. Schriver, K., and Rohrschneider, L. (1981): Organization of $pp60^{src}$ and selected cytoskeletal proteins within adhesion plaques and junctions of Rous sarcoma virus-transformed rat cells. *J. Cell Biol.*, 89:525–535.
221. Seif, R., and Cuzin, F. (1977): Temperature-sensitive growth regulation in one type of transformed rat cells induced by the *tsa* mutant of polyoma virus. *J. Virol.*, 24:721–728.
222. Seif, R., and Martin, R. G. (1979): Growth state of the cell early after infection with simian virus 40 determines whether the maintenance of transformation will be A gene dependent or independent. *J. Virol.*, 31:350–359.
223. Seif, R., and Martin, R. G. (1979): Simian virus 40 small t antigen is not required for the maintenance of transformation but may act as a promoter (cocarcinogen) during establishment of transformation in resting rat cells. *J. Virol.*, 32:979–988.
224. Seif, R., Seif, I., and Wantyghem, J. (1983): Rat cells transformed by SV40 give rise to tumor cells which contain no viral proteins and often no viral DNA. *Mol. Cell. Biol.*, 3:1138–1145.
225. Seifert, S. C., and Gelehrter, T. D. (1978): Mechanism of dexamethasone inhibition of plasminogen activator in rat hepatoma cells. *Proc. Natl. Acad. Sci. USA*, 75:6130–6133.
226. Sela, B. A., Lis, H., Sharon, N., and Sachs, L. (1971): Quantitation of N-acetyl D-galactosamine

like sites on the surface membrane of normal and transformed mammalian cells. *Biochim. Biophys. Acta*, 249:564–572.

227. Shin, S., Freedman, V. H., Risser, R., and Pollack, R. (1975): Tumorigenicity of virus-transformed cells in nude mice is correlated specifically with anchorage independent growth *in vitro*. *Proc. Natl. Acad. Sci. USA*, 72:4435–4439.
228. Shin, S., and Freedman, V. H. (1977): Neoplastic growth of animal cells in nude mice. In: *Proceedings of the Second International Workshop on Nude Mice*, pp. 337–349. University of Tokyo Press, Tokyo.
229. Shiu, R. P. C., Pouyssegur, J., and Pastan, I. (1977): Glucose deprivation accounts for the induction of two transformation-sensitive membrane proteins in Rous sarcoma virus transformed chick embryo fibroblasts. *Proc. Natl. Acad. Sci. USA*, 74:3840–3844.
230. Smith, H. S., Scher, C. D., and Todaro, G. J. (1971): Induction of cell division in medium lacking serum growth factor by SV40. *Virology*, 44:359–370.
231. Soeda, E., Arrand, J. R., and Griffin, B. E. (1979): Polyoma virus: The early region and its T antigens. *Nucleic Acids Res.*, 7:839–857.
232. Stenman, S., and Vaheri, A. (1978): Distribution of a major connective tissue protein, fibronectin, in normal human tissues. *J. Exp. Med.*, 147:1054–1064.
233. Stephenson, J. R., and Todaro, G. J. (1982): Viral-encoded transforming proteins and transforming growth factors. In: *Advances In Viral Oncology, Vol. 1*, edited by G. Klein, pp. 59–82. Raven Press, New York.
234. Stoker, M. G. P. (1973): Role of diffusion boundary layer in contact inhibition of growth. *Nature*, 246:200–203.
235. Stoker, M., and McPherson, I. (1961): Studies on transformation of hamster cells by polyoma virus *in vitro*. *Virology*, 14:359–370.
236. Stoker, M., and Piggott, D. (1974): Shaking 3T3 cells: Further studies on diffusion boundary effects. *Cell*, 3:207–215.
237. Stone, K. R., Smith, R. E., and Joklik, W. K. (1974): Changes in membrane polypeptides that occur when chick embryo fibroblasts and NRK cells are transformed with avian sarcoma viruses. *Virology*, 58:86–100.
238. Strickland, S., and Beers, W. H. (1976): Studies on the role of plasminogen activator in ovulation: In vitro response of granulosa to gonadotropins, cyclic nucleotides and prostaglandins. *J. Biol. Chem.*, 251:5694–5702.
239. Strickland, S., Reich, E., and Sherman, M. I. (1976): Plasminogen activator in early embryogenesis: Enzyme production by trophoblast and parietal endoderm. *Cell*, 9:231–240.
240. Strickland, S., and Mahdavi, V. (1978): The induction of differentiation in teratocarcinoma stem cells by retinoic acid. *Cell*, 15:393–403.
241. Tegtmeyer, P. (1975): Function of simian virus 40 gene A in transforming infection. *J. Virol.*, 15:613–618.
242. Temin, H. M., and Rubin, H. (1958): Characteristics of an assay for Rous sarcoma virus cells in tissue culture. *Virology*, 6:669–688.
243. Todaro, G. J., and Green, H. (1963): Quantitative studies of the growth of mouse embryo cells in culture and their development into established lines. *J. Cell Biol.*, 17:299–313.
244. Todaro, G. J., and Green, H. (1964): An assay for cellular transformation by SV40. *Virology*, 23:117–119.
245. Todaro, G. J., and Green, H. (1964): Enhancement by thymidine analogs of susceptibility of cells to transformation by SV40. *Virology*, 24:393–403.
246. Todaro, G. J., De Larco, J. E., and Cohen, S. (1976): Transformation by murine and feline sarcoma viruses specifically blocks binding of epidermial growth factor (EGF) to cells. *Nature*, 264:26–31.
247. Tooze, J. (editor) (1981): *Molecular Biology of Tumor Viruses, DNA Tumor Viruses*. Cold Spring Harbor Laboratory, Cold Spring Harbor, New York.
248. Troll, W., Meyn, M. S., and Rossmann, T. G. (1978): Mechanisms of protease action in carcinogenesis. In: *Carcinogenesis, Vol. 2*, edited by T. J. Slaga, A. Sivak, and R. K. Boutwell, pp. 301–312. Raven Press, New York.
249. Troll, W., Klassen, A., and Janoff, A. (1970): Tumorigenesis in mouse skin: Inhibition by synthetic inhibitors of proteases. *Science*, 169:1211–1213.
250. Tsai, R. L., and Green, H. (1972): Study of intracellular collagen precursors using DNA-cellulose chromatography. *Nature*, 237:171–173.

251. Ukena, T. E., Borysenko, J. Z., Karnovsky, M. J., and Berlin, R. D. (1974): Effects of colchicin, cytochalasin B and 2 deoxyglucose on the topographical organization of surface-bound concanavalin A in normal and transformed fibroblasts. *J. Cell Biol.*, 61:70–82.
252. Umezawa, H. (1976): Structures and activities of protease inhibitors of microbial origin. *Methods Enzymol.*, 45:678–695.
253. Umezawa, H. (1978): Metabolites under preclinical development for cancer treatment. *Recent Results Cancer Res.*, 63:120–134.
254. Unkeless, J. C., Tobia, A., Ossowski, L., Quigley, J. P., Rifkin, D. B., and Reich, E. (1973): An enzymatic function associated with transformation of fibroblasts by oncogenic viruses. I. Chick embryo fibroblast cultures transformed by avian RNA tumor viruses. *J. Exp. Med.*, 137:85–111.
255. Unkeless, J. C., Danø, K., Kellerman, G. M., and Reich, E. (1974): Fibrinolysis associated with oncogenic transformation. Partial purification and characterization of the cell factor, a plasminogen activator. *J. Biol. Chem.*, 249:2295–2305.
256. Vaheri, A., and Ruoslahti, E. (1975): Fibroblast surface antigen produced but not retained by virus-transformed human cells. *J. Exp. Med.*, 142:530–538.
257. Vaheri, A., Ruoslahti, E., Westermark, B., and Ponten, J. (1976): A common cell type specific surface antigen in cultured human glial cells and fibroblasts: Loss in malignant cells. *J. Exp. Med.*, 143:64–72.
258. Vaheri, A., and Mosher, D. F. (1978): High molecular weight, cell surface associated glycoprotein (fibronectin) lost in malignant transformation. *Biochem. Biophys. Acta*, 516:1–25.
259. Vaheri, A., Kurkinen, M., Letho, V., Linder, E., and Timple, R. (1978): Codistribution of pericellular matrix proteins in cultured fibroblasts and loss in transformation: Fibronectin and procollagen. *Proc. Natl. Acad. Sci. USA*, 75:4944–4948.
260. Vassali, J. D., Hamilton, J., and Reich, E. (1976): Macrophage plasminogen activator: Modulation of enzyme production by anti-inflammatory steroids, mitotic inhibitors and cyclic nucleotides. *Cell*, 8:271–281.
261. Vogt, M., and Dulbecco, R. (1963): Steps in the noeplastic transformation of hamster embryo cells by polyoma virus. *Proc. Natl. Acad. Sci USA*, 49:171–179.
262. Vollet, J. J., Brugge, J. S., Noonan, C. M., and Butel, J. S. (1977): The role of SV-40 gene A in the alteration of microfilaments in transformed cells. *Exp. Cell Res.*, 105:119–126.
263. Wang, K., Ash, J. F., and Singer, S. J. (1975): Filamin, a new high molecular weight protein found in smooth muscle and non-muscle cells. *Proc. Natl. Acad. Sci. USA*, 72:4483–4486.
264. Wang, E., and Goldberg, A. R. (1976): Changes in microfilament organization and surface topography upon transformation of chick embryo fibroblasts with Rous sarcoma virus. *Proc. Natl. Acad. Sci. USA*, 73:4065–4069.
265. Wang, E., and Goldberg, A. R. (1979): Effects of the *src* gene product on microfilament and microtubule organization in avian and mammalian cells infected with the same temperature sensitive mutant of Rous sarcoma virus. *Virology*, 92:201–210.
266. Warburg, O. (1930): *The Metabolism of Tumors*. Constable, London.
267. Wartiovaara, J., Linder, E., Ruoslahti, E., and Vaheri, A. (1974): Distribution of fibroblast surface antigen. Association with fibrillar structures of normal cells and loss upon viral transformation. *J. Exp. Med.*, 140:1522–1533.
268. Weber, M. J. (1973): Hexose transport in normal and in Rous sarcoma virus-transformed cells. *J. Biol. Chem.*, 218:2978–2983.
269. Weber, M. J. (1975): Inhibition of protease activity in cultures of Rous sarcoma virus transformed cells: Effect on the transformed phenotype. *Cell*, 5:253–261.
270. Weber, K., and Groeschel-Stewart, U. (1974): Antibody to myosin: The specific visualization of myosin-containing filaments in nonmuscle cells. *Proc. Natl. Acad. Sci. USA*, 71:4561–4564.
271. Weber, M. J., Hale, A. H., and Roll, D. E. (1975): Role of protease activity in malignant transformation by Rous sarcoma virus. In: *Proteases and Biological Control*, edited by E. Reich, D. B. Rifkin, and E. Shaw, pp. 915–930. Cold Spring Harbor Laboratory, Cold Spring Harbor, New York.
272. Weber, M. J., and Friis, R. R. (1979): Dissociation of transformation parameters using temperature-conditional mutants of Rous sarcoma virus. *Cell*, 16:25–32.
273. Weber, M. J., Salter, D. W., and McNair, T. E. (1982): Increased glucose transport in malignant cells: Analysis of its molecular basis. In: *Molecular Interrelations of Nutrition and Cancer*, edited by M. S. Arnolt, J. Van Eys, and Y. M. Wang, Raven Press, New York.

274. Weil, R. (1978): Viral tumor antigens. A novel type of mammalian regulator protein. *Biochim. Biophys. Acta*, 56:301–309.
275. Weiss, R., Teich, N., Varmus, H., and Coffin, J. (editors) (1982): *Molecular Biology of Tumor Viruses, RNA Tumor Viruses*, Cold Spring Harbor Laboratory, Cold Spring Harbor, New York.
276. Werb, Z. (1978): Biochemical actions of glucocorticoids on macrophages in culture. *J. Exp. Med.*, 147:1695–1712.
277. Wessells, W. K., Speuner, B. S., Ash, J. F., Bradley, M. O., Ludvena, M. A., Taylor, E. L., Wrenn, J. T., and Yamada, K. M. (1971): Microfilaments in cellular and developmental processes. *Science*, 171:135–143.
278. Wiche, G., Lundblad, V. J., and Cole, R. D. (1977): Competence of soluble cell extracts as microtubule assembly systems. Comparison of simian virus 40 transformed and untransformed mouse 3T3 fibroblasts. *J. Biol. Chem.*, 252:794–796.
279. Wiche, G., Furtner, R., Steinhaus, N., and Cole, D. (1979): Expression of simian virus 40 gene A affects tubulin stability. *J. Virol.*, 32:47–51.
280. Wigler, M., and Weinstein, I. B. (1976): Tumor promotor induces plasminogen activator. *Nature*, 259:232–233.
281. Wigley, C. B., and Summerhayes, I. C. (1979): Loss of LETS protein is not a marker for salivary gland or bladder epithelial cell transformation. *Exp. Cell Res.*, 118:394–398.
282. Willingham, M. O., Yamada, K. M., Yamada, S. S., Pouyssegur, J., and Pastan, I. (1977): Microfilament bundles and cell shape are related to adhesiveness to substratum and are dissociable from growth control in cultured fibroblasts. *Cell*, 10:375–380.
283. Wilson, E. L., and Reich, E. (1978): Plasminogen activator in chick fibroblasts: Induction of synthesis by retinoic acid, synergism with viral transformation and phorbol ester. *Cell*, 15:385–392.
284. Wolff, B. A., and Goldberg, A. R. (1976): Rous sarcoma virus transformed fibroblasts having low levels of plasminogen activator. *Proc. Natl. Acad. Sci. USA*, 73:3613–3617.
285. Wolff, B. A., and Goldberg, A. R. (1978): The expression of plasminogen activator in Rous sarcoma virus-transformed cells is controlled both by the virus and the cell. *Virology*, 89:570–577.
286. Wolosewick, J. J., and Porter, K. R. (1976): Stereo high-voltage electron microscopy of whole cells of the human diploid line WI38. *Am. J. Anat.*, 147:303–324.
287. Yamada, K. M., and Weston, J. A. (1974): Isolation of a major cell surface glycoprotein from fibroblasts. *Proc. Natl. Acad. Sci. USA*, 71:3492–3496.
288. Yamada, K. M., and Pastan, I. (1976): The relationship between cell surface protein and glucose and α-aminoisobutyrate transport in transformed chick and mouse cells. *J. Cell. Physiol.*, 89:827–830.
289. Yamada, K. M., Ohanian, S. H., and Pastan, I. (1976): Cell surface protein decreases microvilli and ruffles on transformed mouse and chick cells. *Cell*, 9:241–245.
290. Yamada, K. M., Yamada, S. S., and Pastan, I. (1976): Cell surface protein partially restores morphology, adhesiveness and contact inhibition of movement to transformed fibroblasts. *Proc. Natl. Acad. Sci. USA*, 73:1217–1221.
291. Yamada, K. M., Schlesinger, D. H., Kennedy, D. W., and Pastan, I. (1977): Characterization of a major fibroblast cell surface glycoprotein. *Biochemistry*, 16:5552–5559.
292. Yamada, K. M., Yamada, S. S., and Pastan, I. (1977): Quantitation of a transformation sensitive, adhesive cell surface glycoprotein: Decrease on several-untransformed permanent cell lines. *J. Cell Biol.*, 74:649–654.
293. Yamada, K. M., and Olden, K. (1978): Fibronectins-adhesive glycoproteins of cell surface and blood. *Nature*, 275:179–184.
294. Yamanishi, K., and Rapp, F. (1979): Production of plasminogen activator by human and hamster cells infected with human cytomegalovirus. *J. Virol.*, 31:415–419.
295. Yang, N. S., Kirkland, W., Jorgensen, T., and Furmanski, P. (1980): Absence of fibronectin and presence of plasminogen activator in both normal and malignant human mammary epithelial cells in culture. *J. Cell Biol.*, 84:120–130.
296. Yau, T. M., and Weber, M. J. (1972): Changes in acyl group composition of phospholipids from chicken embryonic fibroblasts after transformation by Rous sarcoma virus. *Biochem. Biophys. Res. Commun.*, 49:114–120.

297. Yerganian, G., and Leonard, M. J. (1961): Maintenance of normal *in situ* chromosomal features in long term tissue cultures. *Science*, 133:1600–1603.
298. Zimmerman, M., Quigley, J. P., Ashe, B., Dorn, C., Goldfarb, R., and Troll, W. (1978): Direct fluorescent assay of urokinase and plasminogen activators of normal and malignant cells: Kinetics and inhibitor profiles. *Proc. Natl. Acad. Sci. USA*, 75:750–753.

Advances in Viral Oncology, Volume 4,
edited by George Klein. Raven Press,
New York © 1984.

Transformation and Reverse Transformation in Mammalian Cells

Theodore T. Puck

The Lita Annenberg Hazen Laboratory for the Study of Human Development, the Florence R. Sabin Laboratories for Genetic and Developmental Medicine, The Matthew Rosenhaus Laboratory of The Eleanor Roosevelt Institute for Cancer Research; Department of Biochemistry, Biophysics and Genetics; and Department of Medicine, University of Colorado Health Sciences Center, Denver, Colorado 80262

Cancer is one of the most challenging of the human afflictions. The following questions embody complex aspects of the disease that have resisted understanding at cellular and molecular levels: (a) Why do the characteristic changes in cell morphology, alterations in structure and behavior of the cell surface, induction of chromosomal defects and karyotype instability, and loss of the power to regulate cell reproduction so frequently occur concurrently when normal cells become malignant? (b) What is the relationship, if any, between these features and the fact that an initiating insult can be followed by time intervals of 20 years or more before a cancerous disease develops? (c) Why do cancers rapidly develop resistance to therapeutic agents that initially produce highly successful responses in a given patient? (d) What is the nature of the process by which cells progress from the precancerous to the fully malignant state? (e) What is the relationship between carcinogenesis caused by cancer viruses and that which occurs spontaneously or through the action of other agents?

It is the purpose of this chapter to present experimental facts, theories, and speculations in an attempt to establish a unified formulation that can afford reasonable explanations for a substantial portion of these manifestations. Space does not permit an exhaustive consideration of the vast literature on the subject. Emphasis is placed on results of our own studies, primarily with the Chinese hamster ovary (CHO) cell, but occasionally with other cells, and to selected contributions from other laboratories. Important phenomena, such as the role of gene amplification (50), protease activation (33), and immune phenomena, which play fundamental roles in various malignancies, are not considered here. The new insights into cancer, which promise fundamental understanding as well as more powerful treatment and prevention, are attributable to the developments made in concepts and techniques of genetic and biochemical analyses of somatic cells in tissue culture and of recombinant DNA studies (41). A most important feature of this chapter is that the new techniques of cellular and molecular biology permit definitive experimental testing of the mechanisms proposed and their consequences. Some of the relationships discussed here have been previously reviewed (36,37,41).

REVERSE TRANSFORMATION REACTION

The transformation of a normal cell to the malignant state involves coordinate changes in several different characteristics. In this chapter, we are concerned with three of these phenomena, which often accompany each other in *in vitro* studies (36,37): (a) Transformed cells lose the capacity to stop growing when a confluent monolayer has been achieved on an appropriate surface or when single cells are placed in standard nutrient medium in suspension. (b) Cell morphology changes in characteristic fashion. Thus fibroblasts, for example, lose their normal, smooth-surfaced, spindle shape and become highly compact and pleomorphic. Occasionally, they exhibit oscillating knobs projecting periodically from the membrane, together with changes in cell surface antigen activity, specific active transport activities, and deposition of specific matrix molecules, such as fibronectin. (c) Cells lose the characteristic integrity of their karyotype, exhibiting variations in chromosome number and structure.

In 1970, we described reversal of these transformation characteristics when the spontaneously transformed CHO cell was treated with cyclic AMP (cAMP) derivatives (17,19,21,42). Other laboratories have described similar behavior in other cells (15,18,23). Cyclic AMP itself is fully active if one simultaneously adds an agent like theophylline to suppress the enzymatic action of phosphodiesterase (19). Finally, agents such as parachlorothio-cAMP and bromo-cAMP are by themselves fully active and, indeed, even more active than dibutyryl cAMP (T. T. Puck and R. K. Robins, *unpublished data*). The phenomena of reverse transformation in the CHO cell are reversible. Removal of the cAMP derivatives from the medium causes the cells to revert to their initial morphologic and behavioral modes.

While cAMP derivatives eliminated growth in suspension of transformed CHO cells in standard nutrient, growth on plastic surfaces was affected only slightly, if at all (36,37). Moreover, in the latter case, the pattern of colonial growth resulting from single cells changed from the random arrangement and three-dimensional pattern typical of the transformed cell to a strictly monolayered growth exhibiting the typical parallel fibroblastic arrangement of the cells (Fig. 1). Cellular morphology changed from a highly compact pleomorphic shape with knob-like structures extruding and retracting from the membrane to a smooth, spindle-shaped, fibroblast-like cell (Fig. 2). Finally, specific cell membrane features were altered by the presence of the cAMP derivatives, including change in the active transport of α-aminoisobutyric acid (36,37), changes in activity of specific cell surface antigens (36,37), restoration of the deposition of fibronectin about the cell surface in a pattern characteristic of the normal fibroblasts (30), and disappearance of the knobbed extrusions from the membrane, so that a smooth, tranquilized cell surface resulted (Fig. 3). Other compounds like retinoic acid with no obvious structural relationships to cAMP also bring about at least some of the reverse transformation phenomena.

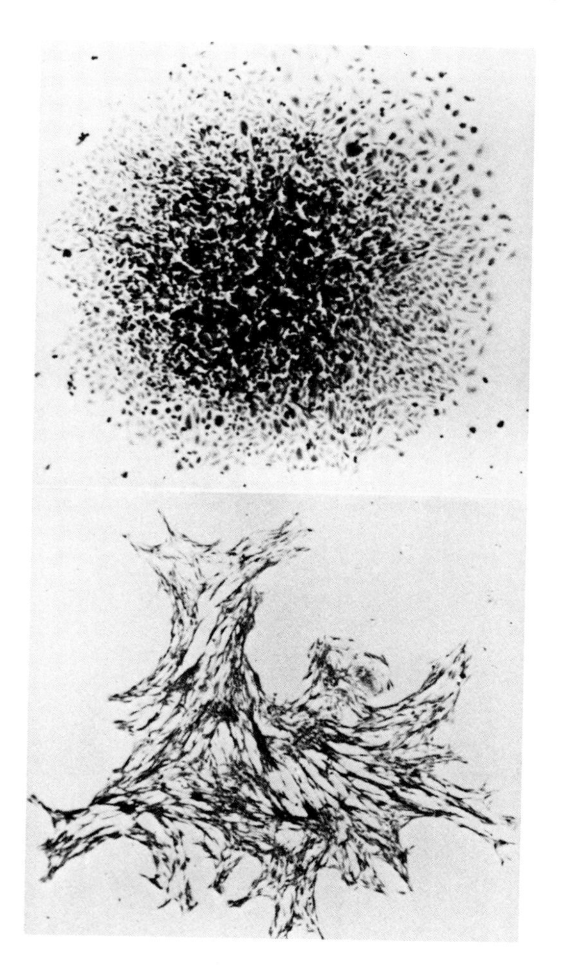

FIG. 1. Top: A colony of CHO cells in normal medium, revealing compact cells, three-dimensional growth in the center of the colony, and the random growth pattern at the edges. **Bottom:** An identical colony to that above but grown in the presence of reverse transformation conditions, so that the cells are elongated, the growth is monolayered, and the cells are fibroblastic in orientation.

ROLE OF THE CYTOSKELETON IN INTERPHASE

The cellular cytoskeleton consists of three types of fibers—microtubules, microfilaments, and intermediate filaments—together with a variety of associated proteins. The work of Porter et al. (34) in particular has indicated the structural interrelatedness of its components at the electron microscopic level. The intimate molecular composition and structure of these components are currently under intense study in various centers.

That the changes in cell morphology are due to changes in the cytoskeleton was demonstrated by the following experimental results. Colcemid in concentrations in which it specifically prevents cytoskeletal organization prevents all the morphologic changes of cAMP derivatives. Moreover, colcemid causes normal fibroblasts to assume the typical morphology of the transformed CHO cell, even reproducing the

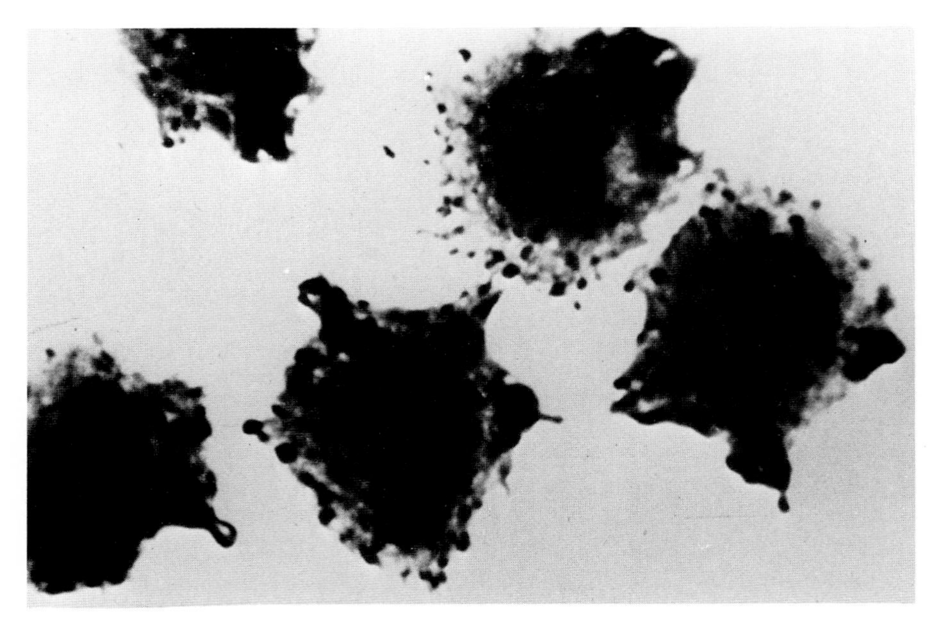

FIG. 2. Top: Cell morphology of the CHO cell showing its highly compact form and knob-like protrusions of the cell membrane. **Bottom:** CHO cells after exposure to cAMP. The cells have changed to a smooth, spindle-shaped form devoid of knob-like structures. (Light microscope.)

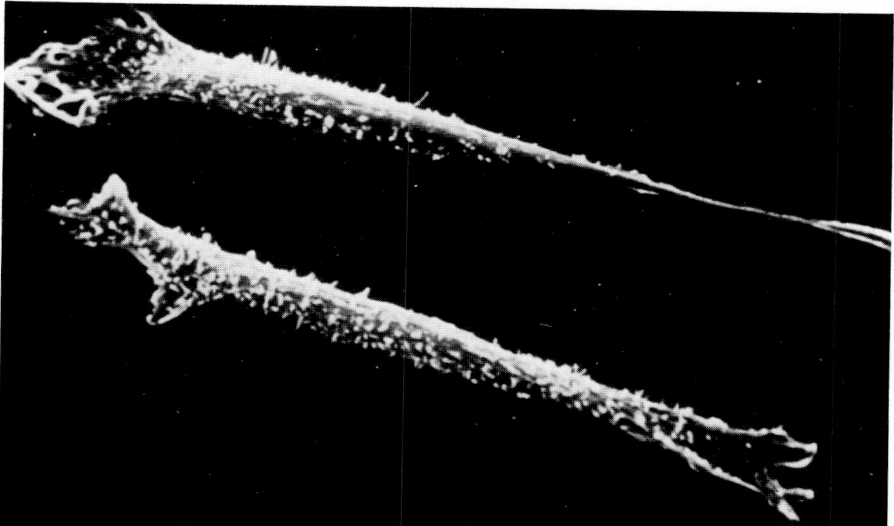

FIG. 3. Scanning electron micrograph of **(top)** CHO cell, revealing many knobs extruding from its surface, and **(bottom)** reverse transformed CHO cells, showing their extended shape and smooth, tranquil surface features.

formation of the pulsating knob-like membrane extrusions (42). This pulsating action of the knobs (or blebs) has been interpreted as contributing to the tendency of the transformed cell to break away from its point of origin and set up new foci of growth at other points in the body (17).

Transmission electron microscopy showed that the transformed CHO cell possessed a sparse and relatively randomly arranged microtubular structure in interphase, whereas after reverse transformation by cAMP derivatives, a dense parallel microtubular matrix similar to that in normal fibroblasts arises. Moreover, the agglutinated bundles of microfilaments at the base of the knob-like structures in the membrane of the transformed cell disappear under the action of reverse transformation agents (29,34).

These experiments were interpreted to mean that the changes in cell morphology and membrane of reverse transformation are attributable to the action of the newly organized cytoskeleton, and that cAMP is necessary for this cytoskeletal organization (29). Experiments also showed that new protein synthesis is not necessary for cytoskeletal organization (32). The only chemical reaction demonstrated to result directly from the presence of cAMP in mammalian cells is protein phosphorylation. Presumably, then, the phosphorylation activity induced by cAMP in the CHO cell acts on already available peptides to cause organization of these cytoskeletal components into an integrated matrix which carries out the necessary metabolic regulatory activities of the normal fibroblast.

The role of the cytoskeleton in maintaining structures and functions of the cell membrane has never been clarified in intimate molecular terms, but ample evidence exists to demonstrate such a connection: (a) We have shown that cell surface antibodies attaching to the CHO cell cause it to assume a spherical form (37,40), a phenomenon we interpreted as indicating that attachment of antibodies to membrane sites associated with cytoskeletal attachment causes displacement of the latter, possibly by conformational changes in specific membrane macromolecules. (b) Lectins behave in a similar fashion, indicating the participation of carbohydrate residues in such reactions (53–55). (c) Other changes in cell surface phenomena, including capping of antigens (24), active transport of amino acids (36,37), and fibronectin deposition (30), were prevented in some cases by colcemid, in other cases by cytochalasin B (a compound that disorganizes cytoskeletal microfilaments), and in some cases by either or both compounds.

Cells that become spherical under the influence of lectins and other agents are released from their surface anchorage. We interpret this as evidence for an interaction between the cytoskeleton and cell attachment sites necessary for normal cell anchorage.

These experiments demonstrated involvement of the cytoskeleton in morphologic and cell surface-related activities of the reverse transformation reaction in the CHO cell but did not prove its relevance to metabolic regulation generally or reproductive control specifically. Indeed, in contrast to our proposal, other workers have concluded that the cytoskeleton exercises a role in cell morphologic maintenance and locomotion but not in growth regulation (64).

As a test of the role of the cytoskeleton in the changed degree of growth regulation of normal and transformed cells, we carried out experiments on normal and transformed cells which had been blocked at the G_0 point of the reproduction cycle by permitting densely confluent cultures to exhaust their growth media. Fresh medium was then added in each case, in both the presence and absence of colcemid in concentrations sufficiently low to ensure action specific to the microtubules. The induction of ornithine decarboxylase (ODC) was measured as an indicator of the resumption of the normal reproductive cycle by the cells blocked in G_o. Under these circumstances, colcemid completely suppressed the induction of ODC in the normal cell. In the transformed cell, however, colcemid not only failed to inhibit ODC induction but caused an even greater than normal induction of this enzyme (48). Other studies have demonstrated that the effects of growth-stimulating factors and hormones on mammalian cells are intensified when the microtubules are disrupted (13,43) and diminished when microtubules are stabilized (9).

Other metabolic effects of the cytoskeleton in interphase cells have also been demonstrated. We have shown that when mouse hepatoma cells are treated with colcemid in concentrations sufficiently low to limit its action to microtubule disorganization, large amounts of tyrosine amino transferase (TAT) activity are induced (Table 1). Similarly, it has been shown that colchicine totally prevents transcription of the β-casein gene but not of ribosomal RNA genes (56). The induction of HMG-CoA reductase enzyme has been demonstrated to display opposite responses to colchicine and cAMP addition, respectively (1,59). An effect of colchicine in preventing gene exposure by cAMP is discussed below. Clearly, the cytoskeleton exerts a variety of regulatory metabolic actions in the biochemical economy of mammalian cells.

ROLE OF THE CYTOSKELETON IN MITOSIS

The structure and function of the mammalian cytoskeleton are demonstrably different during mitosis from that which prevails during interphase. Therefore, one

TABLE 1. *Induction of TAT activity in rat hepatoma cells*[a]

Reagent added	Relative TAT activity
A. None	100.0
B. 10^{-3} M dibutyryl cAMP + 3×10^{-5} M testololactone	389.0 ± 64
C. 5×10^{-6} M colcemid	205.0 ± 20
D. Combined agents of B and C	513.0 ± 44

[a]Demonstration that TAT activity can be induced in rat hepatoma cells either by dibutyryl cAMP or by colcemid in contrast to the mutually antagonistic action of these agents in the reverse transformation reaction. Moreover, the action of the two agents together in TAT activity induction is additive. These actions involve a different set of dynamics from those in the reverse transformation situation of CHO. Testololactone acts as a synergizing agent in the action of cAMP derivatives (17). Cells were incubated overnight at 37°C in basal medium consisting of F12 plus 8% fetal calf serum. The reagents indicated were added for 4 additional hours, after which the cells were harvested and enzyme assays performed.

may expect different consequences from damage to cytoskeletal structures to occur during these different phases of the life cycle. In mitosis, the cytoskeleton takes the form of the spindle, whose function it is to distribute an exact complement of chromosomes to each of the daughter cells. We designed experiments to test whether small damage to the mitotic cytoskeleton might produce a karyotypic change, in contrast to the reversible biochemical changes during interphase described in the preceding section. Consequently, CHO cells were treated with amounts of colcemid far less than that required to cause metaphase arrest. After various times, the drug was removed and the cells allowed to resume normal progression around the reproductive cycle. A second and much higher dose of colcemid then was administered, sufficient to block the arrested mitotic cells in mitosis. The arrested mitotic cells were collected and chromosomal counts determined.

These experiments demonstrated that the presence of the first, small amount of colcemid induced in the CHO cell a degree of karyotype lability that virtually erased the clustering of chromosomes about the modal chromosome number of 22. Cells were produced in almost equal frequencies, with chromosome numbers varying from 20 to 44 (8). Small amounts of colcemid are able to damage but not completely disorganize microtubules, with the result that nondisjunction in mitosis becomes a relatively frequent instead of a rare event. We postulate that genetic damage to the cytoskeletal genes in a transformed cell can produce karyotype disorganization in mitosis similar to what we observed resulting from small amounts of colcemid. Thus such cytoskeletal damage will constantly introduce new chromosomal combinations in the daughter cells, leading to an evolutionary progression at the cellular level in which cells better able to escape the regulatory actions on reproduction will be continuously selected for. This process would explain the karyotype lability exhibited by so many transformed cells, a mechanism that has also been discussed by Heston and White (16) on the basis of epidemiologic considerations. We have also attributed to this effect the propensity of cancer cells to develop resistance to therapeutic agents and thus thwart efforts to treat malignancy in patients.

ROLE OF ONCOGENES AND PHOSPHORYLATION IN REVERSE TRANSFORMATION

Erikson and his associates (12) first identified the gene product of the *src* oncogene as a phosphorylating enzyme. Subsequent studies (22) demonstrated that many if not most cellular phosphorylations carried out by this and other oncogene products act on tyrosine moieties as opposed to cAMP-induced phosphorylations, which tend to be confined to serine and threonine residues (25). These developments gave rise to the experimental question: Can cAMP reverse cellular transformation induced by the *src* oncogene? A definitive experiment was possible by adding cAMP derivatives to cells transformed to malignancy by infection with Rous sarcoma virus (RSV). Such cells, when treated with cAMP derivatives, change their morphology from round, poorly anchoring cells, which are capable of excellent single cell growth in agar suspension, to typical elongated fibroblasts with no capacity for growth in suspension in standard nutrients but excellent growth on surfaces (Table 2). This finding led to the possibility that the decision between normal and malignant behavior might lie in the ratio of tyrosine to serine and threonine phosphorylations in one or more particular proteins[1] (39). Several laboratories have demonstrated an association between *src* gene transformation and disruption of the cell cytoskeleton (5,10,28,52), as would be expected from the considerations presented here.

Identification of the critical phosphorylated proteins involved in transformation is now required, since both cAMP and the *src* gene have been shown to phosphorylate a variety of different proteins, of which only a small fraction affect transformation. In 1972, Hsie and Puck (20) demonstrated that treatment of CHO cells with mutagenic agents could yield cells resistant to the reverse transformation action of cAMP. Studies on the phosphorylation of such resistant mutants under the action of cAMP were undertaken by Bloom and Lockwood (2), LeCam and his colleagues (26), and by ourselves (14). The first two of these groups (2,26) found individual mutant cells resistant to the reverse transformation by cAMP, which failed to yield any of the set of characteristic protein phosphorylations exhibited by the parental transformed cell.

Our own studies (14) produced a different series of resistant mutants. Examination of 2D gels for phosphorylated proteins was carried out in the hope of narrowing the number of such molecules that could serve as candidates for the critical step or steps involved in the reversal of malignant transformation. The parental CHO cell exhibited at least seven clear changes in protein phosphorylations as a result of treatment with cAMP derivatives. Some mutants were isolated which,

TABLE 2. *Typical comparison of growth on treated plastic surfaces and in agarose suspension of normal vole fibroblasts and RSV-transformed vole fibroblasts in the presence and absence of 1 mM DBcAMP*[a]

Cell	Presence of 1 mM DBcAMP	Number of colonies developing	
		In agarose suspension	On treated plastic surfaces
Normal vole fibroblast	Absent	0	182 ± 3
	Present	0	101 ± 3
RSV-transformed vole fibroblast	Absent	206 ± 3	288 ± 8
	Present	0	170 ± 5

[a]Two thousand cells were plated in each case, and the number of colonies developing after 10 days of incubation was scored.

[1]Other workers (47) have reported a Rous sarcoma transformed cell not exhibiting the reverse transformation reaction when treated with cAMP derivatives, a fact we interpret as indicating existence of further complicating features in the cell used, or as due to differences in time and concentration of cAMP administration.

like those of other investigators (2,26), had lost all seven of these phosphorylations in the course of their becoming resistant to cAMP action. However, one mutant has been found which differs in only one of the cAMP-induced phosphorylations exhibited by the wild type CHO cell. While uncertainties in the results of 2D gel analysis hinder the decision as to whether there is indeed only one difference in phosphorylation from the wild type, it is clear that this mutant appears to be defective in a phosphorylation step that is much closer to the end point of the cAMP effect on reverse transformation than the other mutants. Isolation and identification of such proteins should be a definitive step in molecular characterization of the reverse transformation reaction.

These studies led to another aspect of the phosphorylation problem. Small amounts of cAMP can inhibit growth of transformed CHO cells in suspension, and larger amounts can also inhibit growth on plates. Presumably, a phosphorylated protein can block cell multiplication in either case but more effectively in normal cells, which are not anchored to an appropriate surface. The process of surface anchorage in the normal CHO cell results in a cytoskeletal configuration permitting growth to proceed in the presence of cAMP concentrations that are inhibitory in the suspended cell. Studies were undertaken to determine how CHO cell growth might be affected by incorporation of particular human chromosomes into the CHO genome by cell hybridization. Such studies might identify specific chromosomal loci associated with the cAMP growth regulatory response. Such hybrids were compared with CHO for their growth inhibition by cAMP derivatives (T. T. Puck, *in preparation*).

Typical results are shown in Table 3. It is evident that particular human chromosomes greatly enhance while other chromosome combinations suppress the growth inhibitory action of cAMP. Indeed, by the incorporation of specific deletion mutants of human chromosome 11, it was possible to roughly map specific genetic loci for such actions. The genetic alteration in these cells produced by inclusion of particular human chromosomal regions may involve a variety of different phosphorylation and dephosphorylation actions, possibly including that of particular kinases and

TABLE 3. *Demonstration that inhibition of growth of CHO cells in suspension by cAMP derivatives is specifically augmented or suppressed, respectively, in hybrid cells containing particular human chromosomes*

Cell employed	Human chromosomes present	Relative plating efficiency in 10^{-3} M dibutyryl cAMP + 10 μg/ml testololactone, as compared to standard growth medium alone (%)
CHO	0	11.0
CHO hybrid #153-E9A	21	3.0
CHO hybrid #167-E4E	12	0
CHO hybrid #PF1	2, 3, 4, 5, 6, 7, 12, 13, 14, 16, 18, 21, X	90.0

phosphatases. The phosphorylation patterns of such hybrid cells are under study, together with the more precise genetic mapping of the loci responsible for these various actions. This inhibition of growth by cAMP derivatives does not involve irreversible killing, since growth of surface-attached cells is restored when the cAMP is removed from the growth medium.

These experiments indicate that specific chromosomes and chromosomal regions exert specific cell growth regulatory effects under the influence of cAMP derivatives. These phenomena are compatible with the possibility that specific oncogene actions may be involved, particularly since an oncogene has been assigned to chromosome 11 (11). They also provide a possible explanation for the demonstrated association of specific kinds of malignancy with specific chromosomal defects, such as the deletion in chromosome 11 associated with Wilms' tumor (44) and that of chromosome 13 which occurs in retinoblastoma (65). The experiments indicate the wide variety of regulatory actions that may involve different protein phosphorylations.

ROLE OF GENE SEQUESTRATION AND EXPOSURE

In attempting to account for the various types of regulation of gene activity exhibited by mammalian cells, we proposed that separate mechanisms may exist for regulating large regions of the genome as opposed to processes that operate on single or small groups of specific genes. Thus in a tissue such as the liver, for example, large numbers of genes specific for brain, lung, and other tissues would need to be sequestered in fairly permanent fashion. On the contrary, the gene specific for alcohol dehydrogenase must be capable of rapid and specific induction in liver cells when alcohol is encountered in the blood. Hence we postulated the existence of at least two different kinds of states in which mammalian cell genes may exist (37,38). In the one case, genes would be sequestered by a mechanism that operates on relatively large portions of the genome. (The inactivation of the second X chromosome may be an example of this kind.) In the second case, single or small groups of genes would be regulated in response to specific chemical signals. These latter genes would exist in a condition allowing ready and specific expression, a state we provisionally call "exposed." It was worthwhile, therefore, to determine whether a significant difference in gene exposure could be demonstrated in normal and transformed cells, and whether cAMP derivatives, under the conditions that produce reverse transformation, could affect this situation.

A parameter that might be related to gene exposure was developed as follows, based on previous work of Weintraub and Groudine (63). Nuclei were removed from the test cells, suspended in buffer, and treated with DNAse I, a nuclease that randomly hydrolyzes DNA sequences with which it makes effective contact. After incubation under carefully controlled conditions, the enzyme action is stopped. The DNA remaining in the nuclei is quantitatively extracted and subjected to gel electrophoresis, so that the DNA fragments remaining after enzyme treatment are sorted out according to size. These gels can be examined by photometric scanning in a densitometer, so that the distribution of the DNA in the various size ranges can be approximated. Finally, the amount of DNA solubilized as a result of the enzyme treatment of the nuclei is determined by means of perchloric acid precipitation and ultraviolet absorption of the soluble nucleotides that have been produced.

The results obtained by this procedure on nuclei from the transformed CHO cell were compared with those from primary cultures of normal Chinese hamster fibroblasts obtained from ovarian tissue. The following results were obtained: (a) Under standard conditions, normal fibroblasts exhibit considerably greater DNA sensitivity to enzyme hydrolysis than do nuclei from the transformed cell. (b) When the transformed cells are treated with cAMP under conditions that bring about the reverse transformation reaction, their nuclei are restored to a degree of enzyme sensitivity indistinguishable from that of the normal fibroblasts. (c) When the normal cells are treated with cAMP derivatives under identical conditions, no appreciable change in the sensitivity of their nuclei to hydrolysis by DNAse I occurs (51). These facts establish that some of the DNA in nuclei of the transformed CHO cell exist in a different state of exposure from that of the normal cell, and that the situation prevailing in the normal cell can be restored by cAMP addition to the transformed cell, at least as far as the relatively crude measurements carried out so far can detect.

To determine whether the effect of cAMP on the DNAse sensitivity of the transformed cell nuclei is fortuitous or bears a relationship to the phenomenon of reverse transformation, experiments were carried out in which transformed cells treated with cAMP were simultaneously treated with colcemid in a concentration that prevents microtubular organization and the morphologic and other changes of reverse transformation. Under these conditions, the cAMP-induced increase in DNA sensitivity of the nuclei failed to occur (T. T. Puck and F. Ashall, *unpublished data*). It is concluded that the increase of DNA sensitivity to nuclease hydrolysis in the nuclei of transformed cells is dependent on the establishment of a fibroblast-like cytoskeletal organization, as is the reverse transformation reaction itself. It is reasonable to propose, at least as a working hypothesis, that DNA exposure in the CHO cell can exist in alternative states, depending on the integrity of the cytoskeletal system whose organization requires the action of cAMP.

Preliminary experiments demonstrate that if the isolated nuclei from the transformed cell are treated with cAMP, instead of treatment of the whole cells as in the preceding experiments, no effect on DNA sensitivity of the nuclei is achieved. This suggests that the cytoplasmic cytoskeletal structure is involved in these reactions which determine DNA exposure (T. T. Puck and F. Ashall, *unpublished data*).

The data obtained so far do not identify the specific genes whose exposure or sequestration is affected by cAMP in normal and transformed cells, respectively. Experiments are now in progress to determine specifically what regions of the DNA in each cell type are affected by treatment with cAMP derivatives and their agents.

DISCUSSION

The experimental facts described here lend themselves to simple interpretation on the basis of the involvement of the cellular cytoskeleton in all the phenomena of malignancy described above. We assume that a reaction chain exists in normal surface-anchored cells. This can lead to cessation of reproduction when a sufficiently high cell density has been achieved. Thus cell-cell interactions through specific membrane molecules induce a cAMP-triggered reorganization of the membrane-attached cytoskeleton, which in turn induces a series of biochemical reactions that cause growth cessation in a medium otherwise sufficient for cell multiplication. Such normal cells in suspension are growth suppressed, because lack of surface anchorage promotes a cytoskeletal arrangement that continuously keeps cells in the nonreproductive mode. Failure of normal growth suppression can occur by a block in any of the steps of this complex chain of events. Such a metabolic block may activate expression of an oncogene which normally should be inert. However, it is at least conceivable that other kinds of blocks can also lead to unregulated cell division. Thus experiments are necessary to determine how universal is participation of cell membrane elements, cytoskeletal structures, oncogene actions, and protein on phosphorylating events in different types of cancer. Such studies are underway.

Our central postulate (37,38) has been that the cytoskeleton plays several fundamental roles throughout the cell life cycle, that its structure and function in different cells and at different parts of the life cycle are designed to facilitate performance of different specialized metabolic functions, and that a given damage to the cytoskeleton can result in different kinds of metabolic and karyotypic derangement of normal cells.

The concept of reverse transformation implies that a chain of metabolic events leading to regulation of reproduction exists in every cell; that cancer involves a block in one or more steps in this chain; that an agent which restores the integrity of this chain would cause a transformed cell to lose some or all of its characteristics associated with malignancy; that different transformed cells may differ in the identity of the blocked metabolic step(s), and therefore may require different metabolites to effect reverse transformation; and that if karyotype disorganization has proceeded sufficiently far complexities in this relatively straightforward picture may be introduced. The fact that some cancerous cells display phenomena resembling reverse transformation when treated with retinoic acid, but not cyclic AMP derivatives illustrates an expectation of this formulation (Moore, E. E., *in preparation*). The fact that many transformed cells display a syndrome of associated characteristics has been explained by the demonstration that the cell cytoskeleton is required both for normal regulation of a variety of metabolic actions including reproduction and for normal distribution of the chromosomes between the mitotic daughter cells so that damage to the cytoskeleton can cause a group of associated symptoms. Finally, these considerations contain the possibility of a new principle for cancer treatment based on restoration of the blocked pathways rather than the conventional attempts to produce selective cytotoxic effects on the transformed cells.

We have postulated that the molecular action of the cytoskeleton during mitosis can offer a model for some of its molecular actions during interphase (29,37). In mitosis, microtubular elements of the cytoskeleton attach through the mediation of particular proteins to specific sites on the chromosome, i.e., at the chromosomal kinetochores in which particular proteins are bound to the centromeric DNA, which is the seat of highly reiterated sequences (37,58). During interphase as well, cytoskeletal elements may attach to particular chromosomal regions, which in this case may be different from their mitotic attachment sites. This cytoskeletal attachment of microtubular elements in interphase may be associated with chromosomal condensation in specific regions and with gene sequestration, which prevents specific regions of the genome from being acted on by effectors of gene expression in the fluid phase. In addition, some of the reiterated DNA regions may function as the basis of attachment of cytoskeletal elements and related intermediate proteins as part of the gene sequestration and exposure phenomenon (38).

Reports supporting this view have been published, in which electron microscopy has revealed the existence of well-formed cytoskeletal elements in mammalian cell nuclei (37). However, we cannot rule out the participation of cytoskeletal interaction with the nuclear membrane as an additional or alternative mechanism for achievement of cytoskeletal control of the genome. In either case, interaction between cytoskeletal and nuclear structure is supported both by our own demonstration that reverse transformation is accompanied by changes in DNA sensitivity to an endonuclease (51), an event that is antagonized by colchicine, and by the demonstration (31) that chromatin compaction and nuclear morphology are also changed during reverse transformation.

It has been demonstrated that cytoskeletal proteins can be phosphorylated by virtue of cAMP action or through the agency of an oncogene product (46). The question arises as to why phosphorylation at a tyrosine moiety should be so damaging to the process of regulation of reproduction. One may consider that phosphorylation could convert a tyrosine residue, which normally seeks out the hydrophobic interior environment of a protein, into a hydrophilic region, with affinity for the exterior aqueous region. The resulting conformational change, perhaps taking place in a cytoskeletal element, might damage its ability to carry out its normal function in gene sequestration and exposure, and therefore in growth regulation.

The role of the cytoskeleton in transformation may account for certain puzzling aspects of cancer mentioned above. The concurrence of the changes in cell morphology, cell membrane, induction of chromosomal defects, and loss of regulation of cell reproduction that accompany transformation may be due to the fact that all these properties can result from defects in the cell cytoskeleton.

The need for the action of a promotor, such as phorbol ester, to bring into expression a previously induced cancerous mutation would appear to be at least provisionally explainable on the basis of the sequestration reaction which we have shown also to involve the cytoskeleton. A mutation in a gene regulating growth will not be brought into expression as long as the gene forms part of a chromosomal region that is sequestered from contact with the medium. Changes induced by a promotor that can expose such genes will allow expression of such a mutation. We postulate, therefore, that the promoting action of such agents as the phorbol esters involves exposure of transforming mutations in previously sequestered genes by an action involving the cytoskeleton in a fashion similar although not necessarily identical to that discussed here.

Preliminary experiments have shown that retinoic acid, an agent that antagonizes the carcinogenesis-promoting action of phorbol ester and the *src* gene product, behaves like cAMP derivatives in inhibiting the growth of CHO cells in suspension under conditions in which surface growth is unaffected. This behavior suggests that retinoic acid has an action similar to that of cAMP in altering gene sequestration so as to return growth control to that of a normal fibroblast (T. T. Puck, *unpublished data*).

Although this general picture of cytoskeletal involvement is consistent with various effects demonstrated to be produced as a result of treatment of mammalian cells with phorbol esters (62a–e) on the one hand and retinoic acid (27) on the other, the multiplicity of actions that have been described makes necessary highly detailed studies before reliable conclusions can be reached (27,62a–e). Some of the biochemical and morphologic actions of phorbol esters in various cells act like those of colcemid and cytochalasin B (45) in opposing effects induced by cAMP in the CHO cell; however, others may be different. Obviously, detailed molecular study of the actions of those agents in the same cell is required. Such studies are now in progress with the CHO cell. It is possible that the gene amplification studied by Schimke et al. (50), which is accompanied by changes such as formation of double minute chromosomal entities, is also a reflection of cytoskeletal damage of the kind that produces the chromosomal instability in transformed cells. Other experimental facts and theoretical considerations about the interactions of promoting agents, cell membrane elements, and oncogenes have been discussed by other workers, of whom a few are cited (62a–e).

The progression of cells from the precancerous to the cancerous state may be explained as a consequence of the genetic drift imposed by defects in the mitotic cytoskeleton, since selection will always favor more resistant genotypes. Finally, cancer could be produced either by activation of a specific oncogene by the dual action of a mutation in its control genetic region followed by exposure of the previously sequestered region or, in the case of cancer viruses, by incorporation of new oncogenes in regions of the genome where they would not be subject to the regulatory effects of controlling DNA sequences. Intensive study is in progress in several laboratories of the structural and functional interrelationships between oncogenes, promotors of gene expression, and transposable elements. Currently available techniques in somatic cell genetics, gene mapping, and measurement of gene exposure, sequestration, and expression offer hope for elucidation of the molecular nature of malignancy (38).

The effects of cAMP derivatives which we have observed in the CHO cell disappear when such reagents are removed from the medium; cAMP continues to be needed for normal continuing organization of the cytoskeleton. However, some interesting experiments carried out by Prasad and Hsie (35) with retinoblastoma cells and by Cho-Chung and co-workers (6) with mammary tumor cells indicate that changes initiated by cAMP derivatives remain after removal of the drug, so that an irreversible elimination of malignant properties results. Such behavior suggests that the critical cell may have been blocked by a mutation that prevents its traverse of a differentiation pathway which normally would have produced a fully mature, nonmultiplying form. In that case, the reverse transformation reaction could permit the cell to go on through its previously forbidden state of differentiation so that it can no longer revert to a malignant form. This hypothesis must be tested experimentally in accordance with the studies on leukemia of Sachs (49).

Several other agents have been described as capable of carrying out reverse transformation reactions in malignant cells. These include retinoic acid (27), thioproline and 2-amino-2-thiazoline chlorhydrate (4), and inhibitors of prostaglandins and antiinflammatory drugs (60,61). These agents may restore steps in blocked reaction chains at points beyond that at which cAMP operates. Careful analysis of such actions is vitally needed. Theoretically, cancer might be possible without demonstrable defects in the cytoskeleton. Such an occurrence would be compatible with the formulations proposed here, provided that a block in the regulation of cell multiplication occurred at a point beyond those steps that involve action of the cytoskeleton, if such steps exist. This question, which is now under study, would appear to be resolvable by examination of different cancers for changes in cytoskeletal structure, growth regulation, and other properties after addition of cAMP and other reverse transformation agents.

The reverse transformation concept offers a new theoretical possibility for treatment of cancer, i.e., induction of restoration of malignant cells to normal form rather than approaches designed to kill the cancer cell with as much selectivity as possible. It is interesting to speculate whether some spontaneous remissions in cancer may not result from changed metabolic conditions producing effects such as those described here.

A number of experiments have used the approach of reverse transformation to control malignancies in animals and even in some human patients. Results of definite interest have been reported (4,6,7,57,60). If, indeed, important therapeutic results can be achieved, these will undoubtedly become even more valuable when further molecular elucidation of the effects described here has been reached so that maximal beneficial effects of such treatment can be secured.

SUMMARY

Older evidence is reviewed and new experiments described supporting the previous proposal that at least one type of cancer is caused by a defect in the cytoskeleton. The tendency of a variety of different cellular characteristics to become altered concomitantly in transformation and reverse transformation is shown to be due to the variety of cell functions that are dependent on cytoskeletal integrity. In interphase in normal cells, the cytoskeleton plays an essential role in maintaining (a) cell morphology, (b) specific membrane structure and functions, including substrate attachment, cell-cell attachment and interaction, cell surface antigen activity, surface deposition of molecules, such as fibronectin, membrane capping, specific active transport, and development of a smooth, relatively tranquil membrane, as opposed to one studded with oscillating knobbed proturberances, and (c) a smooth, compact nuclear structure, regulation of specific gene activity at appropriate points in the life cycle, including genes necessary for reproductive regulation, and control of the degree of exposure of relatively large chromosomal regions. Cyclic AMP is necessary for organization of the normal interphase cytoskeleton, so that a cAMP deficiency can result in cell transformation.

In mitosis, a damaged cytoskeleton can cause unequal distribution of the chromosomes to the daughter cells, resulting in karyotype derangement and genetic drift, which causes cancer cells to evolve in the direction of steadily increasing malignancy and resistance to cancer therapeutic treatments.

Some metabolic effects of cAMP can be demonstrated to occur independently of changes in the cytoskeleton and may be unrelated to transformation. The specific phosphorylation effects of cAMP and at least some oncogenes appear to be antagonistic with respect to their effects on transformation or its reversal. The regulatory effect of cAMP phosphorylation on cell reproduction has been demonstrated and has been shown to involve genetic loci with antagonistic actions that are contained on several different human chromosomes. The actions of phorbol esters and retinoic acid appear to be explainable in a general way on the basis of these considerations, especially the gene exposure reaction, but their specific molecular sites of action remain to be delineated. These considerations offer a model for the nature of cancer that unifies much of its diverse manifestations and appear to be resolvable by the methods of somatic cell biochemical genetics and recombinant DNA technology.

ACKNOWLEDGMENTS

This chapter is contribution no. 437 of the Eleanor Roosevelt Institute for Cancer Research. This work was supported by grants from NIH (HD02080), Monsanto Co., and Reynolds Industries. Dr. Theodore T. Puck is a Research Professor of the American Cancer Society. Technical assistance in the laboratory was rendered by Cyndi Trombly; editorial assistance and help in preparation of the manuscript was provided by Carol Potera.

REFERENCES

1. Beg, Z. H., Slonik, J. A., and Brewer, H. B. (1980): In vitro and in vivo phosphorylation of rat liver 3-hydroxy-3-methyl glutaryl coenzyme A reductase and its modulation by glucagon. *J. Biol. Chem.*, 258:8541–8545.
2. Bloom, G. S., and Lockwood, A. H. (1980): Specific protein phosphorylation during cyclic AMP

induced mediated morphological reversion of transformed cells. *J. Supramol. Struct.*, 14:241–254.

3. Bourne, H. R., Coffino, P., Melmon, K. L., Tomkins, G. M., and Weinstein, Y. (1975): Genetic analysis of cyclic AMP in a mammalian cell. *Adv. Cyclic Nucleotide Res.*, 5:771–786.
4. Brugarolas, A., and Gosalvez, M. (1980): Treatment of cancer by an inducer of reverse transformation. *The Lancet*, 8207:68–70.
5. Burr, J. G., Dreyfuss, G., Penman, S., and Buchanan, J. M. (1980): Association of the src gene products of Rous sarcoma virus with cytoskeletal structures of chicken embryo fibroblasts. *Proc. Natl. Acad. Sci. USA*, 77:3484–3488.
6. Cho-Chung, Y. S., Clair, T., Bodwin, J. S., and Berghoffer, B. (1981): Growth arrest and morphological changes of hamster breast cells by dibutyryl cyclic AMP and L-arginine. *Science*, 214:77–79.
7. Cho-Chung, Y. S., and Gullino, P. M. (1974): In vivo inhibition of growth of two hormone-dependent mammary tumors by dibutyryl cyclic AMP. *Science*, 183:87–88.
8. Cox, D. M., and Puck, T. T. (1969): Chromosomal nondisjunction: The action of colcemid in Chinese hamster cells in vitro. *Cytogenetics*, 8:158–169.
9. Crossin, K. L., and Carney, D. H. (1981): Microtubule stabilization by taxol inhibits initiation of DNA synthesis by thrombin and by epidermal growth factor. *Cell*, 27:341–350.
10. David-Pfeuty, T., and Singer, S. J. (1980): Altered distributions of the cytoskeletal protein vinculin and α-actinin in cultured fibroblasts transformed by Rous sarcoma virus. *Proc. Natl. Acad. Sci. USA*, 77:6687–6691.
11. DeMartinville, B., Giacalone, J., Shih, C., Weinberg, R. A., and Franke, U. (1983): Oncogene from human E J bladder carcinoma is located on the short arm of chromosome 11. *Science*, 219:498–501.
12. Erikson, R. L., Collett, M. S., Erikson, E., and Purchio, A. F. (1979): Evidence that the avian sarcoma virus transforming gene products is a cyclic AMP independent protein kinase. *Proc. Natl. Acad. Sci. USA*, 76:6200–6264.
13. Friedkin, M., Legg, A., and Rozengurt, E. (1979): Antitubulin agents enhance the stimulation of DNA synthesis by polypeptide growth factors in 3T3 mouse fibroblasts. *Proc. Natl. Acad. Sci. USA*, 76:3909–3912.
14. Gabrielson, E. G., Scoggin, C., and Puck, T. T. (1982): Phosphorylation changes induced by cAMP derivatives in the CHO cell and selected mutants. *Exp. Cell Res.*, 142:63–68.
15. Heidrick, M. L., and Ryan, W. L. (1970): Cyclic nucleotides in cell growth in vitro. *Cancer Res.*, 30:376–378.
16. Heston, L. L., and White, J. (1978): Pedigree of 30 families with Alzheimer disease: Association with defective organization of microfilaments and microtubules. *Behav. Genet.*, 8:315–331.
17. Hsie, A. W., Jones, C., and Puck, T. T. (1971): Further changes in differentiation state accompanying the conversion of Chinese hamster cells to fibroblastic form by dibutyryl adenosine cyclic 3′:5′ monophosphate and testosterone. *Proc. Natl. Acad. Sci. USA*, 68:1648–1652.
18. Hsie, A. W., O'Neill, J. P., Li, A. P., Borman, C. S., Schroeder, C. H., and Kawashima, K. (1977): Control of cell shape by adenosine 3′:5′ phosphate in Chinese hamster ovary cells: Studies of cyclic nucleotide analogue action, protein kinase activation and microtubule organization. *Adv. Pathbiol.*, 6:181–191.
19. Hsie, A. W., and Puck, T. T. (1971): Morphological transformation of Chinese hamster cells by dibutyryl adenosine cyclic 3′:5′ monophosphate and testosterone. *Proc. Natl. Acad. Sci. USA*, 68:358–361.
20. Hsie, A. W., and Puck, T. T. (1972): Production of mutants with respect to reverse transformation in cultured Chinese hamster cells. *J. Cell Biol.*, 55:118a.
21. Hsie, A. W., and Waldren, C. A. (1970): The conversion of Chinese hamster cells to fibroblastic form by dibutyryl cyclic AMP. *J. Cell Biol.*, 47:922.
22. Hunter, J., and Sefton, B. (1980): Transforming gene product of Rous sarcoma virus phosphorylates tyrosine. *Proc. Natl. Acad. Sci. USA*, 77:1311–1315.
23. Johnson, G. S., and Pastan, I. (1972): Role of 3′,5′ adenosine monophosphate in regulation of morphology and growth of transformed and normal fibroblasts. *J. Natl. Cancer Inst.*, 48:377–387.
24. Jones, C., Moore, E. E., and Lehmann, D. W. (1979): Genetic and biochemical analysis of the a_1 cell surface antigen associated with human chromosome 11. *Proc. Natl. Acad. Sci. USA*, 76:6491–6495.

25. Krebs, E. G., and Beavo, J. A. (1979): Phosphorylation dephosphorylation of enzymes. *Annu. Rev. Biochem.*, 48:923–959.
26. LeCam, A. N., Nicholas, J. C., Singer, T. J., Cabral, F., Pastan, I., and Gottesman, M. M. (1981): Cyclic AMP-dependent phosphorylation in intact cells and in cell free extracts from Chinese hamster ovary cells. *J. Biol. Chem.*, 256:933–941.
27. Lotan, R. (1981): Effect of vitamin A and its analogs (retinoids) on normal and neoplastic cells. *Biochem. Biophys. Acta*, 605:33–91.
28. McClain, D. A., Maness, P. F., and Edelman, G. M. (1978): Assay for early, cytoplasmic effects of the src gene product of Rous sarcoma virus. *Proc. Natl. Acad. Sci. USA*, 75:2750–2754.
29. Meek, W. D., and Puck, T. T. (1979): Role of the microfibrillar system in knob action of transformed cells. *J. Supramol. Struct.*, 12:335–354.
30. Nielson, S. E., and Puck, T. T. (1980): Surface deposition of fibronectin in the course of reverse transformation of CHO cells by cyclic AMP. *Proc. Natl. Acad. Sci. USA*, 77:985–989.
31. Nicolini, C., and Beltrame, F. (1982): Coupling of chromatin structure to cell geometry during the cell cycle: Transformed versus reverse-transformed cells. *Cell Biol. Int. Rep.*, 6:63–71.
32. Patterson, D., and Waldren, C. A. (1973): The effect of inhibitors of RNA and protein synthesis on dibutyryl cyclic AMP mediated morphological transformation on Chinese hamster ovary cells *in vitro*. *Biochem. Biophys. Res. Commun.*, 50:566–573.
33. Pollack, R., Risser, R., Conlon, S., Freedman, V., Rifkin, D., and Shin, S. (1975): Production of plasminogen activator and colonial growth in semi-solid medium are *in vitro* correlates of tumorigenicity in the immune deficient nude mouse. In: *Proteases and Biological Control*, edited by E. Reich, and D. Rifkin, pp. 885–899. Cold Spring Harbor Laboratory, Cold Spring Harbor, New York.
34. Porter, K., Puck, T. T., Hsie, A. W., and Kelley, D. (1974): An electron microscope study of the effects of cyclic AMP on Chinese hamster ovary cells. *Cell*. 2:145–158.
35. Prasad, K. N., and Hsie, A. W. (1971): Morphologic differentiation of mouse neuroblastoma cells induced *in vitro* by dibutyryl adenosine 3′:5′ cyclic monophosphate. *Nature*, 233:141–142.
36. Puck, T. T. (1977): Cyclic AMP, the microtubular-microfilament system and cancer. *Proc. Natl. Acad. Sci. USA*, 74:4491–4495.
37. Puck, T. T. (1979): Studies on cell transformation. *Somatic Cell Genet.*, 5:973–990.
38. Puck, T. T. (1983): Human gene mapping, exposure and expression. In: *Banbury Rep. 14: Recombinant DNA Applications to Human Disease*. Cold Spring Harbor Laboratory, Cold Spring Harbor, New York.
39. Puck, T. T., Erikson, R. L., Meek, W. D., and Nielson, S. E. (1981): Reverse transformation of vole cells transformed by avian sarcoma virus containing the src gene. *J. Cell. Physiol.*, 107:399–412.
40. Puck, T. T., and Jones, C. (1973): Cyclic AMP and microtubular dynamics in cultured mammalian cells. In: *Cyclic AMP, Cell Growth and the Immune Response*, edited by B. W. Lichtenstein and C. W. Parker, pp. 338–348. Springer-Verlag, New York.
41. Puck, T. T., and Kao, F. T. (1982): Somatic cell genetics and its application to medicine. *Annu. Rev. Genet.*, 16:225–271.
42. Puck, T. T., Waldren, C. A., and Hsie, A. W. (1972): Membrane dynamics in the action of dibutyryl adenosine 3′:5′ cyclic monophosphate and testosterone on mammalian cells. *Proc. Natl. Acad. Sci. USA*, 69:1943–1947.
43. Otto, A. M., Zumbe, A., Gibson, L., Kubler, A. M., and de Asua, L. J. (1979): Cytoskeleton disrupting drugs enhance effect of growth factors and hormones on initiation of DNA synthesis. *Proc. Natl. Sci. USA*, 76:6435–6438.
44. Riccardi, V. M., Sujansky, E., Smith, A., and Francke, U. (1978): Chromosomal imbalance in the aniridia-Wilms' tumor association: 11p interstitial deletion. *Pediatrics*, 61:604–610.
45. Rifkin, D. B., Crowe, R. M., and Pollack, R. (1979): Tumor promoters induce changes in the chick embryo fibroblast cytoskeleton: *Cell*, 18:361–368.
46. Rohrschneider, L. R. (1980): Adhesion plaques of Rous sarcoma virus-transformed cells contain the src gene product. *Proc. Natl. Acad. Sci. USA*, 77:3514–3518.
47. Roth, C. W., Singhr, T., Pastan, I., and Gottesman, M. M. (1982): Rous sarcoma virus transformed cells are resistant to cyclic AMP. *J. Cell. Physiol.*, 111:42–48.
48. Rumsby, G., and Puck, T. T. (1982): Ornithine decarboxylase induction and the cytoskeleton in normal and transformed cells. *J. Cell. Physiol.*, 111:133–139.

49. Sachs, L. (1978): Control of normal cell differentiation and the phenotypic reversion of malignancy in myeloid leukemia. *Nature*, 274:535–539.
50. Schimke, R. T., Kaufman, R. J., Alt, F., and Kellems, R. F. (1978): Gene amplification and drug resistance in cultural mammalian cells. *Science*, 202:1051–1055.
51. Schonberg, S., Patterson, D., and Puck, T. T. (1983): Resistance of Chinese hamster ovary cell chromatin to endonuclease digestion. *Exp. Cell Res.*, 145:57–62.
52. Sen, A., and Todaro, G. J. (1979): A murine sarcoma virus associated protein kinase: Interaction with actin and microtubular protein. *Cell*, 17:347–356.
53. Storrie, B. (1973): Antagonism by dibutyryl adenosine cyclic 3′,5′ monophosphate and testosterone of cell rounding reactions. *J. Cell Biol.*, 59:471–479.
54. Storrie, B. (1974): Effect of dibutyryl adenosine cyclic 3′,5′ monophosphate and testololactone on concanavalin A binding and cell killing. *J. Cell Biol.*, 62:247–252.
55. Storrie, B. (1975): Antagonism by dibutyryl adenosine cyclic 3′,5′ monophosphate and testololactone of concanavalin A capping. *J. Cell Biol.*, 66:392–403.
56. Teyssot, B., and Houdebine, L. M. (1980): Effect of colchicine on the transcription rate of β-casein and 28 S-ribosomal RNA genes in the rabbit mammary gland. *Biochim. Biophys. Res. Commun.*, 97:463–473.
57. Tisdale, M. J. (1979): The significance of cyclic AMP and cyclic GMP in cancer treatment. *Cancer Treat. Rev.*, 6:1–15.
58. Villasante, A., Corces, V. G., Manso-Martinez, R., and Avila, J. (1981): Binding of microtubule protein to DNA and chromatin: Possibility of simultaneous linkage of microtubule to nucleic acid and assembly of microtubule structure. *Nucleic Acid Res.*, 9:895–909.
59. Volpe, J. J. (1977): Microtubules and the regulation of 3-hydroxy-3-methylglutaryl coenzyme A reductase *J. Biol. Chem.*, 254:2568–2571.
60. Waddell, W. R. (1975): Treatment of intra-abdominal and abdominal wall desmoid tumors with drugs that affect the metabolism of cyclic 3′,5′-adenosine monophosphate. *Ann. Surg.*, 181:299–302.
61. Waddell, W. R., Gerner, R. E., and Reich, M. P. (1983): Nonsteroid anti-inflammatory drugs for desmoid tumors and carcinoma of the stomach. *J. Surg. Oncol.*, 22:197–211.
62a. Weinstein, I. B., Lee, L. S., Fisher, P. B., Mufson, A., and Yamasaki, H. (1979): Action of phorbol esters in cell culture: Mimicry of transformation, altered differentiation, and effects on cell membranes. *J. Supramol. Struct.*, 12:195–208.
62b. Boutwell, R. K. (1974): The functions and mechanisms of promoters of carcinogenesis. *CRC Crit. Rev. Toxicol.*, 2:419–443.
62c. Wertz, P. W., and Mueller, G. C. (1980): Inhibition of 12-0-tetradecanoylphorbol-13-acetate-accelerated phospholipid metabolism by 5,8,11,14-eicosatetraynoic acid. *Cancer Res.*, 40:776–781.
62d. Soprano, K. J., and Baserga, R. (1980): Reactivation of ribosomal RNA genes in human-mouse hybrid cells by 12-0-tetradecenoylphorbol 13-acetate. *Proc. Natl. Acad. Sci. USA*, 77:1566–1569.
62e. Perrella, F. W., Ashendel, C. L., and Boutwell, R. K. (1982): Specific high-affinity binding of the phorbol ester tumor promotor 12-0-tetradecanoylphorbol-13-acetate ti-isolated nuclei and nuclear macromolecules in mouse epidermis. *Cancer Res.*, 42:3496–3501.
63. Weintraub, H., and Groudine, M. (1976): Chromosomal subunits in active genes have an altered conformation. *Science*, 193:848.
64. Willingham, M. C., Yamada, K. M., Yamada, S. S., Pouyssegur, J., and Pastan, I. (1977): Microfilament bundles and cell shape are related to adhesiveness to substratum and are dissociable from growth control in cultured fibroblasts. *Cell*, 10:375–380.
65. Wilson, M. G., Ebbin, A. J., Touner, J. W., and Spencer, W. H. (1977): Chromosomal anomalies in patients with retinoblastoma. *Clin. Genet.*, 12:1–8.

Advances in Viral Oncology, Volume 4,
edited by George Klein. Raven Press,
New York © 1984.

DNA Methylation: Role in Viral Transformation and Persistence

Walter Doerfler

Institute of Genetics, University of Cologne, Cologne, Germany

One of the important problems in molecular biology and one that pervades all realms of modern biology pertains to the mechanisms involved in eukaryotic gene regulation. Gene activity is presumed to be regulated at different levels. DNA-protein interactions are at the core of these regulatory processes. There is evidence from many different eukaryotic systems that DNA methylation, probably at highly specific sites, can modulate these interactions and thus affect the regulation of eukaryotic gene expression. As a relatively simple example for a modulating function on DNA-protein interactions, the activity of restriction endonucleases in prokaryotes is usually negatively interfered with by DNA methylations at highly specific sequences. Occasionally, however, the activity of a restriction endonuclease is premised on the presence of modified nucleotides. The restriction endonuclease DpnI (from *Diplococcus pneumoniae*), for example, requires an N^6-methyladenine (6-mA) residue in the sequence 5′-GATC-3′[1] (103), whereas the sequence DpnII cannot act at 6-mA-containing sites. Thus it becomes obvious that a methylated nucleotide represents only a signal. The functional significance of this signal is decided on by the surrounding sequences and the type of DNA-protein interactions this signal is able to modulate. It would not be surprising, therefore, to observe that methylated nucleotides could affect different genetic functions and might even exert seemingly opposing effects, depending on the sequence that was modified.

In biochemical terms, the role that DNA methylation can play in gene regulation is not yet understood. The signal may be recognized directly, it may function via structural alterations of DNA due to the modification of specific bases. It has been shown that at high concentrations of salt, poly(dG·dC)·poly(dG·dC) can form a left-handed Z structure of DNA (4,5,128,180). On the other hand, with poly(dG·m^5dC)·poly(dG·m^5dC), the transition from the B to the Z form occurs at close to physiologic salt concentrations (4,5). Thus DNA methylation at specific sequences could entail structural alterations of DNA. The B-Z transition is facilitated by alternating dC·dG, i.e., the predominant site of DNA methylation in

---

[1]Nucleotide sequences will be presented in the 5′-3′ direction. Abbreviations for deoxyribonucleotides are the conventional ones. Only one strand will be given.

eukaryotes, or, more generally, by alternating pyrimidine-purine tracks. The sequence dCdCdGdG has been shown to lead to a transition from the B to the A form of DNA (15). Whatever structural alterations in DNA as a consequence of DNA methylation may have been demonstrated to occur with synthetic nucleotides, it remains to be proven that any of these structural changes has an immediate effect on the genetic activity of DNA.

A regulatory signal, such as DNA methylation, could assume special significance in early development, in differentiation, or in catastrophic events as complex as genetic disease or cancer. In these fundamental processes, differential gene activity, i.e., patterns of the sequential turning on or off of genes, will eventually lead to transitions into more advanced or regressed developmental stages and will have to be explained in molecular terms.

With the recent discovery of so-called *onc* genes, which were previously considered to be viral but in fact constitute cellular genes transduced by viral genomes and which may play a role in certain stages of development or cellular transformation (summarized in ref. 92), the question arises what the states of methylation of these genes are in normal as well as transformed cells. Until now, there have been only occasional reports about the levels of DNA methylation in tumor cells (53,105). Most of these reports suffer from the failure to analyze the states of methylation of specific genes or sets of specific genes. Hence these findings have been difficult to correlate to functional alterations in tumor cells. Only recently could it be shown that substantial hypomethylation existed in specific genes, e.g., in the genes for human growth hormones, α- or γ-globin, in human cancer cells derived from some tumors as compared with control cells from the surrounding tissue of the same patients (48). The hypomethylation was even more pronounced in metastatic tumor tissue from one patient.

Investigations on the functional role that DNA methylation can play in eukaryotic gene regulation have been facilitated by viral systems, particularly by studies on integrated genomes of oncogenic viruses. I have recently summarized the earlier findings on these systems in several reviews (29,31,32). More general aspects of DNA methylation and its functional implications have been extensively treated in a series of review articles (7,29,31,39,74,131,132,183).

This chapter concentrates mainly on findings made in viral systems because of the special advantages offered by these systems. It is not my intention to repeat all the information on DNA methylation that has already been treated in earlier reviews (29,31). I concentrate instead on recent developments in the field but present sufficient background material for the newcomer to DNA methylation.

BASIC FACTS ON DNA METHYLATION

Modified Bases

In naturally occurring DNA, two bases have been found to be specifically modified by DNA methylation: cytosine to 5-methylcytosine (5-mC) and adenine to N^6-methyladenine (6-mA) (78,142,188). Prokaryotic DNA and DNA from lower eukaryotes contain 5-mC and 6-mA; in DNA from higher eukaryotes, only 5-mC can be detected. Other modifications may occur to minute extents. The DNA of the T-even bacteriophages carries glucosylated hydroxymethylcytosine in genetically determined patterns (13,85,112). Modified bases could not be detected in DNA from *Drosophila* (42,102,143,144). Of course, the methods available for the detection of methylated bases would not be sensitive enough to retrieve methylated bases if they occurred at $<0.01\%$. The levels of DNA methylation in prokaryotic and eukaryotic DNA vary widely, depending on the organism or the specific DNA sequences analyzed in a given organism. In *Aedes albopictus*, 5-mC is found at about 0.03 mol%, in mammals at 2 to 8 mol%, and in certain higher plants at 30 to 50 mol% (2,22,35–37,139,162,167,168).

Methodology

Although there are reliable methods to determine modified nucleotides in DNA, all methods are limited in sensitivity. Chemical methods exhibit the decisive drawback of failing to yield information on the specific sites of DNA methylation. Two-dimensional chromatographic and electrophoretic separations (67), gas chromatography (130), mass spectrometry (129,141), and high pressure liquid chromatography (50,51,101) have been successfully applied to the determination of 5-mC after total hydrolysis of DNA. Usually, none of these chemical methods is sensitive enough for detailed genetic analyses. It has been shown that with 5-mC covalently linked to bovine serum albumin, antibodies can be raised that can recognize 5-mC in DNA (38,43,108,137). However, there are questions with respect to the absolute specificity of this recognition.

A direct method is needed to detect all methylated nucleotides in their original sequence position in genomic DNA. Obviously, eukaryotic DNA cannot be cloned in pro- or eukaryotic vectors prior to being sequenced. Such a procedure invariably would alter the original patterns of methylation of genomic eukaryotic DNA. A method that would permit the direct sequencing of genomic DNA and at the same time spot all methylated bases, in particular 5-mC, would be required. Such a method is currently being developed. The presence of 5-mC in DNA can be detected by the Maxam-Gilbert nucleotide-sequencing procedure (113,124a), since 5-mC reacts less efficiently with hydrazine than with cytosine or thymine and hence causes blanks in the usual sequencing ladders (124a).

In recent years, the method of choice for studies on the functional significance of DNA methylation has been based on restriction endonucleases. This method is highly specific, sensitive, and facilitates the sequence-specific localization of methylated bases in genomic DNA. An incidental observation by Waalwijk and Flavell (177) opened the way to this analytic possibility. The restriction endonuclease pair HpaII (from *Haemophilus parainfluenzae*) and MspI (from *Moraxella* species) are isoschizomers and cleave at 5′-CCGG-3′ sites. When the internal C-residue of this sequence is methylated, the endonuclease HpaII cannot cleave, whereas MspI is active. Conversely, methylation of the external C-residue renders MspI refractory, while HpaII remains active (145). The availability of this isoschizomeric pair of restriction endonucleases has facilitated the elucidation of patterns of methylation in specific genes. DNA is cleaved with either enzyme; the fragments are separated by electrophoresis on agarose or polyacrylamide gels; and the distribution of fragments in a specific segment A of cellular or viral DNA is visualized by Southern blotting (146) and hybridization to the (^{32}P)-labeled cloned segment A. Depending on the cleavage patterns generated by the HpaII or MspI restriction endonuclease, methylation patterns in DNA segment A can be deduced. The analysis obviously is restricted to methylations in the sequence 5′-CCGG-3′ and to the cloned segment A in total intracellular DNA.

A large proportion of all 5-mC residues in eukaryotic DNA is located in the dinucleotide sequence 5′-CpG-3′ (35,55,59,110,122), but only a certain percentage of these dinucleotides is shared by the 5′-CCGG-3′ sequence. Thus the HpaII-MspI or SmaI-XmaI isoschizomeric pairs will recognize only part of the 5-mC residues in a given gene. The enzyme pair Sau3A and TaqI can be used for the sequence 5′-GATCGA-3′ in a similar fashion (153) to detect 5-mC. TaqI is not inhibited by 5-mC but by 6-mA. The isochizomeric pair DpnI and DpnII has already been mentioned. A number of additional enzymes are sensitive to methylated nucleotides, but insensitive isoschizomers usually are not available. Such restriction endonucleases still can be used to determine methylated sequence combinations, provided the gene or DNA segment to be analyzed is available as a clone that has been propagated on a methylation-deficient strain of *Escherichia coli*. The unmethylated cloned DNA then can be used as an internal standard for pattern comparison. Similarly, viral DNA can be utilized in this way, since free virion DNA is often not methylated (67,90,163).

Distribution of 5-mC in Eukaryotic DNA

The modified nucleotide 5-mC in eukaryotes occurs predominantly in the sequence 5′-CpG-3′ (55,59,110,121,122) and less frequently in the dinucleotide combinations 5′-CpA-3′, 5′-CpT-3′, and 5′-CpC-3′ (66). The distribution of 5-mC in eukaryotic DNA is nonrandom. It has been recognized in many different systems, notably, in well-characterized eukaryotic objects of molecular biology, such as viral systems, that segments of the genome that are actively transcribed are undermethylated (20,96,156,157,170). Regions that have been turned off have proved to be strongly methylated. In animal cell DNA, about 70% of all 5′-CpG-3′ sequences are methylated. Active nuclear regions of DNA that exhibit increased sensitivity toward DNAseI (54,181) show 30 to 40% of all 5′-CpG-3′ sites to be methylated (122). In this context, it is interesting to note that the nearest neighbor sequence 5′-CpG-3′ is statistically underrepresented in most higher eukaryotes and their viruses (120,155). Perhaps this sequence is located predominantly in functionally crucial positions in DNA sequences that exert a regulatory role. For some genes or groups of genes, clusters of this sequence have been recognized in front of or close to the promotor/leader sequence, e.g., in globin gene or in early regions of adenovirus DNA (29,34,48a,48b,96,97). Thus gene regulation may also depend on the strategic positioning of nucleotide sequences that can be modified by methylation at crucial sites.

A genetic signal, such as a modified nucleotide (e.g., 5-mC) in the DNA of higher eukaryotes may be considered to have different coding potentials, depending on the environment and the type of DNA-protein interactions it may modulate or the nature of the structural change in DNA conformation it may help to stabilize. Thus it would not be surprising if the effect of DNA methylation on gene expression, which has usually been reported to be negative, might occasionally be found positive in special cases. Moreover, one would have to strictly analyze the site of DNA methylations. Inverse correlations between DNA methylation and gene activity have been established for the 5′-CCGG-3′ (HpaII) sites or the 5′-GCGC-3′ (HhaI) sites. In certain genes, methylations of C-residues in other sequences may be decisive. Simplistic rules will not suffice; each system and its functional correlations will have to be studied in depth.

De Novo and Maintenance Methylations

Patterns of DNA methylation are inherited and passed on from cell generation to cell generation (72,149,184). This DNA methylation is thought to be upheld by the action of maintenance DNA-methyltransferases, which require hemimethylated DNA as a substrate. Methylated bases on one strand serve as the signal to methylate bases on the complementary newly synthesized strand in a symmetric position. Hemimethylated DNA has been shown to be a substrate in DNA methylation (65). According to that scheme, demethylations of previously methylated DNA would require interference with the action of maintenance DNA-methyltransferases during or after DNA replication. On the basis of this mechanism, demethylation was possible only passively and dependent on DNA replication. Recently, evidence has been adduced that DNA demethylation also could be an active, enzymatically controlled process in the absence of DNA replication. A DNA demethylating activity has been discovered in the nucleus of murine erythroleukemic cells (57) but this finding has not yet been confirmed.

De novo methylation also can occur. The best evidence for this mechanism is still derived from a viral system (see below). In brief, DNA extracted from adenovirons is not detectably methylated (42,67,176), nor is intracellular free viral DNA in productively or abortively infected cells (170). When adenovirus DNA is covalently inserted into cellular DNA in transformed or tumor cells (27–30,34a,52), specific patterns of viral DNA methylation are observed (96,156,157,170). It is noteworthy that the final patterns of adenoviral DNA methylation are established only many cell generations after the integration event (98), and this establishment proceeds in a nonrandom fashion (99). From these observations, it is apparent that DNA methylation can occur *de novo* and must be subject to tight control. Unfortunately, we know next to nothing about these control mechanisms.

If DNA methylation at specific sites did indeed exert a pivotal role in gene regulation, studies on DNA-methyltransferases—for *de novo* and maintenance methylations—and on the regulation of their activities would assume great importance. Work on prokaryotic and eukaryotic DNA-methyltransferases has been summarized previously (74), and this review will not be repeated here. Little information is available on eukaryotic DNA-methyltransferases. Moreover, this latter class of enzymes, in general, has not been shown to exhibit sequence specificities. Perhaps the experimental conditions under which these enzymes were tested were too artifactual to reveal the natural activity.

Can DNA Methylation Alter the Structure of DNA?

It is an interesting and significant question whether DNA methylation at highly specific sites affects essential biologic functions directly or via structural changes of DNA that are elicited or stabilized by DNA methylation. DNA can assume many different conformations, depending on experimental conditions. Theoretically, several different structures of DNA must be considered (for review, see ref. 13a); three families of DNA helices have been examined in detail: A, B, and Z. In solutions containing high concentrations of salt, poly(dG·dC)·poly(dG·dC) can form a left-handed Z structure. For poly(dG·m^5dC)·poly(dG·m^5dC), the transition from the B to the Z form is observed at close to physiologic salt concentrations (4,5,93,119). Di- and polyvalent ions are effective in causing this transition, as well as a variety of exotic ionic conditions (86).

The sequence dG·m^5dC can have a striking effect on the B-Z transition. The B-Z transition is facilitated by alternating dC·dG or, more generally, by alternating pyrimidine-purine tracks, whereas the sequence dCdCdGdG may favor transition from the B to the A form of DNA (15). It is presumed that the latter sequence might sterically inhibit the B-Z conversion. Since certain DNA sequences can cause specific structural altertions, it will be important to determine under which conditions these transitions can occur in the living cell and which modifications might stabilize certain transitions. DNA methylation at highly specific sites could have this effect. On the other hand, a methylated base with the methyl group protruding into the major groove of B DNA can efficiently provide a signal in its own right.

Thus structural changes as a consequence of DNA methylation do not have to be invoked to explain functional effects of DNA methylation. In keeping with this notion, it has been pointed out that all aspects of DNA-protein interactions, as exemplified by the binding of λ repressor to DNA, can be explained in molecular terms with the classic B form of DNA (76). This interpretation has been documented by X-ray analysis.

Possible Biologic Functions of DNA Methylation

The most extensively studied effect of DNA methylation at highly specific sites is that on the regulation of gene expression. In particular, investigations on viral systems have yielded some of the most convincing results, which are reviewed in the following sections. Evidence has also been adduced for DNA methylation affecting mismatch repair in prokaryotes (179), genetic recombination in bacteriophage λ (95), and in virus latency (189). Modification of DNA may influence other mechanisms as well. DNA replication has been hypothesized to be affected by DNA methylation (159), but convincing evidence is not yet available. There is evidence that DNA damage, caused by ultraviolet irradiation or chemical carcinogens, may lead to heritable loss of DNA methylation at some sites (89).

DNA METHYLATION AND GENE ACTIVITY: A CRITICAL REVIEW OF THE EVIDENCE

Hypotheses correlating DNA methylation and gene activity in eukaryotes evolved from observations on the differential states of DNA methylation in developing organisms and on the basis of theoretical considerations (77,133,135). A series of experimental approaches were designed that have yielded more definite insight into the functional significance of DNA methylation. The main lines of evidence, extensively reviewed earlier (29,31,39,131), are as follows.

Inverse Correlations

Inverse correlations between the extent of DNA methylation and the level of expression of a given gene have been established in different systems analyzing a number of eukaryotic and viral genes. Since these correlations were essentially restricted to 5′-CCGG-3′ and 5′-GCGC-3′ sites, it could not be expected that they applied to any gene without exceptions. Moreover, DNA methylation may not be a decisive factor in the regulation of all genes. Alternative mechanisms are conceivable. In general, it is inappropriate to extrapolate the findings in a given set of systems to all other systems. It also must be considered whether these inverse correlations indicate DNA methylation to be the cause or the consequence of gene inactivation.

Experimental evidence was provided by studies on DNA methylation of a certain gene in different organs of one organism or in viral systems (6,10,20,29,115, 116,156,157). In a certain gene that was actively expressed in one organ, all or most of the 5′-CCGG-3′ sites were found to be unmethylated; the DNA in the same gene in other organs not expressing this gene was completely methylated at the 5′-CCGG-3′ sites. These sites were preferentially investigated because of the availability of the isoschizomeric restriction endonuclease pair HpaII and MspI (177), which permitted determination of the state of methylation at all 5′-CCGG-3′ sites. In many genes, the state of methylation at these sites appears to have special significance. In this way, inverse correlations between the extent of DNA methylation at 5′-CCGG-3′ sites in specific genes and the degree to which these genes are expressed have originally been established in several different systems (12,20,69,96,98,109,144a,156,157,164,165,170,182).

In some cases, these inverse correlations have been extended to other sequences containing 5′-CpG-3′ dinucleotides, e.g., the sequence 5′-GCGC-3′, the recognition sequence of the restriction endonuclease HhaI (96,98).

As mentioned above, inverse correlations between gene activity and DNA methylation at 5′-CCGG-3′ and 5′-GCGC-3′ sites do not invariably exist. In some genes, DNA methylation at sites other than HpaII sites or HhaI sites may be important in regulation; other genes may not respond to that signal at all. Thus unusual methylation patterns with respect to expression have been reported for vitellogenin genes A1 and A2 in hepatocytes (56). These genes are heavily methylated, irrespective of expression. Perhaps hormone inducibility of these genes could offer an explanation. Similarly, the DNA around the start site of transcription of the α2(I) collagen gene in chicken was not methylated; the central and 3′ regions of the gene were methylated, irrespective of transcription (117). For the two rat insulin genes, no evidence is believed to exist for a correlation between DNA methylation and gene expression (8a).

Studies on the Expression of *In Vitro* Methylated Genes

Investigations directed toward the core of the problem used *in vitro* methylated cloned genes that were microinjected into the nucleus of oocytes of *Xenopus laevis* (49,104,171–174) or into mammalian cells in culture (97,148,178). Genes methylated at specific sites usually were not expressed in these experiments. The use of sequence-specific *de novo* DNA-methyltransferases from prokaryotic organisms allowed the determination of highly specific sequences presumably involved in gene regulation (49,97,104,148,171–174,178). Other sequences, e.g., the sequence 5′-GGCC-3′, when methylated *in vitro* did not exhibit any effect on gene expression (34,173). These data, accumulated from work in different systems and in different laboratories, provided strong direct evidence for the notion that methylated sequences at highly specific sites in the promoter region of a gene play an important role in the regulation of gene expression. These observations demonstrated that DNA methylation caused transcriptional inactivation, at least in the genes tested, and was not a consequence of gene inactivation. These findings do not preclude the possibility that investigations on other genes might lead to different conclusions.

From our own contribution to this particular problem, it was of paramount importance that work on the *in vitro* methylation of a certain viral gene was preceded by a thorough investigation of the state of methylation of the same gene in the integrated form. Since we initially had only one DNA-methyltransferase, HpaII, available for these studies, we had to ascertain that the 5′-CCGG-3′ site was actually involved in the regulation of this particular gene. Thus *in vivo* studies were an important prerequisite for the successful initiation of *in vitro* investigations. Moreover, it was essential to ascertain that *in vitro* DNA methylation of a gene was complete before its transcriptional activity was tested.

Inhibition of DNA Methyltransferases by 5-Azacytidine

Further evidence comes from work with the cytidine analog 5′-azacytidine. This compound can be incorporated into replicating DNA (11,18,83,84) and, because of its chemical structure (N instead of C in the 5-position), cannot be methylated. However, the main inhibitory role of 5-azacytidine toward DNA methylation emanates from its ability to inhibit DNA-methyltransferases (11,18,83,84). In a number of different systems, it has been possible to activate previously dormant genes and induce their expression (14,64,73,84,160). The original report on the gene-activating property of 5-azacytidine described the activation of a complex set of cellular functions leading to *in vitro* differentiation of mouse fibroblasts to twitching muscle cells (16).

By 5-azacytidine treatment, specific genes could be turned on in animals (19) and in man (106). The inhibitory drug stimulated fetal hemoglobin synthesis in anemic baboons (19) and selectively increased γ-globin synthesis in patients with β^{+}-thalassemia (106). Concomitant demethylations in the genes involved also could be shown (106). Since 5-azacytidine is a highly toxic agent, extreme caution should be exerted (106).

It is not to be expected that all dormant cellular genes can be turned on by treatment of cells with 5-azacytidine. Absence of DNA methylation is a necessary but not sufficient precondition for gene activation (98,126,164). A crucial mechanism, such as gene activity, may be subject to multifaceted regulatory mechanisms; DNA methylation constitutes one important parameter. Thus, depending on the stringency of inactivation for a given gene, 5-azacytidine treatment may or may not lead to the activation of a certain gene or set of genes.

Interestingly, ultimate chemical carcinogens inhibit the transfer of methyl groups to hemimethylated DNA in *in vitro* studies using DNA-methyltransferase from mouse spleen. Carcinogenic agents may lead to heritable changes in DNA methylation patterns by a variety of mechanisms (187).

For a complete understanding of the role of DNA methylation in gene regulation, methods will be required to determine the state of methylation at possibly all the 5′-CpG-3′ sites in a gene and its adjacent sequences. Improved *in vitro* transcription systems also will be required to evaluate the effect of methylation in different 5′-CpG-3′ sites. Presently available *in vitro* transcription systems do not respond to DNA methylation.

The results obtained by 5-azacytidine induction are interesting but also are subject to some doubt in that a toxic analog could have additional, poorly understood effects, and the interpretation of the induction phenomena may be more complicated than previously thought.

Manipulated Genes and *In Vitro* Methylation

It has been shown in a number of different systems that absence of DNA methylation at the 5′-ends and/or promotor sites of eukaryotic or viral genes correlates with gene expression (48b,96,97,126,136,150,185). In one instance (185), estrogen treatment of chickens caused demethylation of an HpaII site at the 5′-end of the vitellogenin gene. It was an attractive hypothesis that specific DNA methylations close to or at the promotor end of a gene might play a role in the turn-off of transcription. Of course, it is still unknown what site or sites precisely exert the decisive effect.

The methodologic repertoire of recombinant DNA technology opened the way to attempt a direct approach to this problem. In our own laboratory, two sets of experiments have been initiated:

1. The E2a gene of adenovirus type 2 DNA had previously been shown to be transcriptionally inactivated by HpaII methylation after subsequent microinjection into the nuclei of *X. laevis* oocytes, whereas the unmethylated gene was transcribed (171,172). The same gene methylated at 5′-GGCC-3′ sites was not inactivated (173). We have *in vitro* manipulated the gene and methylated the 5′-CCGG-3′ sites either at the 5′-end or at the 3′-region of the gene. Methylation of three 5′-CCGG-3′ sites at the 5′-end inactivated the E2a gene of adenovirus type 2 (Ad2) DNA after microinjection into nuclei of *Xenopus laevis* oocytes (104), whereas methylation of eleven 5′-CCGG-3′ sites at the 3′-end affected the expression of the E2a gene only slightly.
2. The plasmid pSVO CAT carries the prokaryotic chloramphenicol acetyltransferase (CAT) gene and an HindIII site just in front of that gene for the insertion of foreign promotor sequences (58). Various adenovirus promoters were inserted at the HindIII site and proved active in facilitating expression of the CAT gene. Promoters that carried HpaII sites upstream from the TATA signal could be inactivated by methylation at 5′-CCGG-3′ sites (97). Promoters with no HpaII sites, e.g., the early SV40 promoter in the plasmid pSV2 CAT, or with HpaII sites downstream from the TATA signal were insensitive to DNA methylation (97).

It has also been shown that methylations of the globin gene between −700 and +100 relative to the cap site of the gene inactivate the gene, methylations between +100 and +1950 do not affect transcription (7a).

These results support the concept that DNA methylation at specific sites (5′-CCGG-3′) in the promoter region, at least of adenovirus genes and of the globin gene, causes transcriptional inactivation. Thus for some genes, a causal relationship between DNA methylation and transcriptional inactivation seems to be established.

VIRUS-TRANSFORMED CELL SYSTEMS

With respect to studies on the regulation of gene expression, virus-transformed cells or cells latently infected with viruses present an interesting model in that some of the viral genes continue to be expressed while others are permanently turned off. Inverse correlations between gene activity and levels of DNA methylation could first be established using virus-transformed cells (20,156,157). The next sections concentrate on the adenovirus, herpes virus, and retrovirus systems but refer also to other viral systems.

The Adenovirus System

Virion DNA and Free Viral DNA in Infected Cells are not Detectably Methylated

At present, there is no evidence that virion DNA extracted from purified adenovirus particles or free intracellular viral DNA extracted from productively or abortively infected cells is detectably methylated (42,67,156–158,170,176). This statement is supported by the results of experiments employing two-dimensional thin-layer chromatography (67), differential cleavage of viral DNA (156–158) with the restriction endonuclease pair HpaII and MspI (177), or high performance liquid chromatography (42). In particular, the latter method is highly sensitive. Ad2 virion DNA yielded <0.04% of 5-mC among the cytidine residues of Ad2 DNA, if any (42). The amount of 6-mA was found to be <0.01% among the adenine residues of Ad2 DNA (67). Trace amounts of 5-mC were occasionally found in Ad2 DNA preparations extracted from virions that had not been treated with DNAse prior to extraction of DNA. It is likely that these DNA preparations contained trace amounts of cellular DNA from KB host cells in which Ad2 virus had been propagated. Thus the low levels of 5-mC in these viral DNA preparations (42) might be explained by contamination with tiny amounts of cellular DNA.

Human cell DNA contains 3.6% (KB DNA) to 4.4% human embryonic kidney cell DNA (HEK DNA) 5-mC (67). Ad2 DNA grown on these host cells seems to be devoid of 5-mC or 6-mA as determined by the most sensitive techniques available. Thus the interesting problem arises of how viral DNA escapes methylation, whereas the DNA of the host is extensively methylated. Several explanations can be considered (see below).

Human KB cellular DNA inserted into virion DNA and encapsidated into complete Ad12 particles does not appear to be significantly methylated, whereas the same cellular DNA sequences in human KB cells are strongly methylated (23). Symmetric recombinants (SYREC) of Ad12 DNA are full length molecules identical in size with viral DNA and constitute recombinants between the left terminus of Ad12 DNA and approximately 26 to 30 kb of human cellular KB DNA (23,24). The length of the adenovirus terminal fragment in the recombinant can vary (24). The example of the recombinant SYREC molecule provides direct evidence for the notion that cellular DNA sequences escape DNA methylation when they are inserted into free viral or virus-like genomes, which replicate independently of cellular DNA. Thus transfer of methyl groups to DNA may depend on the location in a certain nuclear compartment rather than on the origin of the sequence to be modified.

There are several additional examples of virion DNA not being methylated to detectable levels (90,163). In contrast, the DNA of the frog virus FV3 is extensively methylated; >20% of all the cytidine residues in FV3 DNA are 5-mC (186).

Experiments using the isoschizomeric restriction enzyme pair HpaII and MspI (177) demonstrated that the free intracellular viral DNA sequences in productively or abortively infected cells are not extensively methylated either (170), in that no difference was observed in the cleavage patterns of viral DNA by the two enzymes. Ad2 DNA replicates to large copy numbers in human cells, whereas Ad12 DNA cannot replicate in hamster (BHK21) cells (25,26,33,47,170). There is no evidence in either system that free parental virus DNA or newly synthesized free viral DNA in productively infected cells becomes methylated at 5′-CCGG-3′ sites. Levels of methylation at other 5′-CpG-3′-containing sites in free intracellular viral DNA have not yet been investigated. Similarly, it will be mandatory to exclude that extremely low levels of 5-mC could exist even in free viral DNA at 5′-CCGG-3′ sites. We have initiated a study on this topic using small Ad2 DNA fragments as plasmid clones in blot hybridization experiments (U. Wienhus and W. Doerfler, *unpublished experiments*).

We have shown previously (46,125) that in BHK21 cells abortively infected with Ad12, early viral functions but not late viral genes are expressed. Moreover, the regulation of early versus late gene expression of adenovirus DNA in productively infected cells functions meticulously without apparent recourse to DNA methylation as a possible regulatory signal. Of course, we cannot yet rule out the presence of very few methyl groups at decisive regulatory sites in free viral DNA in productively or abortively infected cells. Viral DNA methylation at a level found in integrated viral genomes in transformed cells (96,156–158) does not exist in free adenovirus DNA. Thus it is likely that regulatory principles independent of DNA methylation are operative in free viral DNA. Teleologically speaking, it could be considered inopportune for adenovirus DNA to succumb to long-term cellular regulatory signals. As can be deduced from work on integrated adenovirus DNA sequences, methylation of integrated viral sequences leads to the long-term, perhaps irreversible silencing of viral genes. Such a mechanism would be impractical for the regulation of viral genes, and adenovirus DNA must have developed different ways for the temporal regulation of its expression schedule. Of course, the questions remain of how adenovirus DNA regulates expression and how free viral DNA manages to escape DNA methylation.

Inverse Correlations Between the Extent of DNA Methylation of Integrated Adenoviral DNA and the Level of Expression of Adenoviral Genes in Transformed Cells

The DNA of hamster cell lines transformed by Ad12 has been shown to contain about 40% more 5-mC than the DNA of nonvirus transformed baby hamster kidney cells (67). Similarly, polyoma virus-transformed BHK21 cells contained twice the amount of 5-mC than normal BHK21 cells (134). In this instance, it has been shown that the level of 5-mC in cellular DNA was dependent on cell density and the rate of cellular DNA synthesis.

In the course of intensive investigations on the mode of integration of adenovirus DNA in transformed and tumor cells (for recent reviews, see refs. 30,34a,52), we have discovered that, in contrast to free intracellular viral or virion DNA, integrated Ad12 DNA in hamster cells is strongly methylated at 5′-CCGG-3′ (HpaII) and 5′-

CCCGGG-3′ (SmaI) sites (158). Prior to that finding, evidence was adduced that 5′-GTCGAC-3′ (SalI) sites were more strongly methylated in DNA from the two Ad12-transformed hamster cell lines T637 and HA12/7 than in DNA from BHK 21 cells, the parent line of T637 cells (61). It had also been demonstrated (67) and recently confirmed (42) that the total 5-mC content of DNA from Ad12-transformed T637 and HA12/7 cells (3.7% of 5-mC) was higher than that of DNA from BHK21 cells (2.5% of 5-mC). Furthermore, it was shown that both the 5′-CCGG-3′ (MspI) sites and the 5′-TCGA-3′ (TaqI) sites were more markedly methylated in the DNA from T637 cells than BHK21 cells. MspI sites were more intensely methylated than TaqI sites (42). Thus evidence is accumulating from different types of analyses that a number of different 5′-CpG-3′-containing sequences are more strongly modified in Ad12-transformed cell lines than in nonvirus transformed cells. Interestingly, the levels of DNA methylation in Ad12-induced tumors are strikingly low (see below) and increase with continuing passage of these tumor cells in culture (98).

The finding that integrated viral DNA sequences are extensively methylated, whereas virion DNA or free viral DNA in infected cells is not methylated, provides one of the best lines of evidence for the occurrence of *de novo* methylation of DNA in mammalian cell systems. With the advent of restriction endonuclease pairs diagnostic for DNA methylation at certain nucleotides in specific nucleotide sequences (177), it became possible to directly demonstrate methylation of adenovirus DNA sequences in transformed cells (156–158).

Since we had previously determined the patterns of expression of the early adenoviral genes in some of the Ad12-transformed hamster lines (125) and in Ad12-induced rat brain tumor lines (80), we were in a good position to investigate relationships between adenovirus DNA methylation and adenoviral gene expression. An inverse correlation was discovered between adenoviral gene expression and DNA methylation at 5′-CCGG-3′ sites (156,157). Similar results were subsequently observed in Ad2-transformed hamster cells (170). Early adenovirus genes were undermethylated in cell lines in which they were expressed. Late viral genes usually not expressed in adenovirus-transformed cells were heavily methylated at 5′-CCGG-3′ sites. Conversely, early viral genes not expressed in transformed cell lines, as the E3 region in the Ad12-transformed hamster cell line HA12/7 (96,125,138) or the E2a region in the Ad2-transformed hamster cell line HE2 or HE3 (44,82), are methylated at some or all of the MspI sites (156,157,170). In some of the Ad2-induced rat brain tumor lines, some of the late viral genes are expressed (80); these viral genes are undermethylated at MspI sites (157). Comparable inverse correlations frequently but not invariably involving the 5′-CCGG-3′ sites were subsequently described for genes in many viral and nonviral eukaryotic systems (for reviews, see refs. 29,30,31,39, and 131).

Although the Ad12 sequences integrated in cellular DNA in virus-induced tumors (100) were hardly methylated at all at 5′-CCGG-3′ or 5′-GCGC-3′ sites, viral DNA sequences were not expressed in these tumors, as evidenced by Northern blotting of RNA extracted from the tumors (98). We concluded that the absence of

DNA methylation would probably constitute a necessary but not a sufficient precondition for viral gene expression, as had been demonstrated in other systems (126,164).

In the same series of investigations, we discovered that upon explantation of tumor cells into tissue culture and upon serial passage, the levels of DNA methylation in integrated viral genomes at the 5′-CCGG-3′ and 5′-GCGC-3′ sites increased. This shift in the patterns of methylation was gradual and did not noticeably alter the absence of expression of the persisting viral genomes. It was not random but appeared to follow an ordered schedule, with certain MspI sites becoming methylated earlier than others (99). We do not yet know what factors affect or regulate the extent of viral DNA methylation in tumors or in transformed cells. There is evidence from work on human diploid fibroblasts that methylation patterns can be inherited unstably in clonal isolates (140). There is random drift in patterns of methylation during replication of fibroblasts.

T637 hamster cells are an Ad12-transformed cell line containing about 22 copies of Ad12 DNA per diploid genome (147,158). Morphologic revertants of this cell line have been isolated and characterized (40,41,60,62). Some of these revertants contain only one remaining copy of viral DNA or fragments thereof (40). In comparison to the levels of methylation of viral DNA sequences in cell line T637, viral DNA methylation is markedly increased in the revertants, and the extent of expression is strikingly diminished (40,138).

Methylation at the 5′-Terminus of a Gene Affects Gene Expression

The inverse correlation established in many different eukaryotic systems between the extents of DNA methylation at 5′-CCGG-3′ sites and gene expression does not pertain to DNA methylation along the entire stretch of a gene or group of genes. Evidence has been adduced from studies on Ad12-transformed hamster cells that 5-mC positioned at the 5′-CCGG-3′ sites at the 5′-termini of viral genes and in the promotor/leader sequences correlates with the shut-off of integrated viral genes (96). The data demonstrated that some of the 5′-CCGG-3′ (MspI) sites at the 3′-termini in the E1 regions or in the E2a regions of integrated Ad12 genomes in cell lines T637, HA12/7, and A2497-3 are methylated, although these regions are expressed in all three cell lines (45,125,138). In contrast, in cell line HA12/7, the MspI sites at the 5′-terminus of the E3 region are all methylated (96); consequently, this region fails to be expressed (96,125,138). Similar correlations have been observed for the 5′-GCGC-3′ (HhaI) sites. Evidence derived from several eukaryotic genes, the adenine phosphoribosyl transferase and dihydrofolate reductase genes (150), the globin gene (48b), and the rDNA in *X. laevis* (105a) also points to methylation at the 5′-termini of these genes as the decisive shut-off signal. Additional examples are cited above. It is unknown at present in what way DNA methylation at highly specific sites blocks the initiation of transcription.

We have begun a study using the promoters of several Ad12 early gene segments to regulate the short-term expression of the CAT gene in mammalian cells. So far, we have demonstrated that insertion of the E1a (pAd12 1a CAT) or of the gene IX promotor of Ad12 DNA (pAd12 IX CAT) into the pSVO vector (58) could prime the expression of the CAT gene after transfection of the vector into mouse LTk$^-$ cells (97).

Methylation with the HpaII DNA-methyltransferase of the pAd12 1a-CAT plasmid abolished its activity. There were two HpaII sites upstream from the TATA signal in the E1a promoter. One of these sites was part of an Ad12 enhancer (75a). Methylation of the pSV2 CAT plasmid lacking HpaII sites in the SV40 promoter did not affect the CAT activity of this plasmid. Methylation of the pAd12 IX CAT plasmid slightly enhanced its activity. In this promoter, there was one HpaII site downstream from the TATA signal (97).

The data obtained so far support the notion that methylation of viral promoter sequences at highly specific sites can cause transcriptional inactivation.

The E2a Region of the Ad2 Genome is not Expressed in Some Transformed Cell Lines, Although the Entire Gene Persists in These Cell Lines

The Ad2-transformed hamster cell lines HE1, HE2, and HE3 (17,82) have served a useful role in studying the function of 5-mC at specific sites. It has been demonstrated both by direct immunoprecipitation on SDS-polyacrylamide gels with cell extracts (82) and by *in vitro* translation of hybrid-selected messenger RNA (44) that cell lines HE2 and HE3 do not express the Ad2-specific DNA-binding protein, whereas cell line HE1 does express this 72K protein. The DNA-binding protein of Ad2 (166) is encoded in the E2a region of the viral genome. All three hamster cell lines contain the intact E2a region within the integrated Ad2 genomes (169). Cell lines HE1, HE2, and HE3 contain two to four, two to four, and six to ten genome fragment equivalents per cell, respectively.

We have been able to demonstrate that, in addition to the intact E2a regions, each cell line contains the complete late promoter of the E2a region (173), and that in cell line HE1, in which the DNA-binding protein is synthesized, this late promotor is being utilized (174). Remarkably, the E2a region is controlled by a complex array of promoter sequences, three promoters being used predominantly early and one region being used predominantly late in a productive infection cycle (3,9). The failure of the DNA-binding protein to be expressed in cell lines HE2 and HE3 cannot be due simply to a defective gene or deletions in the promoter region of the E2a segment. In cell lines HE2 and HE3, all 5′-CCGG-3′ sites are methylated; in cell line HE1, the same sites are unmethylated, as determined by restriction enzyme analysis using HpaII and MspI endonucleases (170). Thus with the E2a region of Ad2 DNA in three transformed hamster cell lines, we have discovered an example of a perfect inverse correlation between DNA methylation at 5′-CCGG-3′ sites and the expression of the DNA-binding protein. It is worth noting that the 5′-GGCC-3′ (HaeIII) sites in the E2a region are not methylated at all (173). It appears that the HaeIII site does not have any effect on the E2a gene expression via DNA methylation.

As mentioned earlier, results on the correlation between DNA methylation at specific sites and shut-off of these genes could indicate a causal relationship between DNA methylation and gene expression, or DNA methylation could be a consequence of the absence of gene expression. To determine more precisely the role that DNA methylation can play in the control of gene expression, we have developed systems in which the effects of *in vitro* methylation at specific sites of certain viral segments on gene expression can be investigated. Some of the results are described in the following section.

In Vitro *Methylation of Cloned Adenoviral Genes and Its Effect on Gene Expression*

In order to distinguish between the possibility of DNA methylation representing the cause or the consequence of gene inactivation, the following experiments were designed. The cloned E2a region of Ad2 DNA was *in vitro* methylated using the HpaII DNA-methyltransferase [from *Hemophilus parainfluenzae* (111)], which modified the 5′-CCGG-3′ sequences at the internal C residues. This prokaryotic DNA-methyltransferase was used because, in the Ad2-transformed hamster cell lines HE2 and HE3, the E2a region in the integrated Ad2 genomes was completely methylated at the 5′-CCGG-3′ sites, and the E2a region was not expressed in these cell lines (see above). Conversely, all 5′-CCGG-3′ sites of the E2a region were unmodified in cell line HE1 in which the E2a region was expressed. When the *in vitro* methylated E2a genes of Ad2 DNA were microinjected into nuclei of *X. laevis* (63,161), the microinjected DNA remained methylated for at least 24 hr, and its transcription into mRNA was completely blocked (104,171,172).

On the other hand, unmethylated DNA that was microinjected independently into nuclei of *X. laevis* oocytes was readily expressed as Ad2-specific RNA (171,172). It could be shown that the late E2a promoter of unmethylated DNA was utilized in *Xenopus* oocytes. Viral RNA was not detectably initiated on the methylated E2a fragment. Unmethylated sea urchin histone genes coinjected with methylated Ad2 genes continued to be expressed in *Xenopus* oocytes (172). Thus the transcriptional block of the E2a region of Ad2 DNA was not artifactual but was somehow caused by methylation at the 5′-CCGG-3′ sites (171,172).

We also methylated the 5′-GGCC-3′ sites in the E2a region of Ad2 DNA, i.e., the inverse sequence, by using the BsuRI DNA-methyltransferase [from *Bacillus subtilis* (68)]. After microinjection into nuclei of *X. laevis*, the BsuRI methylated E2a region continued to be expressed (173). In cell lines HE1, HE2, and HE3, the 5′-GGCC-3′ sites in the E2a regions of integrated Ad2 DNA were not methylated (173). We concluded that DNA methylation had to occur at highly specific sequences in order to be functionally relevant.

Furthermore, DNA methylation at specific sites (5′-CCGG-3′) caused gene inactivation. Similar conclusions were reached by workers investigating other experimental systems (49,114,148,178). In the case of the thymidine kinase gene from herpes simplex virus, methylation of adenine at the EcoRI site led to gene inactivation upon microinjection into hamster cells (178). It cannot be decided at present whether the functionally relevant sites for DNA methylation will be the same for all genes. It is conceivable that some genes do respond to methylation at different sites. Moreover, it will have to be determined which parts of a gene or of its regulatory sequences must be methylated for gene inactivation to ensue.

The cloned E2a region of Ad2 DNA was *in vitro* methylated such that either the 5′ end and the promoter with three 5′-CCGG-3′ sites was modified or the 3′ main fragment of the gene with eleven 5′-CCGG-3′ sites. Either construct combination was microinjected into nuclei of *Xenopus laevis* oocytes, and the expression of the Ad2 E2a gene was monitored. Methylations of the three HpaII sites in the promoter segment eliminate transcription of the viral DNA fragment, whereas methylations at the eleven HpaII sites in the main part of the gene appear to decrease transcription only slightly (104). These results demonstrate that the three 5′-CCGG-3′ sites in the promoter and 5′ end are decisive in the shut-off of the E2a region. It is not yet clear to what extent the other eleven such sites in the gene have a modulating effect on gene expression.

DNA-Methyltransferase Activities in Extracts of Adenovirus-Infected and Uninfected Human Cells

The DNA isolated from highly purified adenovirions does not contain detectable amounts of 5-mC or 6-mA (34,42,67,156–158,176). In contrast, viral DNA integrated into cellular DNA is methylated at specific sites in highly specific patterns (96,98,156,157,170). The question is still unresolved of how adenovirus DNA replicating in the nucleus of infected cells can escape becoming methylated. Possibilities include the following: (a) The activity of DNA-methyltransferase(s) in the host cell could be altered after infection. (b) Host DNA-methyltransferase(s) might be sequestered into cellular chromatin and not have access to free viral DNA replicating in the nucleus. It is also important to recall that free viral DNA replicates at a high rate in permissive cell systems (26). More complex explanations also are possible.

In an attempt to investigate some of these possibilities, we have compared the DNA-methyltransferase activities in uninfected and Ad2-infected HeLa cells. In nuclear extracts of these cell systems, no difference in DNA-methyltransferase activities could be observed (34,94). It could be demonstrated clearly that these nuclear extracts were capable of *de novo* methylating adenovirus DNA, KB DNA, or salmon sperm DNA.

Nuclear extracts of uninfected KB cells and of KB cells at 4, 12, or 36 hr after infection with Ad2 were prepared by treatment with 0.3 M NaCl. The DNA-methyltransferase activity in these extracts was determined using adenovirus DNA or various cellular DNA preparations as methyl-acceptors and (^{3}H)-S-adenosyl-methionine as methyl-donor. DNA was *de novo* methylated. The activity was specific for double-stranded DNA. Activities in extracts from infected or uninfected KB cells were similar. Methyltransferase activity also was found in extracts from uninfected HEK cells (94).

Adenovirion or KB DNA thus methylated *in vitro* was analyzed for the bases or deoxyribonucleotides modified in the reaction. Label was found in 5-methylcytosine. The same result was found with nuclear extracts from primary HEK cells. Experiments have been initiated to fractionate and purify the activity from nuclear extracts.

Herpesviruses

Herpesviruses offer another interesting system for studies on the regulation of gene expression (for review, see ref. 88). They can transform cells to tumor cells, and the viral genome persists in these cells, at least in part, in an episomal, nonintegrated form (107). It is likely that a portion of the viral DNA also becomes integrated into the cellular genome (1). An additional highly interesting aspect is that of latent infections of cells or organisms by herpesviruses (for review, see ref. 151). An interesting model system has been described recently in that a lymphoblastoid T-cell line can be persistently infected with herpes simplex virus type 1 (HSV1) (70). The nonproductive state of these cells can be maintained by concanavalin A (ConA) but reversed with phytohemagglutinin.

Particular interest has been devoted to cell lines transformed with herpesviruses. Some of these lines produce infectious virus; others are nonproducer lines. An inverse relationship has been described between the extents of DNA methylation and the expression of the herpesvirus saimiri genome in lymphoid tumor cell lines (20). In the marmoset lymphoid cell line 1670, at least part of the herpesvirus saimiri DNA molecules persist as covalently closed, circular, episomal DNA. This cell line does not produce infectious virus. Accordingly, analyses using the restriction endonucleases HpaII, MspI, SacII, and SmaI have revealed extensive methylation of the persisting viral genomes. More than 80% of all HpaII sites in viral DNA were methylated in this cell line. Similarly, herpesvirus saimiri DNA also persists as covalently closed, circular DNA in virus-producing lymphoid cell lines. In these producer lines, viral DNA is not detectably methylated, as determined by restriction analyses using the same restriction endonucleases (20).

Similar differences in viral DNA methylation between producer and nonproducer lines are apparent in repetitive viral DNA sequences. These data are in agreement with the inverse correlations between DNA methylation at specific sites and the absence of gene expression as reported in other viral and nonviral eukaryotic systems. Of course, the herpesvirus saimiri system is not yet particularly well suited for studies on the regulation of gene expression, as viral functions have not yet been completely mapped. Moreover, it is unclear in what way the replication of the viral genome might be affected by the absence or presence of 5-mC in the viral genome. In any event, these studies demonstrate that nonintegrated viral DNA also can be methylated.

Recently, the locations of specifically unmethylated 5′-CpG-3′ sites in the herpesvirus saimiri genome have been mapped in herpesvirus saimiri-transformed nonproducer lines or virus-induced tumors (21). In the tumors, no evidence was found for methylation of the 5′ terminal C in the sequence 5′-CCGG-3′ or of the internal C in the sequence 5′-GGCC-3′. The 5′-CpG-3′ sequences were extensively methylated. Of 32 sites, 28 were methylated in the viral DNA in tumor cells. Four sites, however, were practically unmethylated. DNA from the nonvirus-producing tumor cell line that has been kept in continuous culture for 7 years also exhibited four specifically unmethylated sites (21). It is unresolved how these specific patterns of methylation arise. Since little information is available on the functional organization of the herpesvirus saimiri genome, functional correlations of the specific patterns of unmethylated sites cannot be drawn.

The DNA from purified virions of HSV1 is not detectably methylated. The DNA can be methylated by long-term *in vitro* incubation with a DNA-methyltransferase preparation from rat liver (176).

The organization of intracellular Epstein-Barr virus DNA in Epstein-Barr virus-transformed human lymphocytes was analyzed (91). It was found that this viral DNA was increasingly methylated with continuous cell passage. Specific cellular phenotypes could not be assigned to a certain DNA methylation pattern (91).

The model of herpes simplex virus latency in a lymphoblastoid T-cell line (70) has already been mentioned. In cells treated with ConA, the nonproductive state is maintained. Under these conditions, the persisting viral genome is extensively methylated. By phytohemagglutinin treatment, the cells are converted to the producer state. In producing cells, the viral DNA is not methylated or undermethylated (189). It will be interesting to investigate further in what way the state of methylation of persisting viral genomes in latently infected cells is responsible for the state of latency and the repression or expression of the viral genome.

In summary, there are similarities between the adenovirus and herpesvirus systems with respect to correlations between DNA methylation and inactive viral genes. Because the herpesvirus system is highly complex, it is not easy to perform detailed functional analyses.

Retroviruses

Interesting observations on proviral DNA and highly significant functional correlations have been reported in the retrovirus field. The 5-mC content of mouse mammary tumor virus (MMTV)-specific proviral DNA sequences in normal and neoplastic tissues of the mouse was determined (12). Proviral DNA was acquired both in milk-borne infection and by genetic transmission. The MMTV proviral DNA sequences originating from germ line infection are heavily methylated at both HpaII and HhaI sites. Exogenously acquired MMTV proviral DNA, however, is not methylated at these sites, neither in infected nor in transformed cells. In transformed cells, cellular DNA sequences are undermethylated. It has been known for some time that the endogenous proviral MMTV DNA is not extensively expressed, whereas acquired proviral DNA is actively expressed as mRNA (175).

Similar conclusions have been reached for proviral sequences of avian sarcoma virus, which are methylated at the 5′-CCGG-3′ sites in a nonpermissive rat cell line but not in permissive chicken cells (69). In the permissive cells, a portion of the endogenous virus DNA is also methylated. Unintegrated virus DNA carries no detectable 5-mC.

Several investigators have previously demonstrated that the DNA of expressed genes is preferentially sensitive to digestion by DNAse I, possibly due to an altered chromatin structure (54,181). Transcriptionally active ev-3 and transcriptionally inactive ev-1 endogenous retroviral genes in chicken cells differ distinctly in a number of parameters (64). The ev-3 DNA is hypomethylated, highly sensitive to DNAse I, and exhibits nuclease hypersensitive sites in its two long terminal repeats (LTRs). Conversely, ev-1 DNA is strongly methylated and is not sensitive to DNAse I digestion. When the chicken cells are exposed to the analog 5-azacytidine for 24 hr (equivalent to two to three cell generations), the ev-1 DNA becomes undermethylated, acquires at least one DNAse-sensitive site, and is transcriptionally activated (64). These data are consistent with the interpretation that the loss of methyl groups from DNA leads to the transcriptional activation of previously dormant genes, and that DNA methylation is somehow associated with the regulation of gene expression. Similar conclusions have been deduced from other virus systems, in particular from *in vivo* and *in vitro* studies with adenoviruses. 5-Azacytidine induction and a concomitant decrease in the level of DNA methylation was also observed with mouse endogenous C type virus (123).

Mechanisms controlling gene expression were also investigated during animal development, using retrovirus genes as models (154). Substrains of mice have been constructed that carry the Moloney murine leukemia virus (M-MuLV) genome in the germ line. These genomes are in part endogenous (i.e., genetically transmitted) and in part exogenous (i.e., somatically acquired by superinfection). The endogenous, genetically silent genomes are highly methylated in the 5′-GCGC-3′ sequences (HhaI sites). These genomes are not infectious in a DNA transfection assay. The exogenous genome copies are hypomethylated and infectious.

The mov-3 substrain of mice carries the M-MuLV as a Mendelian gene in the germ line. The integrated provirus has been molecularly cloned, and the cloned DNA has been tested for methylation and infectivity (71). The genomic proviral and flanking mouse sequences are extensively methylated and not infectious; the cloned mov-3 locus is not methylated, since it was replicated in a prokaryotic host, and is highly infectious. These data further support the notion that methylation of specific genes is causally related to gene inactivation.

The replication of M-MuLV is blocked in murine embryonal carcinoma cells. After infection of embryonal carcinoma cells, up to 100 genome equivalents are detectable in these cells in the integrated state. The proviral genomes are extensively methylated and refractory to cleavage with the restriction endonuclease SmaI. Integrated M-MuLV DNA in differentiated cells, however, is hypomethylated (152). When DNA from infected embryonal carcinoma cells is transfected into permissive cells, viral expression is not induced. The proviral genomes, however, could be induced by 5-azacytidine treatment. It is concluded (152) that *de novo* methylation activity may be a property of early embryonic cells. It is likely from results reported above (94) that actively replicating cells represent an ample source of *de novo* methylating activity.

When cloned retroviral genomes were microinjected into mouse zygotes, or when preimplantation mouse embryos were infected with M-MuLV, the viral genomes became *de novo* methylated and blocked in expression. *De novo* methylation and inhibition of expression failed to occur in postimplantation mouse embryos (81).

Recently, it was reported (124) that the newly acquired M-MuLV was not methylated and yet was not expressed. Within 15 days after infection, the integrated viral DNA in undifferentiated teratocarcinoma cells became methylated. In undifferentiated cells, 5-azacytidine treatment altered the pattern of DNA methylation but did not activate the virus. Viral activation was possible only in differentiated cells. The authors postulate that two independent mechanisms are operative in the regulation of gene expression, of which only one is dependent on DNA methylation (123,124). Similar conclusions had earlier been reached for the expression of adenoviral genomes in hamster tumor cells (98,99).

Interesting interrelationships between DNA methylation and regulation of gene expression have been reported in MMTV-transformed cells (79,118). In a mouse thymoma cell line, MMTV expression can be induced by short-term treatment with dexamethasone. In this cell line, all the regions containing 5′-LTRs are demethylated; other sites of the proviral genome remain methylated (118). Obviously, the LTRs carry promotor elements of the viral genome, and demethylation of these regions upon induction appears to be essential for expression.

In vitro methylation of specific regions of the cloned Moloney sarcoma virus genome has also been shown to inhibit the transforming activity of this virus (114). In this particular system, it has been determined that *in vitro* methylation of the v-*mos* gene (viral oncogene) is more inhibitory to transformation than methylation of the LTR of the viral genome.

Bacteriophage Mu

There are a few examples implicating specific DNA methylation in the regulation of gene expression in prokaryotic systems (reviewed in ref. 31). A notable example is that of the mom gene of bacteriophage Mu. The product of the mom gene modifies adenosine in the sequence $^{C}_{G}A^{G}_{C}NPy$. The mom gene is not expressed in dam^- mutants of *E. coli*, i.e., cells that are deficient in the adenosine DNA-methyltransferase (75,87,127). The modification of adenine residues is unusual [(N^6-1-acetamido)-adenine] (75), and the mechanism of regulation of the mom gene is highly complex.

CONCLUSIONS

The apparently simple modification of DNA by introducing methylated nucleotides has highly complex functional implications. There is an increasing amount of evidence that the biologic meaning of a modification signal may depend on the position at which the signal has been placed. In the present stage of the analysis of this problem, two shortcomings are apparent: (a) Only a limited number of methylated 5′-CpG-3′ sequences can be precisely localized in genomic DNA and can be correlated with the functional state of certain genes. A method to determine complete patterns of methylation in specific segments of genomic DNA will have to be used. (b) Any experimental approach to the methylation problem, e.g., *in vitro* methylation of cloned DNA fragments or constructions of partly methylated genes followed by microinjection or transfection, is necessarily limited. Functional implications can be drawn, which are limited by the experimental design chosen.

Another general precautionary comment is necessary. Observations and the corresponding interpretations gleaned from one gene or one set of genes cannot be extrapolated to any other set of genes. It is apparent from the work summarized in this chapter that the evidence in favor of the notion that DNA methylation at highly specific promoter sites plays a decisive role in the regulation of eukaryotic gene expression has become convincing when one considers certain viral systems. For other genes or systems, the situation is more problematic. In other genes, evidence is lacking, possibly because the right sites of methylation could not be investigated because of a lack of proper restriction endonucleases. I have attempted to describe DNA methylation as a complex regulatory signal, with its significance depending on the specific genetic location. With this admittedly exploratory interpretation in mind, one would not be surprised if methylated nucleotides could have different, sometimes even opposing functions.

The subject of this chapter has been the influence of DNA methylation on viral gene expression. Viral genes become subject to this regulatory signal mainly when they are inserted into the host chromosome, sometimes when they persist in the episomal state under host control. High levels of DNA methylation at 5′-CCGG-3′, 5′-GCGC-3′, and possibly other sites can be correlated with transcriptional inactivation *in vivo*, i.e., in cell cultures, and in *in vitro* experiments. The occurrence of 5-mC in or close to the promoter sequence of a gene is of particular importance. It is conceivable that only certain promoters are methylation sensitive. Some of the inactive and completely methylated genes investigated, but by no means all genes that have been subjected to that test, can be induced by a specific (?) inhibitor of DNA-methyltransferase, i.e., 5-azacytidine. In the course of or as a prerequisite for this induction, these genes also become demethylated. It is not known whether nonspecific toxic effects of the analog may also play a role in the induction event.

There is evidence that after the original insertion event of viral into chromosomal host DNA, unidentified events taking effect over many cell generations are required for the permanent fixation of foreign DNA. During that time, the final pattern of methylation is established. This pattern is not instituted in a randomized fashion; rather, the establishment follows a distinct pattern (99). Moreover, it has been shown in both viral and nonviral eukaryotic systems that absence of DNA methylation is frequently a necessary but not sufficient precondition for gene expression.

Obviously, other factors are also involved, probably in an intricate interplay with signals, such as methylated nucleotides in DNA.

Most of the *in vitro* methylation experiments and constructions summarized herein implicate DNA methylation at specific sites as the cause, not the consequence, of transcriptional inactivation of certain viral genes. The results do not preclude the possibility that in other genes, introduction of methyl groups could also follow transcriptional inactivation.

DNA methylation as a regulatory function will have to be viewed as a long-term signal employed for genes that must be turned off for a long time. Viral genes in free, i.e., nonintegrated, viral genomes do not become subject to methylation. It is unknown at present how free viral DNA escapes the action of DNA-methyltransferases. Perhaps these enzymes are topographically restricted to distinct nuclear compartments, e.g., chromatin, and thus fail to obtain access to free viral DNA. Obviously, it would be inopportune for parasitic genomes in a eukaryotic cell to surrender to long-term shutoff signals that the host cell has developed for its own regulatory repertoire.

In critically examining recent experimental work on DNA methylation and its value as a regulatory signal in higher eukaryotic systems, one is reminded of Humpty Dumpty objecting to Alice: "Now if you had the two eyes on the same side of the nose, for instance—or the mouth at the top—that would be some help" (8). It indeed would be helpful to recognize distinctive features of the different meanings that the methylation signal can assume. Much information substantiating the regulatory function of methylated nucleotides at specific sites in certain genes has been adduced. More work remains to be done before one can understand the involved interactions required between a methylation signal other factors implicated in gene regulation.

ACKNOWLEDGMENTS

I thank Birgit Kierspel for essential help with the list of references and Petra Böhm for typing the manuscript. Work in my laboratory was supported by the Deutsche Forschungsgemeinschaft through SFB74-C1 and by the Ministry of Science and Research of the State of Nordrhein-Westfalen.

REFERENCES

1. Adams, A., Lindahl, T., and Klein, G. (1973): Linear association between cellular DNA and Epstein-Barr virus DNA in a human lymphoblastoid cell line. *Proc. Natl. Acad. Sci. USA*, 70:2888–2892.
2. Adams, R. L. P., McKay, E. L., Craig, L. M., and Burdon, R. H. (1979): Methylation of mosquito DNA. *Biochim. Biophys. Acta*, 563:72–81.
3. Baker, C. C., and Ziff, E. B. (1981): Promoters and heterogeneous 5′ termini of the messenger RNAs of adenovirus serotype 2. *J. Mol. Biol.*, 149:189–221.
4. Behe, M., and Felsenfeld, G. (1981): Effects of methylation on a synthetic polynucleotide: The B-Z transition in poly(dG-m^5dC)·poly(dG-m^5dC). *Proc. Natl. Acad. Sci. USA*, 78:1619–1623.
5. Behe, M., Zimmerman, S., and Felsenfeld, G. (1981): Changes in the helical repeat of poly(dG-m^5dC)·poly(dG-m^5dC) and poly(dG-dC)·poly(dG-dC) associated with the B-Z transition. *Nature*, 293:233–235.

6. Bird, A. P., Taggart, M. H., and Smith, B. A. (1979): Methylated and unmethylated DNA compartments in the sea urchin genome. *Cell*, 17:889–901.
7. Burdon, R. H., and Adams, R. L. P. (1980): Eukaryotic DNA methylation. *Trends Biochem. Sci.*, 5:294–297.
7a. Busslinger, M., Hurst, J., and Flavell, R. A. (1983): DNA methylation and the regulation of globin gene expression. *Cell*, 34:197–206.
8. Carrol, L. (1977): *Through the Looking Glass*. St. Martin's Press, New York.
8a. Cate, R. L., Chick, W., and Gilbert, W. (1983): Comparison of the methylation patterns of the two rat insulin genes. *J. Biol. Chem.*, 258:6645–6652.
9. Chow, L. T., Broker, T. R., and Lewis, J. B. (1979): Complex splicing patterns of RNAs from the early regions of adenovirus-2. *J. Mol. Biol.*, 134:265–303.
10. Christman, J. K., Price, P., Pedrinan, L., and Acs, G. (1977): Correlation between hypomethylation of DNA and expression of globin genes in Friend erythroleukemia cells. *Eur. J. Biochem.*, 81:53–61.
11. Christman, J. K., Weich, N., Schoenbrun, B., Schneiderman, N., and Acs, G. (1980): Hypomethylation of DNA during differentiation of Friend erythroleukemia cells. *J. Cell Biol.*, 86:366–370.
12. Cohen, J. C. (1980): Methylation of milk-borne and genetically transmitted mouse mammary tumor virus proviral DNA. *Cell*, 19:653–662.
13. Cohen, S. S. (1968): *Virus-Induced Enzymes*. Columbia University Press, New York.
13a. *Cold Spring Harbor Symp. Quant. Biol.* (1982): Volume 47.
14. Conklin, K. F., Coffin, J. M., Robinson, H. L., Groudine, M., and Eisenman, R. (1982): Role of methylation in the induced and spontaneous expression of the avian endogenous virus ev-1: DNA structure and gene products. *Mol. Cell. Biol.*, 2:638–652.
15. Conner, B. N., Takano, T., Tanaka, S., Itakura, K., and Dickerson, R. E. (1982): The molecular structure of d(ICpCpGpG), a fragment of right-handed double helical A-DNA. *Nature*, 295:294–299.
16. Constantinides, P. G., Jones, P. A., and Gevers, W. (1977): Functional striated muscle cells from non-myoblast precursors following 5-azacytidine treatment. *Nature*, 267:364–366.
17. Cook, J. L., and Lewis, A. M., Jr. (1979): Host response to adenovirus 2-transformed hamster embryo cells. *Cancer Res.*, 39:1455–1461.
18. Creusot, F., Acs, G., and Christman, J. K. (1982): Inhibition of DNA methyltransferase and induction of Friend erythroleukemia cell differentiation by 5-azacytidine and 5-aza-2′-deoxycytidine. *J. Biol. Chem.*, 257:2041–2048.
19. De Simone, J., Heller, P., Hall, L., and Zwiers, D. (1982): 5-Azacytidine stimulates fetal hemoglobin synthesis in anemic baboons. *Proc. Natl. Acad. Sci. USA*, 79:4428–4431.
20. Desrosiers, R. C., Mulder, C., and Fleckenstein, B. (1979): Methylation of herpesvirus saimiri DNA in lymphoid tumor cell lines. *Proc. Natl. Acad. Sci. USA*, 76:3839–3843.
21. Desrosiers, R. C. (1982): Specifically unmethylated cytidylic-guanylate sites in herpesvirus saimiri DNA in tumor cells. *J. Virol.*, 43:427–435.
22. Demling, B. (1981): Sequence arrangement of a highly methylated satellite DNA of a plant, Scilla: A tandemly repeated inverted repeat. *Proc. Natl. Acad. Sci. USA*, 78:338–342.
23. Deuring, R., Klotz, G., and Doerfler, W. (1981): An unusual symmetric recombinant between adenovirus type 12 DNA and human cell DNA. *Proc. Natl. Acad. Sci. USA*, 78:3142–3146.
24. Deuring, R., and Doerfler, W. (1983): Proof of recombination between viral and cellular genomes in human KB cells productively infected by adenovirus type 12: Structure of the junction site in a symmetric recombinant (SYREC). *Gene*, 25.
25. Doerfler, W. (1968): The fate of the DNA of adenovirus type 12 in baby hamster kidney cells. *Proc. Natl. Acad. Sci. USA*, 60:636–643.
26. Doerfler, W. (1969): Nonproductive infection of baby hamster kidney cells (BHK21) with adenovirus type 12. *Virology*, 38:587–606.
27. Doerfler, W. (1975): Integration of viral DNA into the host genome. *Curr. Top. Microbiol. Immunol.*, 71:1–78.
28. Doerfler, W. (1977): Animal virus-host genome interactions. In: *Comprehensive Virology*, Vol. 10, edited by H. Fraenkel-Conrat and R. R. Wagner, pp. 279–399. Plenum Publishing Co., New York.
29. Doerfler, W. (1981): DNA methylation—A regulatory signal in eukaryotic gene expression. *J. Gen. Virol.*, 57:1–20.

30. Doerfler, W. (1982): Uptake, fixation and expression of foreign DNA in mammalian cells: The organization of integrated adenovirus DNA sequences. *Curr. Top. Microbiol. Immunol.*, 101:127–194.
31. Doerfler, W. (1983): DNA methylation and gene activity. *Annu. Rev. Biochem.*, 52:93–124.
32. Doerfler, W. (1984): DNA methylation and its functional significance: Studies on the adenovirus system. *Curr. Top. Microbiol. Immunol.*, 108:79–98.
33. Doerfler, W., and Lundholm, U. (1970): Absence of replication of the DNA of adenovirus type 12 in BHK21 cells. *Virology*, 40:754–757.
34. Doerfler, W., Kruczek, I., Eick, D., Vardimon, L., and Kron, B. (1982): DNA methylation and gene activity: The adenovirus system as a model. *Cold Spring Harbor Symp. Quant. Biol.*, 47:593–603.
34a. Doerfler, W., Gahlmann, R., Stabel, S., Deuring, R., Lichtenberg, U., Schulz, M., Eick, D., and Leisten, R. (1983): On the mechanism of recombination between adenoviral and cellular DNAs: The structure of junction sites. In: *The Molecular Biology of Adenoviruses*, edited by W. Doerfler. *Current Topics Microbiol. Immunol.*, 109:193–228.
35. Doskočil, J., and Šormova, Z. (1965): The occurrence of 5-methylcytosine in bacterial deoxyribonucleic acid. *Biochim. Biophys. Acta*, 95:513–514.
36. Drozhdenyuk, A. P., Sulimova, G. E., and Vanyushin, B. F. (1977): Content of 5-methylcytosine in different families of repeating sequences of some higher plant DNA. *Biokhimiya*, 42:1439–1444.
37. Dunn, D. B., and Smith, J. D. (1958): The occurrence of 6-methyl-aminopurine in deoxyribonucleic acid. *Biochem. J.*, 68:627–636.
38. Eastman, E. M., Goodman, R. M., Erlanger, B. F., and Miller, O. J. (1980): 5-Methylcytosine in the DNA of the polytene chromosomes of the diptera sciara coprophila, drosophila melanogaster and D. persimilis. *Chromosoma*, 79:225–239.
39. Ehrlich, M., and Wang, R. Y.-H. (1981): 5-Methylcytosine in eukaryotic DNA. *Science*, 212:1350–1357.
40. Eick, D., Stabel, S., and Doerfler, W. (1980): Revertants of adenovirus type 12-transformed hamster cell line T637 as tools in the analysis of integration patterns. *J. Virol.*, 36:41–49.
41. Eick, D., and Doerfler, W. (1982): Integrated adenovirus type 12 DNA in the transformed hamster cell line T637: Sequence arrangements at the termini of viral DNA and mode of amplification. *J. Virol.*, 42:317–321.
42. Eick, D., Fritz, H.-J., and Doerfler, W. (1983): Quantitative determination of 5-methylcytosine in DNA by reverse phase high-performance liquid chromatography. *Anal. Biochem.*, 135.
43. Erlanger, B. F., and Beiser, S. M. (1964): Antibodies specific for ribonucleosides and ribonucleotides and their reaction with DNA. *Proc. Natl. Acad. Sci. USA*, 52:68–74.
44. Esche, H. (1982): Viral gene products in adenovirus type 2-transformed hamster cells. *J. Virol.*, 41:1076–1082.
45. Esche, H., and Siegmann, B. (1982): Expression of early viral gene products in adenovirus type 12 infected and transformed cells. *J. Gen. Virol.*, 60:99–113.
46. Esche, H., Schilling, R., and Doerfler, W. (1979): In vitro translation of adenovirus type 12-specific mRNA isolated from infected and transformed cells. *J. Virol.*, 30:21–31.
47. Fanning, E., and Doerfler, W. (1976): Intracellular forms of adenovirus DNA. V. Viral DNA sequences in hamster cells abortively infected and transformed with human adenovirus type 12. *J. Virol.*, 20:373–383.
48. Feinberg, A. P., and Vogelstein, B. (1983): Hypomethylation distinguishes genes of some human cancers from their normal counterparts. *Nature*, 301:89–92.
48a. Felsenfeld, G., Nickol, J., McGhee, J., and Behe, M. (1982): Chromatin structure and DNA methylation. *Cold Spring Harbor Symp. Quant. Biol.*, 47:577–584.
48b. Flavell, R. A., Grosveld, F., Busslinger, M., de Boer, E., Kioussis, D., Mellor, A. L., Golden, L., Weiss, E., Hurst, J., Bud, H., Bullmann, H., Simpson, E., James, R., Townsend, A. R. M., Taylor, P. M., Schmidt, W., Ferluga, J., Leben, L., Santamaria, M., Atfield, G., and Festenstein, H. (1982): Structure and expression of the human globin genes and murine histocompatibility antigen genes. *Cold Spring Harbor Symp. Quant. Biol.*, 47:1067–1078.
49. Fradin, A., Manley, J. L., and Prives, C. L. (1982): Methylation of simian virus 40 HpaII site affects late, but not early, viral gene expression. *Proc. Natl. Acad. Sci. USA*, 79:5142–5146.
50. Fritz, H.-J., Belagaje, R., Brown, E. L., Fritz, R. H., Jones, R. A., Lees, R. G., and Khorana,

H. G. (1978): High-pressure liquid chromatography in polynucleotide synthesis. *Biochemistry*, 17:1257–1267.

51. Fritz, H.-J., Eick, D., and Werr, W. (1982): Analysis of synthetic oligodeoxyribonucleotides. In: *Chemical and Enzymatic Synthesis of Gene Fragments*, edited by H. G. Gassen and A. Lang. pp. 199–223. Verlag Chemie, Weinheim.
52. Gahlmann, R., and Doerfler, W. (1983): Integration of viral DNA into the genome of the adenovirus type 2-transformed hamster cell line HES without loss or alteration of cellular nucleotides. *Nucleic Acids Res.*, 11:7347–7361.
53. Gautsch, J. W., and Wilson, M. C. (1983): Delayed *de novo* methylation in teratocarcinoma suggests additional tissue-specific mechanisms for controlling gene expression. *Nature*, 301:32–37.
54. Garel, A., Zolan, M., and Axel, R. (1977): Genes transcribed at diverse rates have a similar conformation in chromatin. *Proc. Natl. Acad. Sci. USA*, 74:4876–4871.
55. Gautier, F., Bünemann, H., and Grotjahn, L. (1977): Analysis of calf-thymus satellite DNA: Evidence for specific methylation of cytosine in C-G sequences. *Eur. J. Biochem.*, 80:175–183.
56. Gerber-Huber, S., May, F. E. B., Westley, B. R., Felber, B. K., Hosbach, H. A., Andres, A.-C., and Ryffel, G. U. (1983): In contrast to other Xenopus genes the estrogen-inducible vitellogenin genes are expressed when totally methylated. *Cell*, 33:43–51.
57. Gjerset, R. A., and Martin, D. W., Jr. (1982): Presence of a DNA demethylating activity in the nucleus of murine erythroleukemia cells. *J. Biol. Chem.*, 257:8581–8583.
58. Gorman, C. M., Moffat, L. F., and Howard, B. H. (1982): Recombinant genomes which express chloramphenicol acetyltransferase in mammalian cells. *Mol. Cell. Biol.*, 2:1044–1051.
59. Grippo, P., Iacarrino, M., Parisi, E., and Scarano, E. (1968): Methylation of DNA in developing sea urchin embryos. *J. Mol. Biol.*, 36:195–208.
60. Groneberg, J., and Doerfler, W. (1979): Revertants of adenovirus type 12-transformed hamster cells have lost part of the viral genomes. *Int. J. Cancer*, 24:67–74.
61. Groneberg, J., Chardonnet, Y., and Doerfler, W. (1977): Integrated viral sequences in adenovirus type 12-transformed hamster cells. *Cell*, 10:101–111.
62. Groneberg, J., Sutter, D., Soboll, H., and Doerfler, W. (1978): Morphological revertants of adenovirus type 12-transformed hamster cells. *J. Gen. Virol.*, 40:635–645.
63. Grosschedl, R., and Birnstiel, M. L. (1980): Identification of regulatory sequences in the prelude sequences of an H2A histone gene by the study of specific deletion mutants in vivo. *Proc. Natl. Acad. Sci. USA*, 77:1432–1436.
64. Groudine, M., Eisenman, R., and Weintraub, H. (1981): Chromatin structure of endogenous retroviral genes and activation by an inhibitor of DNA methylation. *Nature*, 292:311–317.
65. Gruenbaum, Y., Cedar, H., and Razin, A. (1982): Substrate and sequence specificity of an eukaryotic DNA methylase. *Nature*, 295:620–622.
66. Gruenbaum, Y., Naveh-Many, T., Cedar, H., and Razin, A. (1981): Sequence specificity of methylation in higher plant DNA. *Nature*, 292:860–862.
67. Günthert, U., Schweiger, M., Stupp, M., and Doerfler, W. (1976): DNA methylation in adenovirus, adenovirus transformed cells, and host cells. *Proc. Natl. Acad. Sci. USA*, 73:3923–3927.
68. Günthert, U., Jentsch, S., and Freund, M. (1981): Restriction and modification in Bacillus subtilis: Two DNA methyltransferases with BsuRI specificity. II. Catalytic properties, substrate specificity, and mode of action. *J. Biol. Chem.*, 256:9346–9351.
69. Guntaka, R. V., Rao, P. Y., Mitsialis, S. A., and Katz, R. (1980): Modification of avian sarcoma proviral DNA sequences in nonpermissive XC cells but not in permissive chicken cells. *J. Virol.*, 34:569–572.
70. Hammer, S. M., Richter, B. S., and Hirsch, M. S. (1981): Activation and suppression of herpes simplex virus in a human T lymphoid cell line. *J. Immunol.*, 127:144–148.
71. Harbers, K., Schnieke, A., Stuhlmann, H., Jähner, D., and Jaenisch, R. (1981): DNA methylation and gene expression: Endogenous retroviral genome becomes infectious after molecular cloning. *Proc. Natl. Acad. Sci. USA*, 78:7609–7613.
72. Harland, R. M. (1982): Inheritance of DNA methylation in microinjected eggs of Xenopus laevis. *Proc. Natl. Acad. Sci. USA*, 79:2323–2327.
73. Harris, M. (1982): Induction of thymidine kinase in enzyme-deficient Chinese hamster cells. *Cell*, 29:483–492.
74. Hattman, S. (1981): DNA methylation. In: *The Enzymes*, edited by P. D. Boyer, pp. 517–547. Academic Press, New York.

75. Hattman, S. (1982): DNA methyltransferase-dependent transcription of the phage Mu mom gene. *Proc. Natl. Acad. Sci. USA*, 79:5518–5521.
75a. Hearing, P., and Shenk, T. (1983): The adenovirus type 5 E1a transcriptional control region contains a duplicated enhancer element. *Cell*, 33:695–703.
76. Hochschild, A., Irwin, N., and Ptashne, M. (1983): Repressor structure and the mechanism of positive control. *Cell*, 32:319–325.
77. Holliday, R., and Pugh, J. E. (1975): DNA modification mechanisms and gene activity during development. *Science*, 187:226–232.
78. Hotchkiss, R. D. (1958): The quantitative separation of purines, pyrimidines and nucleosides by paper chromatography. *J. Biol. Chem.*, 175:315–332.
79. Hynes, N. E., Rahmsdorf, U., Kennedy, N., Fabiani, L., Michalides, R., Nusse, R., and Groner, B. (1981): Structure, stability, methylation, expression and glucocorticoid induction of endogenous and transfected proviral genes of mouse mammary tumor virus in mouse fibroblasts. *Gene*, 15:307–317.
80. Ibelgaufts, H., Doerfler, W., Scheidtmann, K. H., and Wechsler, W. (1980): Adenovirus type 12-induced rat tumor cells of neuroepithelial origin: Persistence and expression of the viral genome. *J. Virol.*, 33:423–437.
81. Jähner, D., Stuhlmann, H., Stewart, C. L., Harbers, K., Löhler, J., Simon, I., and Jaenisch, R. (1983): *De novo* methylation and expression of retroviral genomes during mouse embryogenesis. *Nature*, 298:623–628.
82. Johansson, K., Persson, H., Lewis, A. M., Pettersson, U., Tibbetts, C., and Philipson, L. (1978): Viral DNA sequences and gene products in hamster cells transformed by adenovirus type 2. *J. Virol.*, 27:628–639.
83. Jones, P. A., and Taylor, S. M. (1981): Hemimethylated duplex DNAs prepared from 5-azacytidine-treated cells. *Nucleic Acids Res.*, 9:2933–2947.
84. Jones, P. A., Taylor, S. M., Mohandas, T., and Shapiro, L. J. (1982): Cell cycle-specific reactivation of an inactive X-chromosome locus by 5-azadeoxycytidine. *Proc. Natl. Acad. Sci. USA*, 79:1215–1219.
85. Josse, J., and Kornberg, A. (1962): Glucosylation of deoxyribonucleic acid. III. α- and β-glucosyl transferases from T4-infected Escherichia coli. *J. Biol. Chem.*, 237:1968–1976.
86. Jovin, T. M., van de Sande, J. H., Zarling, D. A., Arndt-Jovin, D. J., Eckstein, F., Füldner, H. H., Greicer, C., Grieger, I., Hamori, E., Kalisch, B., McIntosh, L. P., and Robert-Nicoud, M. (1982): Generation of left-handed Z DNA in solution and visualization in polytene chromosomes by immunofluorescence. *Cold Spring Harbor Symp. Quant. Biol.*, 47:639–646.
87. Kahmann, R. (1982): Methylation regulates the expression of a DNA-modification function encoded by bacteriophage Mu. *Cold Spring Harbor Symp. Quant. Biol.*, 47:143–154.
88. Kaplan, A. S. (editor) (1973): *The Herpesviruses*. Academic Press, New York.
89. Kastan, M. B., Gowans, B. J., and Lieberman, M. W. (1982): Methylation of deoxycytidine incorporated by excision-repair synthesis of DNA. *Cell*, 30:509–516.
90. Kaye, A. M., and Winocour, E. (1967): On the 5-methyl-cytosine found in the DNA extracted from polyomavirus. *J. Mol. Biol.*, 24:475–478.
91. Kintner, C., and Sugden, B. (1981): Conservation and progressive methylation of Epstein-Barr viral DNA sequences in transformed cells. *J. Virol.*, 38:305–316.
92. Klein, G. (editor) (1982): *Advances in Viral Oncology, Vol. 1: Oncogene Studies*. Raven Press, New York.
93. Klysik, J., Stirdivant, S. M., Larson, J. E., Hart, P. A., and Wells, R. D. (1981): Left-handed DNA in restriction fragments and a recombinant plasmid. *Nature*, 290:672–677.
94. Knust-Kron, B., Eick, D., and Doerfler, W. (1984): *Unpublished observations*.
95. Korba, B. E., and Hays, J. B. (1982): Partially deficient methylation of cytosine in DNA at $CC^{A}_{T}GG$ sites stimulates genetic recombination of bacteriophage lambda. *Cell*, 28:531–541.
96. Kruczek, I., and Doerfler, W. (1982): The unmethylated state of the promoter/leader and 5′-regions of integrated adenovirus genes correlates with gene expression. *EMBO J.*, 1:409–414.
97. Kruczek, I., and Doerfler, W. (1983): Expression of the chloramphenicol acetyltransferase gene in mammalian cells under the control of adenovirus type 12 promoters: Effect of promoter methylation on gene expression. *Proc. Natl. Acad. Sci. USA*, 80.
98. Kuhlmann, I., and Doerfler, W. (1982): Shifts in the extent and patterns of DNA methylation upon explantation and subcultivation of adenovirus type 12-induced hamster tumor cells. *Virology*, 118:169–180.

99. Kuhlmann, I., and Doerfler, W. (1983): Loss of viral genomes from hamster tumor cells and nonrandom alterations in the patterns of methylation of integrated adenovirus type 12 DNA. *J. Virol.*, 47:631–636.
100. Kuhlmann, I., Achten, S., Rudolph, R., and Doerfler, W. (1982): Tumor induction by human adenovirus type 12 in hamsters: Loss of the viral genome from adenovirus type 12-induced tumor cells is compatible with tumor formation. *EMBO J.*, 1:79–86.
101. Kuo, K. C., McCune, R. A., and Gehrke, C. W. (1980): Quantitative reversed-phase high performance liquid chromatographic determination of major and modified deoxyribonucleosides in DNA. *Nucleic Acids Res.*, 8:4763–4776.
102. Kurnick, N. B., and Herskowitz, I. H. (1952): The estimation of polyteny in drosophila salivary gland nuclei based on determination of deoxyribonucleic acid content. *J. Cell Comp. Physiol.*, 39:281–299.
103. Lacks, S., and Greenberg, B. (1975): A deoxyribonuclease of Diplococcus pneumoniae specific for methylated DNA. *J. Biol. Chem.*, 250:4060–4066.
104. Langner, K.-D., Vardimon, L., Renz, D., and Doerfler, W. (1984): DNA methylations of three 5′-CCGG-3′ sites in the promoter and 5′ regions inactivate the E2a gene of adenovirus type 2. *(Submitted)*.
105. Lapeyre, J. N., and Becker, F. F. (1979): 5-Methylcytosine content of nuclear DNA during chemical hepatocarcinogenesis and in carcinomas which result. *Biochem. Biophys. Res. Commun.*, 87:698–705.
105a. La Volpe, A., Taggart, M., Macleod, D., and Bird, A. (1982): Coupled demethylation of sites in a conserved sequence of Xenopus ribosomal DNA. *Cold Spring Harbor Symp. Quant. Biol.*, 47:585–592.
106. Ley, T. J., De Simone, J., Anagnou, N. P., Keller, G. H., Humphries, R. K., Turner, P. H., Young, N. S., Heller, P., and Nienhuis, A. W. (1982): 5-Azacytidine selectively increases γ-globin synthesis in a patient with β^+ thalassemia. *N. Engl. J. Med.*, 307:1469–1475.
107. Lindahl, T., Adams, A., Bjursell, G., Bornkamm, G. W., Kaschka-Dierich, C., and Jehn, U. (1976): Covalently closed circular duplex DNA of Epstein-Barr virus in human lymphoid cell line. *J. Mol. Biol.*, 102:511–530.
108. Lubit, B. W., Pham, T. D., Miller, O. J., and Erlanger, B. F. (1976): Localization of 5-methylcytosine in human metaphase chromosomes by immunoelectron microscopy. *Cell*, 9:503–509.
109. Mandel, J. L., and Chambon, P. (1979): DNA methylation: Organ specific variations in the methylation pattern within and around ovalbumin and other chicken genes. *Nucleic Acids Res.*, 7:2081–2103.
110. Manes, C., and Menzel, P. (1981): Demethylation of CpG sites in DNA of early rabbit trophoblast. *Nature*, 293:589–590.
111. Mann, M. B., and Smith, H. O. (1977): Specificity of HpaII and HaeIII methylases. *Nucleic Acids Res.*, 4:4211–4221.
112. Mathews, C. K., Brown, F., and Cohen, C. S. (1964): Virus-induced acquisition of metabolic function. VII. Biosynthesis de novo of deoxycytidylate hydroxymethylase. *J. Biol. Chem.*, 239:2957–2963.
113. Maxam, A. M., and Gilbert, W. (1977): A new method for sequencing DNA. *Proc. Natl. Acad. Sci. USA*, 74:560–564.
114. McGeady, M. L., Jhappan, C., Ascione, R., and van de Woude, G. F. (1983): In vitro methylation of specific regions of the cloned Moloney sarcoma virus genome inhibits its transforming activity. *Mol. Cell. Biol.*, 3:305–314.
115. McGhee, J. D., and Ginder, G. D. (1979): Specific DNA methylation sites in the vicinity of the chicken β-globin genes. *Nature*, 280:419–420.
116. McGhee, J. D., and Felsenfeld, G. (1980): Nucleosome structure. *Annu. Rev. Biochem.*, 49:1115–1156.
117. McKeon, C., Ohkubo, H., Pastan, I., and de Crombrugghe, B. (1982): Unusual methylation patterns of the α2(I) collagen gene. *Cell*, 29:203–210.
118. Mermod, J.-J., Bourgeois, S., Defer, N., and Crépin, M. (1983): Demethylation and expression of murine mammary tumor proviruses in mouse thymoma cell lines. *Proc. Natl. Acad. Sci. USA*, 80:110–114.
119. Möller, A., Nordheim, A., Nichols, S. R., and Rich, A. (1981): 7-Methylguanine in poly(dG-dC)·poly(dG-dC) facilitates Z-DNA formation. *Proc. Natl. Acad. Sci. USA*, 78:4777–4781.
120. Morrison, J. M., Keir, H. M., Subak-Sharpe, H., and Crawford, L. V. (1967): Nearest neighbour

base sequence analysis of the deoxyribonucleic acids of a further three mammalian viruses: Simian virus 40, human papilloma virus and adenovirus type 2. *J. Gen. Virol.*, 1:101–108.

121. Naveh-Many, T., and Cedar, H. (1982): Topographical distribution of 5-methylcytosine in animal and plant DNA. *Mol. Cell. Biol.*, 2:758–762.
122. Naveh-Many, T., and Cedar, H. (1981): Active gene sequences are undermethylated. *Proc. Natl. Acad. Sci. USA*, 78:4246–4250.
123. Niwa, O., and Sugahara, T. (1981): 5-Azacytidine induction of mouse endogenous type C virus and suppression of methylation. *Proc. Natl. Acad. Sci. USA*, 78:6290–6294.
124. Niwa, O., Yokota, Y., Ishida, H., and Sugahara, T. (1983): Independent mechanisms involved in suppression of the Moloney leukemia virus genome during differentiation of murine teratocarcinoma cells. *Cell*, 32:1105–1113.

124a. Ohmori, H., Tomizawa, J. I., and Maxam, A. M. (1978): Detection of 5-methylcytosine in DNA sequences. *Nucleic Acids Res.*, 5:1479–1485.

125. Ortin, J., Scheidtmann, K.-H., Greenberg, R., Westphal, M., and Doerfler, W. (1976): Transcription of the genome of adenovirus type 12. III. Maps of stable RNA from productively infected human cells and abortively infected and transformed hamster cells. *J. Virol.*, 20:355–372.
126. Ott, M.-O., Sperling, L., Cassio, D., Levilliers, J., Sala-Trepat, J., and Weiss, M. C. (1982): Undermethylation at the 5′ end of the albumin gene is necessary but not sufficient for albumin production by rat hepatoma cells in culture. *Cell*, 30:825–833.
127. Plasterk, R. H. A., Vrieling, H., and van de Putte, P. (1983): Transcription initiation of Mu mom depends on methylation of the promoter region and a phage-coded transactivator. *Nature*, 301:344–347.
128. Pohl, F. M., and Jovin, T. M. (1972): Salt-induced cooperative conformational change of a synthetic DNA: Equilibrium and kinetic studies with poly(dG-dC). *J. Mol. Biol.*, 67:375–396.
129. Razin, A., and Cedar, H. (1977): Distribution of 5-methylcytosine in chromatin. *Proc. Natl. Acad. Sci. USA*, 74:2725–2728.
130. Razin, A., and Sedat, J. (1977): Analysis of 5-methylcytosine in DNA. *Anal. Biochem.*, 77:370–377.
131. Razin, A., and Riggs, A. D. (1980): DNA methylation and gene function. *Science*, 210:604–610.
132. Razin, A., and Friedman, J. (1981): DNA methylation and its possible biological roles. *Prog. Nucleic Acids Res. Mol. Biol.*, 25:33–52.
133. Riggs, A. D. (1975): X inactivation, differentiation and DNA methylation. *Cytogenet. Cell Genet.*, 14:9–25.
134. Rubery, E. D., and Newton, A. A. (1973): DNA methylation in normal and tumor virus-transformed cells in tissue culture. I. The level of DNA methylation in BHK21 cells and in BHK21 cells transformed by polyoma virus (Py Y Cells). *Biochim. Biophys. Acta*, 324:24–36.
135. Sager, R., and Kitchin, R. (1975): Selective silencing of eukaryotic DNA. *Science*, 189:426–433.
136. Sanders Haigh, L., Blanchard Owens, B., Hellewell, S., and Ingram, V. M. (1982): DNA methylation in chicken α-globin gene expression. *Proc. Natl. Acad. Sci. USA*, 79:5332–5336.
137. Sano, H., Royer, H. D., and Sager, R. (1980): Identification of 5-methylcytosine in DNA fragments immobilized on nitrocellulose paper. *Proc. Natl. Acad. Sci. USA*, 77:3581–3585.
138. Schirm, S., and Doerfler, W. (1981): Expression of viral DNA in adenovirus type 12-transformed cells, in tumor cells, and in revertants. *J. Virol.*, 39:694–702.
139. Shapiro, H. S., and Chargaff, E. (1960): Studies on the nucleotide arrangement in DNA. VI. Patterns of nucleotide sequence in the DNA of rye germ and its fractions. *Biochim. Biophys. Acta*, 39:68–82.
140. Shmookler Reis, R. J., and Goldstein, S. (1982): Variability of DNA methylation patterns during serial passage of human diploid fibroblasts. *Proc. Natl. Acad. Sci. USA*, 79:3949–3953.
141. Singer, J., Roberts-Ems, J., Luthardt, F. W., and Riggs, A. D. (1979): Methylation of DNA in mouse early embryos, teratocarcinoma cells and adult tissues of mouse and rabbit. *Nucleic Acids Res.*, 7:2369–2385.
142. Sinsheimer, R. (1955): The action of pancreatic deoxyribonuclease. II. Isomeric dinucleotides. *J. Biol. Chem.*, 215:579–583.
143. Smith, S. S., and Thomas, C. A., Jr. (1981): Two dimensional restriction analysis of Drosophila DNAs: Males and females. *Gene*, 13:395–408.
144. Smith, S. S., Reilly, J. G., and Thomas, C. A., Jr. (1981): High resolution two-dimensional restriction analysis of methylation in complex genomes. *ICN-UCLA Symp. Mol. Cell. Biol.*, edited by D. Brown, and C. F. Fox, pp. 635–645.

144a. Smith, S. S., Yu, J. C., and Chen, C. W. (1982): Different levels of DNA modification at 5′-CCGG in murine erythroleukemia cells and the tissues of normal mouse spleen. *Nucleic Acids Res.*, 10:4305–4320.
145. Sneider, T. W. (1980): The 5′cytosine in CCGG is methylated in two eukaryotic DNAs and MspI is sensitive to methylation at this site. *Nucleic Acids Res.*, 8:3829–3840.
146. Southern, E. M. (1975): Detection of specific sequences among DNA fragments separated by gel electrophoresis. *J. Mol. Biol.*, 98:503–517.
147. Stabel, S., Doerfler, W., and Friis, R. R. (1980): Integration sites of adenovirus type 12 DNA in transformed hamster cells and hamster tumor cells. *J. Virol.*, 36:22–40.
148. Stein, R., Razin, A., and Cedar, H. (1982): In vitro methylation of the hamster adenine phosphoribosyltransferase gene inhibits its expression in mouse L cells. *Proc. Natl. Acad. Sci. USA*, 79:3418–3422.
149. Stein, R., Gruenbaum, Y., Pollak, Y., Razin, A., and Cedar, H. (1982): Clonal inheritance of the pattern of DNA methylation in mouse cells. *Proc. Natl. Acad. Sci. USA*, 79:61–65.
150. Stein, R., Sciaki-Gallili, N., Razin, A., and Cedar, H. (1983): Pattern of methylation of two genes coding for housekeeping functions. *Proc. Natl. Acad. Sci. USA*, 80:2422–2426.
151. Stevens, J. G. (1975): Latent herpes simplex virus and the nervous system. *Curr. Top. Microbiol. Immunol.*, 70:31–50.
152. Stewart, C. L., Stuhlmann, H., Jähner, D., and Jaenisch, R. (1982): *De novo* methylation, expression and infectivity of retroviral genomes introduced into embryonal carcinoma cells. *Proc. Natl. Acad. Sci. USA*, 79:4098–4102.
153. Streeck, R. E. (1980): Single-strand and double-strand cleavage at half-modified and fully modified recognition sites for the restriction nucleases Sau 3A and TaqI. *Gene*, 12:267–275.
154. Stuhlmann, H., Jähner, D., and Jaenisch, R. (1981): Infectivity and methylation of retroviral genomes is correlated with expression in the animal. *Cell*, 26:221–232.
155. Subak-Sharpe, H., Burk, R. R., Crawford, L. V., Morrison, J. M., Hay, J., and Keir, H. M. (1966): An approach to evolutionary relationships of mammalian DNA viruses through analysis of the pattern of nearest neighbour base sequences. *Cold Spring Harbor Symp. Quant. Biol.*, 31:737–748.
156. Sutter, D., and Doerfler, W. (1979): Methylation of integrated viral DNA sequences in hamster cells transformed by adenovirus 12. *Cold Spring Harbor Symp. Quant. Biol.*, 44:565–568.
157. Sutter, D., and Doerfler, W. (1980): Methylation of integrated adenovirus type 12 DNA sequences in transformed cells is inversely correlated with viral gene expression. *Proc. Natl. Acad. Sci. USA*, 77:253–256.
158. Sutter, D., Westphal, M., and Doerfler, W. (1978): Patterns of integration of viral DNA sequences in the genomes of adenovirus type 12-transformed hamster cells. *Cell*, 14:569–585.
159. Taylor, J. H. (editor) (1979): Enzymatic methylation of DNA: Pattern and possible regulatory roles. Chromosome structure. In: *Molecular Genetics*, part 3, pp. 89–115. Academic Press, New York.
160. Taylor, S. M., and Jones, P. A. (1979): Multiple new phenotypes induced in 10T1/2 and 3T3 cells treated with 5-azacytidine. *Cell*, 17:771–779.
161. Telford, J. L., Kressmann, A., Koski, R. A., Grosschedl, R., Müller, F., Clarkson, S. G., and Birnstiel, M. L. (1979): Delimitation of a promoter for RNA polymerase II by means of a functional test. *Proc. Natl. Acad. Sci. USA*, 76:2590–2594.
162. Thomas, A. J., and Sherratt, H. S. A. (1956): The isolation of nucleic acid fractions from plant leaves and their purine and pyrimidine composition. *Biochem. J.*, 62:1–4.
163. Tjia, S., Carstens, E. B., and Doerfler, W. (1979): Infection of Spodoptera frugiperda cells with Autographa californica nuclear polyhedrosis virus. II. The viral DNA and the kinetics of its replication. *Virology*, 99:399–409.
164. Van der Ploeg, L. H. T., and Flavell, R. A. (1980): DNA methylation in the human γδβ-globin locus in erythroid and non erythroid tissues. *Cell*, 19:947–958.
165. Van der Ploeg, L. H. T., Groffen, J., and Flavell, R. A. (1980): A novel type of secondary modification of two CCGG residues in human β-globin gene locus. *Nucleic Acids Res.*, 8:4563–4574.
166. Van der Vliet, P. C., and Levine, A. J. (1973): DNA-binding proteins specific for cells infected by adenovirus. *Nature*, 246:170–174.
167. Vanyushin, V. L., Belozersky, A. N., Kokurina, N. A., and Kadirova, D. X. (1968): 5-Methylcytosine and 6-methylaminopurine in bacterial DNA. *Nature*, 218:1066–1067.

168. Vanyushin, B. F., Tkacheva, S. G., and Belozersky, A. N. (1970): Rare bases in animal DNA. *Nature*, 225:948–949.
169. Vardimon, L., and Doerfler, W. (1981): Patterns of integration of viral DNA in adenovirus type 2-transformed hamster cells. *J. Mol. Biol.*, 147:227–246.
170. Vardimon, L., Neumann, R., Kuhlmann, I., Sutter, D., and Doerfler, W. (1980): DNA methylation and viral gene expression in adenovirus-transformed and -infected cells. *Nucleic Acids Res.*, 8:2461–2473.
171. Vardimon, L., Kuhlmann, I., Cedar, H., and Doerfler, W. (1981): Methylation of adenovirus genes in transformed cells and in vitro: Influence on the regulation of gene expression. *Eur. J. Cell Biol.*, 25:13–15.
172. Vardimon, L., Kressmann, A., Cedar, H., Maechler, M., and Doerfler, W. (1982): Expression of a cloned adenovirus gene is inhibited by in vitro methylation. *Proc. Natl. Acad. Sci. USA*, 79:1073–1077.
173. Vardimon, L., Günthert, U., and Doerfler, W. (1982): In vitro methylation of the BsuRI (5′-GGCC-3′) sites in the E2a region of adenovirus type 2 DNA does not affect expression in Xenopus laevis oocytes. *Mol. Cell. Biol.*, 2:1574–1580.
174. Vardimon, L., Renz, D., and Doerfler, W. (1983): Can DNA methylation regulate gene expression? *Recent Results Cancer Res.*, 84:90–102.
175. Varmus, H. E., Quintrell, N., Medeiros, E., Bishop, J. M., Nowinski, R. C., and Sarkars, N. H. (1973): Transcription of mouse mammary tumor virus genes in tissues from high and low tumor incidence mouse strains. *J. Mol. Biol.*, 79:663–679.
176. Von Acken, U., Simon, D., Grunert, F., Döring, H. P., and Kröger, H. (1979): Methylation of viral DNA in vivo and in vitro. *Virology*, 99:152–157.
177. Waalwijk, C., and Flavell, R. A. (1978): MspI, an isoschizomer of HpaII which cleaves both unmethylated and methylated HpaII sites. *Nucleic Acids Res.*, 5:3231–3236.
178. Waechter, D. E., and Baserga, R. (1982): Effect of methylation on expression of microinjected genes. *Proc. Natl. Acad. Sci. USA*, 79:1106–1110.
179. Wagner, R., Jr., and Meselson, M. (1976): Repair tracts in mismatched DNA heteroduplexes. *Proc. Natl. Acad. Sci. USA*, 73:4135–4139.
180. Wang, A. H.-J., Quigley, G. J., Kolpak, F. J., Crawford, J. L., van Boom, J. H., van der Marel, G., and Rich, A. (1979): Molecular structure of a left-handed double helical DNA fragment at atomic resolution. *Nature*, 282:680–686.
181. Weintraub, H., and Groudine, M. (1976): Chromosomal subunits in active genes have an altered conformation. *Science*, 193:848–856.
182. Weintraub, H., Larsen, A., and Groudine, M. (1981): α-Globin gene switching during the development of chicken embryos: Expression and chromosome structure. *Cell*, 24:333–344.
183. Wigler, M. H. (1981): The inheritance of methylation patterns in vertebrates. *Cell*, 24:285–286.
184. Wigler, M., Levy, D., and Perucho, M. (1981): The somatic replication of DNA methylation. *Cell*, 24:33–40.
185. Wilks, A. F., Cozens, P. J., Mattaj, I. W., and Jost, J.-P. (1982): Estrogen induces a demethylation at the 5′ end region of the chicken vitellogenin gene. *Proc. Natl. Acad. Sci. USA*, 79:4252–4255.
186. Willis, D. B., and Granoff, A. (1980): Frog virus 3 DNA is heavily methylated at CpG sequences. *Virology*, 107:250–257.
187. Wilson, V. L., and Jones, P. A. (1983): Inhibition of DNA methylation by chemical carcinogens in vitro. *Cell*, 32:239–246.
188. Wyatt, G. R. (1951): Recognition and estimation of 5-methylcytosine in nucleic acids. *Biochem. J.*, 48:581–584.
189. Youssoufian, H., Hammer, S. M., Hirsch, M. S., and Mulder, C. (1982): Methylation of the viral genome in an *in vitro* model of herpes simplex virus latency. *Proc. Natl. Acad. Sci. USA*, 79:2207–2210.

Advances in Viral Oncology, Volume 4,
edited by George Klein. Raven Press,
New York © 1984.

Malignant Transformation by Rous Sarcoma Virus: From Phosphorylation to Phenotype

Michael J. Weber

Department of Microbiology, University of Illinois, Urbana, Illinois 61801

When chicken embryo fibroblasts become transformed by Rous sarcoma virus (RSV), they undergo numerous alterations, some of which are summarized in Table 1. These are termed, collectively, "the transformed phenotype." Transformation by RSV results in the appearance of approximately 1,000 new RNA sequences (51) and a quantitative or qualitative alteration in as many as 4% of the proteins displayed on a two-dimensional gel (17). Therefore, it is probable that, complex as the "transformed phenotype" may appear to be, the known functional alterations in cellular phenotype constitute only a tiny fraction of the total virus-induced changes in cellular properties.

Both the initiation and maintenance of all the known manifestations of transformation are dependent on the expression of the *src* gene (113). The only protein known to be coded by this gene is $pp60^{src}$ (15,91), a tyrosine-specific protein kinase (11,26,27,40,48,56,57,69,70,74). Other proteins may be found to be coded for by *src* (73), and $pp60^{src}$ may be found to have activities in addition to its tyrosine kinase activity (11,50). At present, however, the problem of determining the mechanism by which $pp60^{src}$ causes the diverse manifestations of transformation is focused on the analysis of proteins that become phosphorylated on tyrosine in RSV-transformed cells.

Other *onc* genes in addition to *src* code for tyrosine-specific protein kinases, including the transforming proteins of several avian sarcoma viruses (Fujinami

TABLE 1. *The transformed phenotype*

Rounded morphology
Disruption of microfilament bundles
Decreased extracellular matrix (e.g., fibronectin and collagen)
Decreased adhesiveness
Increased production of plasminogen activator
Increased hexose transport
Decreased membrane microviscosity and fatty acid unsaturation
Loss of density-dependent growth inhibition
Decreased requirement for mitogenic hormones
Acquisition of anchorage-independent growth
Tumorigenicity

sarcoma virus, PRCII, Y73, Esh sarcoma virus, and UR2), the Abelson murine leukemia virus, and the Snyder-Theilen and Gardner-Arnstein strains of feline sarcoma virus (4,7,30,41,47,53,77,83,85,95,102,114,120). In addition, the cellular receptors for the mitogenic polypeptide hormones epidermal growth factor (EGF), platelet-derived growth factor (PDGF), transforming growth factors (TGF), and insulin all have associated tyrosine protein kinase activity (28,32,38,55, 62,63,76,79,86,87,94,111). The focus of this chapter is on RSV, since this is the best understood oncogenic agent with respect to both the molecular and cellular biology of malignant transformation. The purpose of this chapter is to review current approaches to investigating the mechanism by which $pp60^{src}$ phosphorylation of cellular proteins can induce the transformed phenotype.

THE PATHWAY FROM $pp60^{src}$ TO THE TRANSFORMED PHENOTYPE

Three possible models for the interaction of $pp60^{src}$ with cellular targets are outlined in Fig. 1. The models are not mutually exclusive but rather present different modes of behavior for $pp60^{src}$. The weight of the evidence favors the last model (#3). In this discussion, a distinction is made between "targets" and "substrates" for $pp60^{src}$. The former term has a deliberately teleologic connotation. "Targets" are the cellular proteins that, when acted on by $pp60^{src}$, cause measurable alterations in cellular growth or metabolism (i.e., the transformed phenotype). "Substrates," on the other hand, are cellular proteins that can be acted on by $pp60^{src}$ but may or may not play a role in transformation.

The first model in Fig. 1 indicates that the transformed phenotype could arise from a linear sequence of molecular events triggered by interaction of $pp60^{src}$ with a single target. Evidence indicating that various parameters of transformation can

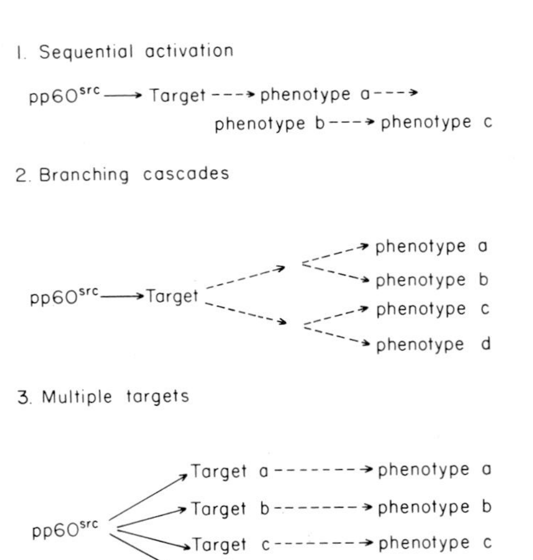

FIG. 1. Models of transformation by RSV.

be dissociated from each other makes this hypothesis untenable (9,10,18, 52,110,115,117). For example, increased hexose transport can be induced by RSV in the absence of detectable plasminogen activator activity (116), and increased plasminogen activator activity can be induced by RSV in cells whose hexose uptake is limited to a near normal rate (D. W. Salter, B. Webb, and M. J. Weber, *unpublished data*). Neither increased hexose transport, nor plasminogen activator, nor gross cytoskeletal alterations are necessary for the loss of growth control; nor is the loss of growth control necessary for the appearance of any of these parameters of transformation (2,66,115,122). It is not possible to arrange these observations in a way that is consistent with model 1. This is not to say that all the manifestations of transformation are expressed independently of the expression of other transformation phenotypes. For example, cyclic AMP analogs can partially lower the hexose transport rate and plasminogen activator levels of RSV-transformed cells (117; K. D. Nakamura and M. J. Weber, *unpublished observations*), and addition of fibronectin or inhibitors of fibrinolysis to transformed cultures can have a modest effect in normalizing the cellular morphology (2,115,122). Nonetheless, it is clear that the pathway leading from pp60^{src} to the transformed phenotype must have some branch-points.

If a biochemical cascade or pathway has branch-points, there must be some proteins in it which are pleiotropic. Models 2 and 3 in Fig. 1 differ only in that pp60^{src} is designated as a pleiotropic protein in model 3.

Partial Transformation Mutants of RSV

Strong evidence that pp60^{src} has more than one physiologically significant primary target (i.e., that model 3 is correct) comes from analysis of various *src* mutants of RSV which induce only a partially transformed phenotype. Friis and his associates (6,44) were the first to point out that the occurrence of these partially transforming virus mutants could provide evidence for a multitarget model of transformation, the partially transforming characteristics of these mutants being a consequence of their ability to alter some targets better than others.

These workers (6,44) isolated eight temperature-conditional transformation mutants of RSV. One (tsGI201) induced only a small increase in plasminogen activator activity or loss of cell-surface fibronectin at the permissive temperature (36° C) and did not cause a fully rounded morphology. Nonetheless, cells infected with this mutant were fully transformed with respect to loss of growth control. TsGI202, 203, 204, and 205 were temperature-conditional with respect to all the measured transformation parameters, and thus were not "partially transforming" in the sense used here. TsGI251, 252, and 253 were all capable of inducing transformed growth properties at the high temperature (42° C) but did not induce fully transformed levels of plasminogen activator or glucose transport and did not cause cells to lose all fibronectin, become agglutinable by concanavalin A, or assume a round morphology at this temperature.

These data are consistent with the notion that $pp60^{src}$ has more than one primary target. According to this hypothesis, the failure of certain virus mutants to induce particular manifestations of transformation would be attributable to a failure of the mutant $pp60^{src}$ to phosphorylate certain targets. However, an alternative hypothesis was that these mutants differed only quantitatively from the wild type virus (i.e., they were simply "leaky"), and the partially transformed phenotype resulted from different transformation parameters differing in their sensitivity to $pp60^{src}$ activity or differing in the sensitivity and accuracy of the assays used to detect them. For example, most of the data in the original paper from the Friis group (6) could be explained by hypothesizing that the assays for altered growth properties were more sensitive than assays for alterations in biochemical properties. Thus cells could appear to be transformed with respect to their growth properties even though they were apparently near normal with respect to biochemical measures of transformation, such as hexose transport and plasminogen activator.

To test this alternative hypothesis, Weber and Friis (116) compared the biologic properties of cells infected with tsGI201 and tsGI251 to those of cells infected with various *src* mutants, which were temperature-conditional for all the measured parameters of transformation when the infected cells were cultured at various temperatures between the permissive and restrictive temperatures. The idea behind this experimental plan was to utilize mutants that were coordinately temperature-conditional for all the measured parameters of transformation and make them "leak" to various degrees by holding them at intermediate temperatures. Thus a "hierarchy" of degrees of transformation could be established, and the phenotypic properties of intermediate (i.e., leaky) mutants could be determined. It then would be possible to determine whether the properties of the putative partial transformation mutants could be mimicked by other mutants held at the intermediate temperatures. Weber and Friis (117) found that the properties of cells infected with tsGI201 and held at 36° C could not be distinguished from those cells infected with fully temperature-conditional mutants, such as tsGI202 and tsNY68, which were held at temperatures intermediate between 36 and 42° C. However, the properties of cells infected with tsGI251 could not be mimicked by making any of the other mutants leak: cells infected with tsGI251 were thermosensitive for most of the biochemical parameters of transformation but were cold-sensitive for their ability to grow in soft agar or in low serum. These findings were taken as evidence that $pp60^{src}$ has at least two physiologically significant targets.

Anderson et al. (3) have also isolated and characterized partial transformation mutants of RSV. The biologic properties of these CU mutants are summarized in Table 2. It is clear that these mutants fulfill the criteria for partial (as opposed to leaky) mutants, since the transformation parameters they induce do not fit into a consistent hierarchy of degrees of transformation. For example, cells infected with CU2 have little surface fibronectin but do not grow well in soft agar. On the other hand, cells infected with CU12 retain much of their fibronectin but grow well in soft agar.

TABLE 2. *Dissociation of transformation parameters*

Parameter expression	CU2	tsCU11 (36° C)	CU12
Transformed	Fibronectin Density inhibition	Fibronectin Anchorage dependence Casein plaques Plasminogen activator	Anchorage dependence Density inhibition Casein plaques Plasminogen activator
Intermediate or abnormal	Anchorage dependence Casein plaques Blebby morphology Plasminogen activator	Density inhibition Hexose transport	Hexose transport Fibronectin Fusiform morphology Adhesiveness[b]
Untransformed	Hexose transport Adhesiveness[b] Focus formation	Morphology Adhesiveness Focus formation	Adhesiveness[b] Focus formation

[a]Value varied with assay conditions.
[b]Adhesiveness even greater than normal cell control.

Another *src* mutant that induces a partially transformed phenotype has been reported by Fujita et al. (45). This mutant, ST529, codes for a thermosensitive pp60^{src} protein, and the cells it infects are normal at 42° C. At 36° C, the infected cells appear similar to CU12 in that they are fully transformed with respect to growth control and plasminogen activator but have a fusiform morphology and retain a great deal of surface fibronectin.

Calothy, Pessac, and their collaborators (18,19,88) have isolated several mutants of RSV which confer on neuroretinal cells the ability to grow in cell culture but fail to transform fully either neuroretinal cells or fibroblasts with respect to morphology, hexose transport, tumor-specific surface antigen (TSSA), or anchorage-independent growth. The stimulation of neuroretinal cell growth depends on the activity of the *src* gene, since *src* deletion mutants do not cause the growth stimulation. These findings are consistent with the notion that different cellular targets are involved in growth stimulation of neuroretinal cells than are required for the appearance of other transformation parameters. It is still possible, however, that the growth stimulation is simply a manifestation of transformation which is sensitive to pp60^{src} activity. It would require the isolation of a mutant that fails to stimulate the growth of neuroretinal cells but still transforms by other criteria to provide evidence that growth stimulation of neuroretinal cells requires a unique cellular target. It is clear from studies on fibroblasts that loss of growth control can be dissociated from other parameters of transformation.

Direct evidence that partially transforming virus mutants have an altered specificity for phosphorylation of cellular substrate proteins has come from analysis of the phosphotyrosine-containing proteins in cells infected with these mutants (see below).

Cellular Transformation Mutants

Somatic cell genetics is a potentially powerful tool for elucidating the sequence of events whereby $pp60^{src}$ modifies the infected cell. Isolation of cellular mutants that fail to respond fully to $pp60^{src}$ activity may be essential to distinguish some substrates of $pp60^{src}$ from physiologically significant targets. In addition, isolation of cellular mutants that fail to express certain parameters of transformation could help determine the role of these parameters (if any) in tumorigenicity and metastasis.

An example of the utility of somatic cell genetics in dissecting the pathways leading to the transformed phenotype comes from recent work of P. L. Kaplan and B. Ozanne *(personal communication)*. These workers analyzed transformation by Kirsten sarcoma virus of a variant 3T3 cell line which lacks receptors for EGF (89). They confirmed earlier reports that these cells were capable of being transformed morphologically by Kirsten sarcoma virus but found, in addition, that the cells could not be transformed with respect to anchorage-independent growth. This finding provides convincing evidence that morphologic transformation and anchorage-independent growth are dissociable manifestations of transformation. In addition, it supports the notion that production of growth factors by transformed cells is an essential component of their altered growth regulation.

Isolation of cell variants that do not respond to the activity of $pp60^{src}$ has generally involved transformation of an established mammalian cell line by a RSV subgroup which has a mammalian tropism (e.g., subgroup D), since there are no well-characterized established cell lines of untransformed chicken fibroblasts. The transformed cell lines then are subjected to a selection procedure to isolate "revertants" which have a more normal phenotype. In the past, selection has been for cells with normal growth control or increased adhesiveness.

Reversion of RSV-transformed cells has generally been associated with loss of the viral genome, decreased proviral transcription, or mutations in the proviral *src* gene (24,81,121). Although revertants of these types provide interesting biologic material for studies on RSV integration, transcription, and genome structure, they are not useful for analysis of the interaction of $pp60^{src}$ with the cell.

Lau, Krzyzek, Faras, and their collaborators (25,65–67,118) have isolated a revertant of a RSV-transformed vole cell line, which may be mutated in a target of $pp60^{src}$. This variant, termed revertant 866, is morphologically near normal but produces plasminogen activator, forms colonies in soft agar, and is tumorigenic in nude mice. The suggestion that this revertant may be mutated in a cellular target for $pp60^{src}$ is based on the following lines of evidence: First, the cells contain the RSV genome in amounts roughly equivalent to the parental level. Second, the cells produce parental levels of $pp60^{src}$, which is active as a kinase. Third, virus can be rescued from these cells by fusion with chicken embryo cells; this virus is capable of transforming chicken cells, suggesting that the revertant properties of the cell line are not attributable to alterations in the proviral *src*.

Although strongly suggestive, these data do not prove that 866 is mutated in a cellular target for pp60^{src}. For example, the precise number of copies of provirus per cell in this revertant and in the parental transformed cell is not clear (118). If these cells contain multiple copies of the RSV genome, it is possible that the virus rescued from these cells by fusion was not the same virus that was initially responsible for transformation. Alternatively, the rescued provirus could be the provirus expressed in the revertant cells, but it could code for a pp60^{src} able to transform chicken cells but not vole cells. Just such a *src* mutant has been described by Varmus and his co-workers (81,112). A revertant of a RSV-transformed rat cell was found to carry a mutated *src* which could cause morphologic transformation of chicken cells but not rat cells. Thus the notion that the revertant cell line 866 has picked up mutations in the proviral *src* has not been excluded. Supporting this suggestion is the fact that the peptide maps of pp60^{src} from the 866 revertant differ from those of the parental pp60^{src}.

To prove that a phenotypically normal revertant of a transformed cell line is actually mutated in a target of pp60^{src}, one must demonstrate that the revertant is totally refractory to retransformation by wild type virus, and that the superinfecting wild type virus enters the cell, is integrated, and expressed. There are no cases in which this has been done.

IDENTIFICATION OF PHOSPHOTYROSINE-CONTAINING PROTEINS

Since pp60^{src} has a protein kinase activity that phosphorylates on tyrosine, the first step in determining the biochemical steps leading to the transformed phenotype is the identification of cellular proteins which become phosphorylated on tyrosine during RSV transformation. The involvement of phosphotyrosine in transformation was not appreciated until recently; this phosphoamino acid is a minor component (1% of the phosphorylhydroxy amino acids in transformed cells, and less than 0.1% in normal cells), and because phosphotyrosine is not easily separated from other phosphorylated cellular constituents (56,57,101).

The following is a brief review of work that identifies proteins that become phosphorylated *in vivo* during transformation. Studies performed *in vitro* are not discussed since it is clear that, under currently used assay conditions, pp60^{src} will phosphorylate a large number of nonphysiologic substrates *in vitro* (39,71). Attempts to identify phosphotyrosine-containing proteins have utilized either immunologic reagents to precipitate proteins that might contain phosphotyrosine or gel electrophoretic techniques to separate the proteins whose phosphorylation state changes upon transformation.

The first protein found to be phosphorylated on tyrosine *in vivo* was pp60^{src} itself. pp60^{src} is phosphorylated on tyrosine 416 (105) (in the carboxyl terminal half of the molecule) as well as on serine number 17 (near the amino terminus of the molecule). It is not clear whether the tyrosine phosphorylation is due to autophosphorylation by pp60^{src} or whether this phosphorylation is a result of some other (cellular) phosphotyrosine protein kinase.

Purchio (90) has reported that purified $pp60^{src}$ is capable of autophosphorylation, whereas Levinson et al. (70) disagree. *Src* expressed in *Echerichia coli* codes for a $p60^{src}$ which is unphosphorylated (48,74), suggesting that $p60^{src}$ cannot phosphorylate itself. However, some aspect of the *E. coli* environment (e.g., the lack of cyclic AMP-dependent kinases to phosphorylate the serine residue in $pp60^{src}$, or the lack of enzymes required for fatty acylation) may have blocked the ability of the enzyme to autophosphorylate. Arguing in favor of autophosphorylation is the finding that the degree of tyrosine phosphorylation in $pp60^{src}$ correlates with the kinase activity of the protein in cells infected with *src* mutants (103,105). It could be, however, that the mutations in *src* affect the ability of the protein to function as a substrate as well as an enzyme.

The appearance of phosphotyrosine on tyrosine-specific protein kinases is not uncommon. The transforming proteins of those tumor viruses that code for tyrosine kinases can be phosphorylated on tyrosine (7,41,47,53,57,64,77,78,83,85,102,120), as can the hormone receptors which possess tyrosine kinase activity (38,62,63,79,111).

$pp60^{src}$ forms a complex with an 85,000 molecular weight (MW) protein, which is phosphorylated on serine, and with a 50,000 MW protein, which is phosphorylated on tyrosine (13,14,16,35,49,80,82). The 85,000 MW protein is a major cellular protein whose amount is regulated by heat shock and glucose. The 50,000 MW protein is a minor species of unknown function. The complex, which may serve as an intermediate in the transit of $pp60^{src}$ from polysomes to the membrane (16,35), was discovered based on the ability of the 50,000 and 85,000 MW proteins to coimmunoprecipitate with $pp60^{src}$. The PRCII transforming protein forms a similar complex (1).

Sefton et al. (100) surveyed the phosphorylation state of various cytoskeletal proteins by immunoprecipitating the proteins from extracts of normal or transformed cells which had been labeled *in vivo* with ^{32}P. They found that filamin, myosin, α-actinin, tubulin, actin, and vimentin were not phosphorylated on tyrosine in RSV-transformed cells, but that vinculin was. Vinculin is a protein found in "adhesion plaques," which are assumed to be the sites that attach a cell to the substratum and that may anchor actin-containing microfilaments (58). Adhesion plaques are also a locale of $pp60^{src}$ (96,106). Vinculin also can be phosphorylated *in vitro* by purified $pp60^{src}$ in a reaction that is enhanced by anionic phospholipids, such as phosphoinositol (59).

An attempt to develop a general immunologic procedure for detection of phosphotyrosine-containing proteins was described by Ross et al. (98), who raised antisera against *p*-azobenzyl phosphonate, linked to keyhole limpet hemocyanin. The specific antibodies could bind to a column of phosphotyramine-Sepharose and could be eluted with phenyl phosphate. The antibody was capable of precipitating the phosphotyrosine-containing pp120 from cells infected with the Abelson leukemia virus and also could precipitate a phosphoprotein of 110,000 MW from RSV-transformed cells. Pp60^{src} and a 36,000 MW protein (see below) are the predominant phosphotyrosine-containing proteins in RSV-transformed cells (72), but the antibody did not precipitate these proteins in amounts proportional to their concentrations in cells. Indeed, these proteins were not visible in the published autoradiogram (98). Thus antiphosphotyrosine antibodies may be useful for identifying tyrosine phosphorylation of some proteins, but they do not provide quantitative information about that phosphorylation and are likely to give false negative results. Monoclonal antibodies against phosphotyrosine have recently been developed (H. Eisen, *personal communication*) and may prove to be useful reagents for the qualitative identification of some phosphotyrosine-containing proteins.

Radke and Martin (93) examined the phosphorylation state of cellular proteins by labeling cells *in vivo* with ^{32}P and separating the proteins from a cell extract on a two-dimensional gel, using isoelectric focusing in the first dimension and SDS-polyacrylamide gel electrophoresis in the second. A major cellular protein of approximately 36,000 MW and of unknown function was found to become phosphorylated upon transformation by RSV. This phosphorylation was subsequently shown to be on tyrosine or on tyrosine and serine (39,92,93). Purified pp60^{src} *in vitro* phosphorylates the protein exclusively on tyrosine. Reports on the *in vivo* labeling of the 36,000 MW protein, however, give varying values for the extent of phosphorylation on phosphoserine, apparently dependent on the methods used to isolate the protein. The transformation-related phosphorylation on serine *in vivo* may occur because the phosphorylation on tyrosine makes the protein more susceptible to phosphorylation by some other kinase.

The 36,000 MW protein also becomes phosphorylated in cells infected with other avian sarcoma viruses, the Abelson virus, or Snyder-Theilen feline sarcoma virus (30). In addition, it becomes phosphorylated in A431 cells treated with EGF (28,32,55). This cell line contains large numbers of EGF receptors, and treatment of these cells with EGF leads to a concomitantly large increase in total cellular phosphotyrosine. Interestingly, treating normal chick embryo fibroblasts or 3T3 cells with EGF does not lead to a detectable phosphorylation of the 36,000 MW protein (76), even though EGF is mitogenic for these cells.

Fractionation experiments indicate that the 36,000 MW protein can be associated either with membranes or with the detergent-insoluble "framework" of the cells; or it may be soluble (23,31,36,61). These findings are not necessarily in conflict with each other, since many membrane proteins can appear in the detergent-insoluble framework, either through linkage to the cytoskeleton or because of their solubility properties. The 36,000 MW protein copurifies with a mitochondrial malic dehydrogenase activity (99), but there is some question whether this protein is in fact a malic dehydrogenase (36). It is possible that the co-purification of the two proteins reflects some functional association, but this remains to be determined.

Cooper and Hunter (29) have refined the isoelectric focusing procedure used by Radke and Martin (93) by treating the two-dimensional gel with 1 N KOH following electrophoresis. This treatment preferentially hydrolyzes the phosphoserine phosphoester bond and also reduces the background due to nucleic acid, thus making phosphorylations on tyrosine considerably more prominent. Autoradiograms of the gels showed several radioactive spots increased in the transformed cells, seven of which were shown to contain phosphotyrosine. Although the two-dimensional gel technique has excellent resolving power, its sensitivity is limited because of the limited amount of material that can be applied to the gels, and because many proteins do not focus tightly in the isoelectric focusing dimension. In addition, some proteins contain alkali-sensitive, tyrosine-phosphate linkages, which are hydrolyzed by the KOH treatment (e.g., the 50,000 MW protein mentioned above). Also, even though alkali treatment enriches phosphotyrosine linkages severalfold, the enormous excess of phosphoserine linkages may obscure detection of additional tyrosine-phosphorylated proteins. For these reasons, the seven phosphotyrosine-containing proteins detected by Cooper and Hunter (29) represent a minimum value.

Martinez et al. (72) electrophoresed total extracts of RSV-transformed cells on SDS-polyacrylamide tube gels, sliced the gels into 60 slices, and determined that phosphotyrosine-containing proteins were present in every slice. Qualitatively similar results were obtained by Beemon et al. (8), who subdivided gels into 10 slices. Although the one-dimensional procedure has considerably less resolving power than the two-dimensional procedures used by Cooper and Hunter (29), the increased sensitivity revealed the existence of a minimum of 30 phosphotyrosine-containing proteins. The major protein was the 36,000 MW protein described by Radke and Martin (93), which accounted for approximately 10% of the total phosphotyrosine in these cells. In addition, peaks of phosphotyrosine were noted at around 60,000 MW (presumably $pp60^{src}$ itself) and <20,000 MW, each of which amounted to about 5% of the total cellular phosphotyrosine (72).

TARGETS VERSUS SUBSTRATES

The existence of these numerous phosphotyrosine-containing proteins is consistent with a multitarget model of transformation, as predicted by the properties of the partial transformation mutants described above. Direct evidence that the partial transformation mutants carry mutations that alter the specificity of $pp60^{src}$ for various cellular proteins has come from analysis of the relative degree of phosphorylation of different proteins in cells infected with these mutants. For example, cells infected with CU2 display less than 10% of the wild type level of phosphorylation of the 36,000 MW protein but still contain 50% of the wild type level of total phosphotyrosine (75). Phosphorylation of several individual proteins can be dissociated one from the other, as can the biologic effects of these mutants (34). Thus it is most likely that the ability of the mutants to induce some parameters of transformation better than others is a consequence of their ability to cause the phosphorylation of some proteins better than others.

Although many of the phosphotyrosine-containing proteins found in RSV-transformed cells may be substrates for $pp60^{src}$ (or for cellular kinases activated by

$pp60^{src}$), they may not be physiologically significant targets. For purposes of comparison, it is worth noting that when the Abelson-transforming protein is expressed in *E. coli*, numerous *E. coli* proteins become phosphorylated on tyrosine (114). These proteins may be substrates for the Abelson protein, but, presumably, they are not physiologically significant targets.

The first protein shown to contain a phosphotyrosine that is unnecessary for transformation was $pp60^{src}$ itself. Snyder et al. (108) mutagenized cloned *src in vitro* converting tyrosine 416 (which is the major site of phosphorylation in wild type $pp60^{src}$) into a phenylalanine. The mutated *src* was still capable of coding for a $pp60^{src}$ with protein kinase activity. Moreover, the mutated *src* could transform cells morphologically, and the transformed cells could grow in soft agar and were tumorigenic. Since the mutant $pp60^{src}$ still was phosphorylated on tyrosine to a low level at sites other than the major one, it is not possible to conclude with certainty that no tyrosine phosphorylation of $pp60^{src}$ is involved in transformation; nor can one conclude that phosphorylation of tyrosine 416 plays no role in regulating the enzymic activity of $pp60^{src}$ or even $pp60^{v\text{-}src}$. It is clear, however, that phosphorylation of tyrosine 416 is not necessary for transformation or tumorigenicity. We thus would refer to $pp60^{src}$ as a candidate substrate for $pp60^{src}$, but probably not a target.

Phosphorylation of the 50,000 MW protein, which forms a complex with $pp60^{src}$, is also unlikely to play a role in transformation. This protein is poorly phosphorylated in RSV-transformed mammalian cells relative to its phosphorylation in RSV-transformed chick cells (14), yet the phenotypic expression of transformation is similar (22). Moreover, phosphorylation of the 50,000 MW protein is generally not decreased when cells are infected with a temperature-conditional mutant and held at the restrictive temperature (42° C), where the cells are not transformed (14,16,80,82). Presumably, this is because the 50,000 MW protein forms a stable complex with the weakly active form of $pp60^{src}$, and the stability of the complex compensates for the lower activity of the enzyme, allowing the 50,000 MW protein to become phosphorylated to at least as great an extent at 42° C as at 36° C. In addition, the increased stability of the complex at 42° C may help protect the phosphotyrosine from the action of phosphatases. Another possibility is that the binding of $pp60^{src}$ increases the susceptibility of the 50,000 MW protein to phosphorylation by a cellular tyrosine kinase. In any event, the degree of phosphorylation of the 50,000 MW protein does not correlate with the expression of the transformed phenotype, suggesting that it is perhaps a substrate for $pp60^{src}$ but not a target whose phosphorylation is involved in generating the transformed phenotype.

The possible role of phosphorylation of vinculin in transformation has been examined by Rohrschneider and Rosok (97) using partially transforming mutants to correlate vinculin phosphorylation with the expression of various transformation parameters. These workers found that vinculin was poorly phosphorylated in cells infected with the partially transforming mutant CU2. Cells infected with this mutant grow poorly in soft agar but have lost much of their fibronectin. Conversely, cells infected with CU11 and CU12 displayed high levels of vinculin phosphorylation. These cells grow well in soft agar. Cells infected with CU11 have a near normal morphology, and cells infected with CU12 retain much of their fibronectin. These studies indicate that vinculin phosphorylation is not involved in the loss of fibronectin or the acquisition of a rounded morphology, a result that is surprising in view of the role of fibronectin in cellular adhesion and morphology and the presence of vinculin in adhesion plaques (58). On the other hand, vinculin phosphorylation did correlate with the ability of cells to grow in soft agar. Because of the large number of proteins that become phosphorylated on tyrosine in these cells, however, this positive correlation is considerably less definitive than the negative ones.

Rohrschneider and Rosok (97) also noticed that $pp60^{src}$ was found in adhesion plaques of most RSV-transformed cells but was not present in cells infected with CU12. Thus the physical presence of $pp60^{src}$ can have biologic effects apart from its tyrosine kinase activity, and the presence of $pp60^{src}$ in adhesion plaques (rather than its kinase activity) is involved in the loss of surface fibronectin. It is possible that the absence of $pp60^{src}$ in the adhesion plaques of the CU12-infected cells was a secondary consequence of some aspect of the structure of these adhesion plaques. In any event, these correlational studies, while not definitive, are consistent with vinculin being a functionally significant target for $pp60^{src}$.

The 36,000 MW protein described above is the major phosphotyrosine-containing protein in RSV-transformed cells (72). In addition, this protein becomes phosphorylated on tyrosine in cells transformed by other tumor viruses which code for a tyrosine protein kinase (30) and is a substrate for $pp60^{src}$ *in vitro* (39). It is suspected (or at least hoped) that the 36,000 MW protein plays an important role in transformation. Weber and his colleagues (34,60,75) have performed an extensive correlational analysis of the role of this protein in transformation, comparing its degree of phosphorylation to the degree of expression of several transformation parameters in cells infected with partially transforming *src* mutants. They found that phosphorylation of the 36,000 MW protein was neither necessary nor sufficient for loss of fibronectin or gross morphologic change. Similar conclusions have been reached by J. Nawrocki, A. Lau, E. Erickson and A. Faras *(personal communication)*, who analyzed phosphorylation of the 36,000 MW protein in the 866 revertant of the RSV-transformed vole cells. This result is perhaps surprising in view of the reported location(s) of the 36,000 MW protein, in the plasma membrane and/or the cellular "framework" (23,31,36,61). The best correlations with phosphorylation of the 36,000 MW protein were with production of plasminogen activator and with tumorigenicity (34,60,75). Thus phosphorylation of this protein may play a critical role in transformation. However, it will be difficult to perform directly mechanistic studies without some understanding of the physiologic function of the 36,000 MW protein. Whatever its *in vivo* function may be, it appears not to be essential for life, since it is not found in certain lymphoid cell lines (104).

Three enzymes of glycolysis become phosphorylated on tyrosine in RSV-transformed cells (33): enolase, phosphoglycerate mutase, and lactate dehydrogenase. In addition, enolase and phosphoglycerate mutase also become phosphorylated in cells infected with a variety of other tumor viruses that code for tyrosine protein kinases (30). Phosphorylation of enolase and phosphoglycerate mutase was initially seen by Cooper and Hunter (29) on two-dimensional gels as increased phosphorylation of proteins with molecular weights of 46,000 and 28,000. These workers subsequently collaborated with Reiss and Schwartz (33) to identify the two proteins as enolase and phosphoglycerate mutase, respectively, based on immunoprecipitation, gel electrophoresis, and peptide mapping.

The following glycolytic enzymes did not become detectably phosphorylated on tyrosine: triosephosphate isomerase, phosphoglycerate kinase, aldolase, glyceraldehyde phosphate dehydrogenase, and pyruvate kinase. Lactate dehydrogenase can be phosphorylated on tyrosine *in vitro* (C. J. Cooper, *personal communication*). In addition, inspection of the published amino acid sequence of this enzyme (37) reveals a tyrosine preceded by an aspartate three residues away and by a lysine seven residues away. This sequence is quite close to the consensus sequence for phosphorylation by $pp60^{src}$ and related kinases (5,20,21,54,78,84,86,87,107) differing only in that the consensus has a pair of acidic amino acids on the amino terminal side of the tyrosine. Thus, $pp60^{src}$ may directly phosphorylate lactate dehydrogenase.

Cooper and Hunter collaborated with Nakamura and Weber (34) to perform an extensive analysis in which the phosphorylation of enolase and phosphoglycerate mutase was correlated with the expression of various transformation parameters in cells infected with partial transformation mutants of RSV. Phosphorylation of these proteins correlated well overall with the rate of hexose transport (but not with changes in some other manifestations of transformation, such as adhesiveness), consistent with a functional role for the phosphorylations in some aspect of glucose metabolism. On the other hand, although enolase, phosphoglycerate mutase, and lactate dehydrogenase may become phosphorylated on tyrosine during transformation, these three enzymes are generally thought not to be important in the regulation of glycolysis; the reactions they catalyze are completely reversible, and analyses of the distribution of intracellular metabolites do not indicate that these enzymes catalyze rate-limiting reactions.

Thus phosphorylation of these glycolytic enzymes is unlikely to affect their intrinsic enzymic activity in a physiologically relevant way. It is possible, however, that the phosphorylation could alter the intracellular distribution of the enzymes, resulting in the generation of glycolytic products at novel intracellular sites. It is uncertain whether phosphorylation on tyrosine of enolase, lactate dehydrogenase, and phosphoglycerate mutase in RSV-transformed cells is adventitious or physiologically significant. If it is significant, it is likely to be in a surprising way, not directly related to controlling the rates of the individual enzymes.

In assessing the biologic significance of phosphorylations on tyrosine, it should be borne in mind that in every case that has been examined, the stoichiometry of the phosphorylation is low. Only about 10% of the molecules of the 36,000 MW protein have been found to contain phosphotyrosine, and an even lower stoichiometry characterizes phosphorylation on tyrosine in vinculin. Since phosphate on tyrosine turns over rapidly (105), presumably because of the presence of active phosphatases (12,42,43,46,68,109,119), it is possible that, even though the steady-state level of phosphorylation is low, every potential substrate molecule gets phosphorylated within a few hours.

For structural proteins, this transient phosphorylation could have long-lasting effects on cellular physiology and morphology. In other cases, the phosphorylation on tyrosine could render the protein susceptible to some other, more stable modification, which has yet to be identified. If phosphorylation is on a regulatory protein, or if phosphorylation results in the redistribution of a protein, even low level phosphorylations could have large effects. Even though these arguments may seem reasonable, however, they should be recognized for what they are: post-hoc rationalizations for an uncomfortable result. Only correlational evidence exists to support the notion that the tyrosine phosphorylations described above play a role in transformation. It is not inconceivable that every tyrosine phosphorylation detected to date is adventitious, and that we have not yet identified any of the true targets of $pp60^{src}$.

SUMMARY AND CONCLUSIONS

Biologic and genetic evidence indicates that the viral transforming protein $pp60^{src}$ has several primary targets for phosphorylation which play a role in generating the variety of cellular alterations that constitute the transformed phenotype. Indeed, numerous proteins are found to become phosphorylated on tyrosine during RSV transformation. It is doubtful, however, whether all these proteins—or even a majority of them—are functionally significant targets. It is more likely that the high levels of $pp60^{src}$ found in RSV-transformed cells result in the adventitious phosphorylation of numerous proteins that play no functional role in transformation. In future studies on the molecular mechanism of oncogenesis by RSV, the major challenge will be to develop the tools to determine the function of tyrosine phosphorylation in malignant transformation.

ACKNOWLEDGMENTS

This work was supported by USPHS grant CA-12467. Thanks are due to Paula Evans, Ricardo Martinez, and Ken Nakamura for helpful comments on the manuscript and to Marc Collett for first pointing out to us the distinction between substrate and targets.

REFERENCES

1. Adkins, B., Hunter, T., and Sefton, B. M. (1982): The transforming proteins of PRCII virus and Rous sarcoma virus form a complex with the same two cellular phosphoproteins. *J. Virol.*, 32:448–455.
2. Ali, I. V., Mautner, V., Lanza, R., and Hynes, R. O. (1977): Restoration of normal morphology adhesion and cytoskeleton in transformed cells by addition of a transformation-specific surface protein. *Cell*, 11:115–126.
3. Anderson, D. D., Beckmann, R. P., Harms, E. H., Nakamura, K., and Weber, M. J. (1981): Biological properties of "partial" transformation mutants of Rous sarcoma virus and characterization of their $pp60^{src}$ kinase. *J. Virol.*, 37:445–458.
4. Barbacid, M., and Lauver, A. V. (1981): Gene products of McDonough feline sarcoma virus have an in vitro-associated protein kinase that phosphorylates tyrosine residues: Lack of detection of this enzymatic activity in vivo. *J. Virol.*, 40:812–821.
5. Baldwin, G. S., Burgess, A. W., and Kamp, B. E. (1982): Phosphorylation of a synthetic gastrin peptide by the tyrosine kinase of A431 cell membrane. *Biochem. Biophys. Res. Commun.*, 109:656–663.
6. Becker, D., Kurth, R., Critchley, D., Friis, R., and Bauer, H. (1977): Distinguishable transformation-defective phenotypes among temperature-sensitive mutants of Rous sarcoma virus. *J. Virol.*, 21:1042–1055.
7. Beemon, K. (1981): Transforming proteins of some feline and avian sarcoma viruses are related structurally and functionally. *Cell*, 24:145–153.
8. Beemon, K., Ryden, T., and McNelly, E. A. (1982): Transformation by avian sarcoma viruses leads to phosphorylation of multiple cellular proteins on tyrosine residues. *J. Virol.*, 42:742–747.
9. Beug, H., Peters, J. H., and Graf, T. (1976): Expression of virus specific morphological cell transformation induced in enucleated cells. *Z. Naturforsch.*, 31:(11/12):766–768.
10. Beug, H., Claviez, M., Jockusch, B., and Graf, T. (1978): Differential expression of Rous sarcoma virus-specific transformation parameters in enucleated cells. *Cell*, 14:843–856.
11. Blithe, D. L., Richert, N. D., and Pastan, I. H. (1982): Purification of a tyrosine-specific protein kinase from Rous sarcoma virus-induced rat tumor. *J. Biol. Chem.*, 257:7135–7142.
12. Brautigant, D. L., Bornstein, P., and Gallis, B. (1981): Phosphotyrosyl-protein phosphatase. *J. Biol. Chem.*, 256:6519–6522.
13. Brugge, J. S., and Darrow, D. (1982): Rous sarcoma virus-induced phosphorylation of a 50,000-molecular weight cellular protein. *Nature*, 295:250–253.
14. Brugge, J. S., Erikson, E., and Erikson, R. L. (1981): The specific interaction of the Rous sarcoma virus transforming protein, $pp60^{src}$, with two cellular proteins. *Cell*, 25:363–372.
15. Brugge, J. S., and Erikson, R. L. (1977): Identification of a transformation-specific antigen induced by an avian sarcoma virus. *Nature*, 269:346–348.
16. Brugge, J., Yonemoto, W., and Darrow, D. (1983): Interaction between the Rous sarcoma virus transforming protein and two cellular phosphoproteins: Analysis of the turnover and distribution of this complex. *Mol. Cell. Biol.*, 3:9–19.
17. Brzeski, H., and Ege, T. (1980): Changes in polypeptide pattern in ASV-transformed rat cells are correlated with the degree of morphological transformation. *Cell*, 22:513–522.
18. Calothy, G., and Pessac, B. (1976): Growth stimulation of chick embryo neuroretinal cells infected with Rous sarcoma virus: Relationship to viral replication and morphological transformation. *Virology*, 71:336–345.
19. Calothy, G., Poirier, F., Dambrime, G., and Pessac, B. (1978): A transformation defective mutant of Rous sarcoma virus inducing chick embryo neuroretinal cell proliferation. *Virology*, 89:75–84.
20. Casnellie, J. E., Harrison, M. L., Hellstrom, K. E., and Krebs, E. G. (1982): A lymphoma protein with an in vitro site of tyrosine phosphorylation homologous to that in $pp60^{src}$. *J. Biol. Chem.*, 257:13877–13879.
21. Casnellie, J. E., Harrison, M. L., Pike, L. J., Hellstrom, K. E., and Krebs, E. G. (1982): Phosphorylation of synthetic peptides by a tyrosine protein kinase from the particulate fraction of a lymphoma cell line. *Proc. Natl. Acad. Sci. USA*, 79:282–286.
22. Chen, Y. C., Hayman, M. J., and Vogt, P. K. (1977): Properties of mammalian cells transformed by mutants of RSV. *Cell*, 11:513–521.
23. Cheng, Y. S., and Chen, L. B. (1981): Detection of phosphotyrosine-containing 34,000-dalton

protein in the framework of cells transformed with Rous sarcoma virus. *Proc. Natl. Acad. Sci. USA*, 78:2388–2392.

24. Chiswell, D. J., Enrietto, P. J., Evans, S., Quade, K., and Wyke, J. A. (1982): Molecular mechanisms involved in morphological variation of avian sarcoma virus-infected rat cells. *Virology*, 116:428–440.
25. Collett, M. S., Brugge, J. S., Erikson, R. L., Lau, A. F., Krzyzek, R. A., and Faras, A. J. (1979): The *src* gene product of transformed and morphologically reverted ASV-infected mammalian cells. *Nature*, 280:195–198.
26. Collett, M. S., and Erikson, R. L. (1978): Protein kinase activity associated with the avian sarcoma virus src gene product. *Proc. Natl. Acad. Sci. USA*, 75:2021–2024.
27. Collett, M. S., Purchio, A. F., and Erikson, R. L. (1980): Avian sarcoma virus-transforming protein, $pp60^{src}$, shows protein kinase activity specific for tyrosine. *Nature*, 285:167–169.
28. Cooper, J. A., Bowen-Pope, D. F., Raines, E., Ross, R., and Hunter, T. (1982): Similar effects of platelet derived growth factor and epidermal growth factor on the phosphorylation of tyrosine in cellular proteins. *Cell*, 31:263–273.
29. Cooper, J. A., and Hunter, T. (1981): Changes in protein phosphorylation in Rous sarcoma virus-transformed chicken embryo cells. *Mol. Cell. Biol.*, 1:165–178.
30. Cooper, J. A., and Hunter, T. (1981): Four different classes of retroviruses induce phosphorylation of tyrosine present in similar cellular proteins. *Mol. Cell. Biol.*, 1:394–407.
31. Cooper, J. A., and Hunter, T. (1982): Discrete primary locations of a tyrosine-protein kinase and of three proteins that contain phosphotyrosine in virally transformed chick fibroblasts. *J. Cell Biol.*, 94:287–296.
32. Cooper, J. A., and Hunter, T. (1981): Similarities and differences between the effects of epidermal growth factor and Rous sarcoma virus. *J. Cell. Biol.*, 91:878–883.
33. Cooper, J. A., Reiss, N. A., Schwartz, R. J., and Hunter, T. (1983): Three glycolytic enzymes are phosphorylated on tyrosine in cells transformed by Rous sarcoma virus. *Nature*, 302:218–222.
34. Cooper, J. A., Nakamura, K. D., Hunter, T., and Weber, M. J. (1983): Phosphotyrosine-containing proteins and expression of transformation parameters in cells infected with partial transformation mutants of Rous sarcoma virus. *J. Virol.*, 46:15–28.
35. Courtneidge, S. A., and Bishop, M. J. (1982): Transit of $pp60^{v\text{-}src}$ to the plasma membrane. *Proc. Natl. Acad. Sci. USA*, 79:7117–7121.
36. Courtneidge, S., Ralston, R., Alitalo, K., and Bishop, J. M. (1982): Subcellular location of an abundant substrate (p36) for tyrosine-specific protein kinases. *Mol. Cell. Biol.*, 3:340–350.
37. Dayhoff, M. O. (1979): *Atlas of Protein Structure*. National Biomedical Research Foundation, Washington, D.C.
38. Ek, B., Westermark, B., Wasteson, A., and Heldin, C. H. (1982): Stimulation of tyrosine-specific phosphorylation by platelet-derived growth factor. *Nature*, 295:419–420.
39. Erikson, E., and Erikson, R. L. (1980): Identification of a cellular protein substrate phosphorylated by the avian sarcoma virus-transforming gene product. *Cell*, 21:829–836.
40. Erikson, R. L., Collett, M. S., Erikson, E., and Purchio, A. F. (1979): Evidence that the avian sarcoma virus transforming gene product is a cyclic-AMP-independent protein kinase. *Proc. Natl. Acad. Sci. USA*, 76:6260–6264.
41. Feldman, R. A., Wang, L. H., Hanafusa, H., and Balduzzi, P. C. (1982): Avian sarcoma virus UR2 encodes a transforming protein which is associated with a unique protein kinase activity. *J. Virol.*, 42:228–236.
42. Foulkes, J. G., Erikson, E., and Erikson, R. L. (1982): Separation of multiple phosphotyrosyl- and phosphoseryl-protein phosphatases from chicken brain. *J. Biol. Chem.*, 258:431–438.
43. Foulkes, J. G., Howard, R. F., and Ziemiecki, A. (1981): Detection of a novel mammalian protein phosphatase with activity for phosphotyrosine. *FEBS Lett.*, 130:197–200.
44. Friis, R. R., Schwarz, R. T., and Schmidt, M. F. G. (1977): Phenotypes of Rous sarcoma virus-transformed fibroblasts: An argument for a multifunctional *src* gene product. *Med. Microbiol. Immunol.*, 164:155–165.
45. Fujita, D. J., Boschek, C. B., Ziemiecki, A., and Friis, R. R. (1981): An avian sarcoma virus mutant which produces an aberrant transformation affecting cell morphology. *Virology*, 111:223–238.
46. Fukami, Y., and Lipmann, F. (1982): Purification of a specific reversible tyrosine-O-phosphate phosphatase. *Proc. Natl. Acad. Sci. USA*, 79:4275–4279.

47. Ghysdael, J., Neil, J. C., and Vogt, P. K. (1981): A third class of avian sarcoma viruses, defined by related transformation-specific proteins of Yamaguchi 73 and Esh sarcoma viruses. *Proc. Natl. Acad. Sci. USA*, 78:2611–2615.
48. Gilmer, T. M., and Erikson, R. L. (1981): Rous sarcoma virus transforming protein $pp60^{src}$ expressed in E. coli functions as a protein kinase. *Nature*, 294:771–773.
49. Gilmore, T. D., Radke, K., and Martin, G. S. (1982): Tyrosine phosphorylation of a 50K cellular polypeptide associated with the Rous sarcoma virus transforming protein $pp60^{src}$. *Mol. Cell. Biol.*, 2:199–206.
50. Graziani, Y., Erikson, E., and Erikson, R. L. (1982): Evidence that the Rous sarcoma virus transforming gene product is associated with glycerol kinase activity. *J. Biol. Chem.*, 258:2126–2129.
51. Groudine, M., and Weintraub, H. (1980): Activation of cellular genes by avian RNA tumor viruses. *Proc. Natl. Acad. Sci. USA*, 77:5351–5354.
52. Hale, A. H., and Weber, M. J. (1975): Hydrolase and serum treatment of normal chick embryo cells: Effects on hexose transport. *Cell*, 5:245–252.
53. Hampe, A., Laprevotte, I., Galibert, F., Fedele, L. A., and Sherr, C. J. (1982): Nucleotide sequences of feline retroviral oncogenes (v-fes) provide evidence for a family of tyrosine-specific protein kinase genes. *Cell*, 30:775–785.
54. Hunter, T. (1982): Synthetic peptide substrates for a tyrosine protein kinase. *J. Biol. Chem.*, 257:4843–4848.
55. Hunter, T., and Cooper, J. A. (1981): Epidermal growth factor induces rapid tyrosine phosphorylation of proteins in A431 human tumor cells. *Cell*, 24:741–752.
56. Hunter, T., and Sefton, B. M. (1980): Transforming gene product of Rous sarcoma virus phosphorylates tyrosine. *Proc. Natl. Acad. Sci. USA*, 77:1311–1315.
57. Hunter, T., Sefton, B. M., and Beemon, K. (1980): Phosphorylation of tyrosine: A mechanism of transformation shared by a number of otherwise unrelated RNA tumor viruses. In: *ICN-UCLA Symposium on Animal Virus Genetics, Vol. 8*, edited by B. M. Fields and R. Jaenisch, pp. 499–514. Academic Press, New York.
58. Hynes, R. (1982): Phosphorylation of vinculin by $pp60^{src}$: What might it mean? *Cell*, 28:437–438.
59. Ito, S., Richert, N., and Pastan, I. (1982): Phospholipids stimulate phosphorylation of vinculin by the tyrosine-specific protein kinase of Rous sarcoma virus. *Proc. Natl. Acad. Sci. USA*, 79:4628–4631.
60. Kahn, P., Nakamura, K., Shin, S., Smith, R. E., and Weber, M. J. (1982): Tumorigenicity of partial transformation mutants of Rous sarcoma virus. *J. Virol.*, 42:602–611.
61. Kaji, A., and Amini, S. (1983): Association of pp 36, a phosphorylated form of the presumed target protein for the src protein of Rous sarcoma virus, with the membrane of chicken cells transformed by Rous sarcoma virus. *Proc. Natl. Acad. Sci. USA*, 80:960–964.
62. Kasuga, M., Zick, Y., Blith, D. L., Karlsson, F. A., Haring, H. U., and Kahn, C. R. (1982): Insulin stimulation of phosphorylation of the beta subunit of the insulin receptor. Formation of both phosphoserine and phosphotyrosine. *J. Biol. Chem.*, 257:9891–9894.
63. Kasuga, M., Zick, Y., Blithe, D. L., Crettaz, M., and Kahn, C. R. (1982): Insulin stimulates tyrosine phosphorylation of the insulin receptor in a cell-free system. *Nature*, 298:667–669.
64. Kawai, S., Yoshida, M., Segawa, K., Sugiyama, H., Ishizaki, R., and Toyoshima, K. (1980): Characterization of Y73, an avian sarcoma virus: A unique transforming gene and its product, a phosphopolyprotein with protein kinase activity. *Proc. Natl. Acad. Sci. USA*, 77:6199–6203.
65. Krzyzek, R. A., Lau, A. F., Vogt, P. K., and Faras, A. J. (1978): Quantitation and localization of RSV-specific RNA in transformed and revertant field vole cells. *J. Virol.*, 25:518–526.
66. Lau, A. F., Krzyzek, R. A., Brugge, J. S., Erikson, R. L., Schollmeyer, J., and Faras, A. J. (1979): Morphological revertants of an avian sarcoma virus-transformed mammalian cell line exhibit tumorigenicity and contain $pp60^{src}$. *Proc. Natl. Acad. Sci. USA*, 76:3904–3908.
67. Lau, A. F., Krzyzek, R. A., and Faras, A. J. (1981): Loss of tumorigenicity correlates with a reduction in $pp60^{src}$ kinase activity in a revertant subclone of avian sarcoma virus-infected field vole cells. *Cell*, 23:815–824.
68. Leis, J. F., and Kaplan, N. O. (1982): An acid phosphatase in the plasma membranes of human astrocytoma showing marked specificity toward phosphotyrosine protein. *Proc. Natl. Acad. Sci. USA*, 79:6507–6511.
69. Levinson, A. D., Opperman, H., Levintow, L., Varmus, H., and Bishop, J. M. (1978): Evidence

that the transforming gene of avian sarcoma virus encodes a protein kinase associated with a phosphoprotein. *Cell*, 15:561–572.

70. Levinson, A. D., Opperman, H., Varmus, H. E., and Bishop, J. M. (1980): The purified product of the transforming gene of avian sarcoma virus phosphorylates tyrosine. *J. Biol. Chem.*, 255:11973–11980.
71. Maness, P. F., and Levy, B. T. (1983): Highly purified pp60^{src} induces the actin transformation in microinjected cells and phosphorylates selected cytoskeletal proteins in vitro. *Mol. Cell. Biol.* 3:102–112.
72. Martinez, R., Nakamura, K. D., and Weber, M. J. (1982): Identification of phosphotyrosine-containing proteins in untransformed and Rous sarcoma virus-transformed chicken embryo fibroblasts. *Mol. Cell. Biol.*, 2:653–665.
73. Mardon, G., and Varmus, H. E. (1983): Frameshift and intragenic suppressor mutations in a Rous sarcoma provirus suggest *src* encodes two proteins. *Cell*, 32:871–879.
74. McGrath, J. P., and Levinson, A. D. (1982): Bacterial expression of an enzymatically active protein encoded by RSV src gene. *Nature*, 295:423–425.
75. Nakamura, K. D., and Weber, M. J. (1982): Phosphorylation of a 36,000 M_r cellular protein in cells infected with partial transformation mutants of rous sarcoma virus. *Mol. Cell. Biol.*, 2:147–153.
76. Nakamura, K. D., Martinez, R., and Weber, M. J. (1983): Tyrosine phosphorylation of specific proteins after mitogen stimulation of chicken embryo fibroblasts. *Mol. Cell. Biol.*, 3:380–390.
77. Neil, J. C., Chysdael, J., and Vogt, P. K. (1981): Tyrosine-specific protein kinase activity associated with P105 of avian sarcoma virus PRCII. *Virology*, 109:223–228.
78. Neil, J. C., Ghysdael, J., Vogt, P. K., and Smart, J. E. (1981): Homologous tyrosine phosphorylation sites in transformation-specific gene products of distinct ASV. *Nature*, 291:675–771.
79. Nishimura, J., Huang, J. S., and Deuel, T. F. (1982): Platelet-derived growth factor stimulates tyrosine-specific protein kinase activity in Swiss mouse 3T3 cell membranes. *Proc. Natl. Acad. Sci. USA*, 79:4303–4307.
80. Oppermann, H., Levinson, A. D., Levintow, L., Varmus, H. E., Bishop, J. M., and Kawai, S. (1981): Two cellular proteins that immunoprecipitate with the transforming protein of Rous sarcoma virus. *Virology*, 113:736–751.
81. Oppermann, H., Levinson, A. D., and Varmus, H. E. (1981): The structure and protein kinase activity of proteins encoded by nonconditional mutants and back mutants in the *src* gene of avian sarcoma virus. *Virology*, 108:47–70.
82. Oppermann, H., Levinson, W., and Bishop, J. M. (1981): A cellular protein that associates with the transforming protein of Rous sarcoma virus is also a heat-shock protein. *Proc. Natl. Acad. Sci. USA*, 78:1067–1071.
83. Patschinsky, T., and Sefton, B. M. (1981): Evidence that there exist four classes of RNA tumor viruses which encode proteins with associated tyrosine protein kinase activities. *J. Virol.*, 39:104–114.
84. Patschinsky, T., Hunter, T., Esch, F. S., Cooper, J. A., and Sefton, B. M. (1983): Analysis of the sequence of amino acids surrounding sites of tyrosine phosphorylation. *Proc. Natl. Acad. Sci. USA*, 79:973–977.
85. Pawson, T., Guyden, J., Kung, T. H., Radke, K., Gilmore, T., and Martin, G. S. (1980): A strain of Fujinami sarcoma virus which is temperature-sensitive in protein phosphorylation and cellular transformation. *Cell*, 22:767–775.
86. Pike, L. J., Gallis, B., Casnellie, J. E., Bornstein, P., and Krebs, E. G. (1982): Epidermal growth factor stimulates the phosphorylation of synthetic tyrosine-containing peptides by A431 cell membranes. *Proc. Natl. Acad. Sci. USA*, 79:1443–1447.
87. Pike, L. J., Marquardt, H., Todaro, G. J., Gallis, B., Casnellie, J. E., Bornstein, P., and Krebs, E. G. (1982): Transforming growth factor and epidermal growth factor stimulate the phosphorylation of a synthetic, tyrosine-containing peptide in a similar manner. *J. Biol. Chem.*, 257:14628–14631.
88. Poirier, F., Calothy, G., Karess, R. E., Erikson, E., and Hanafusa, H. (1982): Role of pp60^{src} kinase activity in the induction of neuroretinal cell proliferation by Rous sarcoma virus. *J. Virol.*, 42:780–789.
89. Pruss, R. M., and Hershman, H. R. (1977): Variants of 3T3 cells lacking mitogenic response to epidermal growth factor. *Proc. Natl. Acad. Sci. USA*, 74:3918–3922.

90. Purchio, A. F. (1982): Evidence that pp60^src^, the product of the Rous sarcoma virus *src* gene, undergoes autophosphorylation. *J. Virol.*, 41:1–7.
91. Purchio, A. F., Erikson, E., Brugge, J. S., and Erikson, R. L. (1978): Identification of a polypeptide encoded by the avian sarcoma virus src gene. *Proc. Natl. Acad. Sci. USA*, 75:1567–1571.
92. Radke, K., Gilmore, T., and Martin, G. S. (1980): Transformation by Rous sarcoma virus: A cellular substrate for transformation-specific protein phosphorylation contains phosphotyrosine. *Cell*, 21:821–828.
93. Radke, K., and Martin, G. S. (1979): Transformation by Rous sarcoma virus: Effects of src gene expression on the synthesis and phosphorylation of cellular polypeptides. *Proc. Natl. Acad. Sci. USA*, 76:5212–5216.
94. Reynolds, F. H., Jr., Todaro, G. H., Fryling, C., and Stephenson, J. R. (1981): Human transforming growth factors induce tyrosine phosphorylation of EGF receptors. *Nature*, 292:259–262.
95. Reynolds, F. J., Jr., Van de Ven, W. J., and Stephenson, J. R. (1980): Feline sarcoma virus P115-associated protein kinase phosphorylates tyrosine. Identification of a cellular substrate conserved during evolution. *J. Biol. Chem.*, 255:11040–11047.
96. Rohrschneider, L. R. (1980): Adhesion plaques of Rous sarcoma virus-transformed cells contain the src gene product. *Proc. Natl. Acad. Sci. USA*, 77:3514–3518.
97. Rohrschneider, L., and Rosok, M. J. (1983): Transformation parameters and $pp60^{src}$ localization in cells infected with partial transformation mutants of Rous sarcoma virus. *Mol. Cell. Biol.*, 3:731–746.
98. Ross, A. H., Baltimore, D., and Eisen, H. N. (1981): Phosphotyrosine-containing proteins isolated by affinity chromatography with antibodies to a synthetic hapten. *Nature*, 294:654–656.
99. Rubsamen, H., Saltenberger, K., Friis, R. R., and Eigenbrocht, E. (1982): Cytosolic malic dehydrogenase activity is associated with a putative substrate for the transforming gene product of Rous sarcoma virus. *Proc. Natl. Acad. Sci. USA*, 79:228–232.
100. Sefton, B. M., Hunter, T., Ball, E. H., and Singer, S. J. (1981): Vinculin: A cytoskeletal target of the transforming protein of Rous sarcoma virus. *Cell*, 24:165–174.
101. Sefton, B. M., Hunter, T., Beemon, K., and Eckhart, W. (1980): Evidence that the phosphorylation of tyrosine is essential for cellular transformation by Rous sarcoma virus. *Cell*, 20:807–816.
102. Sefton, B. M., Hunter, T., and Raschke, W. C. (1981): Evidence that the Abelson virus protein functions in vivo as a protein kinase that phosphorylates tyrosine. *Proc. Natl. Acad. Sci. USA*, 78:1552–1556.
103. Sefton, B. M., Hunter, T., and Beemon, K. (1980): Temperature-sensitive transformation by Rous sarcoma virus and temperature-sensitive protein kinase activity. *J. Virol.*, 33:220–229.
104. Sefton, B. M., Hunter, T., and Cooper, J. A. (1983): Some lymphoid cell lines transformed by Abelson murine leukemia virus lack a major 36,000 dalton tyrosine protein kinase substrate. *Mol. Cell. Biol.*, 3:56–63.
105. Sefton, B. M., Patschinsky, T., Berdot, C., Hunter, T., and Elliott, T. (1982): Phosphorylation and metabolism of the transforming protein of Rous sarcoma virus. *J. Virol.*, 41:813–820.
106. Shriver, K., and Rohrschneider, L. (1981): Organization of $pp60^{src}$ and selected cytoskeletal proteins within adhesion plaques and junctions of Rous sarcoma virus-transformed rat cells. *J. Cell. Biol.*, 89:525–535.
107. Smart, J. E., Oppermann, H., Czernilofsky, A. P., Purchio, A. F., Erikson, R. L., and Bishop, J. M. (1981): Characterization of sites for tyrosine phosphorylation in the transforming protein of Rous sarcoma virus ($pp60^{v\text{-}src}$) and its normal cellular homologue ($pp60^{c\text{-}src}$). *Proc. Natl. Acad. Sci. USA*, 78:6013–6017.
108. Snyder, M. A., Bishop, J. M., Colby, W. W., and Levinson, A. D. (1983): Phosphorylation of tyrosine 416 is not required for the transforming properties and kinase activity of $pp60^{v\text{-}src}$. *Cell*, 32:891–901.
109. Swarup, G., Speeg, K. V., Jr., Cohen, S., and Garbers, D. L. (1982): Phosphotyrosyl-protein phosphatase of TCRC-2 cells. *J. Biol. Chem.*, 257:7298–7301.
110. Tanaka, A., Parker, C., and Kaji, A. (1980): Stimulation of growth rate of chondrocytes by Rous sarcoma virus is not coordinated with other expressions of the *src* gene phenotype. *J. Virol.*, 35:531–541.
111. Ushiro, H., and Cohen, S. (1980): Identification of phosphotyrosine as a product of epidermal growth factor-activated protein kinase in A-431 cell membranes. *J. Biol. Chem.*, 255:8363–8365.
112. Varmus, H. E., Quintrell, N., and Wyke, J. (1981): Revertants of an ASV-transformed rat cell line have lost the complete provirus or sustained mutations in src. *Virology*, 108:28–46.

113. Vogt, P. K. (1977): The genetics of RNA tumor viruses. In: *Comprehensive Virology, Vol. 9*, edited by H. Fraenkel-Conrat and R. R. Wagner, pp. 341–455. Plenum, New York.
114. Wang, J. Y., Queen, C., and Baltimore, D. (1982): Expression of an Abelson murine leukemia virus-encoded protein in Escherichia coli causes extensive phosphorylation of tyrosine residues. *J. Biol. Chem.*, 257:13181–13184.
115. Weber, M. J. (1975): Inhibition of protease activity in cultures of Rous sarcoma virus-transformed cells: Effect on the transformed phenotype. *Cell*, 5:253–261.
116. Weber, M. J., and Friis, R. R. (1979): Dissociation of transformation parameters using temperature-conditional mutants of Rous sarcoma virus. *Cell*, 16:25–32.
117. Weber, M. J., Hale, A. H., Yau, T. M., Buckman, T., Johnson, M., Brady, T. M., and LaRossa, D. D. (1976): Transport changes associated with growth control and malignant transformation. *J. Cell. Physiol.*, 89:711–722.
118. Woods, W. G., Lau, A. F., Krzyzek, R. A., Cervenka, J., and Faras, A. J. (1981): Karyotype analysis and quantitation of viral transforming genes in Rous sarcoma virus transformed, revertant, and retransformed field vole cells. *Cytogenet. Cell Genet.*, 29:153–165.
119. Witt, D. P., and Gordon, J. A. (1980): Specific dephosphorylation of membrane proteins in Rous sarcoma virus-transformed chick embryo fibroblasts. *Nature*, 287:241–243.
120. Witte, O. N., Dasgupta, A., and Baltimore, D. (1980): Abelson murine leukemia virus protein is phosphorylated in vitro to form phosphotyrosine. *Nature*, 283:826–831.
121. Wyke, J. A., and Quade, K. (1980): Infection of rat cells by avian sarcoma virus: Factors affecting transformation and subsequent reversion. *Virology*, 106:217–233.
122. Yamada, K. M., and Olden, K. (1978): Fibronectins-adhesive glycoproteins of cell surface and blood. *Nature*, 275:179–184.

Advances in Viral Oncology, Volume 4,
edited by George Klein. Raven Press,
New York © 1984.

Subcellular Locations of Retroviral Transforming Proteins Define Multiple Mechanisms of Transformation

L. R. Rohrschneider and L. E. Gentry

Fred Hutchinson Cancer Research Center, Tumor Virology Program, Seattle, Washington 98104

Oncogenic retroviruses capable of inducing the rapid onset of neoplasia in animals carry within their genomes genetic information responsible for neoplastic transformation. These transformation-specific nucleotide sequences have been referred to as v-*onc* genes and have been shown to be related to highly conserved sequences (c-*onc* genes) found in normal cells. Consequently, it has been proposed that the oncogenic retroviruses have arisen by a recombination event involving virus and cellular information, and that neoplastic transformation could result from a qualitative or quantitative difference in the expression of the v-*onc* protein. To date, 21 distinct retroviral *onc* genes have been detected, and the transforming proteins synthesized from these genes have been identified. Although much is known concerning the structures of the *onc* genes and their encoded polypeptides, relatively little is known about how these *onc* proteins act to produce the many cellular changes that accompany oncogenic transformation.

To understand the mechanism of transformation at the cellular level, it is important to define the potential cellular sites at which these *onc* polypeptides may act. Studies concerning the intracellular distribution of these proteins may provide insights into the transformation mechanism and may identify crucial cellular structures intimately involved in oncogenesis. In addition, results from these studies may offer clues as to the functions and roles of the c-*onc* gene products in the control of growth and differentiation in normal cells.

Three independent approaches have been utilized to study the subcellular distribution of these transformation-specific proteins: (a) cell fractionation, (b) indirect immunofluorescence, and (c) immunocytochemical electron microscopy. Although these techniques are independent in nature, they all require immunologic reagents necessary for identification of the transforming proteins. Each approach possesses inherent advantages and disadvantages. Cellular fractionation, although providing a gross quantitative overview of location, is highly susceptible to artifactual measurements due to the physical destruction and separation of cellular structures. Indirect immunofluorescence and immunocytochemical electron microscopy are somewhat milder approaches; yet they yield primarily qualitative data and are highly dependent on sensitive immunologic reagents. Nevertheless, it has been possible, using a combination of these approaches, to provide a subcellular analysis of almost all the *onc* gene products.

This chapter provides a brief description of oncogenic retroviruses as well as their *onc*-encoded polypeptides, followed by a more detailed account of the intracellular distribution of these transforming proteins. We present the *onc* gene proteins in groups derived from avian sarcoma viruses, mammalian tumor viruses, and avian acute leukemia viruses. Among the viruses, the cellular location and biochemical properties of the *src* gene product of Rous sarcoma virus (RSV) have been more extensively studied; therefore, we describe this transforming protein first and with greater detail than the other transforming proteins. In our discussion, we identify several subcellular compartments that host distinct *onc* gene products and discuss these subcellular locations in regard to the transformation mechanism.

TRANSFORMING PROTEINS OF RSV

src Gene Protein

RSV was the first retrovirus isolated that was capable of inducing solid tumors in susceptible animals. It is a replication-competent virus, a fact that has facilitated studies on its transformation mechanism. The transforming gene sequence in RSV is termed *src* (for sarcomagenic) and codes for the synthesis of the $pp60^{src}$ transforming protein (27), which is a 60K protein phosphorylated at both a serine and tyrosine residue and possesses a tyrosine-specific kinase activity (48). Thus far, $pp60^{src}$ is the only protein product detected from the *src* gene; however, a second potential coding sequence within the *src* gene has been identified, but no additional protein product has yet been found (92). The interaction of $pp60^{src}$ with multiple targets has been proposed as a mechanism to explain the induction of the total transformed phenotype by RSV (8,14,17).

Cytoplasm

Results of immunofluorescence, immunocytochemical, and cell fractionation experiments have illuminated an increasingly complex spectrum of interactions of $pp60^{src}$ with target cells. Initial reports on the immunofluorescent localization of $pp60^{src}$ revealed that in contrast to the then recognized nuclear location of papova virus large T antigen (138), the *src* gene product of RSV is situated in the cytoplasm (28,113). Although this was somewhat surprising, it was consistent with earlier findings that nucleus-independent cytoplasmic targets exist for the *src* gene product (17,94). Cell fractionation studies also detected from 10 to 30% of $pp60^{src}$ in the soluble cytoplasm (38,87,88). The immunochemical localization techniques consistently indicated that a higher proportion of $pp60^{src}$ was in the soluble cytoplasm (28,113,148). These results could have been biased, however, by the presence of low amounts of autoantibodies for cytoplasmic constituents. This is likely in view of the observation that lower levels of diffuse cytoplasmic fluorescence are seen in RSV transformed cells when highly specific monoclonal or polyclonal antibodies to defined peptide regions of $pp60^{src}$ are employed for immunofluorescence (60,97; L. E. Gentry and L. R. Rohrschneider, *in preparation*).

The soluble $pp60^{src}$ is believed to represent newly synthesized protein in transit to the plasma membrane and is complexed with a 90,000 dalton heat shock protein and a 50,000 dalton cellular protein (37,151). The precise function of this complex is not known, but because the inactive *src* protein in tsNY68-infected cells maintained at the nonpermissive temperature accumulates in the soluble 60K:90K:50K complex, this form probably does not function in the transformation mechanism. The complexed form of $pp60^{src}$ contains little or no tyrosine-specific kinase activity (37,151).

Plasma Membrane

The results of several experimental approaches indicate that the majority of $pp60^{src}$ in RSV transformed cells is situated on the cytoplasmic surface of the plasma membrane. Cell fractionation experiments estimate that approximately 80 to 90% of $pp60^{src}$ is associated with the membrane fraction. However, differences were observed between avian and mammalian cells and depended on the fractionation technique utilized (38,87,88). The hydrophobic interaction of $pp60^{src}$ with the plasma membrane may be attributed to the incorporation of a long chain fatty acid in the amino terminal 8000 dalton domain (56,86,124).

Although these studies indicate the association of the *src* protein with plasma membranes, finer details of that organization have been obtained from techniques that disrupt the cellular structure as little as possible. An immunocytochemical staining technique utilizing ferritin as an electron dense marker detected the pervasive presence of $pp60^{src}$ in the plasma membranes of RSV transformed rat cells (148) (see Figs. 1–3). Specific concentrations of $pp60^{src}$ were also observed in such specialized membrane regions as ruffles (Fig. 2A) and particularly at junctions between cells (113,148) (Figs. 1D, 2B, and 3). Neither immunocytochemical nor immunofluorescence techniques detected $pp60^{src}$ on the external surface of the cells; likewise, lactoperoxidase-catalyzed iodination of the surface of RSV transformed cells failed to iodinate detectable $pp60^{src}$ (88).

By immunofluorescence, the labeling of $pp60^{src}$ in cell-cell junctions often represents the most intense and obvious staining pattern (see Fig. 3). In RSV transformed chicken cells, these junctions can appear as either relatively straight, smooth membrane boundaries between cells (Fig. 1C) or as jagged "stitch-work" (Fig. 1D) that often seems to represent the extensions of fibers that connect cells. In RSV-transformed rat epithelial cells, extensive junctional formation occurs between cells in contact; these junctions all stain with antibodies to $pp60^{src}$ (Fig. 3). Although the junctions displayed here are disparate in structure, the *src* proteins in all these junctions share a common association with vinculin, actin, and α-actinin (131,132). This observation suggests that these are probably not gap junctions, because these

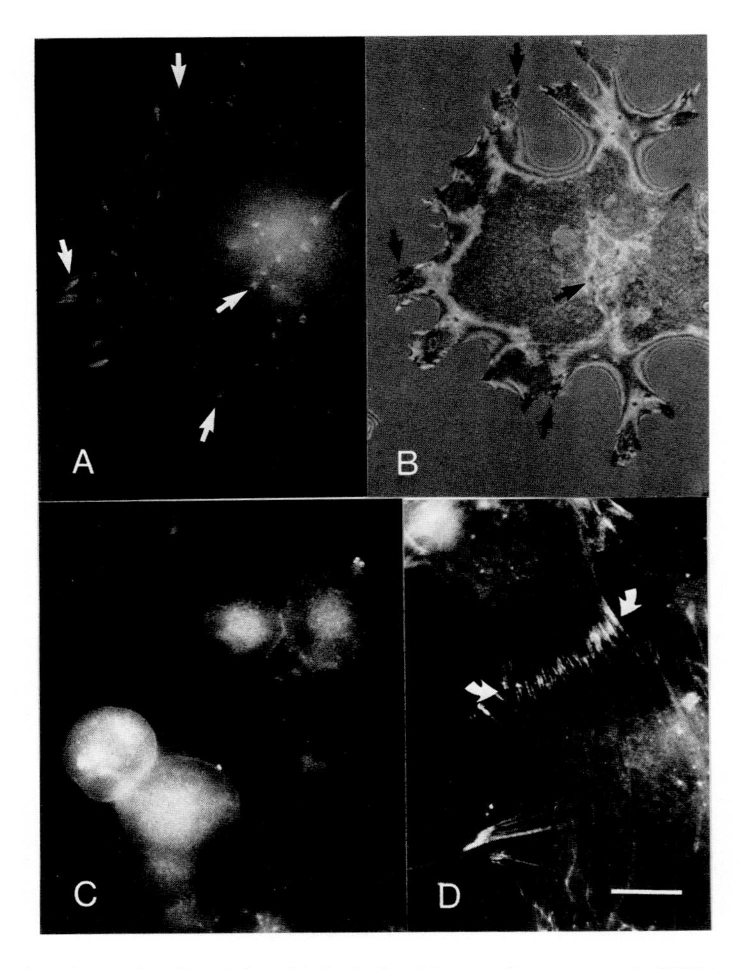

FIG. 1. Locations of $pp60^{src}$ detectable by indirect immunofluorescence in RSV transformed chicken embryo cells (CEC). Affinity-purified antibodies to a peptide region (peptide 1) of $pp60^{src}$ (amino acids 498–512, described in ref. 60) were used to detect the *src* transforming protein. Indirect immunofluorescence with these antibodies reveals the presence of $pp60^{src}$ as speckles on the ventral cell surface **(A)**. This fluorescence pattern for $pp60^{src}$ corresponds exactly with the dark focal contact areas (adhesion plaques) detected under the same cell **(B)** by interference-reflection microscopy. *Arrows*, representative adhesion plaques and the location of corresponding $pp60^{src}$-specific fluorescence. In more rounded cells, $pp60^{src}$ is detectable in the plasma membrane and at junctions that form between these cells **(C)**. In flattened RSV transformed cells, these junctions can display an intricate stitch-work of fluorescence, indicating the location of $pp60^{src}$ **(D)**. Bar = 10 μm, except for **C**, where it represents 16 μm.

cytoskeletal proteins are not known to be associated with gap junctions (59,82). Rather, the junctions containing $pp60^{src}$ more closely resemble the fascia adherens of heart intercalated discs, the zonula adherens of intestinal epithelial cells, and the membrane-associated dense plaques of smooth muscle cells (59,131,137). All

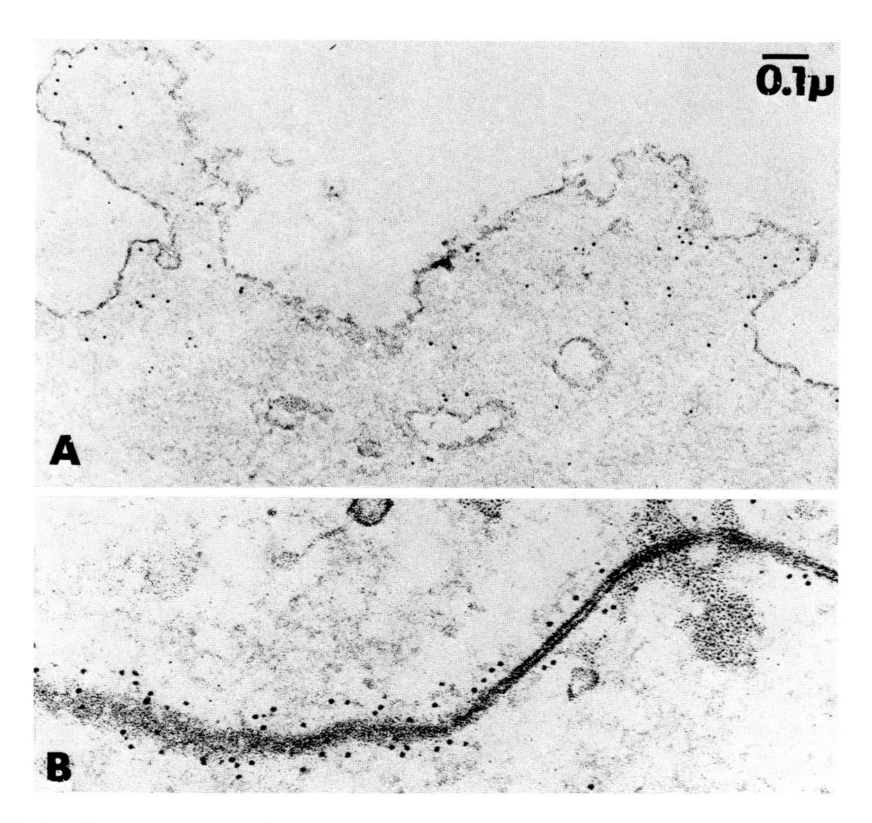

FIG. 2. Electron microscopic immunocytochemical localization of pp60^{src}. The transforming protein, pp60^{src}, in SR-NRK was detected by an immunoferritin-labeling procedure described in refs. 148 and 149. The location of the dark ferritin cores indicates that pp60^{src} is concentrated on the inner face of the plasma membrane, especially in ruffles **(A)** and at junctions between SR-NRK cells **(B)**. (Photograph kindly provided by M. Willingham and I. Pastan, from ref. 148.)

three of these junctions contain vinculin *in vivo*, and microfilament bundles anchor to the plasma membrane at these regions. It has been suggested that vinculin may link the microfilament bundles to the membrane at these sites (59,137).

It is not clear how or why pp60^{src} becomes concentrated in such specialized plasma membrane regions as the cell-cell junctions and adhesion plaques (described below). Although it could enter these sites independently, it is possible that migration along the cytoplasmic face of the plasma membrane would allow access to both these specialized regions. It is also likely that something in either the structure or functioning of these membranes and junctions is acting as a site for pp60^{src} association. In this regard, it is significant that a 36K dalton protein has been detected on the cytoplasmic side of the membrane and at cell junctions, and is a substrate for the tyrosine kinase activity of pp60^{src} (39; J. Cooper and S. Martin, *personal communication*).

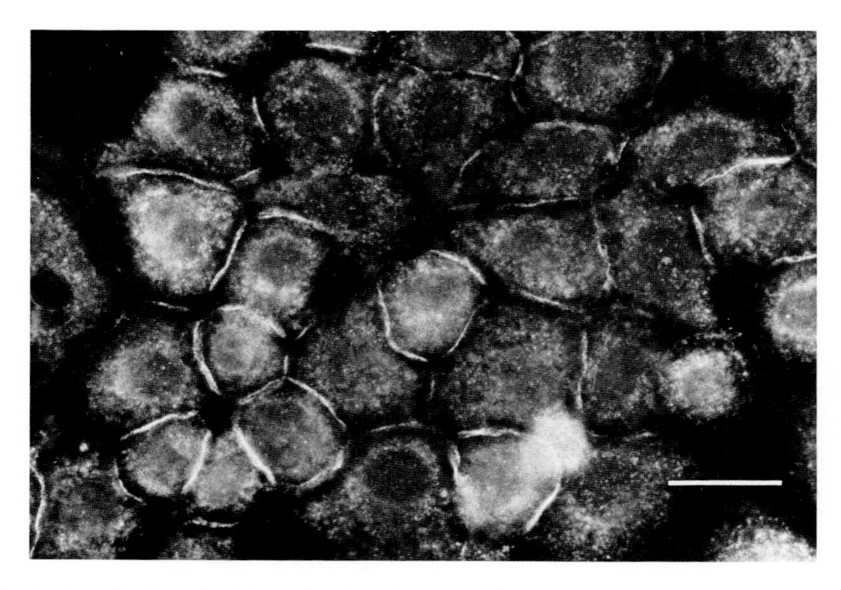

FIG. 3. Localization of $pp60^{src}$ in junctions between SR-NRK cells. SR-NRK cells form dense junctions between cells when confluent, and these junctions contain actin, α-actinin, and vinculin (131). Antibodies to $pp60^{src}$ also stain these junctions, as shown here by indirect immunofluorescence. Bar = 20 μm.

Adhesion Plaques

In fibroblasts and most cells grown in culture, specialized plasma membrane regions participate in attachment to the noncellular substratum. These regions, called focal contacts or adhesion plaques, represent the closest points of approach between the ventral cell membrane and the substratum on which the cell grows (1,79,80). These adhesion regions were first identified in normal, uninfected cells and described by a technique termed interference-reflection microscopy (42). It was later demonstrated that the adhesion plaques of normal cells represented the termini of microfilament bundles within the cell (1,2,76,80,147). The microfilament bundles traverse the cytoplasmic face of the ventral plasma membrane. Their termini within adhesion plaques serve to anchor the plasma membrane to the substratum at those points. The cell membrane of the adhesion plaques is situated within 150Å of the substratum. Extracellular matrix proteins, such as fibronectin, are thought to participate in the anchorage mechanism (133), although the exact nature of that mechanism is not understood. Vinculin also is contained within the adhesion plaques (32,58), and the organization of actin filaments and vinculin closely resembles the structural organization of these proteins in the cell junctions described above.

Using a combination of immunofluorescence and interference-reflection microscopy, it was demonstrated that the adhesion plaques of RSV transformed cells contained the *src* gene product (114) (see Fig. 1A and B). Tumor-bearing rabbit serum to $pp60^{src}$ stained transformed cells at junctions between cells and at membranes in highly rounded cells and also gave a diffuse cytoplasmic staining. When the focal plane of the fluorescence microscope was set on the ventral surface of RSV transformed cells, however, intense speckles of fluorescence were evident. By comparing the location of the fluorescent speckles with the distribution of adhesion plaques seen in the interference-reflection image of the same cell, it was evident that the fluorescence corresponded almost exactly to the position of the adhesion plaques (see Fig. 1A and B). The presence of $pp60^{src}$ at these sites was demonstrated further by isolating the adhesion plaques of RSV transformed normal rat kidney (NRK) cells and demonstrating that $pp60^{src}$ was present and showed an overall enrichment. Interestingly, $pp60^{src}$ in the isolated adhesion structure could not be readily solubilized in nonionic detergents (132). Approximately 12% of the total cellular $pp60^{src}$ was contained in the isolated adhesion plaques of RSV-NRK cells (132). Recently, highly specific polyclonal antibodies to defined peptide regions of $pp60^{src}$ and monoclonal antibodies to $pp60^{src}$ have confirmed the existence of $pp60^{src}$ at the adhesion plaque sites (60,61,97,123). An adhesion plaque location for $pp60^{src}$ has been observed in cells transformed by each different strain of RSV (L. E. Gentry and L. R. Rohrschneider, *unpublished observation*).

The location of $pp60^{src}$ within the adhesion plaque structures is temperature dependent within cells transformed by a virus containing a temperature-sensitive mutation in the *src* gene product (i.e., tsNY68), indicating the functional association of $pp60^{src}$ with the adhesion plaque sites (114). Two sets of experiments suggest that the tyrosine-specific kinase activity of $pp60^{src}$ is active in the adhesion plaques. First, when isolated adhesion plaques of SR-NRK cells were placed in a kinase buffer containing [γ-^{32}P]-ATP, $pp60^{src}$ became phosphorylated on tyrosine within the appropriate peptide region (132; K. Shriver and L. R. Rohrschneider, *in preparation*); several additional proteins also were phosphorylated on tyrosine. When adhesion plaques were isolated from NRK cells infected with a RSV mutant containing a temperature-sensitive defect in the *src* gene product, the phosphorylation of $pp60^{src}$ and other proteins was temperature-sensitive (132).

The second set of experiments supporting the view that $pp60^{src}$ is active as a kinase in the adhesion plaques deals with the adhesion plaque protein, vinculin. Both vinculin and $pp60^{src}$ exhibit a remarkable colocalization within adhesion plaques and cell-cell junctions of RSV-transformed cells (97,131). This colocalization prompted the examination of vinculin as a substrate for the tyrosine kinase activity of $pp60^{src}$. It was subsequently demonstrated that vinculin in RSV transformed cells contains a five- to 10-fold increase in abundance of phosphotyrosine relative to that found in uninfected cells (121). *In vitro* studies demonstrated that purified vinculin could be phosphorylated on tyrosine with the purified *src* gene product (77), and that the phosphorylation *in vivo* and *in vitro* occurred on the same sites (I. Pastan, *personal communication*). It is not known in which cellular compartment vinculin becomes phosphorylated. However, the experimental findings of both colocalization and an enzyme-substrate relationship *in vivo* and *in vitro* indicate that vinculin is a genuine substrate for the enzymatic action of $pp60^{src}$ and suggest that the adhesion plaques and functions associated with or governed by those structures are involved in the mechanism of transformation by RSV.

In addition to the kinase activity of $pp60^{src}$, there are indications that other actions of the *src* protein are at work in adhesion plaque localization. Analysis of several mutants of RSV has indicated that mutants exist that do not express detectable $pp60^{src}$ in the adhesion plaques yet express a functional kinase (115,116). Unlike wild type RSV, these mutants elicit a fusiform transformed morphology, and infected cells contain fibronectin in their extracellular matrix. This suggests a relationship between the presence of $pp60^{src}$ in adhesion plaques and both fibronectin matrix formation and cell morphology. The exact nature of that relationship is not yet understood.

Cytoskeleton

It has been reported that the majority of $pp60^{src}$ in RSV transformed cells is associated in some way with the cytoskeletal framework (30,31). About 90% of the $pp60^{src}$ remains with the detergent-insoluble matrix after extraction, while 10% is obtained in the soluble extract. This result seems inconsistent with other cellular locations reported for $pp60^{src}$. However, it is more likely that this finding reflects the extensive interactions of the cytoskeleton with various cellular structures. Clearly, the adhesion plaques and cell junctions are part of the cytoskeletal network. We have found that $pp60^{src}$ remains in these structures when extracted with nonionic detergent-containing buffers, suggesting that the *src* protein at these sites is cytoskeletal associated (31) (L. R. Rohrschneider, *unpublished observation*). Although other areas of the plasma membrane do not exhibit an obvious connection to the cytoskeleton, many membrane proteins of cultured fibroblasts as well as intact erythrocytes are also components of the cellular matrix (16,24). Hence, in these regions of the cell, it is possible to imagine a coassociation of $pp60^{src}$ with the plasma membrane and the cytoskeletal framework.

***yes* Gene Proteins**

Two replication-defective strains of avian sarcoma virus, Yamaguchi 73 (Y73) and Esh sarcoma virus (ESV), have recently been isolated from spontaneous tumor tissue in chickens (78,143). Like RSV, these virus isolates are capable of inducing sarcomas *in vivo* and transforming fibroblasts *in vitro*; however, they carry within their genomes transforming sequences which, by nucleic acid hybridization, appeared unrelated to the *src* gene (62,81). These transformation-specific sequences have been termed *yes*. The *yes* genes of Y73 and ESV are expressed as *gag*-linked phosphoproteins of approximately 90,000 ($P90^{gag\text{-}yes}$) and 80,000 ($P80^{gag\text{-}yes}$) daltons, respectively. They exhibit associated kinase activities specific for tyrosine residues similar to $pp60^{src}$ and to the transforming proteins of other defective sarcoma viruses (62,63,81). Both the *gag-yes* polyproteins are structurally related to each other, containing peptides and immunologic determinants of $p19^{gag}$ (but not

$p27^{gag}$) and sharing a majority of methionine- and cysteine-containing non-*gag* tryptic peptides. The methionine peptide maps of $P80^{gag\text{-}yes}$ and $P90^{gag\text{-}yes}$, however, bear no relationship to those of $pp60^{src}$ or to the *fps* portions of $P105^{gag\text{-}fps}$ or $P140^{gag\text{-}fps}$ (15,62,63).

The complete nucleotide sequence of Y73 has recently been reported and has provided some important insights into the structure and possible function of the *yes* transforming protein (83). Although early reports indicated that the transforming genes *yes* and *src* were unrelated, close examination of their nucleotide sequences has shown otherwise. At the level of the nucleotide sequence, only 31% of the transforming sequences are homologous; however, the protein products predicted from these genes exhibit a strikingly high degree of homology, approaching more than 95% in some regions. The greatest extent of homology exists in the carboxy-terminal half of the protein molecules, the region most likely responsible for the tyrosine kinase activity, a fact consistent with their similar enzyme specificities, metal ion requirements, and pH optima (63,110). This close structural similarity between the gene products of *yes* and *src* suggests that they may elicit similar mechanisms for cellular transformation.

In an attempt to more thoroughly understand the mechanism of transformation by Y73 and ESV, the subcellular location of the v-*yes* gene products was investigated by indirect immunofluorescence microscopy (61). Due to the absence of stable nonproducer clones of Y73 and ESV transformed cells, it was necessary to utilize antibodies that were specific for the transforming regions of $P90^{gag\text{-}yes}$ and $P80^{gag\text{-}yes}$. Site-specific antibodies to a defined peptide segment of $pp60^{src}$ were found to be cross reactive with the corresponding region in the v-*yes* transforming proteins (60); thus these antibodies were employed for immunofluorescence localization studies (61).

Figure 4 shows the results that were obtained when Y73 and ESV transformed chicken embryo fibroblasts were stained with these cross reactive antipeptide antibodies. Using a combination of indirect immunofluorescence and interference-reflection microscopy, it is clear that the *yes* transformation-specific polyproteins are found within adhesion plaque structures and needle-like interdigitating cellular junctions. In addition to these specialized cellular structures, fluorescence was observed in the cytoplasm of cells transformed with these viruses. A distinct membranous fluorescence was not observed in these cells; however, such flat cells are not readily amenable to detection of low abundance membrane proteins by immunofluorescence. Cell fractionation would be better suited for that analysis. Nevertheless, at the immunofluorescence level, the intracellular distribution of the *yes* proteins appears to be remarkably similar to that of $pp60^{src}$, providing further support that *yes* and *src* are functionally alike.

fps Gene Proteins

Several independent stocks of virus have been isolated from spontaneous sarcomas in chickens which possess homologous transforming sequences unrelated to

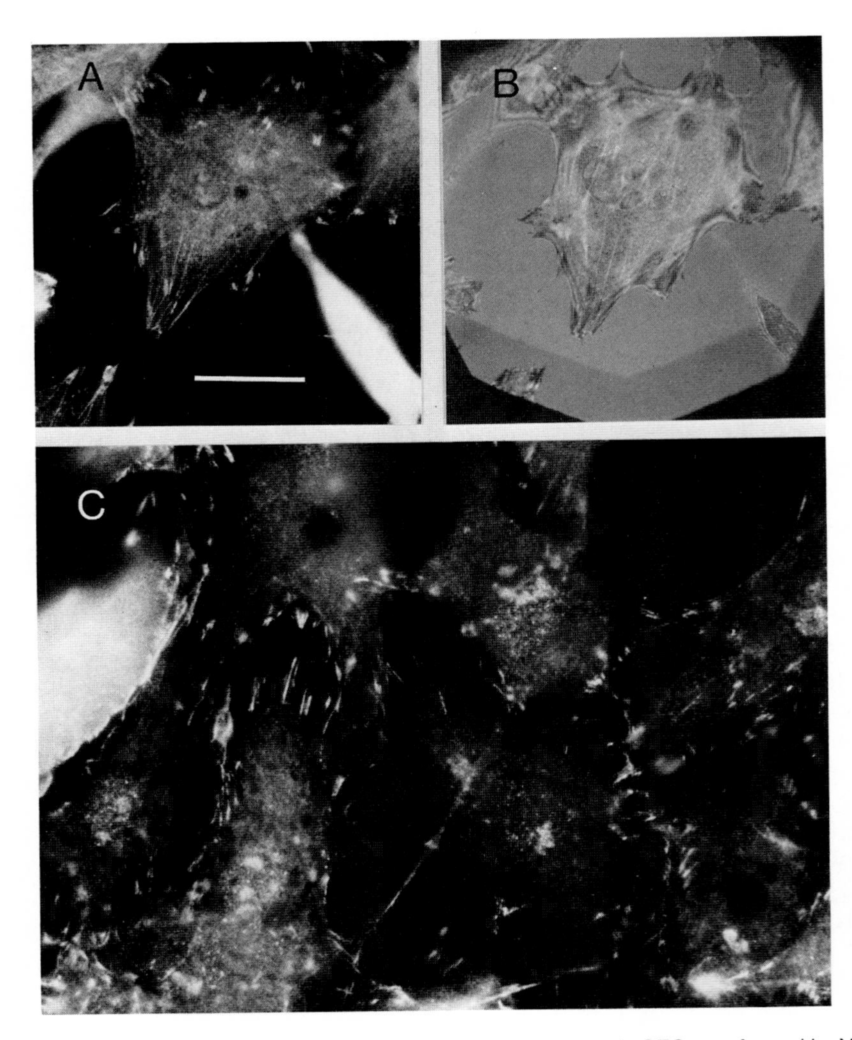

FIG. 4. Location of the *yes* gene products $P90^{gag\text{-}yes}$ and $P80^{gag\text{-}yes}$ in CEC transformed by Y73 virus and ESV, respectively. Affinity-purified antibodies to the peptide 1 region of $pp60^{src}$ cross reacted with the $P90^{gag\text{-}yes}$ protein of Y73 and were used to examine the intracellular distribution of $P90^{gag\text{-}yes}$ in Y73 transformed CEC (61). **A:** Immunofluorescent distribution of $P90^{gag\text{-}yes}$. **B:** Interference-reflection image of the adhesion plaques under the same cell. $P90^{gag\text{-}yes}$ is detectable in the cytoplasm and also concentrated in membrane areas that correspond to the adhesion plaques. Immunofluorescence photograph **(C)** presents a more extensive view of the distribution of $P80^{gag\text{-}yes}$ in ESV transformed CEC. All fluorescence, except a nonspecific nucleolar stain, could be eliminated by preabsorbing the antibodies with excess peptide. Bar = 20 μm.

the *src* gene of RSV. This group of viruses is composed of FuSV (54), several isolates from the Poultry Research Center [PRCII, PRCIIp, and PRCIV (33)], and the most recently discovered UR1 (11). All are replication-defective forms requiring associated helper virus for infectious particles (11,25,26,91), and all carry transformation-specific sequences called *fps*.

Viruses of this group have a target cell specificity similar to that of RSV. When injected intramuscularly into young chickens, however, they induce sarcomas with a preponderance of spindle-shaped cells (11,25,26,54,91) rather than the typical round cell morphologies produced by RSV infection. In tissue culture, the results are generally the same. Agents bearing *fps* induce characteristic fusiform transformation of fibroblasts (25,26,91). UR1, however, is unique; in culture, it produces not only foci consisting of fusiform cells but also foci containing round cells (11).

All the viral genomes containing *fps* sequences synthesize a single protein varying in size from 105K to 170K daltons containing both *gag* and *fps* encoded portions (26,70,91,95,145). These nonstructural, *gag*-related proteins have associated with them a kinase activity which, like the *src* gene product of RSV, phosphorylates substrate proteins at tyrosine residues (26,49,95,145). Support for the view that this tyrosine-specific kinase activity is encoded by the viral genome and is not due to adventitiously associated proteins has come from the analysis of viral mutants temperature-sensitive for transformation (69,102). These viruses produce *gag-fps* transforming proteins which exhibit thermolabile enzymatic activities when assayed either *in vitro* or *in vivo*. Thus it appears that cellular transformation by the *fps*-containing viruses requires the inherent tyrosine-specific kinase activity of the *fps* transforming protein.

With respect to the intracellular distribution of their *gag*-related, transformation-specific polyproteins, the two most thoroughly studied viruses of this group are FuSV and PRCII. These viruses produce *gag-fps* polyproteins of 140K daltons ($P140^{gag\text{-}fps}$) and 105K daltons ($P105^{gag\text{-}fps}$), respectively, which contain determinants of $p27^{gag}$ as well as $p19^{gag}$ (15,91,95). The presence of these *gag* determinants has allowed for not only the identification of these polyproteins but also studies examining their structure and function.

The PRCII transforming protein, although initially synthesized as a 105K dalton polypeptide ($P105^{gag\text{-}fps}$) is posttranslationally modified to yield a protein of 110K daltons ($P110^{gag\text{-}fps}$). This modification apparently involves phosphorylation and yields roughly equimolar amounts of the two proteins at steady state (4,5). The intracellular fate of these proteins has been studied. It was found that the precursor and its modified counterpart are localized in different subcellular compartments of PRCII transformed chicken fibroblasts (5). $P110^{gag\text{-}fps}$ was found to reside almost entirely in the cytoskeletal matrix fraction, whereas $P105^{gag\text{-}fps}$ fractionated with the cytoplasm and plasma membrane components soluble in nonionic detergents. When the two individual fractions were tested for tyrosine-specific kinase activity in the immune complex, it was found that the cytoskeletal-bound form, $P110^{gag\text{-}fps}$, was highly active. The soluble form of this transforming protein ($P105^{gag\text{-}fps}$), on the other hand, was essentially devoid of active kinase (5). These results are similar to the findings of others that the transforming proteins of $pp60^{src}$ and $P120^{gag\text{-}abl}$ and their associated kinase activities are contained in cytoskeletal preparations of virally transformed cells (22,30).

The intracellular distribution of the FuSV transforming protein has been determined by indirect immunofluorescence and quantitative cell fractionation (51). Using rat tumor antisera specific for the *fps* portion of $P140^{gag\text{-}fps}$ and indirect immunofluorescence analysis, it was demonstrated that this transformation-specific polyprotein was located predominantly within the cytoplasm of FuSV-infected cells and also concentrated at the plasma membrane where two cells would make contact. The $P140^{gag\text{-}fps}$ transforming protein was not detectable in adhesion plaques or close contact areas (51; L. R. Rohrschneider, *unpublished observation*). A similar immunofluorescence pattern has been observed using antibodies specific for a peptide region of $pp60^{src}$ which cross react with the *fps* transforming protein (L. R. Rohrschneider and L. E. Gentry, *unpublished observations*). Typical immunofluorescence staining patterns of FuSV and PRCIV transformed cells produced by these cross-reactive, peptide-specific antibodies are shown in Fig. 5.

Quantitative cellular fractionation of FuSV transformed cells showed $P140^{gag\text{-}fps}$ to be associated with plasma membrane and cytoplasmic components, confirming the immunofluorescence results (51). The association of this transforming protein with the membrane fraction was not affected by ionic or nonionic detergents, yet it was readily solubilized when extracted by salts. Thus, in contrast to $pp60^{src}$, which appears to be tightly bound to membrane, $P140^{gag\text{-}fps}$ is only weakly associated; this interaction is governed by salt-sensitive, protein-protein interactions. The fact that the PRCII *gag-fps* transforming protein resides in the cytoskeletal elements makes it plausible that $P140^{gag\text{-}fps}$ may also be associated with these structures. In fact, some of the conditions employed for the extraction studies of $P140^{gag\text{-}fps}$ from

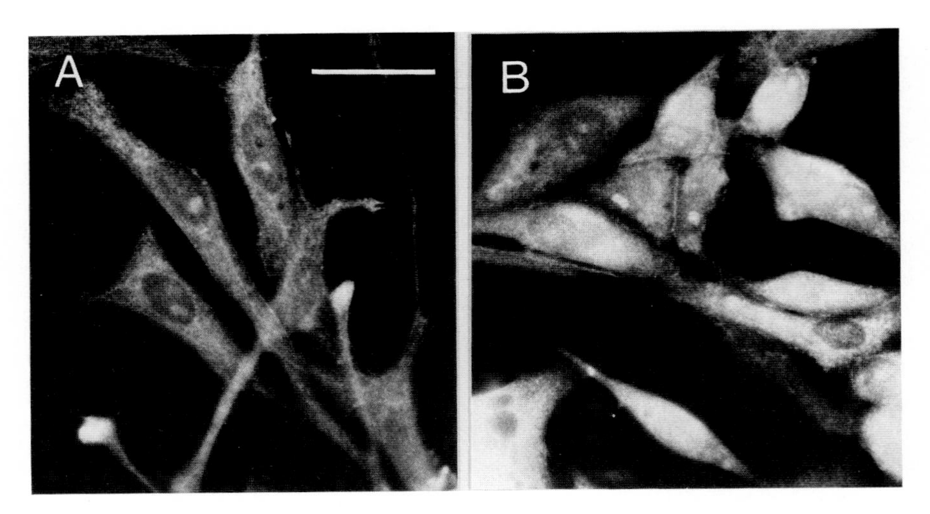

FIG. 5. Immunofluorescent localization of *gag-fps* gene products in transformed CEC. The same affinity-purified antibodies to peptide 1 of $pp60^{src}$ cross reacted with *fps* gene products and were employed for these indirect immunofluorescence experiments (60). **A:** Fluorescent location of $P140^{gag\text{-}fps}$ in FuSV transformed CEC. **B:** Fluorescent location of $P170^{gag\text{-}fps}$ in PRCIV transformed CEC. Again, the nucleolar stain was nonspecific. Bar = 20 μm.

whole cells (51) are similar to procedures utilized for the preparation of cytoskeletons (5,22,31). Under these conditions, the majority of $P140^{gag\text{-}fps}$ remained bound to the insolubilized cytoskeletal structures. Therefore, it becomes likely that $P140^{gag\text{-}fps}$ may also be associated with the cytoskeleton network of the cell similarly to the PRCII transforming protein.

ros Gene Proteins

The transforming virus UR2 represents a recently isolated avian sarcoma virus, which contains transformation-specific genetic information different from *src*, *fps*, or *yes* (11,128). The unique transforming gene sequence present in this virus has been designated as *ros* (146). To date, UR2 is the only RNA tumor virus that has been shown to contain *ros*-related information (128). UR2 is replication defective; when associated with an appropriate helper virus, it is able to transform fibroblasts *in vitro* and induce sarcomas *in vivo* (11).

The UR2 genome encodes a 68K dalton *gag-ros* gene product which, like the *gag*-containing, transformation-specific polyproteins of other defective avian sarcoma viruses, is associated with a protein kinase activity that phosphorylates acceptor proteins at tyrosine residues (50). Although this protein appears to share a similar specificity for tyrosine residues, its enzymatic properties distinguish it from the transforming gene products of other avian sarcoma viruses, a fact consistent with its lack of homology at the DNA level (11,50,128). Cells transformed by UR2 display an accentuated fusiform shape. However, other cellular alterations that usually accompany transformation have not been examined, and subcellular localization studies on $P68^{gag\text{-}ros}$ remain to be reported.

TRANSFORMING PROTEINS OF MAMMALIAN TUMOR VIRUSES

fes (and *fms*) Gene Proteins

Three independent viral isolates comprise the family of feline sarcoma viruses (FeSV). Each virus in this group induces fibrosarcomas on injection into susceptible cats and transforms a broad spectrum of cells in culture. Two of the FeSV isolates are related in that their acquired *onc* genes and protein products are homologous (12,13,141). These two genes are designated v-*fes* and are contained within the genome of both Snyder-Theilen (ST) and Gardner-Arnstein (GA) strains of FeSV. The v-*fes* gene and its protein product also are related to the v-*fps* gene carried by some avian sarcoma viruses such as PRCII and FuSV (12,15,68,127,129). The third [McDonough (SM)] strain of FeSV contains an *onc* gene distinct from that found in ST-FeSV and GA-FeSV and is called v-*fms*. The transforming proteins of these viruses are synthesized as *gag-onc* polyproteins and designated $P85^{gag\text{-}fes}$ (ST-FeSV), $P95^{gag\text{-}fes}$ (GA-FeSV), and $gP180^{gag\text{-}fms}$ (SM-FeSV) (13,109,118,141). SM-FeSV transformed cells contain additional *fms*-related proteins of 140K and 120K daltons, and all three are glycosylated (9). Both the ST- and GA-FeSV transforming proteins have an associated tyrosine kinase activity. However, the transforming proteins from SM-FeSV do not appear to have a similar activity, and SM-FeSV transformed cells do not contain elevated levels of phosphotyrosine (109).

A number of monoclonal antibodies to both the *gag*- and *onc*-specific portions of all three strains of FeSV has allowed experiments to be carried out on the intracellular localization of these transforming proteins (9,142). In Fig. 6 are presented the results of immunofluorescence experiments performed on rat cells transformed with either ST-FeSV or SM-FeSV (L. R. Rohrschneider, F. H. Reynolds, J. R. Stephenson, *in preparation*). The cells are nonproductively transformed (109); therefore, in the ST-FeSV transformed cells, a monoclonal antibody to the amino terminal *gag* protein, p15, was used to detect the location of the $P85^{gag\text{-}fes}$. The fluorescence results indicated that some $P85^{gag\text{-}fes}$ was situated within the cytoplasm around the nucleus, and that a distinctly membranous component also was present. Ruffled membranes near the periphery of the cells stained with the monoclonal antibody and junctions between cells contained prominent fluorescent stain. Both fluorescent staining patterns are indicative of membranous localizations. $P85^{gag\text{-}fes}$ was not found in adhesion plaques.

The detection of $P85^{gag\text{-}fes}$ in the cytoplasm and membrane of ST-FeSV transformed cells did not change when other antibodies to different epitopes were tested. When monoclonal antibodies to the *fes*-specific portion of $P85^{gag\text{-}fes}$ were used in

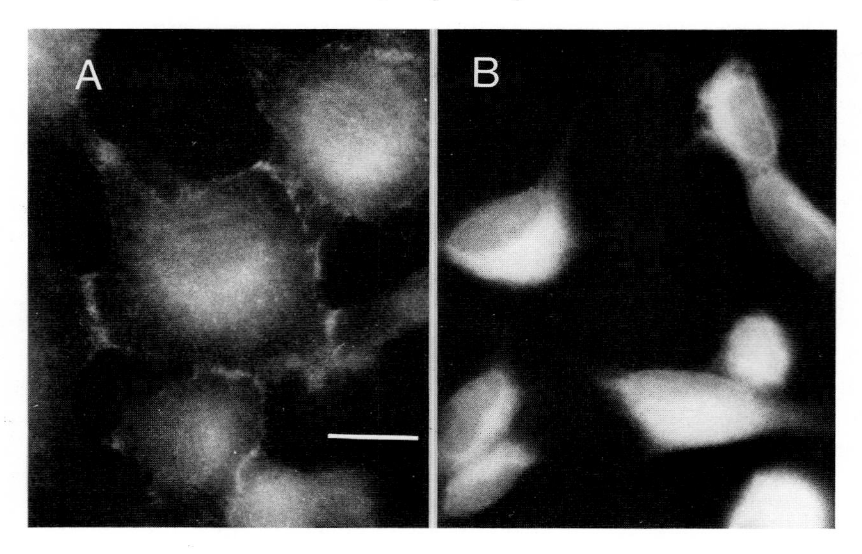

FIG. 6. Immunofluorescent localization of $P85^{gag\text{-}fes}$ and $gP180^{gag\text{-}fms}$ in transformed rat cells. Fischer rat embryo (FRE) cells nonproductively transformed with ST-FeSV were reacted with monoclonal antibodies to the *gag* protein p15 of $P85^{gag\text{-}fes}$ (F. Reynolds, J. Stephenson, and L. R. Rohrschneider, *in preparation*). The cytoplasmic and membrane distribution of $P85^{gag\text{-}fes}$ detected by indirect immunofluorescence is shown in **A**. The fluorescent distribution of $gP180^{gag\text{-}fms}$ in FRE cells nonproductively transformed with SM-FeSV was similarly revealed using monoclonal antibodies to the p30 *gag* protein **(B)**. An intense fluorescence is detectable in the cytoplasm of the SM-FeSV transformed cells. Membrane ruffles also stained, but less intensely (not visible in photograph). Bar = 10 μm.

these experiments, an identical localization was found (L. R. Rohrschneider, *unpublished observation*). This suggests that these epitopes are not eclipsed at these locations. Analysis of GA-FeSV transformed nonproducer cells showed that the $P95^{gag\text{-}fes}$ protein is cytoplasmic and also exhibited an identical, albeit less dramatic, membrane localization. The localization of both these *gag-fes* proteins is qualitatively similar to that reported for *gag-fps*. This is consistent with the fact that these three independently acquired *onc* genes are homologous and synthesize highly related gene products (68,127).

The SM-FeSV carries the distinctly different *fms* transforming gene coding for the $gP180^{gag\text{-}fms}$ (52,111). Monoclonal antibodies to the *gag* protein p30 (142) were used to locate the *gag-fms* gene product within nonproducer transformed rat cells. The immunofluorescence results presented in Fig. 6B demonstrate an intense fluorescence in the cytoplasm that appears to avoid the nucleus. This is similar to previous perinuclear fluorescence obtained (9) using a monoclonal antibody to the v-*fms*-specific segment of $gP180^{gag\text{-}fms}$, which also detected the more abundant $gP120^{fms}$ and a $gP140^{gag\text{-}fms}$ protein. Using either antibody, however, the SM-FeSV transformed cells, unlike ST- or GA-FeSV cells, never exhibited fluorescence at cell-cell junctions. This different pattern of fluorescence suggests that the v-*fms* gene products may share a somewhat different intracellular location than the v-*fes* product. This difference also is reflected in the finding that, of the FeSV strains, only the v-*fms* products are glycosylated and, in addition, lack comparable tyrosine kinase activity (109). The v-*fms* proteins are not exposed on the external cell surface, and cell fractionations have placed their location in the particulate cytoplasm (9). Recently, the v-*fms* gene products have been detected in the cytoskeleton of SM-FeSV transformed cells in association with intermediate filaments (S. Anderson, M. Gonda, and C. J. Sherr, *personal communication*). The SM strain of FeSV, therefore, may transform cells by a mechanism and at an intracellular location different from other known sarcoma viruses.

It is worth mentioning that the ruffled lamellipodia of SM-FeSV transformed cells also stained with antibodies to the p30 of $gP180^{gag\text{-}fms}$ (L. R. Rohrschneider, *unpublished observation*). The intensity was considerably less than that seen in the cytoplasm and, therefore, was difficult to record on film. It is not certain whether this fluorescence pattern is due to cytoplasmic $gP180^{gag\text{-}fms}$ within the ruffle or some minor membrane location.

abl Gene Protein

The Abelson murine leukemia virus (A-MuLV) originally acquired its v-*abl* transforming gene by recombination of M-MuLV with an endogenous sequence in a Balb/c mouse. The acquisition allowed A-MuLV to induce *in vivo* a distinctly different leukemia than the parent MuLV and to transform bone marrow and fibroblast cells *in vitro*. The v-*abl* gene product responsible for this activity is a 120K dalton phosphoprotein containing some MuLV *gag* sequences and exhibiting a tyrosine-specific kinase activity similar to that contained in $pp60^{src}$ (122,144).

The transforming protein of the most common strain of A-MuLV is designated $P120^{gag\text{-}abl}$.

Antibodies to the *abl*-specific portion of $P120^{gag\text{-}abl}$ have been produced by immunizing a specific substrain (C57L/J) of mice with syngeneic A-MuLV transformed lymphoid cells (150). These antibodies stain the surface of viable A-MuLV transformed lymphoid cells, suggesting that at least a portion of $P120^{gag\text{-}abl}$ is exposed on the surface of these tumor cells. This was consistent with the demonstration that viable A-MuLV lymphoid cells could absorb out antibodies to the 120K dalton transforming protein (150).

It was initially thought that the *gag* determinants were not expressed on the external cell surface (150). Recently, however, monoclonal antibodies have detected p15 but not p12 determinants of the protein on the cell surface (N. Rosenberg and O. Witte, *personal communication*). Unlike classic cell surface proteins, however, $P120^{gag\text{-}abl}$ is not glycosylated, and the exposure of $P120^{gag\text{-}abl}$ on the cell surface is somewhat cryptic because neither lactoperoxidase-catalyzed iodination nor trypsin treatment is able to modify cellular $P120^{gag\text{-}abl}$ (150). The lack of trypsin sensitivity and inability to iodinate with lactoperoxidase could be due to the absence of target amino acids in the exposed segment. Alternatively, the exposure of portions of $P120^{gag\text{-}abl}$ on the cell surface may simply indicate that we do not understand everything about the dynamic nature of membrane proteins and their interaction with antibodies.

Other intracellular interactions of $P120^{gag\text{-}abl}$ have been described in addition to, or perhaps together with, the plasma membrane location. When A-MuLV transformed fibroblasts or lymphoid cells are extracted with nonionic detergents, about half the $P120^{gag\text{-}abl}$ can be recovered in the detergent-insoluble matrix (22). This indicates that at least a portion of the v-*abl* protein is associated with the cytoskeleton of transformed cells.

Using monoclonal antibodies to the *gag* proteins p12 and p15 of $P120^{gag\text{-}abl}$, the intracellular distribution of this protein in NRK cells nonproductively transformed with A-MuLV was recently reexamined (L. R. Rohrschneider, *unpublished observation*). This analysis was done because several similarities have emerged between $P120^{gag\text{-}abl}$ and $pp60^{src}$. Both proteins are in the plasma membrane, are associated with the cytoskeletal framework, and contain covalently bound long chain fatty acid (22,31,124,150). In addition, both transforming proteins cause the increased phosphorylation of vinculin on tyrosine; and a 1.2 kb sequence in the cloned v-*abl* gene shows strong nucleotide sequence homology with the C-terminal half of v-*src* (D. Baltimore, *personal communication*). Therefore, it was of interest to determine if $P120^{gag\text{-}abl}$ exhibited a cellular localization similar to $pp60^{src}$. The results in Fig. 7 demonstrate the broad distribution of $P120^{gag\text{-}abl}$ in the membrane of NRK-*abl* cells and the concentration of this protein at junctions between cells. Clearly, the membrane is the most prominent location for $P120^{gag\text{-}abl}$. The ventral surface of these cells, however, also exhibits specific concentrations of $P120^{gag\text{-}abl}$ in regions that probably correspond to substratum adhesion sites (Fig. 7, inset). A similar localization of $pp60^{src}$ has been described in other RSV transformed NRK cells

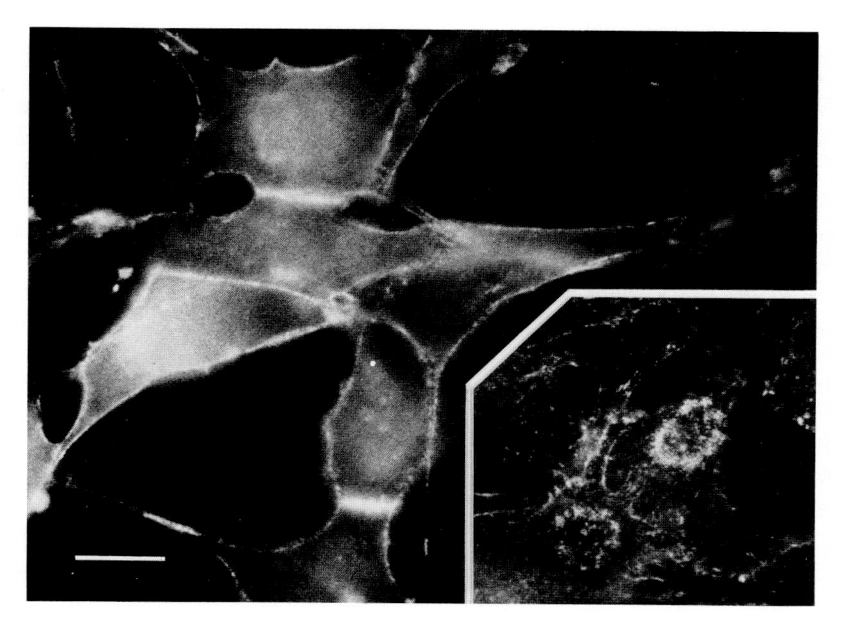

FIG. 7. Cellular distribution of $P120^{gag\text{-}abl}$ in A-MuLV transformed NRK cells. Nonproductively transformed cells were examined for the location of $P120^{gag\text{-}abl}$ by indirect immunofluorescence using monoclonal antibodies to the p12 and p15 *gag* proteins (obtained from B. W. Chesebro through O. Witte). The results illustrate a strong classic membrane fluorescence pattern that outlines the periphery of the cell at each focal plane. These cells probably contain very little cytoplasmic $P120^{gag\text{-}abl}$, because there is no definition of the nucleus within the cytoplasm. *Inset,* clusters of fluorescence at the ventral surface of the cells. These areas most likely represent cell-substratum adhesion sites. Bar = 10 μm.

(97). Therefore, by analogy with $pp60^{src}$, we suspect that the portion of $P120^{gag\text{-}abl}$ related to $pp60^{src}$ is situated on the cytoplasmic face of the plasma membrane and in substratum adhesion sites. The protein sequences of $P120^{gag\text{-}abl}$ unrelated to $pp60^{src}$ may show a more extensive membrane interaction, with some portions exposed on the outside of the cell membrane. The *src*-like sequence of $P120^{gag\text{-}abl}$ appears necessary for fibroblast transformation, whereas both the *src*-like and *gag* portions are necessary for lymphocyte transformation (106a). We speculate that the exposure of some portions of the $P120^{gag\text{-}abl}$ on the cell surface may be necessary for lymphocyte transformation.

ras Gene Proteins

The *ras* gene proteins have been identified as the transforming gene products synthesized from four independent strains of murine sarcoma virus (MuSV). The Harvey (Ha-MuSV) and Kirsten (Ki-MuSV) strains have acquired related but distinct *ras* genes (termed Ha-*ras* and Ki-*ras*), and the RaSV and B-MuSV strains carry Ha-*ras*-related genes originally picked up from a rat and mouse, respectively

(46). The viruses exhibit similar pathology, inducing sarcomas, splenomegaly, and erythroblastosis *in vivo* and transforming fibroblasts in culture.

The transforming protein in the case of Ha-MuSV is synthesized as a precursor of approximately 22,000 daltons (called pro-p21^{ras}) that is processed (perhaps proteolytically) to give p21^{ras} and finally phosphorylated to give the pp21^{ras} product (130). Synthesis of pro-p21^{ras} occurs on nonmembrane-bound ribosomes in the cytoplasm, and the nonphosphorylated product is then transported to the plasma membrane. The transfer does not occur as rapidly as found for pp60^{src}, and even a greater lag follows before the phosphorylated pp21^{ras} accumulates in the membrane fraction (up to 24 hr). The pp21^{ras} is found exclusively in the membrane fraction. At some point in the processing sequence, presumably before or during plasma membrane deposition, a long chain fatty acid is covalently attached to p21^{ras} and pp21^{ras} (124). This posttranslational modification may play a role in anchoring the v-*ras* gene products to the plasma membrane. The *ras* gene products also bind guanine nucleotides and possess an apparent autophosphorylating activity specific for threonine residues. It is not known whether one or all of the *ras* gene protein species contain these activities.

The location of the vast majority of the v-Ha-*ras* gene products in the plasma membrane of Ha-MuSV transformed cells was demonstrated initially by immunofluorescence and immunocytochemical techniques (149). Tumor-bearing rat anti-p21^{ras} sera stained the plasma membranes of transformed cells and did not detect *ras* protein determinants on the external cell surface. These results indicate that both p21^{ras} and pp21^{ras} are situated on the cytoplasmic face of the plasma membrane. Rat monoclonal antibodies to the v-*ras* gene products have been produced and confirm the plasma membrane location of the p21^{ras} proteins (55). It is interesting that the monoclonal antibodies detect more cytoplasmic p21^{ras} than the polyclonal antibodies (compare ref. 149 with ref. 55); this could result from the partial eclipse of the p21^{ras} epitope while in the plasma membrane site.

Other Mammalian *onc* Proteins

Several other mammalian tumor viruses exist with distinct *onc* genes, such as Simian sarcoma virus (SSV), Moloney murine sarcoma virus (Mo-MSV), and the Finkel, Biskis, Jinkins isolate of murine sarcoma virus (FBJ-MSV). These viruses are of interest because they contain distinct *onc* genes and may transform cells by mechanisms different than the other known viruses. At least in the case of the SSV v-*sis* gene product, some similarities with the intracellular location and properties of other transforming proteins have been detected. Recently, antibodies have been prepared to a synthetic peptide of the predicted v-*sis* gene product amino acid sequence derived from the known nucleotide sequence (112). These antibodies identify a 28K dalton protein as the product of the v-*sis* gene (gp28^{sis}). No associated kinase activity is found with this transforming protein. Cell fractionation studies have tentatively indicated that the gp28^{sis} protein is plasma membrane associated and not released into the culture supernatant (K. Robbins and S. Aaronson, *personal communication*). Results from the same laboratory also indicate that $gp28^{sis}$ is glycosylated at a tunicamycin-sensitive site. It is not known whether $gp28^{sis}$ transforms by mechanisms related to that of other glycosylated transforming proteins (e.g., the *erb* B protein and the $gP180^{gag\text{-}fms}$).

The transforming protein synthesized from the v-*mos* gene of Mo-MSV is designated $pp37^{mos}$ and was detected with antibodies prepared to a C-terminal peptide of the protein (101). Experiments with these antibodies indicate that $pp37^{mos}$ is phosphorylated at six serine sites and represents only 0.0005% of the total cell protein in Mo-MSV transformed cell lines (100). (For comparison, $pp60^{src}$ represents about 0.02% of the total cell protein.) This low level of expression precludes detection by fluorescence methods; however, cell fractionation indicates that $pp37^{mos}$ is a soluble cytoplasmic protein and not associated with other proteins or with cytoskeletal elements (100). Acutely infected cells express 30- to 100-fold higher levels of $pp37^{mos}$ that appears to reside in the soluble cytoplasm by both cell fractionation and fluorescence techniques (100). No enzymatic activity has yet been ascribed to this transforming protein.

The localization work on $pp37^{mos}$, especially in stable transformed cell lines, suffers from the low level of expression, and $pp37^{mos}$ at sites other than the soluble cytoplasm may not be detectable. In the stable transformed cell lines, a small percentage (10 to 20%) of active $pp37^{mos}$ could reside at intracellular sites other than the soluble cytoplasm.

The FBJ virus was originally isolated from a spontaneous tumor in a CFI mouse. Serial passage of cell-free extracts resulted in virus that induced osteosarcomas in mice with latent periods as short as 3 weeks. The transforming gene, v-*fos*, has been cloned and characterized, and tumor-bearing rat antibodies to the v-*fos* gene product have provisionally identified the product as a 55K dalton phosphoprotein ($pp55^{fos}$) lacking any viral *gag* determinants (40,41). The phosphorylation occurs on serine residues, and a 39K dalton host cell protein is associated with $pp55^{fos}$. Immunofluorescence and cell fractionation studies indicate that $pp55^{fos}$ is in the nucleus of FBJ transformed cells (T. Curren, A. D. Miller, and I. Verma, *personal communication*). The distribution within the nucleus appears diffuse and excludes the nucleoli. The similarities in phosphorylation and intracellular location between $pp55^{fos}$ and $P110^{gag\text{-}myc}$ are intriguing and could suggest at least some common features in their transformation mechanisms.

One other virus that deserves mention here is the spleen focus-forming viral (SFFV) component of the Friend virus complex. SFFV causes the abnormal growth of erythroid precursor cells both in culture and in susceptible mice yet does not possess a conventional type of *onc* locus (for review, see ref. 139). Instead, this virus contains gene sequences encoding an altered envelope protein, which appears to be involved in virus-mediated oncogenesis. This envelope protein is glycosylated, has a molecular weight of approximately 55K daltons, and is expressed on the external cell surface membrane (120). Its precise function in oncogenesis is not known.

THE TRANSFORMING PROTEINS OF AVIAN ACUTE LEUKEMIA VIRUSES

myc Gene Proteins

The *myc* transforming gene and its protein product are currently of interest because of their possible involvement in certain human tumors (36,43,44). This interest was initially stimulated by the finding that the c-*myc* gene could be activated in B-cell lymphomas of chickens by the integration of an avian leukosis virus long terminal repeat in the vicinity of this locus (75). This indicated that oncogenesis was associated with the activation of c-*myc* by a viral promotor sequence. Recently, in certain nonviral-induced tumors, genomic translocations around the endogenous c-*myc* locus point to the possibility that various alterations in the expression of this gene may be associated with specific types of tumors (35,72,126). The intracellular location of the v- and c-*myc* gene products, therefore, is of considerable interest.

The study of the *myc* locus has its origins within the avian tumor virus system. Among eight independent isolates of defective avian leukemia viruses, four have been found to carry the *myc* sequence in their genome (20). Also, no *myc*-containing mammalian tumor viruses have yet been described. The prototype avian virus with the v-*myc* transforming gene is myelocytomatosis virus (MC29), and viruses with related transforming genes are CMII, OK10, and MH2. The organization of the *myc* sequence within the defective genome of each virus is distinct and has been reviewed recently (20). An interesting oncogenic potential is displayed by this class of viruses in that they are capable of inducing myelocytomatosis as well as carcinomas, sarcomas, mesotheliomas, and endotheliomas (65). Some differences in oncogenic potential among these viruses are apparent *in vivo*. Within the hematopoietic lineage, macrophage-like cells represent the target for each of the *myc*-containing viruses.

In MC29 transformed cells, the transforming protein is expressed as a *gag-myc* fusion protein (termed $P110^{gag\text{-}myc}$) and is phosphorylated within the *gag* and *myc* sequences but lacks any autophosphorylation activity by the immune complex assay (108). The isolation of nonproducer cells transformed with the defective MC29 genome, along with the availability of antibodies to the N-terminal $p19^{gag}$ protein of $P110^{gag\text{-}myc}$, has permitted the localization of this transforming protein by indirect immunofluorescence. In complete contrast to the cytoplasmic and membrane locations for several other retrovirus transforming proteins, MC29 $P110^{gag\text{-}myc}$ protein was detected in the nucleus, somewhat analogous to the large T antigens of papova viruses (3,45). The fluorescence results indicated that $P110^{gag\text{-}myc}$ was inside the nuclear structure (excluding the nucleoli) and not confined to the nuclear lamina. Bone marrow cells, containing natural target cells for MC29 (macrophages), and cells derived from MC29-induced tumors also contained a high level of $P110^{gag\text{-}myc}$ in the nucleus (3,29). This localization was supported by cell fractionation studies. Pulse-chase labeling indicated that this protein was rapidly transported into the nucleus after synthesis (29). Evidence has been presented that $P110^{gag\text{-}myc}$ binds to double stranded DNA (29); however, it is not known whether it functions as such *in vivo*.

Several spontaneous deletion mutants of MC29 have been derived from a transformed cell line (MC29-Q10) producing virus (107). These deletions overlap and map within the v-*myc* sequence of $P110^{gag\text{-}myc}$ (47). This region also contains the v-*myc* phosphorylation sites (108) and would correspond to the 5′ side of the 3′ exon of the v-*myc* genome (7). *In vivo*, these deletion mutants produce none of the tumors normally associated with MC29; *in vitro*, they fail to induce anchorage-independent growth in either macrophages or fibroblasts (107; C. Tachibana and R. Eisenman, *personal communication*). The mutants are capable of inducing focus formation in fibroblasts in culture (107). The location of the transformation-defective *gag-myc* proteins was examined in quail fibroblasts (29). The results of both immunofluorescence and cell fractionation studies indicated that the proteins are in the nucleus in all cases. In addition, each mutant or wild type *gag-myc* protein could be isolated in a chromatin fraction (29). The lack of any discernible difference in location of the wild type and mutant *gag-myc* proteins in transformed fibroblasts is surprising, because the mutants are defective for transformation. The results do indicate that molecular factors (deletions) modifying the ability of *gag-myc* to transform cells do not necessarily alter the intracellular location of this protein. This suggests that one domain may anchor $P110^{gag\text{-}myc}$ to the nuclear structure, and another independent domain may be required to induce neoplastic transformation. Further experiments are required to rule out an alternative dosage effect or the possibility that more subtle changes in nuclear localization may be responsible.

To analyze directly the v-*myc* protein sequences in the mechanism of transformation by MC29, site-specific antibodies to a defined peptide segment of v-*myc* have been prepared (71). The peptide segment chosen was the C-terminal dodecapeptide based on the nucleotide sequence analysis of v-*myc* (7). Affinity-purified antibodies to this dodecapeptide, immune-precipitated $P110^{gag\text{-}myc}$ from MC29 transformed cells and competition with the peptide blocked recognition of that transforming protein. The antipeptide antibodies recognized authentic v-*myc*-containing proteins from the translation of hybrid-selected mRNA. This was demonstrated further by the immune precipitation of the $P90^{gag\text{-}myc}$ from CMII transformed cells and the $P200^{gag\text{-}pol\text{-}myc}$ from OK10 transformed cells (71). In addition, proteins of 60K daltons, presumably products of subgenomic v-*myc* mRNAs, were specifically immune-precipitated from OK10 and MH2 transformed cells. Similar results have been obtained with *myc*-protein specific antibodies prepared against *myc* protein expressed from cloned *myc* segments in *E. coli* (K. Alitalo, *personal communication*).

With the antibodies to a peptide region of *myc* protein, immunofluorescence experiments were performed; the results are presented in Fig. 8. As previously reported, the antipeptide antibodies to *myc* detected $P110^{gag\text{-}myc}$ in the nucleus (Fig. 8A) of cells transformed by MC29. Fluorescence on viable MC29 transformed cells indicated a complete lack of expression of any C-terminal *myc*-related peptides

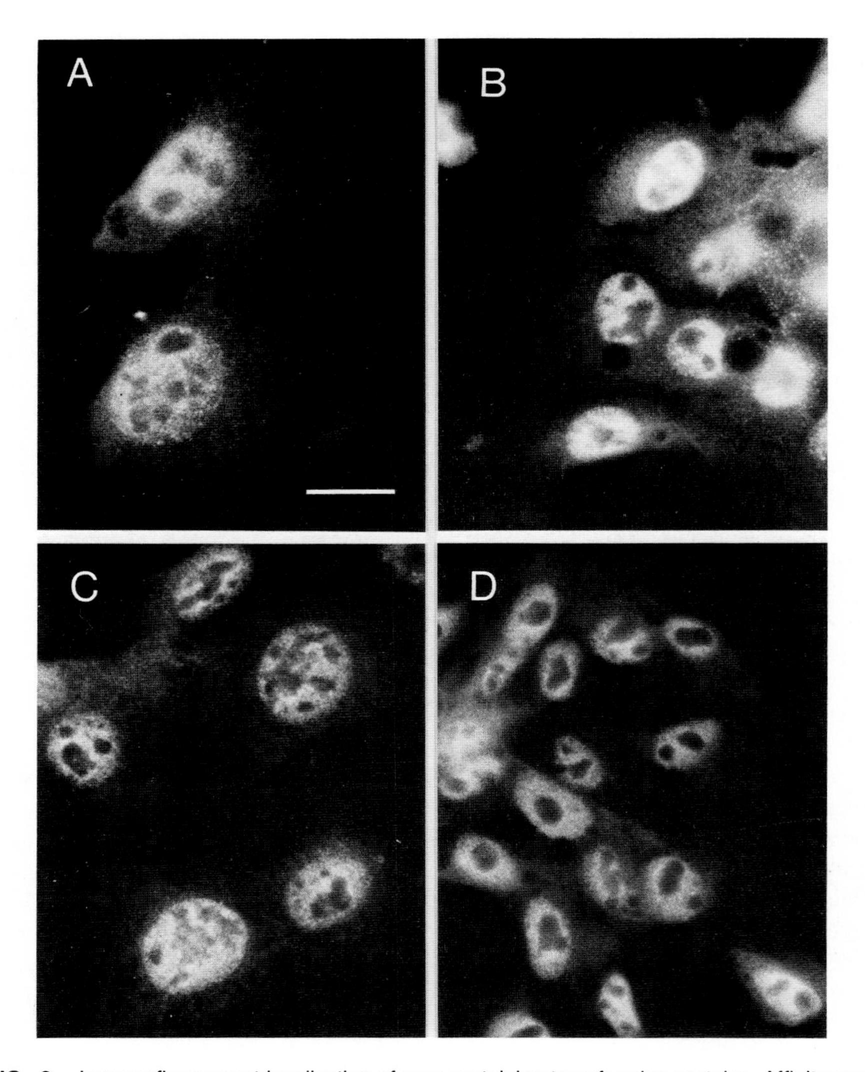

FIG. 8. Immunofluorescent localization of *myc*-containing transforming proteins. Affinity-purified antibodies to the C-terminal dodecapeptide of the *myc* gene product were used to locate the MC29 class of transforming proteins by indirect immunofluorescence. **A:** Location of $P110^{gag-myc}$ in MC29 transformed quail fibroblasts (Q8). **B:** Location of $P90^{gag-myc}$ in CMII transformed quail fibroblasts. **C:** Location of $P200^{gag-pol-myc}$ and $p60^{myc}$ in OK10 transformed cells. **D:** Location of $p60^{myc}$, the only protein detectable with these antibodies in MH2 transformed cells (for details, see ref. 71). In all cases, the transforming proteins are nuclear, and no fluorescence is seen if the antibodies are preabsorbed with the appropriate peptide. Bar = 10 μm.

on the cell surface (L. R. Rohrschneider, *unpublished observation*). Likewise, the $P90^{gag-myc}$ in CMII transformed cells was nuclear (Fig. 8B), and the only *myc*-related protein detectable in MH2 transformed cells ($p60^{myc}$) was localized again within the nucleus (Fig. 8D). In OK10 transformed cells, antibodies to the v-*myc* dodecapeptide stained mainly the nucleus, suggesting that both $P200^{gag\text{-}pol\text{-}myc}$ and $p60^{myc}$ in these cells were nuclear (Fig. 8C). In all cases, competition experiments with excess peptide indicated that the antibodies were specifically recognizing the *myc*-related proteins detected by immune precipitation. Cell fractionation experiments performed in parallel with the immunofluorescence studies were in complete agreement with the nuclear localization of these *myc* proteins (71). These results consistently demonstrate that v-*myc*-related proteins are nuclear.

The v-*myc* gene is composed of two exons from the 3′ end of c-*myc* (7). Therefore, the antibodies to the C-terminal amino acid sequence of v-*myc* should also recognize the c-*myc* gene product. This notion is confirmed by the finding that the antipeptide antibodies specifically recognize a 60K dalton protein from uninfected quail fibroblasts (71) as well as from certain lymphoid tumors of the chicken (71). A protein of somewhat lower molecular weight is also detected by the anti-*myc* antibodies. Although the 60K dalton protein is present at only a fraction of the level of the *gag-myc* transforming proteins, it has been detectable in the nucleus of both quail fibroblasts and lymphoid cell lines (71). The nuclear fluorescence appears similar to that described for the other *myc* proteins but may be slightly more granular in texture. The protein recognized in these cells is the product of the c-*myc* gene ($p60^{c\text{-}myc}$), and its detection in the nucleus represents the first instance of an endogenous *onc* protein being localized by these methods.

erbA and *erbB* Gene Proteins

Avian erythroblastosis virus (AEV) is a replication-defective acute leukemia virus which, upon inoculation into chickens, induces acute erythroblastosis and, in some cases, slowly developing sarcomas (65,66). *In vitro*, this virus is able to transform immature erythroid cells from cultured bone marrow as well as fibroblasts (64,67). Chicken fibroblasts transformed with AEV display an accentuated fusiform shape and produce a number of characteristic transformation parameters (117). Infected cells exhibit an increased rate of hexose uptake, lose intact microfilament bundles, show elevated levels of protease activity, and have little cell surface fibronectin, all of which are also observed upon transformation by RSV (117).

The AEV oncogene is called *erb* and, unlike the transforming regions of other oncogenic retroviruses, is composed of two nonoverlapping domains, each coding for a different protein (10,90,103,125,152). The *erbA* domain is located near the 5′ end of the genome (19,89) and encodes a *gag*-linked polyprotein. Owing to the presence of *gag* determinants, this protein was readily identified in AEV transformed cells and was shown to be approximately 75K daltons in size: $P75^{gag\text{-}erbA}$ (74). The *erbB* domain, on the other hand, is located at the 3′ end of the inserted *erb* and produces a protein encoded entirely by *erbB* sequences. Initial experiments to identify the protein specified by this region involved *in vitro* translation of partially degraded virion RNA. This showed a 40K dalton protein (P40) which was unrelated to either virus structural proteins or the *erbA* gene product (90,103,152).

More recently, carefully controlled hybrid-arrested translations of AEV virion RNA, as well as hybrid-selected translations of AEV mRNA, have indicated that a 61K dalton molecular weight protein, not P40, is the authentic product of the *erbB* domain (105). Apparently, the polypeptide product (P40) seen in earlier studies represented artifactual translation of virion RNA (105).

The role that each region plays in the transformation process was, until recently, somewhat obscure. Initially, it was speculated that both may be transforming proteins, with one responsible for erythroid and the other for fibroblast transformation. In light of genetic studies, however, this seemed unlikely, since all temperature-sensitive mutants of AEV that have been isolated are temperature sensitive for both fibroblast and erythroblast transformation (18,98). Confirmation that only one domain of *erb* is responsible for the transforming activity of AEV came from recent studies utilizing *in vitro* generated mutants of AEV deleted in *erbA* or *erbB* (53). AEV mutants containing deletions in *erbA*, although displaying a reduced transforming potential, were able to elicit the transformation of both erythroblasts and fibroblasts. Mutants of AEV with defects in *erbB*, on the other hand, possessed no transforming capabilities. These results demonstrated that *erbB* encodes the major transforming activity of AEV, and that the *erbA* gene enhances this effect (53).

Following assignment of the *erbB* gene as the essential domain for AEV-induced transformation, it became clear that antisera that would recognize and define the *erbB* gene product in AEV transformed cells were necessary for a better understanding of the transformation process. Two recent independent approaches have been utilized to generate antisera reactive with the *erbB* gene product. One method has used tumor-bearing rat sera (73), whereas another has employed for immunization purposes the *erbB* gene product expressed in *E. coli* (106). Using these antisera, it has been possible to identify and partially characterize the *erbB* gene product as a polypeptide of 62K daltons (p62), which is posttranslationally modified to yield species of 66K ($gp66^{erbB}$) and 68K ($gp68^{erbB}$) daltons, and that both species appear to be glycosylated forms of $p62^{erbB}$ (73,106). Conflicting results concerning the phosphorylation state of the *erbB* proteins have been reported.

The intracellular distribution of the *erb* gene products has been examined by indirect immunofluorescence and cellular fractionation experiments (3,73). Subcellular fractionation studies indicated that the *erbB* transforming proteins $gp66^{erbB}$ and $gp68^{erbB}$ were predominantly found in the cell membrane fraction. In contrast, $P75^{gag\text{-}erbA}$ was located primarily within the cytoplasm (3,73). Indirect immunofluorescence studies provided further details on the location of the *erbB* proteins. Using viable erythroblasts and fibroblasts transformed by AEV, the *erbB* transforming proteins were detected on the surface of the infected cells (73). Evidence that the cell surface fluorescence was attributable to *erbB* proteins came from studies using chicken fibroblasts infected with various deletion mutants. In these cases, the antisera produced a surface fluorescence only with cells synthesizing an intact *erbB* product. These results imply that one or more of the *erbB* proteins is on the cell surface; however, direct evidence that the antibodies were in fact reacting with an *erbB* protein was not presented.

myb Gene Protein

Two virus isolates carry the *myb* transforming sequence in their genome: avian myeloblastosis virus (AMV) and E26. The better characterized strain, AMV, induces exclusively an acute myeloblastic leukemia and does not transform fibroblasts in culture. The v-*myb* gene in AMV has been cloned and sequenced (119), and antibodies to peptide regions predicted from the nucleotide sequence have identified a 48K dalton protein in leukemic cells that is specific for AMV (23). Antibodies produced to *myb* expressed from a cloned segment in *E. coli* also identify the same size protein (85). This putative $p48^{myb}$ is neither glycosylated nor detectable when cells are labeled with radioactive phosphorus and does not exhibit a kinase activity in the immune complex. Both cell fractionation and immunofluorescence indicate that $p48^{myb}$ is in the nucleus of AMV transformed cells (M. Baluda, *personal communication*).

GENERAL SYNTHESIS

A summary of the intracellular location of the transforming proteins covered in this chapter is presented in Table 1. The transforming proteins are listed along with their respective *onc* gene and virus designation. We have attempted to indicate the multiple cellular locations of each transforming protein and have denoted whether the proteins at these sites are glycosylated and/or phosphorylated. The information must be considered incomplete, because data are presently not available concerning all the posttranslational modifications. Further research should reveal more subtle and varied locations. Nevertheless, the current locations thus presented emphasize not only the diversity of intracellular targets that may exist but also some common features in the mechanisms of neoplastic transformation.

Distinct and Common Groups of Transforming Proteins Based on Subcellular Localization

Clear examples of a common location for distinct *onc* gene protein products include the nuclear locations for the $P110^{gag\text{-}myc}$, $pp55^{fos}$, and $p48^{myb}$ synthesized from MC29, FBJ-MSV, and AMV, respectively. By immunofluorescence, all are detectable within the nucleus as a diffuse or sometimes granular stain that is excluded from the nucleoli. Both $P110^{gag\text{-}myc}$ and $pp55^{fos}$ are phosphorylated on serine residues (40,108); $p48^{myb}$, however, is not phosphorylated. No homology exists between the v-*myc*, v-*myb*, and v-*fos* genes (20,140). On the basis of localization, it is reasonable that the transformation mechanisms of these protein products may be related.

The *gag-fes* and *gag-fps* protein products are another example of different isolates of transforming genes whose protein products share similar, discrete, subcellular

TABLE 1. *Summary of subcellular locations of the* onc *gene proteins*

Transforming gene (virus)	Subcellular location[a]					
	Membrane					
	Cell surface	Cytoplasmic side	Cytoplasm	Nucleus	Cytoskeleton	Cell adhesion sites
src (RSV)		$pp60^{\text{src}}$	$pp60^{\text{src}}$		$pp60^{\text{src}}$	$pp60^{\text{src}}$
yes (Y73)		$P90^{\text{gag-yes}}$	$P90^{\text{gag-yes}}$[c]		$P90^{\text{gag-yes}}$[c]	$P90^{\text{gag-yes}}$
abl (A-MuLV)	$P120^{\text{gag-abl}}$	$P120^{\text{gag-abl}}$			$P120^{\text{gag-abl}}$	$P120^{\text{gag-abl}}$
fps (PRCII)			$P105^{\text{gag-fps}}$		$P110^{\text{gag-fps}}$	
fps (FuSV)		$P140^{\text{gag-fps}}$	$P140^{\text{gag-fps}}$			
fes (ST-FeSV)		$P85^{\text{gag-fes}}$	$P85^{\text{gag-fes}}$			
fms (SM-FeSV)			$gP180^{\text{gag-fes}}$		$gP180^{\text{gag-fms}}$	
erb (AEV)	$gp68^{\text{erbB}}$[d] $gp66^{\text{erbB}}$[b,d]		$P75^{\text{gag-erbA}}$			
fos (FBJ virus)				$pp55^{\text{fos}}$		
myc (MC29)				$P110^{\text{gag-myc}}$		
(CMII)				$P90^{\text{gag-myc}}$		
(MH2)			$P100^{\text{gag-mil}}$[e]	$p60^{\text{myc}}$		
(OK10)				$P200^{\text{gag-myc}}$ $p60^{\text{myc}}$		
myb (AMV)				$p48^{\text{myb}}$[b]		
mos (Mo-MSV)			$pp37^{\text{mos}}$			
ras (Ha-MSV)		$pp21^{\text{ras}}$	pro-$p21^{\text{ras}}$			
sis (SSV)		$gp28^{\text{sis}}$[d]				

[a]Refer to the text for details and references. The nomenclature of Coffin et al. (34) has been followed for naming the transforming genes. Brackets at left denote groups of transforming proteins with common locations; *g*, glycoprotein.
[b]Phosphate has not been detected in these proteins.
[c]L. E. Gentry and L. R. Rohrschneider, *unpublished results.*
[d]Tentative location.
[e]See ref. 80a.

localizations. This is not surprising, since nucleotide sequence data indicate that v-*fes* and v-*fps* are highly related, and both are associated with tyrosine kinase activities (12,15,68,119). Their localization is interesting because it differs from other transforming proteins, including those that also contain tyrosine kinase activity. P140$^{gag\text{-}fps}$, for example, is cytoplasmic, with some plasma membrane association that is salt dependent in cell fractionation and exhibits a different affinity for membranes than pp60^{src}. In addition, the *gag-fps* protein product is not detectable in adhesion plaques; although it is found at junctions between cells, these junctions may not be the same as the pp60^{src}-containing junctions in RSV transformed cells (L. R. Rohrschneider, *unpublished observation*). The *gag-fps* and *gag-fes* gene products represent a single class of transforming proteins, based on their cellular distribution. Although they contain a tyrosine-specific kinase activity and are found in the plasma membrane, they may transform cells by a means distinct from other *onc* proteins.

Another class of transforming proteins whose members may share a common or related transformation mechanism (at least in fibroblasts) is represented by pp60^{src} of RSV, P120$^{gag\text{-}abl}$ of A-MuLV, and P90$^{gag\text{-}yes}$ of Y73 virus. The ESV P80$^{gag\text{-}yes}$ is a separate isolate containing the *yes* gene and is also in this category (20,48). All three of these transforming proteins have tyrosine-specific kinase activities, and their principal location is on the cytoplasmic side of the plasma membrane of transformed cells. In addition, nucleotide sequence analyses of the *src*, *abl*, and *yes* genes have indicated that extensive homology exists in certain regions of these genes (152; D. Baltimore, *personal communication*). Most revealing, however, is the localization of each of these transforming proteins within adhesion plaques and close contact areas of their respective transformed cell. This particular localization is distinct from all other transforming proteins and is consistent with the fact that each is able to induce the increased phosphorylation of the adhesion plaque protein vinculin on tyrosine (121). The A-MuLV, RSV, and Y73 all transform fibroblasts. The common localization and enzymatic properties among the transforming proteins of these three viruses could dictate some related fibroblast transformation mechanism.

Other transforming proteins from distinct viruses also are plasma membrane associated but in a manner unrelated to the membrane-associated *onc* proteins described above. For example, the guanine nucleotide binding protein pp21^{ras} of Ha-MSV is on the cytoplasmic side of the membrane but is phosphorylated in threonine only, contains no detectable tyrosine kinase activity, and does not cause the increased phosphorylation of vinculin (99,130,149); neither does gp28^{sis}, which is glycosylated and may be on the plasma membrane. The *erbB* protein also is glycosylated, does not cause the increased phosphorylation of vinculin, and may be exposed on the surface of transformed cells. Therefore, these transforming gene products represent a class of proteins distinct from the P140$^{gag\text{-}fps}$ class or the pp60^{src} class described above.

The large number of transforming proteins found in the plasma membrane highlight it as a general target for transformation. The variations in form and function of the *onc* proteins at this site are reminiscent of some known normal cell membrane polypeptides. The class of membrane proteins that the *onc* transformation-specific polypeptides most closely resemble is receptors that regulate growth and transduce chemical signals across the membrane in normal cells (see ref. 6). Membrane receptors for growth factors, such as EGF, PDGF, and insulin, exhibit tyrosine-specific kinase activity on the cytoplasmic side of the plasma membrane. This class of receptors, as well as others, contains subunits that are glycosylated and exposed on the cell surface. Even a GTP-binding protein (the G protein) that is postulated to couple adenylate cyclase to its activated membrane receptor is found in the plasma membrane. Also, a membrane transport ATPase was found to share limited amino acid sequence homology with one *onc* protein (57). In addition, cell surface receptors for acetylcholine are distributed uniformly throughout the plasma membrane and are clustered on the ventral myotube surface in association with vinculin in cell-substratum adhesion areas (21). The properties and locations of these normal cell membrane proteins are analogous to several of the *onc* gene products, which may suggest similarities in their functions as well.

Determinants of Membrane Localization

The plasma membrane represents a major target site for many types of transforming proteins (e.g., tyrosine kinases). Is there any given determinant that prescribes this membrane localization? The answer is not a simple one. A scan of Table 1, for example, indicates that glycosylation per se does not ordain cell surface expression (at least from current data); and although phosphorylation occurs for $pp60^{src}$ and $pp21^{ras}$ in the plasma membrane, it is more likely that this step is used to regulate the activity of the *onc* protein in the membrane rather than to determine this location. In the case of $pp60^{src}$, it is known that an 8000 dalton segment at the amino terminal end mediates membrane attachment and is necessary for efficient transformation (86). Although no hydrophobic amino acid domain exists in this region, a long chain fatty acid is covalently attached to it and may serve to anchor $pp60^{src}$ to the plasma membrane (56,124). Thus far, three transforming proteins have been found to incorporate radioactive long chain fatty acid. For $pp60^{src}$ and $P120^{gag\text{-}abl}$, the lipid is presumably bound via an amide linkage, whereas $pp21^{ras}$ may covalently couple the lipid through an ester linkage (124). This posttranslational modification suggests that the covalently attached lipid could mediate membrane anchorage, although it is not a universal means for membrane attachment. A similar modification was not readily detectable on $P140^{gag\text{-}fps}$ or $P90^{gag\text{-}yes}$ (B. Sefton, *personal communication*). Perhaps the covalent attachment of lipid delineates a further membrane protein subclass. It will be of interest to know what other *onc* proteins and normal proteins contain this same modification.

Role of the Cytoskeleton in Transformation

Not all transforming proteins have been investigated for a possible cytoskeletal association. In all cases (except one) where this has been studied, however, a direct affiliation with the cytoskeletal framework was demonstrated. The one exception is the pp37^{mos} protein of Mo-MSV found in the soluble cytoplasm of acutely infected and transformed cell lines (100). Other proteins, such as pp60^{src}, P120$^{gag\text{-}abl}$, P105$^{gag\text{-}fps}$, P140$^{gag\text{-}fps}$, and gP180$^{gag\text{-}fms}$, are all associated with the detergent-insoluble matrix after extraction of cells with nonionic, detergent-containing buffers. This is rather remarkable, since each of these transforming proteins is found at discrete subcellular locations. P120$^{gag\text{-}abl}$ is perhaps transmembrane, with the major portion on the cytoplasmic side; pp60^{src} also is tightly bound to the cytoplasmic face of the plasma membrane, whereas the *gag-fps* protein products are less firmly attached to the plasma membrane in addition to being cytoplasmic. gP180$^{gag\text{-}fms}$ is found primarily in the cytoplasm. Thus a diversity of intracellular locations exists, with the only common feature being an association with cytoskeletal elements. This raises the possibility that the cytoskeleton ultimately may be the underlying vehicle that integrates the various cellular compartments and regulates functions, such as growth and differentiation.

Influence of *gag* Determinants on Localization of Transforming Proteins

It is likely that the *gag* determinants found on the amino terminal end of many transforming proteins have nothing to do with determining intracellular location of *gag-onc* proteins. The *src* gene product (pp60^{src}) contains no *gag* protein sequences, whereas P90$^{gag\text{-}yes}$ does; yet each is found at similar intracellular locations. Likewise, the FBJ virus pp55^{fos}, the MC29 P110$^{gag\text{-}myc}$, or p48^{myb}, and the MH2 p60^{myc} are all nuclear proteins, regardless of the presence or absence of *gag*-protein determinants. Converse examples also exist, where *onc* proteins contain related *gag* determinants yet exhibit distinct cellular locations. The P110$^{gag\text{-}myc}$ of MC29 is nuclear, and the P90$^{gag\text{-}yes}$ of Y73 virus is in the adhesion plaques and probably plasma membrane, like pp60^{src}. Also, the P75$^{gag\text{-}erbA}$ of AEV is a cytoplasmic protein, and the P90$^{gag\text{-}myc}$ of CMII virus is detected in the nucleus of transformed cells. Thus *gag* proteins do not appear to dictate intracellular locations of the *gag-onc* fusion proteins.

The exact purpose of the *gag*-protein determinants fused to *onc* gene products is not entirely understood, but results of recent experiments on c-*myc* suggest a different reason for the evolution of such proteins. Results indicate that the c-*myc* gene contains an additional nontranslated 5′ exon not found in the v-*myc* gene. This 5′ exon may regulate translation; therefore, the *gag* sequences may merely substitute for this 5′ controlling exon of c-*myc*, thereby altering translational controls for the c-*myc* mRNA (135). It is not known whether a similar situation exists for other *gag-onc* proteins.

Deletions in the *gag* sequences of the various *gag-onc* fusion proteins have not been generated and studied; thus it is difficult to determine the exact function of these sequences. In the A-MuLV system, however, deletion of almost all *gag* sequences in P120$^{gag\text{-}abl}$ was found to dramatically reduce the ability of A-MuLV to transform lymphocytes (106a). Therefore, although the *gag* protein determinants may not direct a *gag-onc* protein to a particular intracellular location, its presence (at least in the Abelson system) seems to be necessary for transformation of certain target cells.

Intracellular Location of v-*onc* Versus c-*onc* Proteins

The intracellular location of the normal c-*onc* proteins relative to the respective v-*onc* proteins is of interest in studying the mechanism of transformation by these proteins. Unfortunately, only a few examples can be cited. In each case, the c-*onc* protein exhibits an intracellular distribution similar or identical to the respective v-*onc* protein. Both $pp60^{c\text{-}src}$ and $pp60^{v\text{-}src}$ are associated with the plasma membrane of cells (38), and a nuclear location is found for the $p60^{c\text{-}myc}$ as well as for the MC29 class of transforming proteins typified by $P110^{gag\text{-}myc}$ (71). Similarly, both the HaSV $pp21^{v\text{-}ras}$ and its related $pp21^{c\text{-}ras}$ product are situated in the plasma membrane, presumably on the cytoplasmic face (99); and both $p48^{v\text{-}myb}$ and a related normal thymocyte protein ($p110^{c\text{-}myb}$) are nuclear (M. Baluda, *personal communication*). Perhaps further studies will detect more subtle differences in cellular distribution; at this time, however, no major intracellular differences are detectable between the viral proteins and their normal cellular counterparts.

Biochemical differences are readily apparent between these two classes of cellular and viral proteins. In some cases, different sites and even different amino acids are phosphorylated on c-*onc* versus their respective v-*onc* proteins (96,131,134), and some c-*onc* proteins do not contain detectable phosphate, whereas their v-*onc* protein products do (93,104). The inverse situation also has been found, where the c-*onc*, but not the v-*onc*, protein is phosphorylated (23). Even different exons can be used in composing the amino acid sequence of the c-*onc* protein relative to the v-*onc* protein (84,136,140). Thus, although the intracellular location of these cellular and viral proteins may be similar or identical, structural differences exist between the v-*onc* and c-*onc* proteins, implying that they may have altered functions.

These results fit the hypothesis that transformation results from the normal expression of an abnormal c-*onc* protein. The alternative hypothesis envisions the overexpression of a normal c-*onc* protein as the primary pathologic event. Both mechanisms may contribute to neoplastic transformation, and results of the localization studies suggest that regardless of whether or not each c-*onc* protein and its respective v-*onc* protein are functionally identical, they are situated at similar intracellular targets.

Static Versus Dynamic Locations for *onc* Proteins

In all the localization studies on the various viral transforming proteins, one receives the distinct impression (especially from immunofluorescence photographs) that these proteins function by their static location at a particular cellular site. Even pulse-chase radiolabeling and cell fractionation experiments that have attempted to capture various *onc* proteins in transit leave the impression that these proteins migrate to one permanent site where they may function. This may be the case, but it should be kept in mind that cells are by nature dynamic, molecular assemblages requiring continuous synthesis, assembly, contraction, translocation, degradation, and regeneration for normal survival. Therefore, it may not be extraordinary to expect that some of the *onc* proteins may participate in these cellular dynamics; in fact, this could be an essential part of the transformation mechanism. Such movement could depend on the physiologic state of the cell and involve intra- or intercompartmental transitions. Thus our current static outlook on the transformation mechanism may be naive, and more sophisticated means might be required to completely understand neoplastic transformation.

ACKNOWLEDGMENTS

We thank the many colleagues cited in the text for discussion, opinions, and permission to mention their results before publication. We especially thank R. Eisenman and F. Yoshimura for critical evaluation of the manuscript. Work in our laboratory was supported by Public Health Service grants from NCI (CA 20551 and CA 28151). L. G. was supported by a postdoctoral fellowship from an NIH training grant.

REFERENCES

1. Abercrombie, M., and Dunn, G. A. (1975): Adhesions of fibroblasts to substratum during contact inhibition observed by interference reflection microscopy. *Exp. Cell Res.*, 92:57–62.
2. Abercrombie, M., Heaysman, J. E. M., and Pegrum, S. M. (1971): The locomotion of fibroblasts in culture. IV. Electron microscopy of the leading lamella. *Exp. Cell Res.*, 67:359–367.
3. Abrams, H. D., Rohrschneider, L. R., and Eisenman, R. N. (1982): Nuclear location of the putative transforming protein of avian myelocytomatosis virus. *Cell*, 29:427–439.
4. Adkins, B., Hunter, T., and Beemon, K. (1982): Expression of the PRCII avian sarcoma virus genome. *J. Virol.*, 41:767–780.
5. Adkins, B., and Hunter, T. (1982): Two structurally and functionally different forms of the transforming protein of PRCII avian sarcoma virus. *Mol. Cell. Biol.*, 2:890–896.
6. Alberts, B., Bray, D., Lewis, J., Raff, M., Roberts, K., and Watson, J. D. (1983): *Molecular Biology of the Cell*, chs. 6 and 13. Garland Publishing, New York.
7. Alitalo, K., Bishop, J. M., Smith, D. H., Chen, E. Y., Colby, W. W., and Levinson, A. D. (1983): Nucleotide sequence of the v-*myc* oncogene of avian retrovirus MC29. *Proc. Natl. Acad. Sci. USA*, 80:100–104.
8. Anderson, D. D., Beckmann, R. P., Harms, E. H., Nakamura, K., and Weber, M. J. (1981): Biological properties of "partial" transformation mutants of Rous sarcoma virus and characterization of their pp60^{src} kinase. *J. Virol.*, 37:445–458.
9. Anderson, S. J., Furth, M., Wolff, L., Ruscetti, S. K., and Sherr, C. J. (1982): Monoclonal antibodies to the transformation-specific glycoprotein encoded by the feline retroviral oncogene v-*fms*. *J. Virol.*, 44:696–702.
10. Anderson, S. M., Hayward, W. S., Neel, B. G., and Hanafusa, H. (1980): Avian erythroblastosis virus produces two mRNAs. *J. Virol.*, 36:676–683.
11. Balduzzi, P. C., Notter, M. F. D., Morgan, H. R., and Shibuya, M. (1981): Some biological properties of two new avian sarcoma viruses. *J. Virol.*, 40:268–275.
12. Barbacid, M., Breitman, M. L., Lauver, A. V., Long, L. K., and Vogt, P. K. (1981): The transformation-specific proteins of avian (Fujinami and PRCII) and feline (Snyder-Theilen and Gardner-Arnstein) sarcoma viruses are immunologically related. *Virology*, 110:411–419.
13. Barbacid, M., Lauver, A. V., and Devare, S. G. (1980): Biochemical and immunological characterization of polyproteins coded for by the McDonough, Gardner-Arnstein, and Snyder-Theilen strains of feline sarcoma virus. *J. Virol.*, 33:196–207.

14. Becker, D., Kurth, R., Critchley, D., Friis, R., and Bauer, H. (1977): Distinguishable transformation-defective phenotypes among temperature-sensitive mutants of Rous sarcoma virus. *J. Virol.*, 21:1042–1055.
15. Beemon, K. (1981): Transforming proteins of some feline and avian sarcoma viruses are related structurally and functionally. *Cell*, 24:145–153.
16. Ben-Ze'ev, A., Duerr, A., Solomon, F., and Penman, S. (1979): The outer boundary of the cytoskeleton: A lamina derived from plasma membrane proteins. *Cell*, 17:859–865.
17. Beug, H., Claviez, M., Jockusch, B. M., and Graf, T. (1978): Differential expression of Rous sarcoma virus-specific transformation parameters in enucleated cells. *Cell*, 14:843–856.
18. Beug, H., and Graf, T. (1980): Transformation parameters of chicken embryo fibroblasts infected with ts34 mutant of avian erythroblastosis virus. *Virology*, 100:348–356.
19. Bister, K., and Duesberg, P. H. (1979): Structure and specific sequences of avian erythroblastosis virus RNA: Evidence for multiple classes of transforming genes among avian tumor viruses. *Proc. Natl. Acad. Sci. USA*, 76:5023–5027.
20. Bister, K., and Duesberg, P. H. (1982): Genetic structure and transforming genes of avian retroviruses. In: *Advances in Viral Oncology, Vol. 1*, edited by G. Klein, pp. 3–42. Raven Press, New York.
21. Bloch, R. J., and Geiger, B. (1980): The localization of acetylcholine receptor clusters in areas of cell-substrate contact in cultures of rat myotubes. *Cell*, 21:25–35.
22. Boss, M. A., Dreyfuss, G., and Baltimore, D. (1981): Localization of the Abelson murine leukemia virus protein in a detergent-insoluble subcellular matrix: Architecture of the protein. *J. Virol.*, 40:472–481.
23. Boyle, W. J., Lipsick, J. S., Reddy, E. P., and Baluda, M. A. (1983): Identification of the leukemogenic protein of avian myeloblastosis virus and of its normal cellular homologue. *Proc. Natl. Acad. Sci. USA*, 80:2834–2838.
24. Branton, D., Cohen, C. M., and Tyler, J. (1981): Interaction of cytoskeletal proteins on the human erythrocyte membrane. *Cell*, 24:24–32.
25. Breitman, M. L., Neil, J. C., Moscovici, C., and Vogt, P. K. (1981): The pathogenicity and defectiveness of PRCII: A new type of avian sarcoma virus. *Virology*, 108:1–12.
26. Breitman, M. L., Hirano, A., Wong, T., and Vogt, P. K. (1981): Characteristics of avian sarcoma virus strain PRCIV and comparison with strain PRCIIp. *Virology*, 114:451–462.
27. Brugge, J. S., and Erikson, R. L. (1977): Identification of a transformation-specific antigen induced by an avian sarcoma virus. *Nature*, 269:346–348.
28. Brugge, J. S., Steinbaugh, P. J., and Erikson, R. L. (1978): Characterization of the avian sarcoma virus protein $p60^{src}$. *Virology*, 91:130–140.
29. Bunte, T., Greiser-Wilke, I., Donner, P., and Moelling, K. (1982): Association of *gag-myc* proteins from avian myelocytomatosis virus wild-type and mutants with chromatin. *EMBO J.*, 1:919–927.
30. Burr, J. G., and Buchanan, J. M. (1984): Cytoskeletal-associated substrates for the *src* gene product of Rous sarcoma virus. *Mol. Cell. Biol. (submitted)*.
31. Burr, J. G., Dreyfuss, G., Penman, S., and Buchanan, J. M. (1980): Association of the *src* gene product of Rous sarcoma virus with cytoskeletal structures of chicken embryo fibroblasts. *Proc. Natl. Acad. Sci. USA*, 77:3484–3488.
32. Burridge, K., and Feramisco, J. R. (1980): Microinjection and localization of a 130K protein in living fibroblasts: A relationship to actin and fibronectin. *Cell*, 19:587–595.
33. Carr, J. G., and Campbell, J. G. (1958): Three new virus-induced fowl sarcomata. *Br. J. Cancer*, 12:631–635.
34. Coffin, J. M., Varmus, H. E., Bishop, J. M., Essex, M., Hardy, W. D., Martin, G. S., Rosenberg, N. E., Scolnick, E. M., Weinberg, R. A., and Vogt, P. K. (1981): Proposal for naming host cell-derived inserts in retrovirus genomes. *J. Virol.*, 40:953–957.
35. Calame, K., Kim, S., Lalley, P., Hill, R., Davis, M., and Hood, L. (1982): Molecular cloning of translocations involving chromosome 15 and the immunoglobulin C_α gene from chromosome 12 in two murine plasmacytomas. *Proc. Natl. Acad. Sci. USA*, 79:6994–6998.
36. Collins, S. J., and Groudine, M. (1982): Amplification of endogenous *myc*-related DNA sequences in a human myeloid leukemia cell line. *Nature*, 298:679–681.
37. Courtneidge, S. A., and Bishop, J. M. (1982): Transit of $pp60^{v\text{-}src}$ to the plasma membrane. *Proc. Natl. Acad. Sci. USA*, 79:7117–7121.
38. Courtneidge, S. A., Levinson, A. D., and Bishop, J. M. (1980): The protein encoded by the transforming gene of avian sarcoma virus ($pp60^{src}$) and a homologous protein in normal cells

(pp60$^{proto\text{-}src}$) are associated with the plasma membrane. *Proc. Natl. Acad. Sci. USA*, 77:3783–3787.
39. Courtneidge, S., Ralston, R., Alitalo, K., and Bishop, J. M. (1983): Subcellular location of an abundant substrate (p36) for tyrosine-specific protein kinases. *Mol. Cell. Biol.*, 3:340–350.
40. Curran, T., Peters, G., Van Beveren, C., Teich, N. M., and Verma, I. M. (1982): FBJ murine osteosarcoma virus: Identification and molecular cloning of biologically active proviral DNA. *J. Virol.*, 44:674–682.
41. Curran, T., and Teich, N. M. (1982): Candidate product of the FBJ murine osteosarcoma virus oncogene: Characterization of a 55,000 dalton phosphoprotein. *J. Virol.*, 42:114–122.
42. Curtis, A. S. G. (1964): The mechanism of adhesion of cells to glass. A study by interference reflection microscopy. *J. Cell Biol.*, 20:199–215.
43. Dalla-Favera, R., Martinotti, S., Gallo, R. C., Erikson, J., and Croce, C. M. (1983): Translocation and rearrangements of the c-*myc* oncogene locus in human undifferentiated B-cell lymphomas. *Science*, 219:963–967.
44. Dalla-Favera, R., Wong-Staal, F., and Gallo, R. C. (1982): *Onc* gene amplification in promyelocytic leukaemia cell line HL-60 and primary leukaemic cells of the same patient. *Nature*, 299:61–63.
45. Donner, P., Greiser-Wilke, I., and Moelling, K. (1982): Nuclear localization and DNA binding of the transforming gene product of avian myelocytomatosis virus. *Nature*, 296:262–266.
46. Ellis, R. W., Lowy, D. R., and Scolnick, E. M. (1982): The viral and cellular p21 *(ras)* gene family. In: *Advances in Viral Oncology, Vol. 1*, edited by G. Klein, pp. 107–126. Raven Press, New York.
47. Enrietto, P. J., and Hayman, M. J. (1982): Restriction enzyme analysis of partially transformation-defective mutants of acute leukemia virus MC29. *J. Virol.*, 44:711–715.
48. Erikson, R. L., and Purchio, A. F. (1982): Avian sarcoma viruses, protein kinases, and cell transformation. In: *Advances in Viral Oncology, Vol. 1*, edited by G. Klein, pp. 43–58. Raven Press, New York.
49. Feldman, R. A., Hanafusa, T., and Hanafusa, H. (1980): Characterization of protein kinase activity associated with the transforming gene product of Fujinami sarcoma virus. *Cell*, 22:757–765.
50. Feldman, R. A., Wang, L.-H., Hanafusa, H., and Balduzzi, P. C. (1982): Avian sarcoma virus UR2 encodes a transforming protein which is associated with a unique protein kinase activity. *J. Virol.*, 42:228–236.
51. Feldman, R. A., Wang, E., and Hanafusa, H. (1983): Cytoplasmic localization of the transforming protein of Fujinami sarcoma virus: Salt-sensitive association with subcellular components. *J. Virol.*, 45:782–791.
52. Frankel, A. E., Gilbert, J. H., Porzig, K. J., Scolnick, E. M., and Aaronson, S. A. (1979): Nature and distribution of feline sarcoma virus nucleotide sequences. *J. Virol.*, 30:821–827.
53. Frykberg, L., Palmieri, S., Beug, H., Graf, T., Hayman, M. J., and Vennstrom, B. (1982): Transforming capacities of avian erythroblastosis virus mutants deleted in the erbA of erbB oncogenes. *Cell*, 32:227–238.
54. Fujinami, A., and Inamoto, K. (1914): Uber geschwulste bei japanischen haushuhnern, insbesondere uber einen transplantablen tumor. *Z. Krebsforsch.*, 14:94–119.
55. Furth, M. E., Davis, L. J., Fleurdelys, B., and Scolnick, E. M. (1982): Monoclonal antibodies to the p21 products of the transforming gene of Harvey murine sarcoma virus and of the cellular *ras* gene family. *J. Virol.*, 43:294–304.
56. Garber, E. A., Krueger, J. G., Hanafusa, H., and Goldberg, A. R. (1983): Only membrane-associated RSV *src* proteins have amino-terminally bound lipid. *Nature*, 300:161–163.
57. Gay, N. J., and Walker, J. E. (1983): Homology between human bladder carcinoma oncogene product and mitochondrial ATP-synthase. *Nature*, 301:262–264.
58. Geiger, B. (1979): A 130K protein from chicken gizzard: Its localization at the termini of microfilament bundles in cultured chicken cells. *Cell*, 18:193–205.
59. Geiger, B., Dutton, A. H., Tokuyasu, K. T., and Singer, S. J. (1981): Immunoelectron microscope studies of membrane-microfilament interactions: Distribution of α-actinin, tropomyosin, and vinculin in intestinal epithelial brush border and chicken gizzard smooth muscle cells. *J. Cell Biol.*, 91:614–628.
60. Gentry, L. E., Rohrschneider, L. R., Casnellie, J. E., and Krebs, E. G. (1983): Antibodies to a defined region of pp60^{src} neutralize the tyrosine specific kinase activity. *J. Biol. Chem.*, 258:11219–11228.

61. Gentry, L. E., and Rohrschneider, L. R. (1984): Common features of the *yes* and *src* gene products defined by peptide-specific antibodies. *Mol. Cell. Biol. (submitted).*
62. Ghysdael, J., Neil, J. C., Wallbank, A. M., and Vogt, P. K. (1981): Esh avian sarcoma virus codes for a *gag*-linked transformation-specific protein with an associated protein kinase activity. *Virology*, 111:386–400.
63. Ghysdael, J., Neil, J. C., and Vogt, P. K. (1981): A third class of avian sarcoma viruses, defined by related transformation-specific proteins of Yamaguchi 73 and Esh sarcoma viruses. *Proc. Natl. Acad. Sci. USA*, 78:2611–2615.
64. Graf, T. (1975): In vitro transformation of chicken bone-marrow cells with avian erythroblastosis virus. *Z. Naturforsch.*, 30c:847.
65. Graf, T., and Beug, H. (1978): Avian leukemia viruses. Interaction with their target cells *in vivo* and *in vitro*. *Biochim. Biophys. Acta*, 516:269–299.
66. Graf, T., Fink, D., Beug, H., and Royer-Pokora, B. (1977): Oncornavirus-induced sarcoma formation obscured by rapid development of lethal leukemia. *Cancer Res.*, 37:59–63.
67. Graf, T., Royer-Pokora, B., Schubert, G. E., and Beug, H. (1976): Evidence for the multiple oncogenic potential of cloned leukemia virus: *In vitro* and *in vivo* studies with avian erythroblastosis virus. *Virology*, 71:423–433.
68. Hampe, A., Laprevotte, I., Galibert, F., Fedele, L. A., and Sherr, C. J. (1982): Nucleotide sequences of feline retroviral oncogenes (v-fes) provide evidence for a family of tyrosine-specific protein kinase genes. *Cell*, 30:775–785.
69. Hanafusa, T., Mathey-Prevot, B., Feldman, R. A., and Hanafusa, H. (1981): Mutants of Fujinami sarcoma virus which are temperature-sensitive for cellular transformation and protein kinase activity. *J. Virol.*, 38:347–355.
70. Hanafusa, T., Wang, L.-H., Anderson, S. M., Karess, R. E., Hayward, W. S., and Hanafusa, H. (1980): Characterization of the transforming gene of Fujinami sarcoma virus. *Proc. Natl. Acad. Sci. USA*, 77:3009–3013.
71. Hann, S., Abrams, H., Rohrschneider, L., and Eisenmann, R. (1983): Proteins encoded by the v-*myc* and c-*myc* oncogenes: Identification and localization in acute leukemia virus transformants and bursal lymphoma cell lines. *Cell*, 34:789–798.
72. Harris, L. J., Lang, R. B., and Marcu, K. B. (1982): Non-immunoglobulin-associated DNA rearrangements in mouse plasmacytomas. *Proc. Natl. Acad. Sci. USA*, 79:4175–4179.
73. Hayman, M. J., Ramsay, G. M., Savin, K., Kitchener, G., Graf, T., and Beug, H. (1983): Identification and characterization of the avian erythroblastosis virus *erbB* gene product as a membrane glycoprotein. *Cell*, 32:579–588.
74. Hayman, M. J., Royer-Pokora, B., and Graf, T. (1979): Defectiveness of avian erythroblastosis virus: Synthesis of a 75k *gag*-related protein. *Virology*, 92:31–45.
75. Hayward, W. S., Neel, B. G., and Astrin, S. M. (1981): Activation of a cellular *onc* gene by promoter insertion in ALV-induced lymphoid leukosis. *Nature*, 290:475–480.
76. Heath, J. P., and Dunn, G. A. (1978): Cell-to-substratum contacts of chicken fibroblasts and their relation to the microfilament system: A correlated interference-reflexion and high voltage electron-microscope study. *J. Cell Sci.*, 29:197–212.
77. Ito, S., Richert, N., and Pastan, I. (1982): Phospholipids stimulate phosphorylation of vinculin by the tyrosine-specific protein kinase of Rous sarcoma virus. *Proc. Natl. Acad. Sci. USA*, 79:4628–4631.
78. Itohara, S., Hirata, K., Inoue, M., Hatsuoka, M., and Sato, A. (1978): Isolation of a sarcoma virus from a spontaneous chicken tumor. *Gann*, 69:825–830.
79. Izzard, C. S., and Lochner, L. R. (1976): Cell-to-substrate contacts in living fibroblasts: An interference-reflexion study with an evaluation of the technique. *J. Cell Sci.*, 21:129–159.
80. Izzard, C. S., and Lochner, L. R. (1980): Formation of cell-to-substrate contacts during fibroblast mobility: An interference-reflexion study. *J. Cell Sci.*, 42:81–116.
80a. Jansen, H. W., Reuckert, B., and Bister, K. (1983): Two unrelated cell-derived sequences in the genome of avian leukemia and carcinoma inducing retrovirus MH2. *EMBO J.*, 2:1969–1975.
81. Kawai, S., Yoshida, M., Segawa, K., Sugiyana, H., Ishizaki, R., and Toyoshima, K. (1980): Characterization of Y73, an avian sarcoma virus: A unique transforming gene and its product, a phosphopolyprotein with protein kinase activity. *Proc. Natl. Acad. Sci. USA*, 77:6199–6203.
82. Kensler, R. W., and Goodenough, D. A. (1980): Isolation of mouse myocardial gap junctions. *J. Cell Biol.*, 86:755–764.
83. Kitamura, N., Kitamura, A., Toyoshima, K., Hirayama, Y., and Yoshida, M. (1982): Avian

sarcoma virus Y73 genome sequence and structural similarity of its transforming gene product to that of Rous sarcoma virus. *Nature*, 297:205–208.

84. Klempnauer, K.-H., Gonda, T. J., and Bishop, J. M. (1982): Nucleotide sequence of the retroviral leukemia gene v-*myb* and its cellular progenitor c-*myb*: The architecture of a transduced oncogene. *Cell*, 31:453–463.
85. Klempnauer, K.-H., Ramsay, G., Bishop, J. M., Moscovici, M. G., Moscovici, C., McGrath, J. P., and Levinson, A. D. (1983): The product of the retroviral transforming gene v-*myb* is a truncated version of the protein encoded by the cellular oncogene c-*myb*. *Cell*, 33:345–355.
86. Krueger, J. G., Garber, E. A., Goldberg, A. R., and Hanafusa, H. (1982): Changes in amino-terminal sequences of $pp60^{src}$ lead to decreased membrane association and decreased *in vivo* tumorigenicity. *Cell*, 28:889–896.
87. Krueger, J. G., Wang, E., and Goldberg, A. R. (1980): Evidence that the *src* gene product of Rous sarcoma virus is membrane associated. *Virology*, 101:25–40.
88. Krzyzek, R. A., Mitchell, R. L., Lau, A. F., and Faras, A. J. (1980): Association of $pp60^{src}$ and *src* protein kinase activity with the plasma membrane of nonpermissive and permissive avian sarcoma virus-infected cells. *J. Virol.*, 36:805–815.
89. Lai, M. M. C., Hu, S. S. F., and Vogt, P. K. (1979): Avian erythroblastosis virus: Transformation-specific sequences form a contiguous segment of 3.25 Kb located in the middle of the 6-Kb genome. *Virology*, 97:366–377.
90. Lai, M. M. C., Neil, J. C., and Vogt, P. K. (1980): Cell-free translation of avian erythroblastosis virus RNA yields two specific and distinct proteins with molecular weights of 75,000 and 40,000. *Virology*, 100:475–483.
91. Lee, W.-H., Bister, K., Pawson, A., Robins, T., Moscovici, C., and Duesberg, P. H. (1980): Fujinami sarcoma virus: An avian RNA tumor virus with a unique transforming gene. *Proc. Natl. Acad. Sci. USA*, 77:2018–2022.
92. Mardon, G., and Varmus, H. E. (1983): Frameshift and intragenic suppressor mutations in a Rous sarcoma provirus suggest *src* encodes two proteins. *Cell*, 32:871–879.
93. Mathey-Prevot, B., Hanafusa, H., and Kawai, S. (1982): A cellular protein is immunologically crossreactive with and functionally homologous to the Fujinami sarcoma virus transforming protein. *Cell*, 28:897–906.
94. McClain, D. A., Maness, P. F., and Edelman, G. M. (1978): Assay for early cytoplasmic effects of the *src* gene product of Rous sarcoma virus. *Proc. Natl. Acad. Sci. USA*, 75:2750–2754.
95. Neil, J. C., Breitman, M. L., and Vogt, P. K. (1981): Characterization of a 105,000 molecular weight *gag*-related phosphoprotein from cells transformed by the defective avian sarcoma virus PRCII. *Virology*, 108:98–110.
96. Neil, J. C., Ghysdael, J., and Vogt, P. K. (1981): Tyrosine-specific protein kinase activity associated with p105 of avian sarcoma virus PRCII. *Virology*, 109:223–228.
97. Nigg, E. A., Sefton, B. M., Hunter, T., Walter, G., and Singer, S. J. (1982): Immunofluorescent localization of the transforming protein of Rous sarcoma virus with antibodies against a synthetic *src* peptide. *Proc. Natl. Acad. Sci. USA*, 79:5322–5326.
98. Palmieri, S., Beug, H., and Graf, T. (1982): Isolation and characterization of four new temperature-sensitive mutants of avian erythroblastosis virus (AEV). *Virology*, 123:296–311.
99. Papageorge, A., Lowy, D., and Scolnick, E. M. (1982): Comparative biochemical properties of p21 *ras* molecules coded for by viral and cellular *ras* genes. *J. Virol.*, 44:509–519.
100. Papkoff, J., Nigg, E. A., and Hunter, T. (1983): The transforming protein of Moloney murine sarcoma virus is a soluble cytoplasmic protein. *Cell*, 33:161–172.
101. Papkoff, J., Verma, I. M., and Hunter, T. (1982): Detection of a transforming gene product in cells transformed by Moloney murine sarcoma virus. *Cell*, 29:417–426.
102. Pawson, T., Guyden, J., Kung, T.-H., Radke, K.,Gilmore, T., and Martin, G. S. (1980): A strain of Fujinami sarcoma virus which is temperature-sensitive in protein phosphorylation and cellular transformation. *Cell*, 22:767–775.
103. Pawson, T., and Martin, G. S. (1980): Cell-free translation of avian erythroblastosis virus RNA. *J. Virol.*, 34:280–284.
104. Ponticelli, A. S., Whitlock, C. A., Rosenberg, N., and Witte, O. N. (1982): *In vivo* tyrosine phosphorylations of the Abelson virus transforming protein are absent in its normal cellular homolog. *Cell*, 29:953–960.
105. Privalsky, M. L., and Bishop, J. M. (1982): Proteins specified by avian erythroblastosis virus:

Coding region localization and identification of a previously undetected *erb-B* polypeptide. *Proc. Natl. Acad. Sci. USA*, 79:3958–3962.

106. Privalsky, M. L., Sealy, L., Bishop, J. M., McGrath, J. P., and Levinson, A. D. (1983): The product of the avian erythroblastosis virus *erbB* locus is a glycoprotein. *Cell*, 32:1257–1267.

106a. Prywes, R. Foulkes, J. G., Rosenberg, N., and Baltimore, D. (1983): Sequences of the A-MuLV protein needed for fibroblast and lymphoid cell transformation. *Cell*, 34:569–579.

107. Ramsay, G., Graf, T., and Hayman, M. J. (1980): Mutants of avian myelocytomatosis virus with smaller *gag* gene related proteins have an altered transforming ability. *Nature*, 288:170–172.

108. Ramsay, G., Hayman, M. J., and Bister, K. (1982): Phosphorylation of specific sites in the *gag-myc* polyproteins encoded by MC29-type viruses correlates with their transforming ability. *EMBO J.*, 1:1111–1116.

109. Reynolds, F. H., Van deVen, W. J. M., Blomberg, J., and Stephenson, J. R. (1981): Differences in mechanisms of transformation by independent feline sarcoma virus isolates. *J. Virol.*, 38:1084–1089.

110. Richert, N. D., Davies, P. J. A., Jay, G., and Pastan, I. H. (1979): Characterization of an immune complex kinase in immunoprecipitates of avian sarcoma virus-transformed fibroblasts. *J. Virol.*, 31:695–706.

111. Robbins, K. C., Barbacid, M., Porzig, K. J., and Aaronson, S. A. (1979): Involvement of different exogenous FeLV subgroups in the generation of independent feline sarcoma virus isolates. *Virology*, 97:1–11.

112. Robbins, K. C., Devare, S. G., Reddy, E. P., and Aaronson, S. A. (1982): *In vivo* identification of the transforming gene product of simian sarcoma virus. *Science*, 218:1131–1133.

113. Rohrschneider, L. R. (1979): Immunofluorescence on avian sarcoma virus-transformed cells: Localization of the *src* gene product. *Cell*, 16:11–24.

114. Rohrschneider, L. R. (1980): Adhesion plaques of Rous sarcoma virus-transformed cells contain the *src* gene product. *Proc. Natl. Acad. Sci. USA*, 77:3514–3518.

115. Rohrschneider, L. R., and Rosok, M. J. (1983): Transformation parameters and $pp60^{src}$ localization in cells infected with partial transformation mutants of Rous sarcoma virus. *Mol. Cell. Biol.*, 3:731–746.

116. Rohrschneider, L. R., Rosok, M. J., and Gentry, L. E. (1983): Molecular interaction of $pp60^{src}$ with cellular adhesion plaques. *Prog. Nucleic Acid Res. Mol. Biol.*, 29:233–244.

117. Royer-Pokora, B., Beug, H., Claviez, M., Winkhardt, H.-J., Friis, R. R., and Graf, T. (1978): Transformation parameters in chicken fibroblasts transformed by AEV and MC29 avian leukemia viruses. *Cell*, 13:751–760.

118. Ruscetti, S. K., Turek, L. P., and Sherr, C. J. (1980): Three independent isolates of feline sarcoma virus code for three distinct gag-x polyproteins. *J. Virol.*, 35:259–264.

119. Rushlow, K. E., Lautenberger, J. A., Papas, T. S., Baluda, M. A., Perbal, B., Chirikjian, J. G., and Reddy, E. P. (1982): Nucleotide sequence of the transforming gene of avian myeloblastosis virus. *Science*, 216:1421–1423.

120. Ruta, M., and Kabat, D. (1980): Plasma membrane glycoproteins encoded by cloned Rauscher and Friend spleen focus-forming viruses. *J. Virol.*, 35:844–853.

121. Sefton, B. M., Hunter, T., Ball, E. H., and Singer, S. J. (1981): Vinculin: A cytoskeletal target of the transforming protein of Rous sarcoma virus. *Cell*, 24:165–174.

122. Sefton, B. M., Hunter, T., and Raschke, W. C. (1981): Evidence that the Abelson virus protein functions *in vivo* as a protein kinase that phosphorylates tyrosine. *Proc. Natl. Acad. Sci. USA*, 78:1552–1556.

123. Sefton, B. M., and Walter, G. (1982): Antiserum specific for the carboxy terminus of the transforming protein of Rous sarcoma virus. *J. Virol.*, 44:467–474.

124. Sefton, B. M., Trowbridge, I. S., Cooper, J. A., and Scolnick, E. M. (1982): The transforming proteins of Rous sarcoma virus, Harvey sarcoma virus and Abelson virus contain tightly bound lipid. *Cell*, 31:465–474.

125. Sheiness, D., Vennstrom, B., and Bishop, J. M. (1981): Virus-specific RNAs in cells infected by avian myelocytomatosis virus and avian erythroblastosis virus: Modes of oncogene expression. *Cell*, 23:291–300.

126. Shen-Ong, G. L. C., Keath, E. J., Piccoli, S. P., and Cole, M. D. (1982): Novel *myc* oncogene RNA from abortive immunoglobulin-gene recombination in mouse plasmacytomas. *Cell*, 31:443–452.

127. Shibuya, M., and Hanafusa, H. (1982): Nucleotide sequence of Fujinami sarcoma virus: Evolu-

tionary relationship of its transforming gene with transforming genes of other sarcoma viruses. *Cell*, 30:787–795.

128. Shibuya, M., Hanafusa, H., and Balduzzi, P. C. (1982): Cellular sequences related to three new *onc* genes of avian sarcoma virus (*fps*, *yes*, and *ros*) and their expression in normal and transformed cells. *J. Virol.*, 42:143–152.
129. Shibuya, M., Hanafusa, T., Hanafusa, H., and Stephenson, J. R. (1980): Homology exists among the transforming sequences of avian and feline sarcoma viruses. *Proc. Natl. Acad. Sci. USA*, 77:6536–6540.
130. Shih, T. Y., Weeks, M. O., Gruss, P., Dhar, R., Oroszlan, S., and Scolnick, E. M. (1982): Identification of a precursor in the biosynthesis of the p21 transforming protein of Harvey murine sarcoma virus. *J. Virol.*, 42:253–261.
131. Shriver, K., and Rohrschneider, L. R. (1981): Organization of $pp60^{src}$ and selected cytoskeletal proteins within adhesion plaques and junctions of Rous sarcoma virus-transformed rat cells. *J. Cell Biol.*, 89:525–535.
132. Shriver, K., and Rohrschneider, L. (1981): Spatial and enzymatic interactions of $pp60^{src}$ with cytoskeletal proteins in isolated adhesion plaques and junctions from RSV-transformed NRK cells. In: *Cold Spring Harbor Conference on Cell Proliferation, Vol. 8*, pp. 1247–1262. Cold Spring Harbor Laboratory, Cold Spring Harbor, New York.
133. Singer, I. I., and Paradiso, P. R. (1981): A transmembrane relationship between fibronectin and vinculin (130Kd protein): Serum modulation in normal and transformed hamster fibroblasts. *Cell*, 24:481–492.
134. Smart, J. E., Oppermann, H., Czernilofsky, A. P., Purchio, A. F., Erikson, R. L., and Bishop, J. M. (1981): Characterization of sites for tyrosine phosphorylation in the transforming protein of Rous sarcoma virus ($pp60^{v\text{-}src}$) and its normal cellular homologue ($pp60^{c\text{-}src}$). *Proc. Natl. Acad. Sci. USA*, 78:6013–6017.
135. Stanton, L. W., Watt, R., and Marcu, K. B. (1983): Translocation, breakage and truncated transcripts of c-myc oncogene in murine plasmacytomas. *Nature*, 303:401–406.
136. Takeya, T., and Hanafusa, H. (1983): Structure and sequence of the cellular gene homologous to the RSV *src* gene and the mechanism of generating the transforming virus. *Cell*, 32:881–890.
137. Tokuyasu, K. T., Dutton, A. H., Geiger, B., and Singer, S. J. (1981): Ultrastructure of chicken cardiac muscle as studied by double immunolabeling in electron microscopy. *Proc. Natl. Acad. Sci. USA*, 78:7619–7623.
138. Topp, W. C., Lane, D., and Pollack, R. (1980): Transformation by SV40 and polyoma virus. In: *DNA Tumor Viruses. Molecular Biology of Tumor Viruses*, second edition, part 2, edited by John Tooze, pp. 205–296. Cold Spring Harbor Laboratory, Cold Spring Harbor, New York.
139. Troxler, D. H., Ruscetti, S. K., and Scolnick, E. M. (1980): The molecular biology of Friend virus. *Biochim. Biophys. Acta*, 605:305–324.
140. Van Beveren, C., van Straaten, F., Curran, T., Muller, R., and Verma, I. M. (1983): Analysis of FBJ-MuSV Provirus and c-*fos* (mouse) gene reveals that viral and cellular *fos* gene products have different carboxy termini. *Cell*, 32:1241–1255.
141. Van deVen, W. J. M., Khan, A. S., Reynolds, F. H., Mason, K. T., and Stephenson, J. R. (1980): Translational products encoded by newly acquired sequences of independently derived feline sarcoma virus isolates are structurally related. *J. Virol.*, 33:1034–1045.
142. Veronese, F., Kelloff, G. J., Reynolds, F. H., Hill, R. W., and Stephenson, J. R. (1982): Monoclonal antibodies specific to transforming polyproteins encoded by independent isolates of feline sarcoma virus. *J. Virol.*, 43:896–904.
143. Wallbank, A. M., Sperling, F. G., Hubben, K., and Stubbs, E. L. (1966): Isolation of a tumor virus from a chicken submitted to a Poultry Diagnostic Laboratory: Esh sarcoma virus. *Nature*, 209:1265.
144. Wang, J. Y. J., Queen, C., and Baltimore, D. (1982): Expression of an Abelson murine leukemia virus-encoded protein in *Escherichia coli* causes extensive phosphorylation of tyrosine residues. *J. Biol. Chem.*, 257:13181–13184.
145. Wang, L.-H., Feldman, R., Shibuya, M., Hanafusa, H., Notter, M. F. D., and Balduzzi, P. C. (1981): Genetic structure, transforming sequence, and gene product of avian sarcoma virus UR1. *J. Virol.*, 40:258–267.
146. Wang, L.-H., Hanafusa, H., Notter, M. F. D., and Balduzzi, P. C. (1982): Genetic structure and transforming sequence of avian sarcoma virus UR2. *J. Virol.*, 41:833–841.
147. Wehland, J., Osborn, M., and Weber, K. (1979): Cell-to-substratum contacts in living cells: A

direct correlation between interference-reflexion and indirect immunofluorescence microscopy using antibodies against actin and α-actinin. *J. Cell Sci.*, 37:257–273.
148. Willingham, M. C., Jay, G., and Pastan, I. (1979): Localization of the ASV *src* gene product to the plasma membrane of transformed cells by electron microscopic immunocytochemistry. *Cell*, 18:125–134.
149. Willingham, M. C., Pastan, I., Shih, T. Y., and Scolnick, E. M. (1980): Localization of the *src* gene product of the Harvey strain of MSV to plasma membrane of transformed cells by electron microscopic immunocytochemistry. *Cell*, 19:1005–1014.
150. Witte, O. N., Rosenberg, N., and Baltimore, D. (1979): Preparation of syngeneic tumor regressor serum reactive with the unique determinants of the Abelson murine leukemia virus-encoded P120 protein at the cell surface. *J. Virol.*, 31:776–784.
151. Yonemoto, W., Lipsich, L. A., Darrow, D., and Brugge, J. S. (1982): An analysis of the interaction of the Rous sarcoma virus transforming protein, $pp60^{src}$, with a major heat-shock protein. In: *Heat Shock: From Bacteria to Man*, edited by M. J. Schlesinger, M. Ashburner, and A. Tissieres, pp. 289–298. Cold Spring Harbor Laboratory, Cold Spring Harbor, New York.
152. Yoshida, M., and Toyoshima, K. (1980): *In vitro* translation of avian erythroblastosis virus RNA: Identification of two major polypeptides. *Virology*, 100:484–487.

Advances in Viral Oncology, Volume 4,
edited by George Klein. Raven Press,
New York © 1984.

The Reversibility of Neoplastic Transformation: Regulation of Clonal Growth and Differentiation in Hematopoiesis and the Normalization of Myeloid Leukemic Cells

Leo Sachs

Department of Genetics, Weizmann Institute of Science, Rehovot, Israel

The change of normal into malignant cells involves a sequence of genetic changes, including specific chromosome changes (2,38,78,79). Evidence has been obtained with various types of tumors, including sarcomas (79), myeloid leukemias (79,80,82), and teratocarcinomas (14,63), that malignant cells have not lost the genes that control normal growth and differentiation. This was first shown in sarcomas by the finding that it was often possible to reverse the malignant phenotype to a nonmalignant one with a high frequency in cloned sarcoma cells (73–75,79). A comparison of sarcomas with myeloid leukemias showed that reversion of the malignant phenotype can be achieved by two mechanisms.

In one mechanism found with sarcomas (Fig. 1) (73), this reversion was obtained by chromosome segregation, resulting in a change in gene dosage attributable to a change in the balance of specific chromosomes (Fig. 2) (74). This return to the gene balance required for expression of the nonmalignant phenotype occurred without hybridization between different types of cells (29,73–75,79,96). The nonmalignant cells thus were derived from the malignant ones by genetic segregation. Reversion of the malignant phenotype associated with chromosome changes has also been found after hybridization between different types of cells (39,76,95).

In the second mechanism, found in myeloid leukemia, reversion to a nonmalignant phenotype was also obtained in certain clones with a high frequency. In contrast to sarcomas, this reversion was not associated with chromosome segregation. Phenotypic reversion of malignancy in these leukemic cells was obtained by induction of the normal sequence of cell differentiation by the physiologic inducer of differentiation (79,80,82–84).

IN VITRO CLONING AND CLONAL DIFFERENTIATION OF NORMAL HEMATOPOIETIC CELLS

The cloning and clonal differentiation of normal hematopoietic cells in culture made it possible to study the controls that regulate growth (multiplication) and

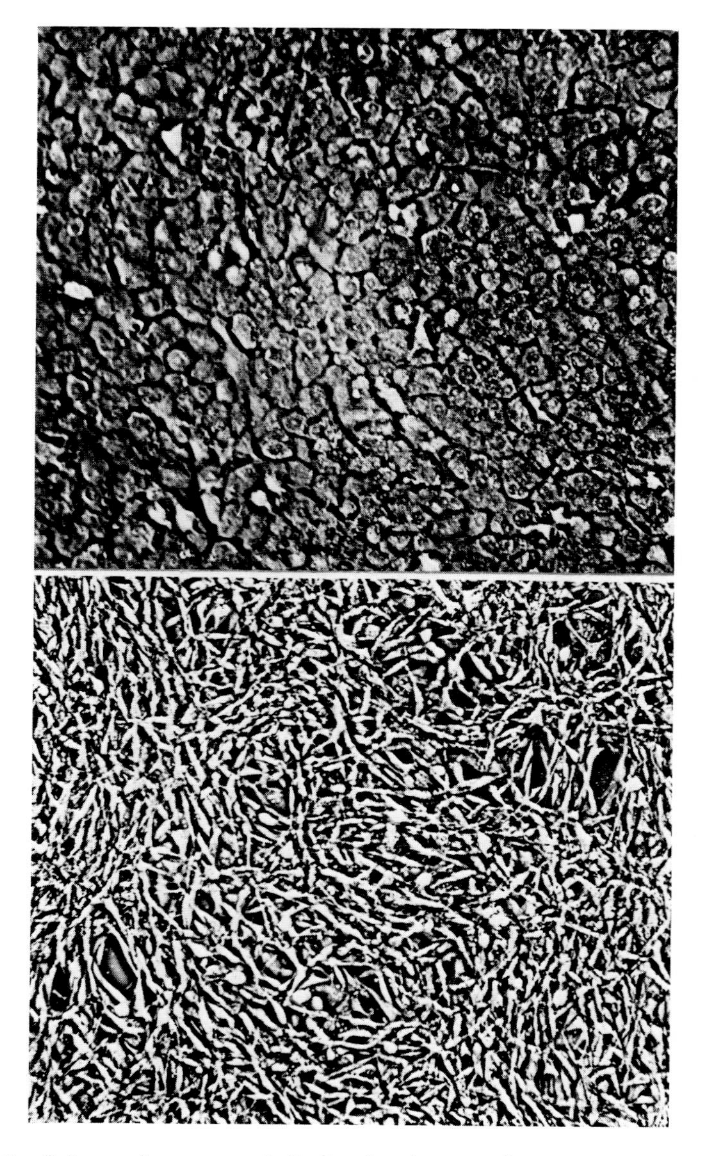

FIG. 1. Cultures of sarcoma cells **(bottom)** and a nonmalignant revertant **(top)** (73).

differentiation of different hematopoietic cell types (79,80,82). We first showed (23,71), as was confirmed by others (4), that normal mouse myeloid precursor cells cultured with a feeder layer of other cell types can form clones of granulocytes and macrophages in culture. We also found that the formation of these clones is due to secretion by cells of the feeder layer of specific inducers that induce the

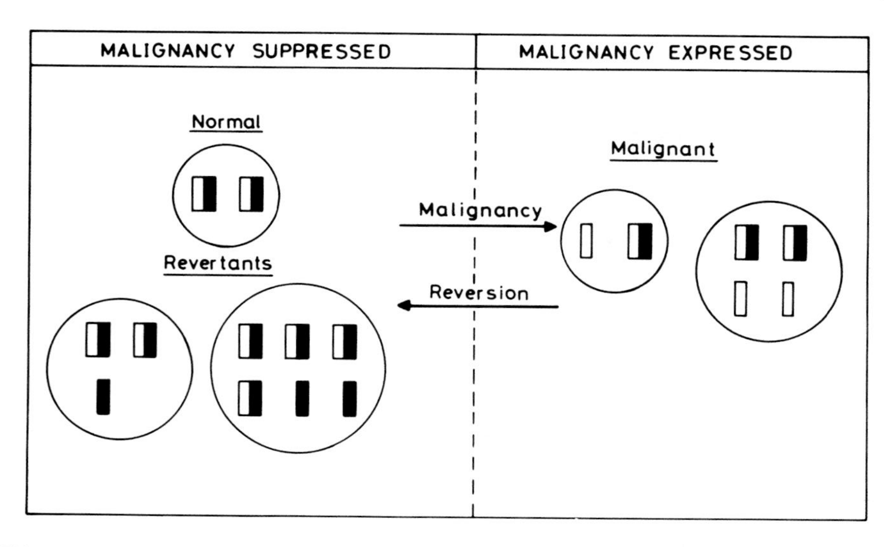

FIG. 2. Gene dosage and the expression and suppression of malignancy from experiments with normal fibroblasts, sarcoma cells, and their nonmalignant revertants (74).

TABLE 1. In vitro *cloning and clonal differentiation of normal hematopoietic cells*

Cloning and differentiation in liquid medium (mast cells and granulocytes) (23)
Cloning and differentiation in agar (macrophages) (71)
Inducer for cloning and differentiation secreted by cells (71)
Inducer in conditioned medium from cells (for macrophages and granulocytes) (36,72)
Different inducer for macrophage and granulocyte clones (36)
Cloning and differentiation of macrophages and granulocytes in methylcellulose (36)
Confirmation of cloning and differentiation in agar (4)
Production of inducer for cloning by some leukemic cells (66)
Cloning and differentiation of human cells (68,70)
Protein inducer of differentiation that does not induce cloning (21)
Terminology used for proteins that induce cloning and differentiation of normal macrophages and granulocytes
Mashran gm (37)
Colony stimulating factor (CSF) (61)
Colony stimulating activity (CSA) (1)
Macrophage and granulocyte inducer (MGI) (41)
MGI-1 (= mashran gm, CSF, CSA) for cloning: MGI-2 for differentiation (45,48,58,82)

formation of clones and the differentiation of cells in these clones to macrophages or granulocytes in mice (36,71,72) and in humans (68). After we first detected their presence in culture supernatants (36,72), these protein inducers were referred to by a number of names; I use the term macrophage and granulocyte inducers (MGI) (Table 1). These proteins can be produced and secreted by various normal and malignant cells in culture and *in vivo* (79). Their production can be induced by a variety of compounds (16,20,55,94), and some cells produce these proteins constitutively (1,36,41,47,87). MGI are a family of proteins that exist in a number of molecular forms that have different biologic activities.

PROTEINS THAT INDUCE CELL GROWTH AND DIFFERENTIATION

The family of MGI proteins includes some that induce cell growth (multiplication) and others that induce differentiation. Those that induce growth, which are also required for normal cell viability, we call MGI-1 (48,82), of which there are different forms. These include proteins that induce the formation of macrophage clones (MGI-1M) (36,48,87), granulocyte clones (MGI-1G) (36,48), or both types of clones (MGI-1GM) (7,41,47). MGI-1 has previously been referred to as *mashran gm* (37), colony-stimulating factor (CSF) (61), colony-stimulating activity (CSA) (1), and MGI (41) (Table 1). The existence of an antibody that does not react with all forms of MGI-1M has shown that there can be different antigenic sites on molecules that belong to the same form of MGI-1 (48). The other main type of MGI, which we call MGI-2 (45,48,82), induces the differentiation of myeloid precursor cells, either leukemic (18) or normal (45,82), without inducing colony formation. This differentiation-inducing protein (18,21) has also been referred to as MGI (18), D factor (59), and GM-DF (8). It has been suggested that there are different forms of MGI-2 for differentiation to macrophages or granulocytes (45). The regulation of MGI-1 and MGI-2 appears to be under the control of different genes (16). Differentiation-inducing protein MGI-2, but not the growth-inducing protein, is a DNA-binding protein (93).

These MGI can be proteins or glycoproteins, depending on the cells in which they are produced. The presence of carbohydrates does not appear to be necessary for their biologic activity (47). Their molecular weights usually are around 23,000 or multiples of this number (48,65,80). MGI-2 activity is more sensitive to proteolytic enzymes and high temperature than MGI-1 activity (47). MGI-2 has a shorter half-life in serum than MGI-1 (57). The ready separability of the different forms of MGI seems to depend on the cells from which they are derived (48). Further studies should determine whether different forms of MGI are derived from a common precursor, and whether tumor cells with the appropriate gene rearrangements, and possibly even normal cells under certain conditions, may produce hybrid molecules of different forms of MGI, including hybrid molecules with MGI-1 and MGI-2 activity (45).

MECHANISM THAT COUPLES GROWTH AND DIFFERENTIATION IN NORMAL CELLS: INDUCTION OF DIFFERENTIATION-INDUCING PROTEIN BY GROWTH-INDUCING PROTEIN

We have developed a simple procedure for isolating normal myeloid precursor cells from the bone marrow (Fig. 3) (51). Incubation of isolated normal myeloid precursors with MGI-1, either MGI-1M or MGI-1G (48), induces the viability and growth of these normal precursors and results in cell differentiation to macrophages or granulocytes even without adding the differentiation-inducing protein MGI-2.

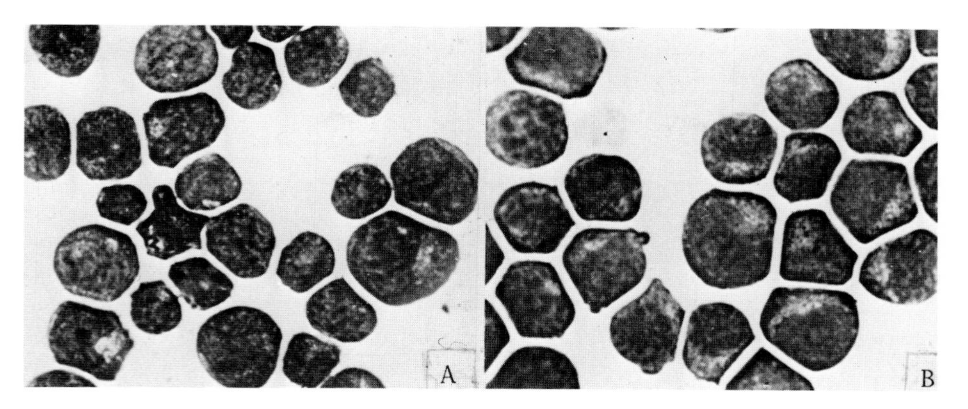

FIG. 3. Photographs of normal myeloid precursor cells isolated from normal bone marrow **(A)** and MGI$^+$D$^+$ myeloid leukemic cells **(B)** (51).

The incubation of normal myeloid precursors with MGI-1 also results in the induction of MGI-2 (45,58,82), which can be detected at 6 hr after the addition of MGI-1. This induction of MGI-2 by MGI-1 thus can account for the induction of differentiation after adding MGI-1 to the normal cells. The induction of differentiation-inducing protein MGI-2 by growth-inducing protein MGI-1 appears to be an effective control mechanism for coupling growth and differentiation in normal cells.

Multiplication of normal cells is regulated at two control points. The first requires MGI-1 to produce more cells that can then differentiate by the MGI-2 induced by MGI-1. The second control is the stopping of cell multiplication that occurs as part of the program of terminal differentiation to mature cells induced by MGI-2. Thus there is a coupling of growth and differentiation in normal cells at both these points. Mature cells can also produce feedback inhibitors that interfere with the induction of growth of the normal precursors by MGI-1 (5,6,37,67).

REGULATION OF GROWTH AND DIFFERENTIATION IN LEUKEMIA

Normal myeloid precursor cells isolated from bone marrow (51) require an external source of MGI-1 for cell viability and growth. There are, however, myeloid leukemic cells that no longer require MGI-1 for viability and growth, so that these leukemic cells can then multiply in the absence of MGI-1 (80,82). This gives the leukemic cells a growth advantage over the normal cells when there is a limiting amount of MGI-1. Starting with a decreased requirement for MGI-1, there are various stages that eventually lead to a complete loss of this requirement. Other myeloid leukemic cells constitutively produce their own MGI-1 (64,66), and these leukemic cells also have a growth advantage compared to normal cells that require an external source of MGI-1 (Fig. 4). A change in the requirement of MGI-1 for growth, either a partial or complete loss of this requirement, or the constitutive production of MGI-1 give a growth advantage to leukemic cells. The growth

Type of myeloid cells	Requirement of MGI-1 for growth
Normal	External source
Leukemic	Decrease ⟶ no requiremet or Constitutive production

FIG. 4. Differences in MGI-1 requirement for growth in normal and leukemic myeloid cells.

advantage to leukemic cells with a constitutive production of MGI-1 would be even greater if MGI-1 does not induce MGI-2 in these leukemic cells.

The existence of myeloid leukemic cells that either no longer require MGI-1 for viability and growth or constitutively produce their own MGI-1, raises the question of whether these leukemic cells can still be induced to differentiate to mature cells by the normal differentiation-inducing protein MGI-2. This question has been answered by showing that there are clones of myeloid leukemic cells that no longer require MGI-1 for growth but can still be induced to differentiate normally to mature macrophages and granulocytes by MGI-2 (Fig. 5) via the normal sequence of gene expression (80,82,84). These mature cells then are no longer malignant *in vivo* (19,57). Many differentiation-associated properties are induced in these cells by MGI-2 (40,42,49,52,84,88). These include the ability to respond to chemotaxis (Fig. 6). Injection of these myeloid leukemic cells into embryos has shown that after such injection, the leukemic cells can participate in hematopoietic differentiation in apparently healthy adult animals (25).

Injection of MGI-2 into animals, or *in vivo* induction of MGI-2 by a compound that induces the production of this differentiation-inducing protein, results in an inhibition of leukemia development in animals with such leukemic cells (Fig. 7) (57). There are also myeloid leukemic cells that constitutively produce their own MGI-1 and that can be induced to differentiate by MGI-2. Our results indicate that induction of normal differentiation in myeloid leukemic cells by MGI-2 can be an approach to therapy based on the induction of normal differentiation in malignant cells (18,54,57,68,79–81).

Leukemic clones that can be induced to differentiate to mature cells by MGI-2 have been found in different strains of mice (8,17,18,34,35) and in humans (55,68).

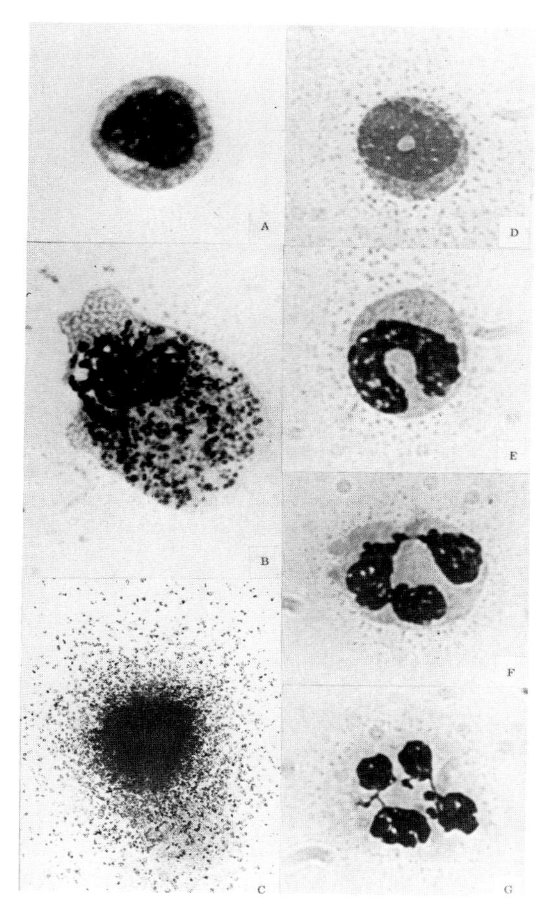

FIG. 5. Differentiation of MGI$^+$-D$^+$ myeloid leukemic cells to mature macrophages and granulocytes by the normal myeloid differentiation-inducing protein MGI-2. Leukemic cell **(A)**, macrophage **(B)**, colony of macrophages **(C)**, stages in differentiation to granulocytes **(D–G)** (18).

They are referred to as MGI$^+$D$^+$ (MGI$^+$, they can be induced to differentiate by MGI-2; D$^+$, differentiation to mature cells). MGI$^+$D$^+$ leukemic cells have specific chromosome changes compared to normal cells (2,27). These changes seem to involve changes in genes other than those involved in the induction of normal differentiation. Other clones of myeloid leukemic cells also can grow without adding MGI-1 but are either partly (MGI$^+$D$^-$) or almost completely (MGI$^-$D$^-$) blocked in their ability to be induced to differentiate by MGI-2 (Fig. 8) (17,30,32,35,48,85,86). These differentiation-defective clones have specific chromosome changes compared to MGI$^+$D$^+$ cells (2,27).

A variety of compounds other than MGI-2 can induce differentiation in MGI$^+$D$^+$ clones; not all are active on the same MGI$^+$D$^+$ clone, and they do not all induce the same differentiation-associated properties. The inducers include certain steroids, lectins, polycyclic hydrocarbons, tumor promoters, lipopolysaccharides, X-irradiation, and compounds used in cancer chemotherapy (80). The existence of

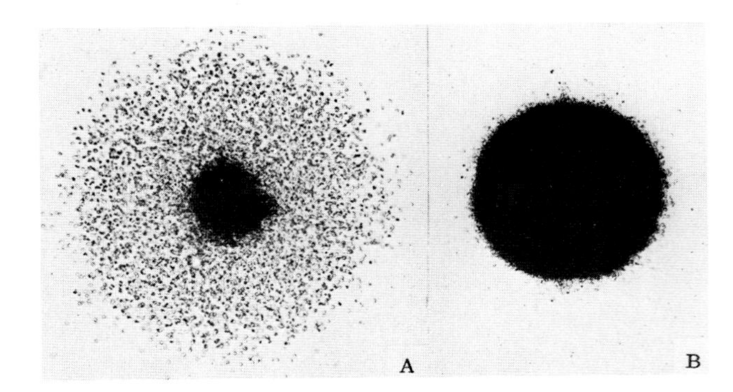

FIG. 6. Colony from a MGI^+D^+ clone **(A)** and from a MGI^-D^- clone **(B)** cultured with MGI-2 (27). The differentiating cells in the MGI^+D^+ clone migrate from the colony when the cells are induced to respond to chemotactic stimuli (88).

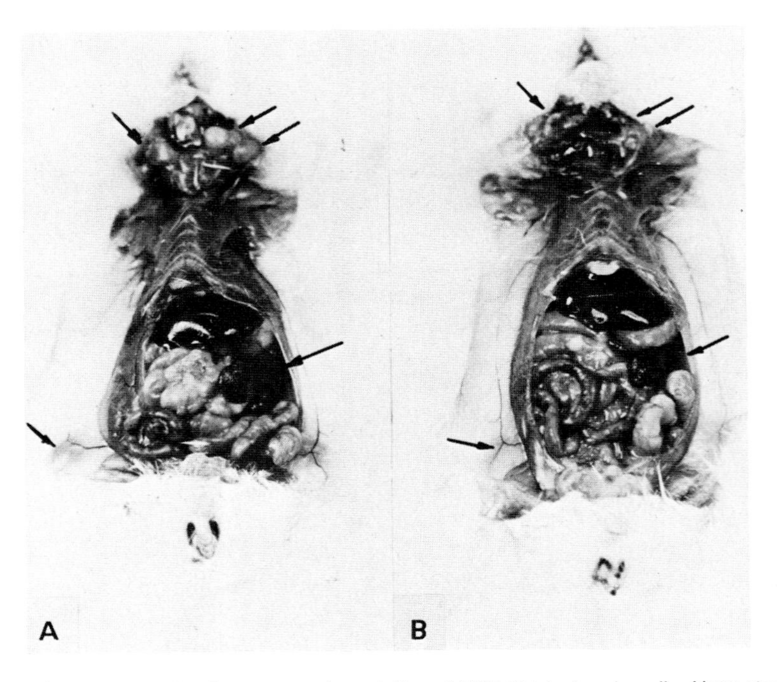

FIG. 7. Mice 38 days after intravenous inoculation of MGI^+D^+ leukemic cells. Untreated leukemic mouse **(A)** and mouse injected with MGI-2 **(B)** (57).

clonal differences in the ability of X-irradiation and cancer chemotherapeutic chemicals to induce differentiation may help to explain differences in response to therapy in different individuals (80). As a result of these experiments, we have suggested that it may be possible to introduce a form of therapy based on induction of differentiation (18,54,56,57,68,79–81). This would include prescreening in culture

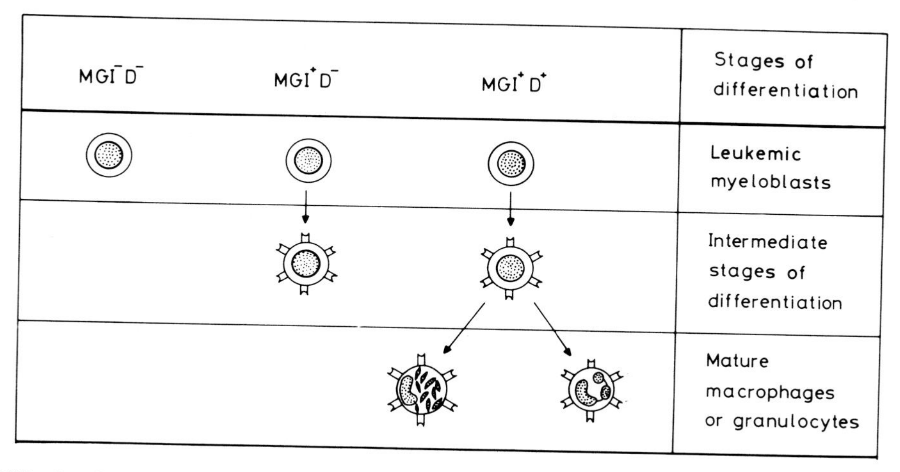

FIG. 8. Classification of different types of myeloid leukemic cell clones according to their ability to be induced to differentiate by MGI-2. MGI$^+$: inducible for differentiation-markers by MGI-2; D$^+$: inducible for mature cells by MGI-2.

to select for the most effective compounds and using these compounds for a low dose chemotherapy protocol aimed at inducing cell differentiation (56). Since different myeloid leukemic clones respond differently to MGI-2 and other compounds, such differences will also occur in leukemic cells from different patients. Based on these suggestions (81), some encouraging clinical results have been obtained with the use of low dose cytosine arabinoside (3,33). Successful treatment of a patient with acute monoblastic leukemia with low dose cytosine arabinoside is shown in Fig. 9 (62).

DIFFERENT PATHWAYS OF GENE EXPRESSION IN DIFFERENTIATION

Some of the compounds that induce differentiation in susceptible clones of MGI$^+$D$^+$ leukemic cells, including lipopolysaccharide, phorbol ester tumor promoters, such as 12-0-tetradecanoyl-phorbol-13-acetate (TPA), and nitrosoguanide, can induce in these clones the production of MGI-2. These compounds thus induce differentiation by inducing in the leukemic cells the endogenous production of the normal differentiation-inducing protein MGI-2 (16,55,94). Other compounds, such as the steroid dexamethasone, can induce differentiation in MGI$^+$D$^+$ clones without inducing MGI-2 (16). This steroid induces differentiation by other pathways of gene expression than MGI-2 (9,52); the same applies to dimethylsulfoxide (DMSO).

In a line of human myeloid leukemic cells, DMSO induces the formation of granulocytes (11), whereas MGI-2 (55,56) and the tumor promoting phorbol ester TPA (55,77), which induces the production of MGI-2 (55), induce the formation of macrophages. Using two-dimensional gel electrophoresis, studies on the protein changes induced by these inducers showed a similar developmental program for

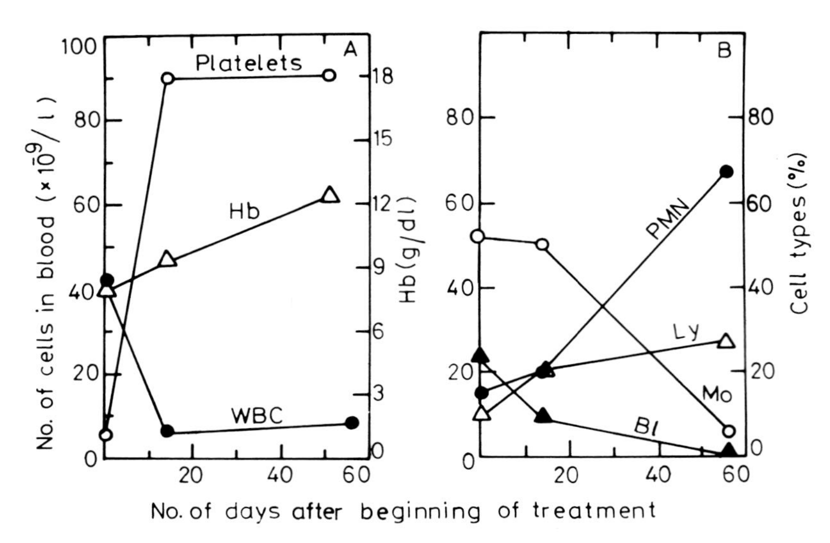

FIG. 9. Blood composition in a patient with acute monoblastic leukemia treated with a low dose of cytosine arabinoside (62).

macrophage differentiation induced by MGI-2 and TPA, which differed from the beginning from the granulocyte program induced by DMSO (Fig. 10) (44). Unlike MGI-2 or DMSO, TPA induces rapid cell attachment of these myeloid leukemic cells to the Petri dish. Combined treatment with TPA and DMSO showed cell attachment, extensive spreading of the cells, the regulation of specific proteins, and expression of the macrophage program. The results indicate that cells in suspension can express either the macrophage or granulocyte program, depending on the inducer, and that changes in cell shape associated with cell attachment can regulate specific proteins and restrict the developmental program to macrophages (Fig. 11) (44). The *in vivo* environment of cells in relation to the possibilities of cell adhesion may play a major role in determining the differentiation program of myeloid and other cell types.

COMPLEMENTATION OF GENE EXPRESSION IN DIFFERENTIATION BY COMBINED TREATMENT WITH DIFFERENT COMPOUNDS

Induction of differentiation in some myeloid leukemic clones requires combined treatment with different compounds (40,53,55,91,92). In these cases, each compound induces changes not induced by others, so that the combined treatment results in new gene expression. This complementation of gene expression can occur at the levels of mRNA production and translation (31). With the appropriate combination of compounds, we have been able to induce all our MGI^-D^- leukemic clones for some differentiation-associated properties (91,92). It will be interesting to determine whether the same applies to differentiation of erythroleukemic cells (22,60). It is possible that all myeloid leukemic cells no longer susceptible to the normal differentiation-inducing protein MGI-2 by itself can be induced to differenti-

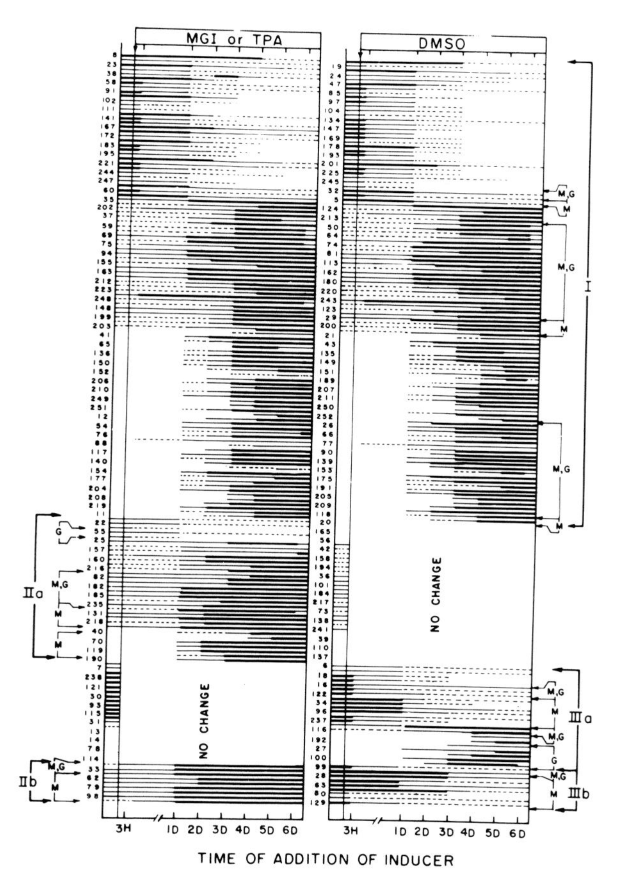

FIG. 10. Schematic summary of cytoplasmic protein changes in human myeloid leukemic cells during macrophage differentiation induced by MGI-2 or TPA and granulocyte differentiation induced by DMSO. The summary is based on analysis of two-dimensional gels of cytoplasmic proteins synthesized in untreated cells and in cells 3 hr (H) and 1 to 6 days (D) after incubation with MGI-2, TPA, or DMSO. *Thick line*, high or maximum synthesis; *thin line*, intermediate synthesis; *dashed line*, low or minimum synthesis. The absence of a line indicates that no synthesis was detected. This summary shows the relative and not the absolute rates of protein synthesis (44).

ate by choosing the appropriate combinations of compounds to give the required complementation. This can include the use of hormones, such as steroids (49,50) or insulin (90,91), and different nonphysiologic compounds (80) with or without MGI-2.

MECHANISMS THAT UNCOUPLE GROWTH AND DIFFERENTIATION IN LEUKEMIA: AUTOINDUCTION OF DIFFERENTIATION IN ONE TYPE OF LEUKEMIA

As mentioned above, there are $MGI^{+}D^{+}$ clones of myeloid leukemic cells that no longer require MGI-1 for growth but still can be induced to differentiate

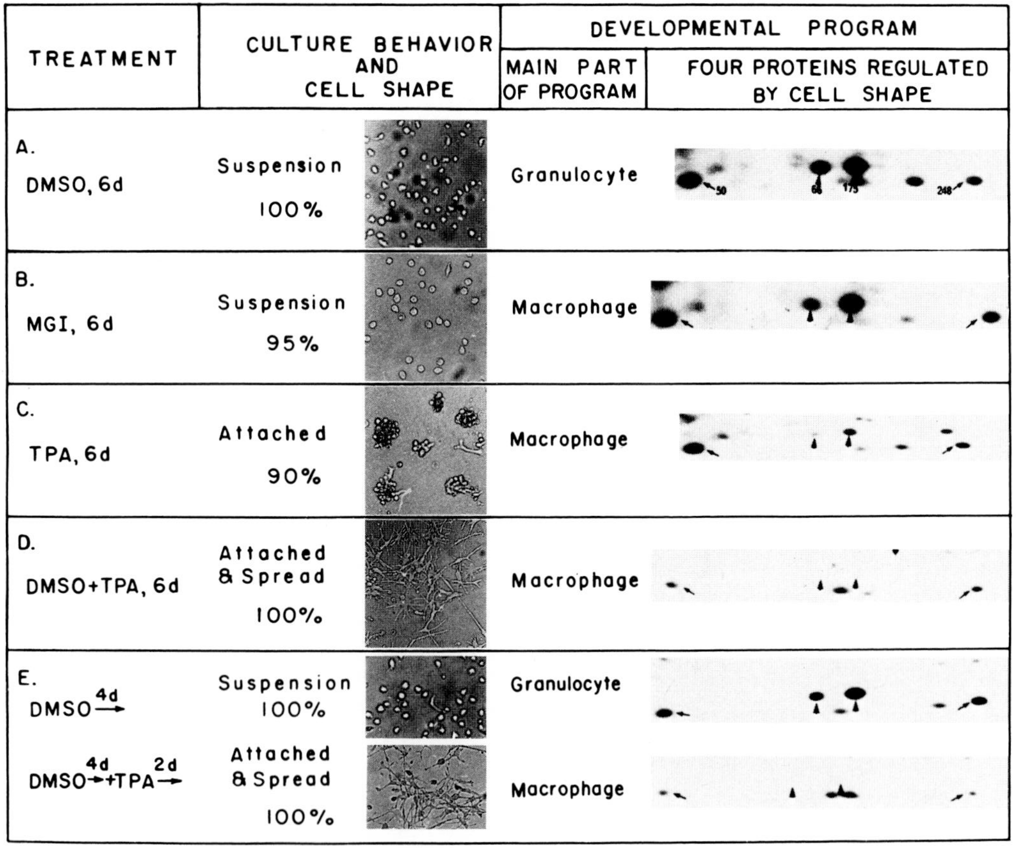

TREATMENT	CULTURE BEHAVIOR AND CELL SHAPE	DEVELOPMENTAL PROGRAM: MAIN PART OF PROGRAM	DEVELOPMENTAL PROGRAM: FOUR PROTEINS REGULATED BY CELL SHAPE
A. DMSO, 6d	Suspension 100%	Granulocyte	50, 68, 175, 248
B. MGI, 6d	Suspension 95%	Macrophage	
C. TPA, 6d	Attached 90%	Macrophage	
D. DMSO+TPA, 6d	Attached & Spread 100%	Macrophage	
E. DMSO $\xrightarrow{4d}$	Suspension 100%	Granulocyte	
DMSO $\xrightarrow{4d}$ +TPA $\xrightarrow{2d}$	Attached & Spread 100%	Macrophage	

FIG. 11. Cell shape and expression of the developmental programs for macrophages and granulocytes and of four class I proteins in human myeloid leukemic cells. The main part of the program refers to expression of class II or III proteins (Fig. 9). The four proteins shown are in class I (Fig. 9). d, Days (44).

normally by MGI-2. These leukemic cells have uncoupled the normal requirement for growth from the normal requirement for differentiation. Experiments on the properties of these cells after induction of differentiation by MGI-2 have shown that the normal requirement for MGI-1 for cell viability and growth is restored in the differentiating leukemic cells (21,58). MGI-1 added to normal myeloid precursors induces the production of MGI-2, so that the cells then can differentiate by the endogenously produced MGI-2. In these leukemic cells, however, MGI-1 did not induce the production of MGI-2 even though, like normal cells, they again required MGI-1 for viability and growth. There was no induction of differentiation after adding MGI-1 (58). Another type of leukemic cell constitutively produces its own MGI-1 and can also show this lack of induction of MGI-2 by MGI-1, so that the cells do not differentiate (89). The absence of induction of MGI-2 by MGI-1 uncouples growth and differentiation in these leukemic cells. The lack of requirement of MGI-1 for growth and the absence of induction of the differentiation-inducing protein MGI-2 by the growth-inducing protein MGI-1 thus are mechanisms that uncouple growth and differentiation in MGI^+D^+ leukemic cells (58,82,89).

In leukemic cells with constitutive production of MGI-1, changes in specific components of the culture medium can result in an autoinduction of differentiation due to restoration of the induction of MGI-2 by MGI-1, which then restores the normal coupling of growth and differentiation (Fig. 12). These changes in the culture medium include the use of mouse or rat serum instead of horse or calf serum, serum-free medium, and removal of transferrin from serum-free medium (89). Autoinduction of differentiation in this type of leukemic cell also may occur under certain conditions *in vivo*.

Type of myeloid cells	Requirement of MGI-1 for growth	Induction of MGI-2 by MGI-1	Differentiation
Normal	+	Production of MGI-2 →	+
Leukemic	+ or −	No production of MGI-2	−
	Constitutive production MGI-1	Production of MGI-2 →	+ *

FIG. 12. Differences in induction of MGI-2 by MGI-1 in normal and leukemic myeloid cells. *Autoinduction of differentiation under specific conditions.

The coupling of growth and differentiation in normal cells is regulated at two control points. The uncoupling of growth and differentiation in MGI^+D^+ leukemic cells is at the first control point; the coupling at the second control in normal cells, between the initiation of differentiation by MGI-2 and the stopping of multiplication in the mature cells, is maintained. There are differentiation-defective MGI^+D^- leukemic cells that, like the MGI^+D^+ leukemic cells, no longer require addition of MGI-1 for growth. In these cells, however, MGI-2 induces only a partial differentiation; mature cells are not produced; and the cells do not stop multiplying. In addition to uncoupling growth and differentiation at the first control point, MGI^+D^- leukemic cells show a second uncoupling between the initiation of differentiation by MGI-2 and the stopping of cell multiplication that occurs as part of the normal program of terminal differentiation. It has been suggested that leukemia originates by uncoupling the first control, and that uncoupling of the second control then results in a futher evolution of leukemia (80,82).

CHROMOSOME CHANGES AND RETROVIRUSES IN LEUKEMIC CELLS

None of the clones of myeloid leukemic cells studied has a completely normal diploid chromosome banding pattern. There are also specific chromosome differences between the normal and leukemic cells and between MGI^+D^+, MGI^+D^-, and MGI^-D^- clones (2,27). Chromosome studies on normal fibroblasts, sarcomas, and revertants from sarcomas which have regained a nonmalignant phenotype have indicated that the difference between malignant and nonmalignant cells is controlled by the balance between genes for expression (E) and suppression (S) of malignancy. When there is enough S to neutralize E, malignancy is suppressed; when the amount of S is not sufficient to neutralize E, malignancy is expressed (Fig. 2) (29,74,79,95). Genes for expression of malignancy (E) have been referred to in present terminology as oncogenes. Studies on the chromosomes of the myeloid leukemic clones suggest that this applies to the origin of malignancy in these leukemias, and that the ability of the leukemic cells to be induced to differentiate by MGI-2 is also dependent on the balance between different genes. It has been shown that MGI^-D^- cells can give rise to MGI^+D^+ progeny by segregation of appropriate chromosomes, and that these chromosome changes then restore the appropriate gene balance required for induction of differentiation by MGI-2 (Table 2) (2). Changes in the balance of specific genes due to changes in gene dosage is a mechanism that could produce the uncoupling of growth and differentiation in myeloid leukemia. Specific changes in gene dosage have also been found in lymphoid leukemia (15,38).

We have suggested from our chromosome data on these mouse myeloid leukemias that inducibility for differentiation by MGI-2 is controlled by the balance between genes on chromosomes 2 and 12, and that these chromosomes also carry genes that control the malignancy of these cells (2). The leukemic cells also had deletions in chromosome 2 and rearrangements with chromosome 12 (2). It has since been

TABLE 2. *Chromosomes of MGI^+D^+ and MGI^-D^- clones*

Cell type	Clone no.	No. of chromosomes in chromosome groups[a]												Modal chromosome no.
		2	3	6	7	12	14	X	T(3;6)	T(3;12)	T(7;15)	T(12;15)	U	
MGI^+D^+	11	2	2	2	2	1+B	2	2	0	0	0	0	0	40
MGI^+D^+	7-M9	1 + F	1	1	1	0	1	1	1	1	1	1	1	38
MGI^+D^+	7-M11	1 + F	1	1	1	0	1	1	1	1	1	1	1	38
MGI^+D^+	7-M16	1 + F	1	1	1	0	1	1	1	1	1	1	1	38
MGI^+D^+	7-M4	1 + F	1	1	1	1	1	1	1	0	1	1	1	38
MGI^+D^+	7-M5	1 + F	1	1	1	1	1	1	1	0	1	1	1	38
MGI^-D^-	7	1 + F	1	1	1	1	1	1	1	1	1	1	1	39

[a]All other chromosome groups had the normal diploid pattern. Translocation (T), deletion (F), insertion (B), unknown (U) (2).

found that (a) the immunoglobulin heavy chain gene is on mouse chromosome 12 (13); (b) the c-*abl* gene is on mouse chromosome 2 (24); and (c) in human chronic myelocytic leukemia, c-*abl* is involved (12) in the translocation of the Philadelphia chromosome (78). In addition to the role of gene dosage (2,79) and the deletions we found in chromosome 2 (2), we are now studying the possible role of DNA rearrangements in the regulation of genes involved in the malignancy of myeloid leukemic cells, cell competence for induction of differentiation, and in the differentiation program.

The origin and further evolution of malignancy can involve, in different cases, changes in gene dosage, deletions, rearrangements, or mutation. Genes for the expression (E) and suppression (S) of malignancy, that include the genes which in present terminology are called oncogenes, are genes involved in the control of normal cell growth and differentiation, and may include genes for proteins like MGI-1 and MGI-2. Genetic changes in the regulation or structure of these genes can produce the uncoupling required for the origin and further evolution of malignancy.

Normal myeloblasts and the different types of myeloid leukemic cells from mice contain endogenous retroviruses (43,46). Superinfection of normal myeloblasts with endogenous ecotropic virus from normal myeloblasts or MGI^+D^+ leukemic cells induces viability and multiplication of these normal cells when MGI-1 is not added, without blocking the ability of these cells to be induced to differentiate by MGI-2. It has been suggested that this promotion of cell multiplication is caused by integration of proviral sequences near regulatory sites that control cell growth (43,46).

This promotion of growth by superinfection with endogenous virus from normal or MGI^+D^+ leukemic cells is not accompanied by differentiation of the cells unless MGI is added. The integration of proviral sequences near growth regulatory sites thus can uncouple growth and differentiation. The cells do not continue to grow for long periods, however, as is the case of leukemic cells. If this stage is followed by the appropriate chromosome changes, this could then produce the more complete uncoupling found in the leukemic cells. Endogenous ecotropic virus from MGI^-D^- leukemic cells did not promote growth of the normal myeloblasts, and it has been suggested that this provirus is not integrated near growth regulatory sites (Fig. 13) (46). The promotion of growth induced by the virus from normal or MGI^+D^+ leukemic cells may increase the probability of these chromosome changes. In human cells, the chromosome changes could produce this uncoupling without the presence of the virus (84).

The finding that infection with mouse myeloblast virus can promote the growth of myeloid precursors without blocking differentiation (43,46) has been confirmed with erythroid precursors infected with the Harvey and Kirsten sarcoma viruses (26). Studies with avian leukosis virus have provided direct evidence that provirus can be integrated near growth regulatory genes, and that this can activate these cellular genes (28,69).

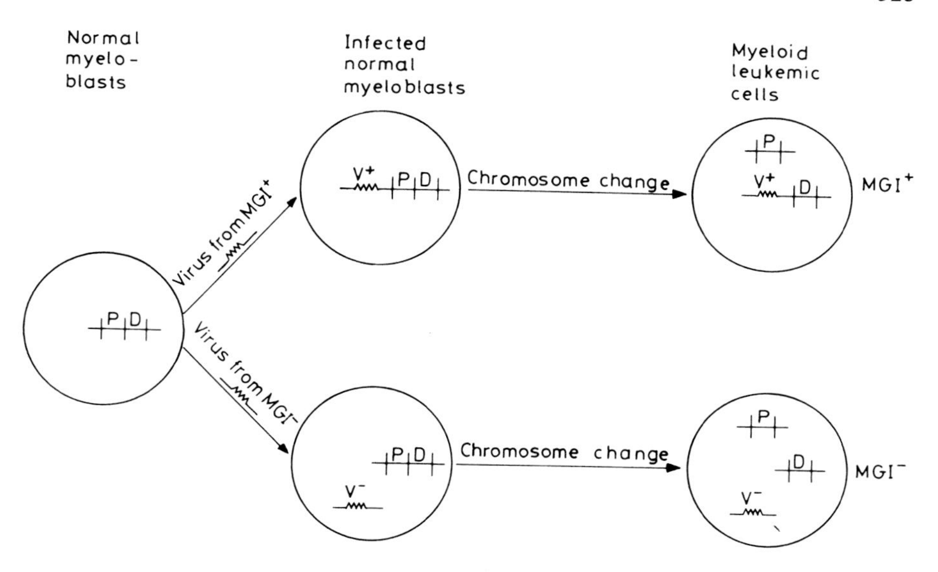

FIG. 13. Model of integration sites for the endogenous virus from normal and MGI^+D^+ leukemic cells (V^+) and from MGI^-D^- leukemic cells (V^-), in relation to the genes for cell viability and growth (proliferation, P) and differentiation (D) (46).

CONSTITUTIVE GENE EXPRESSION AND THE ORIGIN AND FURTHER EVOLUTION OF LEUKEMIA

In leukemic cells that, unlike normal myeloblasts, no longer require MGI-1 for cell viability and growth, the molecular changes required for viability and growth that must be induced in the normal cells are constitutive. This also applies to leukemic cells that constitutively produce their own MGI-1, suggesting that the origin of myeloid leukemia can be a change from an induced to a constitutive expression of genes that control cell viability and growth (80,82).

Studies using two-dimensional gel electrophoresis (42) on changes in the synthesis of specific proteins in normal myeloblasts, MGI^+D^+, MGI^+D^-, and MGI^-D^- leukemic clones at different times after adding MGI-1 and MGI-2 have directly shown that there have been changes from inducible to constitutive gene expression in the leukemic cells. The results also indicate a relationship between constitutive gene expression and uncoupling of the initiation of differentiation by MGI-2 and the stopping of multiplication in mature cells. The leukemic cells were found to be constitutive for changes in the synthesis of specific proteins that were only induced in the normal cells after treatment with MGI-1. These protein changes, which included the appearance of some proteins and the disappearance of others, were constitutive in all the leukemic clones studied derived from different tumors. These have been called C_{leuk} (constitutive for leukemia). Other protein changes were induced by MGI-2 in normal and MGI^+D^+ leukemic cells and were constitutive in MGI^+D^- and MGI^-D^- leukemic cells. There were more of these constitutive changes in MGI$^-$D$^-$ than in MGI$^+$D$^-$ leukemic cells. These have been called C_{def} (constitutive for differentiation defective) (Fig. 14) (42).

These results indicate that changes from an induced to a constitutive expression of certain genes is associated with the uncoupling of growth and differentiation, both at the control that requires MGI-1 to produce more cells and at the control of the stopping of cell multiplication that occurs in the formation of mature cells.

The protein changes during the growth and differentiation of normal myeloblasts are induced by MGI-1 and MGI-2 as a series of parallel multiple pathways of gene expression (42). It can be assumed that the normal developmental program that couples growth and differentiation in normal cells requires synchronous initiation and progression of these multiple parallel pathways. The presence of constitutive gene expression for some pathways can be expected to produce asynchrony in the coordination required for the normal developmental program. Depending on the pathways involved, this asynchrony then could result in an uncoupling of the controls for growth and differentiation and produce different blocks in the ability to be induced for and to terminate the differentiation process.

We have been able to treat MGI$^-$D$^-$ leukemic cells to induce the reversion of specific proteins from the constitutive to the nonconstitutive state. This reversion was associated with a gain of inducibility by MGI-2 for various differentiation-associated properties. Reversion from the constitutive to the nonconstitutive state in these cells restored the synchrony required for induction of differentiation (92).

The suggestion derived from these results (42,80,82) is that myeloid leukemia originates by a change that produces certain constitutive pathways of gene expression, so that cells no longer require MGI-1 for growth or constitutively produce MGI-1 without inducing MGI-2. These leukemic cells can still be induced to differentiate normally by MGI-2 added exogenously or induced in the cells in other ways. The differentiation program induced by MGI-2 can proceed normally when

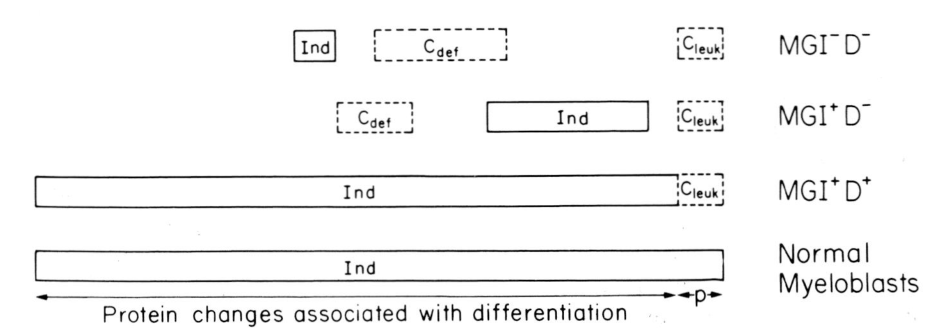

FIG. 14. Schematic summary of protein changes associated with growth (proliferation) and differentiation in normal myeloblasts and different types of myeloid leukemic cells. C_{leuk}, constitutive expression of changes in leukemic compared to normal myeloblasts; C_{def}, constitutive expression of changes in the differentiation-defective MGI$^+$D$^-$ and MGI$^-$D$^-$ clones compared to MGI$^+$D$^+$ clones; Ind, MGI-induced changes; p, protein changes associated with proliferation to produce more cells before differentiation (42).

it is uncoupled from the growth program induced by MGI-1. This can be followed by constitutive expression of other pathways, resulting in the uncoupling of other controls and an asynchrony that interferes with the normal program of terminal differentiation. These second changes then result in the further evolution of leukemia (82).

The MGI-1-independent growth of myeloid leukemic cells seems to proceed through stages (10), probably starting from a small decrease in MGI-1 requirement and ending in complete independence of MGI-1. Presumably, there are stages in the amount of constitutively produced MGI-1 and the degree of lack of inducibility of MGI-2 by MGI-1. These stages would produce different degrees of asynchrony, resulting in differences in the degree of hematologic abnormality; the final stages of asynchrony then result in leukemia.

DEVELOPMENTAL PROGRAMS AND THE REVERSION OF MALIGNANCY IN DIFFERENT TYPES OF TUMOR CELLS

These conclusions on the origin and evolution of myeloid leukemia may be applicable to malignant tumors derived from other types of cells whose viability, growth, and differentiation are induced by other physiologic inducers. Identification of the physiologic inducers of growth and differentiation for different cell types would be a crucial requirement in extending these conclusions to other tumors. Even in the absence of such identification, however, it is likely that teratocarcinoma cells (14,63) may be comparable to MGI^+D^+ myeloid leukemic cells. The presence of fetal proteins in certain tumors may also be due to constitutive gene expression in the tumor of a protein that is induced by the physiologic inducer during the developmental program in the normal fetus (82). There are probably a variety of tumors in which (a) the original malignancy has a normal differentiation program, and the cells are malignant because of uncoupling of the requirement for growth from the requirement for differentiation by changing the gene expression required for growth from inducible to constitutive; and (b) the further evolution of the tumor results from changes from inducible to constitutive of other pathways of gene expression that produce asynchrony in the normal differentiation program, so that mature, nondividing cells are not formed by the physiologic inducer of differentiation. However, even these tumors may still be induced to differentiate to form nonmalignant cells, by treatment with compounds that can reverse the constitutive to the non-constitutive state or induce the differentiation program by other pathways.

REFERENCES

1. Austin, P. E., McCullouch, E. A., and Till, J. E. (1971): Characterization of the factor in L cell conditioned medium capable of stimulating colony formation by mouse marrow cells in culture. *J. Cell. Physiol.*, 77:121–134.
2. Azumi, J., and Sachs, L. (1977): Chromosome mapping of the genes that control differentiation and malignancy in myeloid leukemic cells. *Proc. Natl. Acad. Sci. USA*, 74:253–257.
3. Baccarani, M., and Tura, S. (1979): Correspondence. Differentiation of myeloid leukemic cells: New possibilities for therapy. *Br. J. Haematol.*, 42:485–487.

4. Bradley, T. R., and Metcalf, D. (1966): The growth of mouse bone marrow *in vitro*. *Aust. J. Exp. Biol. Med. Sci.*, 44:287–300.
5. Broxmeyer, H. E., de Sousa, M., Smithyman, A., Ralph, P., Hamilton, J., Kurland, J. I., and Bognacki, J. (1980): Specificity and modulation of the action of lactoferrin, a negative feedback regulator of myelopoiesis. *Blood*, 55:324–333.
6. Broxmeyer, H. E., Smithyman, A., Eger, R. R., Myers, P. A., and de Sousa, M. (1978): Identification of lactoferrin as the granulocyte-derived inhibitor of colony stimulating activity production. *J. Exp. Med.*, 148:1052–1067.
7. Burgess, A. W., Camakaris, J., and Metcalf, D. (1977): Purification and properties of colony-stimulating factor from mouse lung conditioned medium. *J. Biol. Chem.*, 252:1998–2003.
8. Burgess, A. W., and Metcalf, D. (1980): Characterisation of a serum factor stimulating the differentiation of myelomonocytic leukemic cells. *Int. J. Cancer*, 26:647–654.
9. Cohen, L., and Sachs, L. (1981): Constitutive gene expression in myeloid leukemia and cell competence for induction of differentiation by the steroid dexamethasone. *Proc. Natl. Acad. Sci. USA*, 78:353–357.
10. Collins, S. J., Gallo, R. C., and Gallagher, R. E. (1977): Continuous growth and differentiation of human myeloid leukemic cells in culture. *Nature*, 270:347–349.
11. Collins, S. J., Ruscetti, F. W., Gallagher, R. E., and Gallo, R. C. (1978): Terminal differentiation of human promyelocytic leukemia induced by dimethylsulfoxide and other polar compounds. *Proc. Natl. Acad. Sci. USA*, 75:2458–2462.
12. De Klein, A., Van Kessel, A. D., Grosveld, G., Bartman, C. R., Hagemeijer, A., Bootsma, D., Spurr, N. K., Heisterkamp, N., Groffen, J., and Stephenson, J. R. (1982): A cellular oncogene is translocated to the Philadelphia chromosome in chronic myelocytic leukaemia. *Nature*, 300:765–767.
13. D'Eustachio, P., Pravtcheva, D., Marcu, K., and Ruddle, F. H. (1980): Chromosomal location of the structural gene cluster encoding murine immunoglobin heavy chains. *J. Exp. Med.*, 151:1545–1550.
14. Dewey, M. J., Martin, D. W., Jr., Martin, G. R., and Mintz, B. (1977): Mosaic mice with teratocarcinoma-derived mutant cells deficient in hypoxanthine phosphoribosyltransferase. *Proc. Natl. Acad. Sci. USA*, 74:5564–5568.
15. Dofuku, R., Biedler, J. L., Sprengler, B. A., and Old, L. J. (1975): Tirsomy of chromosome 15 in spontaneous leukemia of AKR mice. *Proc. Natl. Acad. Sci. USA*, 72:1515–1517.
16. Falk, A., and Sachs, L. (1980): Clonal regulation of the induction of macrophage and granulocyte inducing proteins for normal and leukemic myeloid cells. *Int. J. Cancer*, 26:595–601.
17. Fibach, E., Hayashi, M., and Sachs, L. (1973): Control of normal differentiation of myeloid leukemic cells to macrophages and granulocytes. *Proc. Natl. Acad. Sci. USA*, 70:343–346.
18. Fibach, E., Landau, T., and Sachs, L. (1972): Normal differentiation of myeloid leukemic cells induced by a differentiation-inducing protein. *Nature*, 237:276–278.
19. Fibach, E., and Sachs, L. (1974): Control of normal differentiation of myeloid leukemic cells. IV. Induction of differentiation by serum from endotoxin treated mice. *J. Cell. Physiol.*, 83:177–185.
20. Fibach, E., and Sachs, L. (1975): Control of normal differentiation of myeloid leukemic cells. VIII. Induction of differentiation to mature granulocytes in mass culture. *J. Cell. Physiol.*, 86:221–230.
21. Fibach, E., and Sachs, L. (1976): Control of normal differentiation of myeloid leukemic cells. XI. Induction of a specific requirement for cell viability and growth during the differentiation of myeloid leukemic cells. *J. Cell. Physiol.*, 89:259–266.
22. Friend, C. (1978): The phenomenon of differentiation in murine erythroleukemic cells. *Harvey Lect.*, 72:253–281.
23. Ginsburg, H., and Sachs, L. (1963): Formation of pure suspension of mast cells in tissue culture by differentiation of lymphoid cells from the mouse thymus. *J. Natl. Cancer Inst.*, 31:1–40.
24. Goff, S. P., D'Eustachio, P., Ruddle, F. H., and Baltimore, D. (1982): Chromosmal assignment of the endogenous proto-oncogene c-abl. *Science*, 218:1317–1319.
25. Gootwine, E., Webb, C. G., and Sachs, L. (1982): Participation of myeloid leukaemic cells injected into embryos in haematopoietic differentiation in adult mice. *Nature*, 299:63–65.
26. Hankins, W. D., and Scolnick, E. M. (1981): Harvey and Kirsten sarcoma viruses promote the growth and differentiation of erythroid precursors *in vivo*. *Cell*, 26:91–97.
27. Hayashi, M., Fibach, E., and Sachs, L. (1974): Control of normal differentiation of myeloid

leukemic cells. V. Normal differentiation in aneuploid leukemic cells and the chromosome banding pattern of D^+ and D^- clones. *Int. J. Cancer*, 14:40–48.
28. Hayward, W. S., Neel, B. G., and Astrin, S. M. (1981): Activation of a cellular onc gene by promoter insertion in ALV-induced lymphoid leukosis. *Nature*, 290:475–479.
29. Hitosumachi, S., Rabinowitz, Z., and Sachs, L. (1971): Chromosomal control of reversion in transformed cells. *Nature*, 231:511–514.
30. Hoffman-Liebermann, B., Liebermann, D., and Sachs, L. (1981): Control mechanism regulating gene expression during normal differentiation of myeloid leukemic cells. Differentiation defective mutants blocked in mRNA production and mRNA translation. *Dev. Biol.*, 81:255–265.
31. Hoffman-Liebermann, B., Liebermann, D., and Sachs, L. (1981): Regulation of gene expression by tumor promoters. III. Complementation of the developmental program in myeloid leukemic cells by regulating mRNA production and mRNA translation. *Int. J. Cancer*, 28:615–620.
32. Hoffman-Liebermann, B., and Sachs, L. (1978): Regulation of actin and other proteins in the differentiation of myeloid leukemic cells. *Cell*, 14:825–834.
33. Housset, M., Daniel, M. T., and Degos, L. (1982): Small doses of Ara-C in the treatment of acute myeloid leukemia: Differentiation of myeloid leukemia cells? *Br. J. Haematol.*, 51:125–129.
34. Ichikawa, Y. (1969): Differentiation of a cell line of myeloid leukemia. *J. Cell. Physiol.*, 74:223–234.
35. Ichikawa, Y., Maeda, N., and Horiuchi, M. (1976): *In vitro* differentiation of Rauscher virus induced myeloid leukemic cells. *Int. J. Cancer*, 17:789–797.
36. Ichikawa, Y., Pluznik, D. H., and Sachs, L. (1966): *In vitro* control of the development of macrophage and granulocyte colonies. *Proc. Natl. Acad. Sci. USA*, 56:488–495.
37. Ichikawa, Y., Pluznik, D. H., and Sachs, L. (1967): Feedback inhibition of the development of macrophage and granulocyte colonies. I. Inhibition by macrophages. *Proc. Natl. Acad. Sci. USA*, 58:1480–1486.
38. Klein, G. (1981): The role of gene dosage and genetic transpositions in carcinogenesis. *Nature*, 194:313–318.
39. Klein, G., Frieberg, S., Wiener, F., and Harris, H. (1973): Studies on hybrid cells derived from the fusion of the TA_3/Ha ascites carcinoma with normal fibroblasts. I. Malignancy, karyotype and formation of isoantigenic variants. *J. Natl. Cancer Inst.*, 50:1259–1268.
40. Krystosek, A., and Sachs, L. (1976): Control of lysozyme induction in the differentiation of myeloid leukemic cells. *Cell*, 9:675–684.
41. Landau, T., and Sachs, L. (1971): Characterization of the inducer required for the development of macrophage and granulocyte colonies. *Proc. Natl. Acad. Sci. USA*, 68:2540–2544.
42. Liebermann, D., Hoffman-Liebermann, B., and Sachs, L. (1980): Molecular dissection of differentiation in normal and leukemic myeloblasts: Separately programmed pathways of gene expression. *Dev. Biol.*, 79:46–63.
43. Liebermann, D., Hoffman-Liebermann, B., and Sachs, L. (1980): Regulation of endogenous type C virus expression during normal myeloid cell differentiation. Evidence for a role in promoting myeloid cell proliferation and differentiation. *Virology*, 107:121–134.
44. Liebermann, D., Hoffman-Liebermann, B., and Sachs, L. (1981): Regulation of gene expression by tumor promoters. II. Control of cell shape and developmental programs for macrophages and granulocytes in human myeloid leukemic cells. *Int. J. Cancer*, 28:285–291.
45. Liebermann, D., Hoffman-Liebermann, B., and Sachs, L. (1982): Regulation and role of different macrophage and granulocyte proteins in normal and leukemic myeloid cells. *Int. J. Cancer*, 29:159–161.
46. Liebermann, D., and Sachs, L. (1979): Increase of normal myeloblast viability and multiplication without blocking differentiation by type C RNA virus from myeloid leukemic cells. *Proc. Natl. Acad. Sci. USA*, 76:3353–3357.
47. Lipton, J., and Sachs, L. (1981): Characterization of macrophage and granulocyte inducing proteins for normal and leukemic myeloid cells produced by the Krebs ascites tumor. *Biochim. Biophys. Acta*, 673:552–569.
48. Lotem, J., Lipton, J., and Sachs, L. (1980): Separation of different molecular forms of macrophage and granulocyte inducing proteins for normal and leukemic myeloid cells. *Int. J. Cancer*, 25:763–771.
49. Lotem, J., and Sachs, L. (1974): Different blocks in the differentiation of myeloid leukemic cells. *Proc. Natl. Acad. Sci. USA*, 71:3507–3511.

50. Lotem, J., and Sachs, L. (1975): Induction of specific changes in the surface membrane of myeloid leukemic cells by steroid hormones. *Int. J. Cancer*, 15:731–740.
51. Lotem, J., and Sachs, L. (1977): Control of normal differentiation of myeloid leukemic cells. XII. Isolation of normal myeloid colony-forming cells from bone marrow and the sequence of differentiation to mature granulocytes in normal and D^+ myeloid leukemic cells. *J. Cell. Physiol.*, 92:97–108.
52. Lotem, J., and Sachs, L. (1977): Genetic dissection of the control of normal differentiation in myeloid leukemic cells. *Proc. Natl. Acad. Sci. USA*, 74:5554–5558.
53. Lotem, J., and Sachs, L. (1978): Genetic dissociation of different cellular effects of interferon on myeloid leukemic cells. *Int. J. Cancer*, 22:214–220.
54. Lotem, J., and Sachs, L. (1978): *In vivo* induction of normal differentiation in myeloid leukemic cells. *Proc. Natl. Acad. Sci. USA*, 75:3781–3785.
55. Lotem, J., and Sachs, L. (1979): Regulation of normal differentiation in mouse and human myeloid leukemic cells by phorbol esters and the mechanism of tumor promotion. *Proc. Natl. Acad. Sci. USA*, 76:5158–5162.
56. Lotem, J., and Sachs, L. (1980): Potential pre-screening for therapeutic agents that induce differentiation in human myeloid leukemic cells. *Int. J. Cancer*, 25:561–564.
57. Lotem, J., and Sachs, L. (1981): *In vivo* inhibition of the development of myeloid leukemia by injection of macrophage and granulocyte inducing protein. *Int. J. Cancer*, 28:375–386.
58. Lotem, J., and Sachs, L. (1982): Mechanisms that uncouple growth and differentiation in myeloid leukemia: Restoration of requirement for normal growth-inducing protein without restoring induction of differentiation-inducing protein. *Proc. Natl. Acad. Sci. USA*, 79:4347–4351.
59. Maeda, M., Horiuchi, M., Numa, S., and Ichikawa, Y. (1977): Characterization of a differentiation stimulating factor for mouse myeloid leukemic cells. *Gann*, 68:435–447.
60. Marks, P., and Rifkind, R. A. (1978): Erythroleukemic differentiation. *Annu. Rev. Biochem.*, 47:419–448.
61. Metcalf, D. (1969): Studies on colony formation *in vitro* by mouse bone marrow cells. I. Continuous cluster formation and relation of clusters to colonies. *J. Cell. Physiol.*, 74:323–332.
62. Michalewicz, R., Lotem, J., and Sachs, L. (1983): Cell differentiation and therapeutic effect of low doses of cytosine arabinoside in human myeloid leukemia. *Leukemia Res.* (*in press*).
63. Mintz, B., and Illmensee, K. (1975): Normal genetically mosaic mice produced from malignant teratocarcinoma cells. *Proc. Natl. Acad. Sci. USA*, 72:3585–3589.
64. Moore, M. A. S. (1982): G-SCF: Its relationship to leukemia differentiation-inducing activity and other hemopoietic regulators. *J. Cell Physiol. [Suppl.]*, 1:53–64.
65. Nicola, N. A., Burgess, A. W., and Metcalf, D. (1979): Similar molecular properties of granulocyte-macrophage colony stimulating factors produced by different organs. *J. Biol. Chem.*, 24:5290–5299.
66. Paran, M., Ichikawa, Y., and Sachs, L. (1968): Production of the inducer for macrophage and granulocyte colonies by leukemic cells. *J. Cell. Physiol.*, 72:251–254.
67. Paran, M., Ichikawa, Y., and Sachs, L. (1969): Feedback inhibition of the development of macrophage and granulocyte colonies. II. Inhibition by granulocytes. *Proc. Natl. Acad. Sci. USA*, 62:81–87.
68. Paran, M., Sachs, L., Barak, Y., and Resnitzky, P. (1970): *In vitro* induction of granulocyte differentiation in hematopoietic cells from leukemic and non-leukemic patients. *Proc. Natl. Acad. Sci. USA*, 67:1542–1549.
69. Payne, G. S., Bishop, J. M., and Varmus, H. E. (1982): Multiple arrangements of viral DNA and an activated host oncogene in bursal lymphomas. *Nature*, 295:209–214.
70. Pike, B., and Robinson, W. A. (1970): Human bone marrow growth in agar gel. *J. Cell. Physiol.*, 76:77–84.
71. Pluznik, D. H., and Sachs, L. (1965): The cloning of normal "mast" cells in tissue culture. *J. Cell. Comp. Physiol.*, 66:319–324.
72. Pluznik, D. H., and Sachs, L. (1966): The induction of clones of normal "mast" cells by a substance from conditioned medium. *Exp. Cell. Res.*, 43:553–563.
73. Rabinowitz, Z., and Sachs, L. (1968): Reversion of properties in cells transformed by polyoma virus. *Nature*, 220:1203–1206.
74. Rabinowitz, Z., and Sachs, L. (1970): Control of the reversion of properties in transformed cells. *Nature*, 225:136–139.

75. Rabinowitz, Z., and Sachs, L. (1970): The formation of variants with a reversion of properties of transformed cells. V. Reversion to a limited life span. *Int. J. Cancer*, 6:388–398.
76. Ringertz, N. R., and Savage, R. E. (1976): *Cell Hybrids*. Academic Press, New York.
77. Rovera, G., Santoli, D., and Samsky, C. (1979): Human promyelocytic leukemic cells in culture differentiate into macrophage-like cells when treated with a phorbol diester. *Proc. Natl. Acad. Sci. USA*, 76:2779–2783.
78. Rowley, J. D. (1977): Mapping of human chromosomal regions related to neoplasia: Evidence from chromosomes 1 and 17. *Proc. Natl. Acad. Sci. USA*, 74:5729–5733.
79. Sachs, L. (1974): Regulation of membrane changes, differentiation, and malignancy in carcinogenesis. *Harvey Lect.*, 68:1–35.
80. Sachs, L. (1978): Control of normal cell differentiation and the phenotypic reversion of malignancy in myeloid leukemia. *Nature*, 274:535–539.
81. Sachs, L. (1978): The differentiation of myeloid leukemia cells. New possibilities for therapy. *Br. J. Haematol.*, 40:509–517.
82. Sachs, L. (1980): Constitutive uncoupling of pathways of gene expression that control growth differentiation in myeloid leukemia: A model for the origin and progression of malignancy. *Proc. Natl. Acad. Sci. USA*, 77:6152–6156.
83. Sachs, L. (1982): Control of growth and differentiation in leukemic cells: Regulation of the developmental program and restoration of the normal phenotype in myeloid leukemia. *J. Cell Physiol. [Suppl.]*, 1:151–164.
84. Sachs, L. (1982): Normal regulatory programmes in myeloid leukemia: Regulatory proteins in the control of growth and differentiation. *Cancer Surveys*, 1:321–342.
85. Simantov, R., and Sachs, L. (1978): Differential desensitization of functional adrenergic receptors in normal and malignant myeloid cells. Relationship to receptor mediated hormone cytotoxicity. *Proc. Natl. Acad. Sci. USA*, 75:1805–1809.
86. Simantov, R., Shkolnik, T., and Sachs, L. (1980): Desensitization of enucleated cells to hormones and the role of cytoskeleton in control of a normal hormone response. *Proc. Natl. Acad. Sci. USA*, 77:4798–4802.
87. Stanley, E. R., and Heard, P. M. (1977): Factors regulating macrophage production and growth. Purification and some properties of the colony stimulating factor from medium conditioned by mouse L cells. *J. Biol. Chem.*, 252:4305–4312.
88. Symonds, G., and Sachs, L. (1979): Activation of normal genes in malignant cells. Activation of chemotaxis in relation to other stages of normal differentiation in myeloid leukemia. *Somat. Cell Genet.*, 5:931–944.
89. Symonds, G., and Sachs, L. (1982): Autoinduction of differentiation in myeloid leukemic cells: Restoration of normal coupling between growth and differentiation in leukemic cells that constitutively produce their own growth-inducing protein. *EMBO J.*, 1:1343–1346.
90. Symonds, G., and Sachs, L. (1982): Cell competence for induction of differentiation by insulin and other compounds in myeloid leukemic clones continuously cultured in serum-free medium. *Blood*, 60:208–212.
91. Symonds, G., and Sachs, L. (1982): Modulation of cell competence for induction of differentiation in myeloid leukemic cells. *J. Cell. Physiol.*, 111:9–14.
92. Symonds, G., and Sachs, L. (1983): Synchrony of gene expression and the differentiation of myeloid leukemic cells: Reversion from constitutive to inducible protein synthesis. *EMBO J.*, 2:663–667.
93. Weisinger, G., and Sachs, L. (1983): DNA-binding protein that induces cell differentiation. *EMBO J.*, 2:2103–2107.
94. Weiss, B., and Sachs, L. (1978): Indirect induction of differentiation in myeloid leukemic cells by lipid A. *Proc. Natl. Acad. Sci. USA*, 75:1374–1378.
95. Wiener, F., Klein, G., and Harris, H. (1974): The analysis of malignancy by cell fusion. V. Further evidence of the ability of normal diploid cells to suppress malignancy. *J. Cell Sci.*, 15:177–183.
96. Yamamoto, T., Rabinowitz, Z., and Sachs, L. (1973): Identification of the chromosomes that control malignancy. *Nature*, 243:247–250.

Subject Index

Subject Index